D0946914

Detection and Estimation for Communication and Radar Systems

Covering the fundamentals of detection and estimation theory, this systematic guide describes statistical tools that can be used to analyze, design, implement, and optimize real-world systems. Detailed derivations of the various statistical methods are provided, ensuring a deeper understanding of the basics. Packed with practical insights, it uses extensive examples from communication, telecommunication, and radar engineering to illustrate how theoretical results are derived and applied in practice. A unique blend of theory and applications, and more than 80 analytical and computational end-of-chapter problems, make this an ideal resource for both graduate students and professional engineers.

Kung Yao is a Distinguished Professor in the Electrical Engineering Department at the University of California, Los Angeles. He received his BS (Highest Honors) and Ph.D. from Princeton University. A Life Fellow of the IEEE, he has worked for or consulted for several leading companies, including AT&T Bell Laboratories, TRW, Hughes Aircraft Company, and Raytheon.

Flavio Lorenzelli received his Ph.D. from the University of California, Los Angeles, and for several years was with ST Microelectronics. The recipient of a Fulbright fellowship in 1989, he has been an engineer at the Aerospace Corporation since 2007 and is a Lecturer in the Electrical Engineering Department at UCLA.

Chiao-En Chen is an Assistant Professor in both the Department of Electrical Engineering and the Department of Communications Engineering at National Chung Cheng University, Taiwan. He received his PhD from the University of California, Los Angeles, in 2008.

"This text is tailor-made for first year graduate students, with its easy-to-follow presentation style, self-contained background materials, and even simulation methods that are perfect for new learners and practitioners."

Zhi Ding, University of California, Davis

"Making things as simple as possible, but not simpler, is an art well mastered by the authors, whose teaching experience shines through the whole book and makes it an ideal text for electrical engineering students, especially those taking courses in wireless communications. The panoply of examples and homework problems included in the book makes it also an invaluable tool for self-study."

Ezio Biglieri, University of California, Los Angeles

"This book strikes a good balance between engineering insight and mathematical rigor. It will make an excellent textbook for either an advanced undergraduate class or a first-year graduate class on detection and estimation theory."

Laurence Milstein, University of California, San Diego

Detection and Estimation for Communication and Radar Systems

KUNG YAO
University of California, Los Angeles

FLAVIO LORENZELLI
The Aerospace Corporation, Los Angeles

CHIAO-EN CHEN
National Chung Cheng University, Taiwan

CAMBRIDGE
UNIVERSITY PRESS

University Printing House, Cambridge CB2 8BS, United Kingdom

One Liberty Plaza, 20th Floor, New York, NY 10006, USA

477 Williamstown Road, Port Melbourne, VIC 3207, Australia

4843/24, 2nd Floor, Ansari Road, Daryaganj, Delhi - 110002, India

79 Anson Road, #06-04/06, Singapore 079906

Cambridge University Press is part of the University of Cambridge.

It furthers the University's mission by disseminating knowledge in the pursuit of education, learning and research at the highest international levels of excellence.

www.cambridge.org
Information on this title: www.cambridge.org/9780521766395

First published 2013

A catalogue record for this publication is available from the British Library

ISBN 978-0-521-76639-5 Hardback

Additional resources for this publication at www.cambridge.org/yaolorenzellichen

To my wife, Mary, and my children, David, Erica, and Roger

K.Y.

To my mom

F.L.

To my family

C.E.C.

Contents

Preface *page* xi

1 Introduction and motivation to detection and estimation 1

1.1 Introduction 1
1.2 A simple binary decision problem 3
1.3 A simple correlation receiver 6
1.4 Importance of SNR and geometry of the signal vectors in detection theory 7
1.5 BPSK communication systems for different ranges 13
1.6 Estimation problems 17
1.6.1 Two simple estimation problems 17
1.6.2 Least-absolute-error criterion 19
1.6.3 Least-square-error criterion 20
1.6.4 Estimation robustness 21
1.6.5 Minimum mean-square-error criterion 23
1.7 Conclusions 28
1.8 Comments 29
References 31
Problems 31

2 Review of probability and random processes 35

2.1 Review of probability 35
2.2 Gaussian random vectors 40
2.2.1 Marginal and conditional pdfs of Gaussian random vectors 41
2.3 Random processes (stochastic processes) 43
2.4 Stationarity 46
2.5 Gaussian random process 51
2.6 Ensemble averaging, time averaging, and ergodicity 53
2.7 WSS random sequence 54
2.8 Conclusions 56
2.9 Comments 57
2.A Proof of Theorem 2.1 in Section 2.2.1 58
2.B Proof of Theorem 2.2 in Section 2.2.1 60

References 60
Problems 61

3 **Hypothesis testing** 65

3.1 Simple hypothesis testing 65
3.2 Bayes criterion 66
3.3 Maximum a posteriori probability criterion 71
3.4 Minimax criterion 72
3.5 Neyman–Pearson criterion 75
3.6 Simple hypothesis test for vector measurements 77
3.7 Additional topics in hypothesis testing (*) 80
3.7.1 Sequential likelihood ratio test (SLRT) 80
3.7.2 Uniformly most powerful test 85
3.7.3 Non-parametric sign test 88
3.8 Conclusions 90
3.9 Comments 91
References 91
Problems 92

4 **Detection of known binary deterministic signals in Gaussian noises** 97

4.1 Detection of known binary signal vectors in WGN 97
4.2 Detection of known binary signal waveforms in WGN 103
4.3 Detection of known deterministic binary signal vectors in colored Gaussian noise 109
4.4 Whitening filter interpretation of the CGN detector 112
4.5 Complete orthonormal series expansion 116
4.6 Karhunen–Loève expansion for random processes 117
4.7 Detection of binary known signal waveforms in CGN via the KL expansion method 124
4.8 Applying the WGN detection method on CGN channel received data (*) 130
4.8.1 Optimization for evaluating the worst loss of performance 132
4.9 Interpretation of a correlation receiver as a matched filter receiver 135
4.10 Conclusions 138
4.11 Comments 139
4.A 139
4.B 140
References 141
Problems 141

5 ***M*-ary detection and classification of deterministic signals** 149

5.1 Introduction 149
5.2 Gram–Schmidt orthonormalization method and orthonormal expansion 150

5.3 M-ary detection 154
5.4 Optimal signal design for M-ary systems 168
5.5 Classification of M patterns 171
5.5.1 Introduction to pattern recognition and classification 171
5.5.2 Deterministic pattern recognition 173
5.6 Conclusions 185
5.7 Comments 186
References 186
Problems 187

6 Non-coherent detection in communication and radar systems 190

6.1 Binary detection of a sinusoid with a random phase 190
6.2 Performance analysis of the binary non-coherent detection system 195
6.3 Non-coherent detection in radar receivers 201
6.3.1 Coherent integration in radar 201
6.3.2 Post detection integration in a radar system 202
6.3.3 Double-threshold detection in a radar system 205
6.3.4 Constant False Alarm Rate (CFAR) 207
6.4 Conclusions 210
6.5 Comments 210
References 211
Problems 211

7 Parameter estimation 214

7.1 Introduction 214
7.2 Mean-square estimation 215
7.2.1 Non-linear mean-square estimation and conditional expectation 218
7.2.2 Geometry of the orthogonal principle and mean-square estimation 220
7.2.3 Block and recursive mean-square estimations 226
7.3 Linear LS and LAE estimation and related robustness and sparse solutions 230
7.3.1 LS estimation 230
7.3.2 Robustness to outlier (*) of LAE solution relative to LS solution 232
7.3.3 Minimization based on l_2 and l_1 norms for solving linear system of equations (*) 234
7.4 Basic properties of statistical parameter estimation 238
7.4.1 Cramér–Rao Bound 243
7.4.2 Maximum likelihood estimator 247
7.4.3 Maximum a posteriori estimator 253
7.4.4 Bayes estimator 255
7.5 Conclusions 258
7.6 Comments 258

7.A Proof of Theorem 7.1 of Section 7.3.3 259
7.B Proof of Theorem 7.3 of Section 7.4.1 260
References 262
Problems 264

8 Analytical and simulation methods for system performance analysis 271

8.1 Analysis of receiver performance with Gaussian noise 272
8.2 Analysis of receiver performance with Gaussian noise and other random interferences 276
8.2.1 Evaluation of P_e based on moment bound method 278
8.3 Analysis of receiver performance with non-Gaussian noises 282
8.3.1 Noises with heavy tails 282
8.3.2 Fading channel modeling and performance analysis 287
8.3.3 Probabilities of false alarm and detection with robustness constraint 293
8.4 Monte Carlo simulation and importance sampling in communication/radar performance analysis 296
8.4.1 Introduction to Monte Carlo simulation 297
8.4.2 MC importance sampling simulation method 299
8.5 Conclusions 304
8.6 Comments 304
8.A Generation of pseudo-random numbers 306
8.A.1 Uniformly distributed pseudo-random number generation 307
8.A.2 Gaussian distributed pseudo-random number generation 309
8.A.3 Pseudo-random generation of sequences with arbitrary distributions 309
8.B Explicit solution of $p_V(\cdot)$ 310
References 312
Problems 314

Index 318

Preface

This publication was conceived as a textbook for a first-year graduate course in the Signals and Systems Area of the Electrical Engineering Department at UCLA to introduce basic statistical concepts of detection and estimation and their applications to engineering problems to students in communication, telecommunication, control, and signal processing. Students majoring in electromagnetics and antenna design often take this course as well. It is not the intention of this book to cover as many topics as possible, but to treat each topic with enough detail so a motivated student can duplicate independently some of the thinking processes of the originators of these concepts. Whenever possible, examples with some numerical values are provided to help the reader understand the theories and concepts. For most engineering students, overly formal and rigorous mathematical methods are probably neither appreciated nor desirable. However, in recent years, more advanced analytical tools have proved useful even in practical applications. For example, tools involving eigenvalue–eigenvector expansions for colored noise communication and radar detection; non-convex optimization methods for signal classification; non-quadratic estimation criteria for robust estimation; non-Gaussian statistics for fading channel modeling; and compressive sensing methodology for signal representation, are all introduced in the book.

Most of the material in the first seven chapters of this book can be covered in a course of 10 weeks of 40 lecture hours. A semester-long course can more thoroughly cover more material in these seven chapters and even some sections of Chapter 8. Homework problems are provided in each chapter. The solutions of odd-numbered problems are available from the Cambridge University Press website. The solutions of the even-numbered problems are available (also from Cambridge University Press) to instructors using this book as a textbook. The prerequisites of this book include having taken undergraduate courses on linear systems, basic probability, and some elementary random processes. We assume the students are familiar with using Matlab for computations and simulations. Indeed, some of the statements in the book and in the homework problems use standard Matlab notations.

Comments and references including bibliographic information are provided at the end of each chapter. The authors of this book certainly appreciate the extensive prior research in journal and book publications on all the topics covered in this book. Omissions of references on some technical topics/methodologies, and even some homework problems that may have appeared elsewhere, are not intentional. In such cases, we seek your understanding and indulgence.

1 Introduction and motivation to detection and estimation

1.1 Introduction

The second half of the twentieth century experienced an explosive growth in information technology, including data transmission, processing, and computation. This trend will continue at an even faster pace in the twenty-first century. Radios and televisions started in the 1920s and 1940s respectively, and involved transmission from a single transmitter to multiple receivers using AM and FM modulations. Baseband analog telephony, starting in the 1900s, was originally suited only for local area person-to-person communication. It became possible to have long-distance communication after using cascades of regeneration repeaters based on digital PCM modulation. Various digital modulations with and without coding, across microwave, satellite, and optical fiber links, allowed the explosive transmissions of data around the world starting in the 1950s–1960s. The emergence of Ethernet, local area net, and, finally, the World Wide Web in the 1980s–1990s allowed almost unlimited communication from any computer to another computer. In the first decade of the twenty-first century, by using wireless communication technology, we have achieved cellular telephony and instant/personal data services for humans, and ubiquitous data collection and transmission using ad hoc and sensor networks. By using cable, optical fibers, and direct satellite communications, real-time on-demand wideband data services in offices and homes are feasible.

Detection and estimation theories presented in this book constitute some of the most basic statistical and optimization methodologies used in communication/telecommunication, signal processing, and radar theory and systems. The purpose of this book is to introduce these basic concepts and their applications to readers with only basic junior/senior year linear system and probability knowledge. The modest probability prerequisites are summarized in Section 2.1. Other necessary random processes needed to understand the material in the rest of this book are also presented succinctly in Sections 2.2–2.4.

The author (KY) has taught a first-year detection and estimation graduate course at UCLA for many years. Given the university's location in southern California, our students have very diverse backgrounds. There are students who have been working for some years in various local aerospace, communications, and signal processing industries, and may have already encountered in their work various concepts in detection and estimation. They may already have quite good intuitions on many of the issues encountered in this course and may be highly motivated to learn these concepts in greater depth. On the other hand, most students (domestic and foreign) in this course have just finished a BS degree in engineering or applied science with little or no prior engineering experience and have not encountered real-life practical information processing systems and technologies. Many of these students may feel the topics covered in the course to be just some applied statistical problems and have no understanding about why one would want to tackle such problems.

The detection problem is one of deciding at the receiver which bit of information, which for simplicity at this point can be assumed to be binary, having either a "one" or a "zero," was sent by the transmitter. In the absence of noise/disturbance, the decision can be made with no error. However, in the presence of noise, we want to maximize the probability of making a correct decision and minimize the probability of making an incorrect decision. It turns out the solution of this statistical problem is based on statistical hypothesis theory already formulated in statistics in the 1930s by Fisher [1] and Neyman–Pearson [2]. However, it was only during World War II, in the analysis and design of optimum radar and sonar systems, that a statistical approach to these problems was formulated by Woodward [3]. Of course, we also want to consider decisions for multiple hypotheses under more general conditions.

The parameter estimation problem is one of determining the value of a parameter in a communication system. For example, in a modern cellular telephony system, the base station as well as a hand-held mobile phone need to estimate the power of the received signal in order to control the power of the transmitted signal. Parameter estimation can be performed using many methods. The simplest one is based on the mean-square estimation criterion, which had its origin in the 1940s by Kolmogorov [4] and Wiener [5]. The related least-square-error criterion estimation method was formulated by Gauss [6] and Laplace [7] in the nineteenth century. Indeed, Galileo even formulated the least-absolute-error criterion estimation in the seventeenth century [8]. All of these estimation methods of Galileo, Gauss, and Laplace were motivated by practical astronomical tracking of various heavenly bodies.

The purpose of this course is to teach some basic statistical and associated optimization methods mainly directed toward the analysis, design, and implementation of modern communication systems. In network jargon, the topics we encounter in this book all belong to the physical layer problems. These methods are equally useful for the study of modern control, system identification, signal/image/speech processing, radar systems, mechanical systems, economic systems, and biological systems. In this course, we will encounter the issue of the modeling of a system of interest. In simple problems, this modeling may be sort of obvious. In complicated problems, the proper modeling of the problem may be its most challenging aspect. Once a model has been formulated, then we can consider the appropriate mathematical tools for the correct and computationally efficient solution of the modeled problem. Often, among many possible solutions, we may seek the theoretically optimized solution based on some analytically tractable criterion. After obtaining this optimum solution, we may consider some solutions that are only slightly sub-optimum but are practical from an engineering point of view (e.g., ease of implementation; low cost of implementation; etc.). If there is not a known optimum solution, how can we determine how good some ad hoc as well as practical solutions are?

In Chapter 1, our purpose is to provide some very simple motivational examples illustrating the concepts of: statistical decision of two possible hypotheses; correlation receiver; relationship of the receiver's detector signal-to-noise ratio (SNR) to the transmitter power, and deterministic and statistical estimations. These examples will relate various simply posed hypothetical problems and human-made and physical phenomena to their possible solutions based on statistical methods. In turn, these methods are useful for characterizing, modeling, analyzing, and designing various engineering problems discussed in the rest of the book.

We will use the notation of denoting a deterministic scalar variable or parameter by a lowercase letter such as z. A deterministic column vector of dimension $M \times 1$ will be denoted by a bold lowercase letter such as $\mathbf{z}$. A scalar random variable (r.v.) will be denoted by an uppercase letter like Z, while a vector random vector will be denoted by a bold uppercase letter like $\mathbf{Z}$. The realizations of the r.v. Z and the random vector $\mathbf{Z}$, being non-random (i.e., deterministic), will be denoted by their corresponding lowercase letters of z and $\mathbf{z}$ respectively. We will use the abbreviation for the "left-hand side" by "l.h.s." and the "right-hand side" by "r.h.s" of an equation. We will also use the abbreviation of "with respect to" by "w.r.t."

At the end of each chapter, for each section, we provide some casual historical background information and references to other relevant journals and books. An asterisk following the section title in a chapter indicates that section may be of interest only to some serious readers. Materials in those sections will not be needed for following chapters. Similarly, an asterisk following an example indicates these materials are provided for the serious readers.

1.2 A simple binary decision problem

In the first motivational example, we consider a simple and intuitively obvious example illustrating the concept of maximum-likelihood (ML) decision. It turns out the ML criterion is the basis of many modern detection, decoding, and estimation procedures. Consider two boxes denoted as Box 0 and Box 1. Each box contains ten objects colored either red (R) or black (B). Suppose we know the "prior distributions" of the objects in the two boxes as shown in (1.1) and in Fig. 1.1.

$$\text{Box 0} \begin{cases} 2 \text{ reds} \\ 8 \text{ blacks} \end{cases}; \quad \text{Box 1} \begin{cases} 8 \text{ reds} \\ 2 \text{ blacks} \end{cases}. \tag{1.1}$$

Furthermore, we define the random variable (r.v.) X by

$$X = \begin{cases} -1, & \text{for a red object}, \\ 1, & \text{for a black object}. \end{cases} \tag{1.2}$$

From these prior probabilities, the conditional probabilities $p_0(x) = p(x|\text{Box } 0)$ and $p_1(x) = p(x|\text{Box } 1)$ of the two boxes are described in Fig. 1.1. For an observed value of x (taking either 1 or -1), these conditional probabilities are called "likelihood" functions. Now, suppose we randomly (with equal probability) pick one object from one

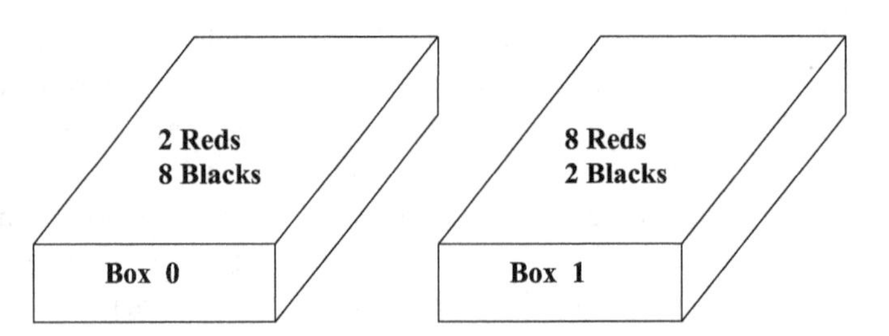

Figure 1.1 Known prior distributions of red and black objects in two boxes.

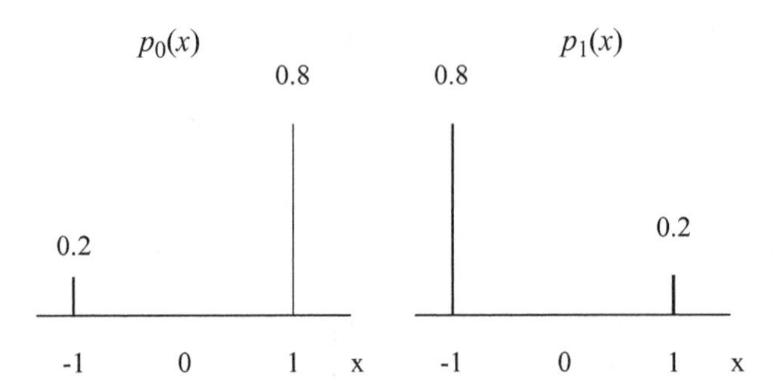

Figure 1.2 Conditional probabilities of $p_0(x)$ and $p_1(x)$.

of the boxes (whose identity is not known to us) and obtain a red object with the r.v. X taking the value of $x = -1$. Then by comparing the likelihood functions, we notice

$$p(x = -1|\text{Box } 0) = 0.2 < 0.8 = p(x = -1|\text{Box } 1). \tag{1.3}$$

After having observed a red object with $x = -1$, we declare Box 1 as the "most likely" box that the observed object was selected from. Thus, (1.3) illustrates the use of the ML decision rule. By looking at Fig. 1.1, it is "intuitively obvious" if a selected object is red, then statistically that object is more likely to come from Box 1 than from Box 0. A reasonably alert ten-year-old child might have come to that decision. On the other hand, if we observed a black object with $x = 1$, then by comparing their likelihood functions, we have

$$p(x = 1|\text{Box } 0) = 0.8 > 0.2 = p(x = 1|\text{Box } 1). \tag{1.4}$$

Thus, in this case with $x = 1$, we declare Box 0 as the "most likely" box that the observed object was selected from, again using the ML decision rule. In either case, we can form the likelihood ratio (LR) function

$$\frac{p_1(x)}{p_0(x)} \begin{cases} > 1 \Rightarrow \text{Declare Box 1}, \\ < 1 \Rightarrow \text{Declare Box 0}. \end{cases} \tag{1.5}$$

Then (1.3) and (1.4) are special cases of (1.5), where the LR function $p_1(x)/p_0(x)$ is compared to the threshold constant of 1. Specifically, for $x = -1$, $p_1(x = -1)/p_0(x = -1) > 1$, we decide for Box 1. For $x = 1$, $p_1(x = 1)/p_0(x = 1) < 1$, we decide for

Box 0. The decision procedure based on (1.5) is called the LR test. It turns out that we will repeatedly use the concepts of ML and LR tests in detection theory in Chapter 3.

Even though the decision based on the ML criterion (or equivalently the LR test) for the above binary decision problem is statistically reasonable, that does not mean the decision can not result in errors. Consider the evaluation of the probability of a decision error, given we have observed a red object with $x = -1$, is denoted by $P_{\mathrm{e}|x=-1}$ in (1.6a):

$$P_{\mathrm{e}|x=-1} = \mathrm{Prob}(\text{Decision error}|x = -1) \tag{1.6a}$$

$$= \mathrm{Prob}(\text{Decide Box } 0|x = -1) \tag{1.6b}$$

$$= \mathrm{Prob}(\text{Object came from Box } 0|x = -1) \tag{1.6c}$$

$$= 2/10 = 1/5 = 0.2. \tag{1.6d}$$

For an observed $x = -1$ the ML criterion makes the decision for Box 1. Thus, the "Decision error" in (1.6a) is equivalent to "Decide Box 0" in (1.6b). But the "Object came from Box 0 given $x = -1$" in (1.6c) means it is a red object from Box 0. The probability of a red object from Box 0 has the relative frequency of 2 over 10, which yields (1.6d). Similarly, the probability of a decision error, given we have observed a black object with $x = 1$, is denoted by $P_{\mathrm{e}|x=1}$ in (1.7a):

$$P_{\mathrm{e}|x=1} = \mathrm{Prob}(\text{Decision error}|x = 1) \tag{1.7a}$$

$$= \mathrm{Prob}(\text{Decide Box } 1|x = 1) \tag{1.7b}$$

$$= \mathrm{Prob}(\text{Object came from Box } 1|x = -1) \tag{1.7c}$$

$$= 2/10 = 1/5 = 0.2. \tag{1.7d}$$

Due to the symmetry of the number of red versus black objects in Box 0 with the number of black versus red objects in Box 1, it is not surprising that $P_{\mathrm{e}|x=-1} = P_{\mathrm{e}|x=1}$.

Next, suppose we want the average probability of a decision error P_{e}. Since the total number of red objects equals the total number of black objects, then $\mathrm{P}(x = -1) = \mathrm{P}(x = 1) = 1/2 = 0.5$. Thus,

$$P_{\mathrm{e}} = P_{\mathrm{e}|x=-1} \times 0.5 + P_{\mathrm{e}|x=1} \times 0.5 = 1/5 = 0.2. \tag{1.8}$$

Equation (1.8) shows on the average the above ML decision rule results in an error 20% of the time. However, suppose we are told there are a total of ten black objects and ten red objects in both boxes, but are not given the prior probability information of (1.1) and Fig. 1.1. Then we can not use the ML decision criterion and the LR test. Given there are a total of ten black objects and ten red objects in both boxes, then regardless of whether the observed object is red or black, we should declare either Box 0 or Box 1 half of the time (by flipping a fair coin and deciding Box 0 if the coin shows a "Head" and Box 1 if the coin shows a "Tail"). In this random equi-probable decision rule, on the average an error occurs 50% of the time. This shows that by knowing the additional conditional probability information of (1.1) and using the ML decision criterion (or the LR test), we can achieve on average a smaller probability of error. This simple example shows the usefulness of statistical decision theory.

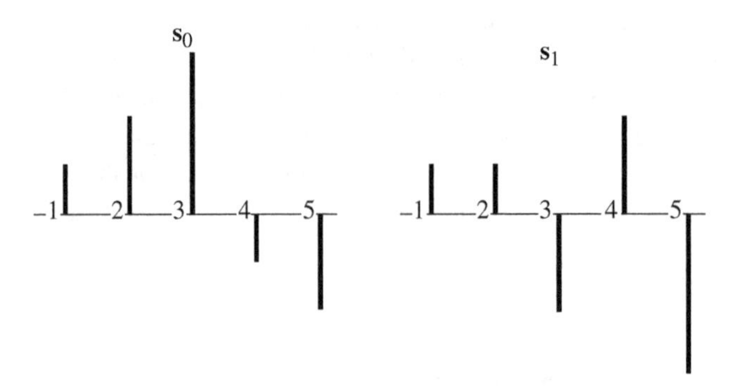

Figure 1.3 Two digital data vectors $\mathbf{s}_0$ and $\mathbf{s}_1$ of length five.

When the two boxes are originally chosen with equal probability in selecting an object, we saw the simple intuitive solution and the LR test solution are identical. However, if the two boxes are not chosen with equal probability, then after an observation the proper decision of the box is less intuitively obvious. As will be shown again in Chapter 3, for arbitrary probability distribution of the two boxes (i.e., hypotheses), the LR test can still be used to perform a statistically optimum decision.

1.3 A simple correlation receiver

In the second motivational example, consider two known digital signal data of length five denoted by the column vectors

$$\mathbf{s}_0 = [1,\ 2,\ 3,\ -1,\ -2]^T,\ \mathbf{s}_1 = [1,\ 1,\ -2,\ 2,\ -3]^T \tag{1.9}$$

as shown in Fig. 1.3. Define the *correlation* of the $\mathbf{s}_i$ vector with the $\mathbf{s}_j$ vector by

$$\mathbf{s}_i^T \mathbf{s}_j = \sum_{k=1}^{5} s_i(k) s_j(k),\ i, j = 0,\ 1. \tag{1.10}$$

Since $\mathbf{s}_i$ is a 5×1 column vector, and $\mathbf{s}_i^T$ is a 1×5 row vector, thus the correlation of the $\mathbf{s}_i$ vector with the $\mathbf{s}_j$ vector, $\mathbf{s}_i^T \mathbf{s}_j$, is a $1 \times 5 \times 5 \times 1 =$ one-dimensional scalar number. In particular, if $i = j$, then with each component of $\mathbf{s}_i$ having the unit of a volt, $\mathbf{s}_i^T \mathbf{s}_i = \|\mathbf{s}_i\|^2$ can be considered as the power of the $\mathbf{s}_i$ vector summed over four equally spaced time intervals. Thus, $\|\mathbf{s}_i\|^2$ represents the energy of the vector $\mathbf{s}_i$. By direct evaluation, we show both vectors have the same energy of

$$\begin{aligned} \|\mathbf{s}_0\|^2 &= 1^2 + 2^2 + 3^2 + (-1)^2 + (-2)^2 = 19 \\ &= 1^2 + 1^2 + (-2)^2 + (-2)^2 + (-3)^2 = \|\mathbf{s}_1\|^2. \end{aligned} \tag{1.11}$$

However, the correlation of $\mathbf{s}_1$ with $\mathbf{s}_0$ by direct evaluation is given by

$$\begin{aligned} \mathbf{s}_1^T \mathbf{s}_0 &= (1 \times 1) + (1 \times 2) + (-2 \times 3) + (-2 \times (-1)) + (-3 \times (-2)) \\ &= 1. \end{aligned} \tag{1.12}$$

Furthermore, suppose the observed vector $\mathbf{x}$ is either $\mathbf{s}_0$ or $\mathbf{s}_1$. That is, we have the model relating the observed data vector $\mathbf{x}$ to the signal vectors $\mathbf{s}_i$ given by

$$\mathbf{x} = \begin{cases} \mathbf{s}_0, \text{ under hypothesis } H_0 \\ \mathbf{s}_1, \text{ under hypothesis } H_1, \end{cases} \tag{1.13}$$

under the two hypotheses of H_0 and H_1. A simple decision rule to determine which vector or hypothesis is valid is to consider the correlation of $\mathbf{s}_1$ with the observed vector $\mathbf{x}$ given by

$$\gamma = \mathbf{s}_1^T \mathbf{x} = \begin{cases} \mathbf{s}_1^T \mathbf{s}_0 = 1 \Leftrightarrow \ \mathbf{x} = \mathbf{s}_0 \Rightarrow \text{Declare hypothesis } H_0 \\ \mathbf{s}_1^T \mathbf{s}_1 = 19 \Leftrightarrow \mathbf{x} = \mathbf{s}_1 \Rightarrow \text{Declare hypothesis } H_1. \end{cases} \tag{1.14}$$

From (1.14), we note if the correlation value is low (i.e., 1), then we know the observed $\mathbf{x} = \mathbf{s}_0$ (or hypothesis H_0 is true), while if the correlation value is high (i.e., 19), then the observed $\mathbf{x} = \mathbf{s}_1$ (or hypothesis H_1 is true). Of course, in practice, the more realistic additive noise (AN) observation model replacing (1.13) may be generalized to

$$\mathbf{X} = \begin{cases} \mathbf{s}_0 + \mathbf{N}, \text{ under hypothesis } H_0 \\ \mathbf{s}_1 + \mathbf{N}, \text{ under hypothesis } H_1, \end{cases} \tag{1.15}$$

where $\mathbf{N}$ denotes the observation noise vector with some statistical properties. The introduction of an AN model permits the modeling of realistic physical communication or measurement channels with noises. Fortunately, an AN channel still allows the use of many statistical methods for analysis and synthesis of the receiver.

As we will show in Chapter 2, if $\mathbf{N}$ is a white Gaussian noise vector, then for the model of (1.15), the decision procedure based on the correlation method of (1.14) is still optimum, except the threshold values of 1 under hypothesis H_0 and 19 under hypothesis H_1 need to be modified under different statistical criteria (with details to be given in Chapter 3). We hope in the statistical analysis of complex systems, as the noise approaches zero, the operations of the noise-free systems may give us some intuitive hint on the optimum general solutions. The fact that the optimum binary receiver decision rule of (1.14) based on the correlation of $\mathbf{s}_1$ with the received vector $\mathbf{x}$ in the noise-free model of (1.13) is still optimum for the white Gaussian AN model of (1.15) is both interesting and satisfying.

1.4 Importance of SNR and geometry of the signal vectors in detection theory

One of the most important measures of the "goodness" of a waveform, whether in the analog domain (with continuous-time values and continuous-amplitude values), or in the digital data domain (after sampling in time and quantization in amplitude) as used in most communication, radar, and signal processing systems, is its signal-to-noise ratio (SNR) value. In order to make this concept explicit, consider the correlation receiver of Section 1.3 with the additive noise channel model of (1.15). In order to show quantitative results with explicit SNR values, we also need to impose an explicit statistical property on the noise vector $\mathbf{N}$ and also assume the two $n \times 1$ signal column vectors $\mathbf{s}_0$ and $\mathbf{s}_1$ have equal energy $\|\mathbf{s}_0\|^2 = \|\mathbf{s}_1\|^2 = E$. The simplest (and fortunately still quite justifiable) assumption of the white Gaussian noise (WGN) property of $\mathbf{N} = [N_1, N_2, \ldots, N_n]^T$ being a column vector of dimension $n \times 1$ having zero mean and a covariance matrix of

$$\mathbf{\Lambda} = \mathrm{E}\{\mathbf{N}\mathbf{N}^T\} = \begin{bmatrix} \sigma^2 & 0 & 0 \\ 0 & \ddots & 0 \\ 0 & 0 & \sigma^2 \end{bmatrix} = \sigma^2 \mathbf{I}_n, \tag{1.16}$$

where σ^2 is the variance of each noise component N_k along the diagonal of $\mathbf{\Lambda}$, and having zero values in its non-diagonal values. Then the SNR of the signal $\mathbf{s}_0$ or $\mathbf{s}_1$ to the noise $\mathbf{N}$ is defined as the ratio of the energy E of $\mathbf{s}_0$ or $\mathbf{s}_1$ to the trace of $\mathbf{\Lambda}$ of (1.16) defined by

$$\mathrm{trace}(\mathbf{\Lambda}) = \sum_{k=1}^{n} \mathbf{\Lambda}_{kk} = n\sigma^2. \tag{1.17}$$

Thus,

$$\mathrm{SNR} = \frac{E}{\mathrm{trace}(\mathbf{\Lambda})} = \frac{E}{n\sigma^2}. \tag{1.18}$$

Due to its possible large dynamic range, SNR is often expressed in the logarithmic form in units of dB defined by

$$\mathrm{SNR(dB)} = 10 \log_{10}(\mathrm{SNR}) = 10 \log_{10}(E/(n\sigma^2)). \tag{1.19}$$

Now, consider the binary detection problem of Section 1.3, where the two 5×1 signal column vectors $\mathbf{s}_0$ and $\mathbf{s}_1$ defined by (1.9) have equal energies of $\|\mathbf{s}_0\|^2 = \|\mathbf{s}_1\|^2 = E = 19$, and the AN channel of (1.15) has a WGN vector $\mathbf{N}$ of zero mean and covariance matrix given by (1.16). Thus, its SNR $= 19/(5\sigma^2)$.

Denote the correlation of $\mathbf{s}_1$ with $\mathbf{X}$ by $\Gamma = \mathbf{s}_1^T \mathbf{X}$. Under the noise-free channel model of (1.12), $\Gamma = \mathbf{s}_1^T \mathbf{x} = \mathbf{s}_1^T \mathbf{s}_0 = 1$ for hypothesis H_0 and $\Gamma = \mathbf{s}_1^T \mathbf{x} = \mathbf{s}_1^T \mathbf{s}_1 = 19$ for hypothesis H_1. But under the present AN channel of model (1.15), $\Gamma = \mathbf{s}_1^T \mathbf{X}$ under the two hypotheses yields two r.v.'s given by

$$\Gamma = \mathbf{s}_1^T \mathbf{X} = \begin{cases} \mathbf{s}_1^T(\mathbf{s}_0 + \mathbf{N}) \\ \mathbf{s}_1^T(\mathbf{s}_1 + \mathbf{N}) \end{cases} = \begin{cases} \mu_0 + \mathbf{s}_1^T \mathbf{N}, \text{ under hypothesis } H_0 \\ \mu_1 + \mathbf{s}_1^T \mathbf{N}, \text{ under hypothesis } H_1 \end{cases}, \tag{1.20}$$

where $\mu_0 = 1$ and $\mu_1 = 19$. Denote the correlation of $\mathbf{s}_1$ with $\mathbf{N}$ by $\tilde{\Gamma} = \mathbf{s}_1^T \mathbf{N}$, which is a Gaussian r.v. consisting of a sum of Gaussian r.v.'s of $\{N_1, \ldots, N_n\}$. Since all the r.v.'s of $\{N_1, \ldots, N_n\}$ have zero means, then $\tilde{\Gamma}$ has a zero mean. Then the variance of $\tilde{\Gamma}$ is given by

$$\sigma_{\tilde{\Gamma}}^2 = \mathrm{E}\{(\mathbf{s}_1^T \mathbf{N})(\mathbf{s}_1^T \mathbf{N})^T\} = \mathrm{E}\{\mathbf{s}_1^T \mathbf{N}\mathbf{N}^T \mathbf{s}_1\} = \sigma^2 \mathbf{s}_1^T \mathbf{s}_1 = 19\sigma^2. \tag{1.21}$$

Using (1.20) and (1.21), we note $\Gamma = \mathbf{s}_1^T \mathbf{X}$ is a Gaussian r.v. of mean 1 and variance $19\sigma^2$ under hypothesis H_0, while it is a Gaussian r.v. of mean 19 and also variance $19\sigma^2$ under hypothesis H_1. The Gaussian pdfs of Γ under the two hypotheses are plotted in Fig. 1.4. Thus, for the additive white Gaussian noise (AWGN) channel, instead of using the decision rule of (1.14) for the noise-free case, under the assumption that both

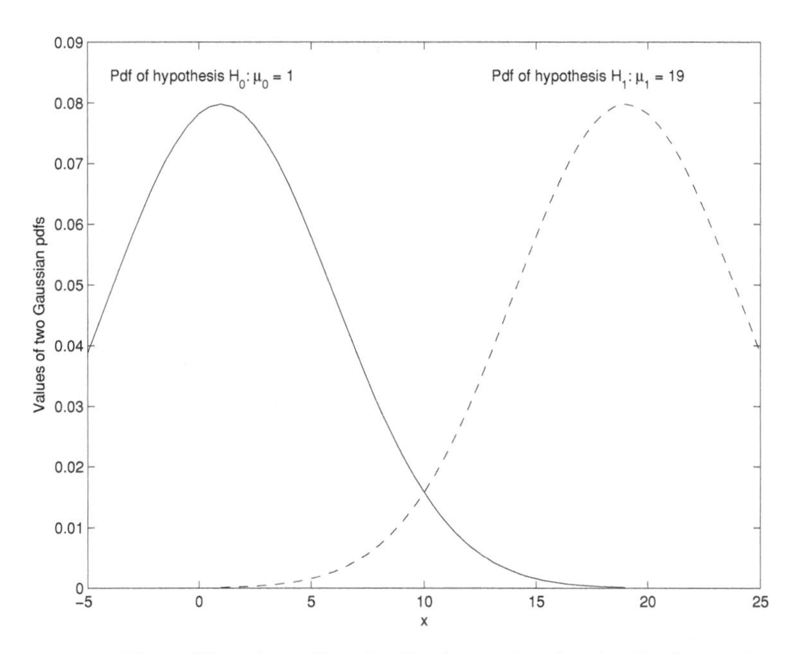

Figure 1.4 Plots of Gaussian pdfs under H_0 of $\mu_0 = 1$ and under H_1 of $\mu_1 = 19$.

hypotheses occur equi-probably and are of equal importance, then we should set the decision threshold at 10, which is the average of 1 and 19. Thus, the new decision rule is given by

$$\Gamma = \mathbf{s}_1^T \mathbf{X} \begin{cases} < 10 \Rightarrow \text{Declare hypothesis } H_0 \\ > 10 \Rightarrow \text{Declare hypothesis } H_1 \end{cases}. \tag{1.22}$$

From (1.20) and Fig. 1.4, under hypothesis H_0, the decision rule in (1.22) states if the noise is such that $\mathbf{s}_1^T \mathbf{X} = (1 + \mathbf{s}_1^T \mathbf{N}) < 10$, then there is no decision error. But if the noise drives $\mathbf{s}_1^T \mathbf{X} = (1 + \mathbf{s}_1^T \mathbf{N}) > 10$, then the probability of an error is given by the area of the solid Gaussian pdf curve under H_0 to the right of the threshold value of 10.

Thus, the probability of an error under H_0, $P_{\mathrm{e}|H_0}$, is given by

$$\begin{aligned} P_{\mathrm{e}|H_0} &= \int_{10}^{\infty} \frac{1}{\sqrt{2\pi}\sigma_\Gamma} e^{-\frac{(\gamma-1)^2}{2\sigma_\Gamma^2}} d\gamma = \int_{\frac{10-1}{\sigma_\Gamma}}^{\infty} \frac{1}{\sqrt{2\pi}} e^{\frac{-t^2}{2}} dt \\ &= \int_{\frac{9}{\sqrt{19}\sigma}}^{\infty} \frac{1}{\sqrt{2\pi}} e^{\frac{-t^2}{2}} dt = \int_{\frac{9\sqrt{5\mathrm{SNR}}}{19}}^{\infty} \frac{1}{\sqrt{2\pi}} e^{\frac{-t^2}{2}} dt \\ &= \mathrm{Q}\left(\left((9/19)\sqrt{5}\right)\sqrt{\mathrm{SNR}}\right) = \mathrm{Q}\left(1.059\sqrt{\mathrm{SNR}}\right). \end{aligned} \tag{1.23}$$

The second integral on the r.h.s. of (1.23) follows from the first integral by the change of variable $t = (\gamma - 1)/\sigma_\Gamma$, the third integral follows from the second integral by using the variance $\sigma_\Gamma^2 = 19\sigma^2$, the fourth integral follows from the third integral by using $\mathrm{SNR} = 19/(5\sigma^2)$. Finally, the last expression of (1.23) follows from the definition of a complementary Gaussian distribution function of the zero mean and unit variance

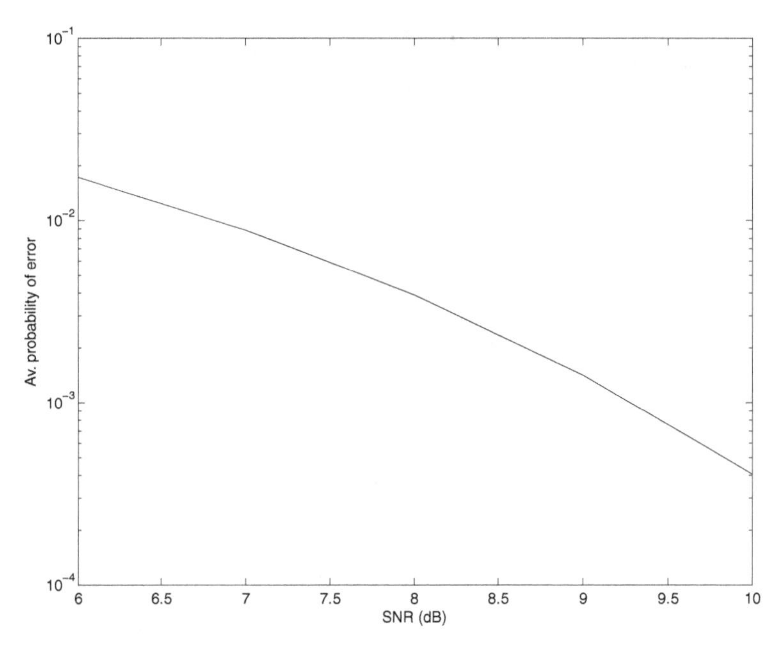

Figure 1.5 Plot of the average probability of error P_e vs. SNR(dB) for detection of $\mathbf{s}_0$ and $\mathbf{s}_1$ of (1.9) in WGN.

Gaussian r.v. defined by

$$\mathrm{Q}(x) = \int_x^{\infty} \frac{1}{\sqrt{2\pi}} e^{\frac{-t^2}{2}} dt. \tag{1.24}$$

By symmetry, under hypothesis H_1, the probability of an error is given by the area of the dashed Gaussian pdf curve under H_1 to the left of the threshold value of 10. Thus, the probability of an error under H_1, $P_{e|H_1}$, is given by

$$\begin{aligned} P_{e|H_1} &= \int_{-\infty}^{10} \frac{1}{\sqrt{2\pi}\sigma_\Gamma} e^{-\frac{(\gamma-19)^2}{2\sigma_r}} d\gamma = \int_{-\infty}^{\frac{10-19}{\sigma_\Gamma}} \frac{1}{\sqrt{2\pi}} e^{\frac{-t^2}{2}} dt \\ &= \int_{-\infty}^{\frac{-9}{\sqrt{19}\sigma}} \frac{1}{\sqrt{2\pi}} e^{\frac{-t^2}{2}} dt = \int_{-\infty}^{\frac{-9\sqrt{5\mathrm{SNR}}}{19}} \frac{1}{\sqrt{2\pi}} e^{\frac{-t^2}{2}} dt \\ &= \int_{\frac{9\sqrt{5\mathrm{SNR}}}{19}}^{\infty} \frac{1}{\sqrt{2\pi}} e^{\frac{-t^2}{2}} dt = \mathrm{Q}\left(1.059\sqrt{\mathrm{SNR}}\right). \end{aligned} \tag{1.25}$$

If hypothesis H_0 and hypothesis H_1 are equi-probable, then the average probability of an error P_e is given by

$$P_e = 1/2 \cdot P_{e|H_0} + 1/2 \cdot P_{e|H_1} = \mathrm{Q}\left(1.059\sqrt{\mathrm{SNR}}\right). \tag{1.26}$$

A plot of the average probability of error P_e of (1.25) is given in Fig. 1.5. Since the Q($\cdot$) function in (1.25) is a monotonically decreasing function, the average probability of error decreases as SNR increases as seen in Fig. 1.5.

While the performance of the binary detection system depends crucially on the SNR factor, some geometric property intrinsic to the two signal vectors $\mathbf{s}_0$ and $\mathbf{s}_1$ is also important.

Now, take the 5×1 signal vectors $\mathbf{s}_0$ and $\mathbf{s}_1$ defined by

$$\mathbf{s}_0 = \left[\sqrt{19/5},\ \sqrt{19/5},\ \sqrt{19/5},\ \sqrt{19/5},\ \sqrt{19/5}\right]^T = -\mathbf{s}_1. \tag{1.27}$$

Clearly, $\|\mathbf{s}_0\|^2 = \|\mathbf{s}_1\|^2 = E = 19$ just as before, except now the correlation of $\mathbf{s}_1$ with $\mathbf{s}_0$ yields

$$\mathbf{s}_1^T \mathbf{s}_0 = -19. \tag{1.28}$$

This means the correlation of $\mathbf{s}_1$ with $\mathbf{X}$ of (1.21) now becomes

$$\Gamma = \mathbf{s}_1^T \mathbf{X} = \begin{cases} \mathbf{s}_1^T(\mathbf{s}_0 + \mathbf{N}) \\ \mathbf{s}_1^T(\mathbf{s}_1 + \mathbf{N}) \end{cases} = \begin{cases} -19 + \mathbf{s}_1^T \mathbf{N}, \text{ under hypothesis } H_0 \\ \;\;\,19 + \mathbf{s}_1^T \mathbf{N}, \text{ under hypothesis } H_1 \end{cases}. \tag{1.29}$$

Thus, $\Gamma = \mathbf{s}_1^T \mathbf{X}$ is a Gaussian r.v. of mean of -19 and variance of $19\sigma^2$ under hypothesis H_0, while it is a Gaussian r.v. of mean of 19 and also with a variance of $19\sigma^2$ under hypothesis H_1. The new decision rule has a threshold of 0 given as the average value of -19 and 19 and is given by

$$\Gamma = \mathbf{s}_1^T \mathbf{X} \begin{cases} < 10 \Rightarrow \text{Declare hypothesis } H_0 \\ > 10 \Rightarrow \text{Declare hypothesis } H_1 \end{cases}. \tag{1.30}$$

Thus, the probability of an error under H_0, $P_{\mathrm{e}|H_0}$, is now given by

$$\begin{aligned} P_{\mathrm{e}|H_0} &= \int_0^\infty \frac{1}{\sqrt{2\pi}\sigma_\Gamma} e^{-\frac{(\gamma+19)^2}{2\sigma_\Gamma^2}} d\gamma = \int_{\frac{0+19}{\sigma_\Gamma}}^\infty \frac{1}{\sqrt{2\pi}} e^{\frac{-t^2}{2}} dt \\ &= \int_{\frac{19}{\sqrt{19}\sigma}}^\infty \frac{1}{\sqrt{2\pi}} e^{\frac{-t^2}{2}} dt = \int_{\frac{19\sqrt{5\mathrm{SNR}}}{19}}^\infty \frac{1}{\sqrt{2\pi}} e^{\frac{-t^2}{2}} dt \\ &= \mathrm{Q}\left(2.236\sqrt{\mathrm{SNR}}\right). \end{aligned} \tag{1.31}$$

Similarly, the probability of an error under H_1, $P_{\mathrm{e}|H_1}$ is given by

$$\begin{aligned} P_{\mathrm{e}|H_1} &= \int_{-\infty}^0 \frac{1}{\sqrt{2\pi}\sigma_\Gamma} e^{-\frac{(\gamma-19)^2}{2\sigma_\Gamma^2}} d\gamma = \int_{-\infty}^{\frac{0-19}{\sigma_\Gamma}} \frac{1}{\sqrt{2\pi}} e^{-\frac{t^2}{2}} dt \\ &= \int_{-\infty}^{\frac{-19}{\sqrt{19}\sigma}} \frac{1}{\sqrt{2\pi}} e^{-\frac{t^2}{2}} dt = \int_{-\infty}^{\frac{-19\sqrt{5\,\mathrm{SNR}}}{19}} \frac{1}{\sqrt{2\pi}} e^{-\frac{t^2}{2}} dt \\ &= \int_{\frac{19\sqrt{5\,\mathrm{SNR}}}{19}}^\infty \frac{1}{\sqrt{2\pi}} e^{-\frac{t^2}{2}} dt = \mathrm{Q}(2.236\sqrt{\mathrm{SNR}}) \end{aligned} \tag{1.32}$$

and the average probability of an error P_{e} is given by

$$P_{\mathrm{e}} = 1/2 \cdot P_{\mathrm{e}|H_0} + 1/2 \cdot P_{\mathrm{e}|H_1} = \mathrm{Q}(2.236\sqrt{\mathrm{SNR}}). \tag{1.33}$$

A plot of the average probability of error P_{e} of (1.33) is given in Fig. 1.6. Comparing (1.33) to (1.26) and Fig. 1.6 to Fig. 1.5, we note that the average probability of error

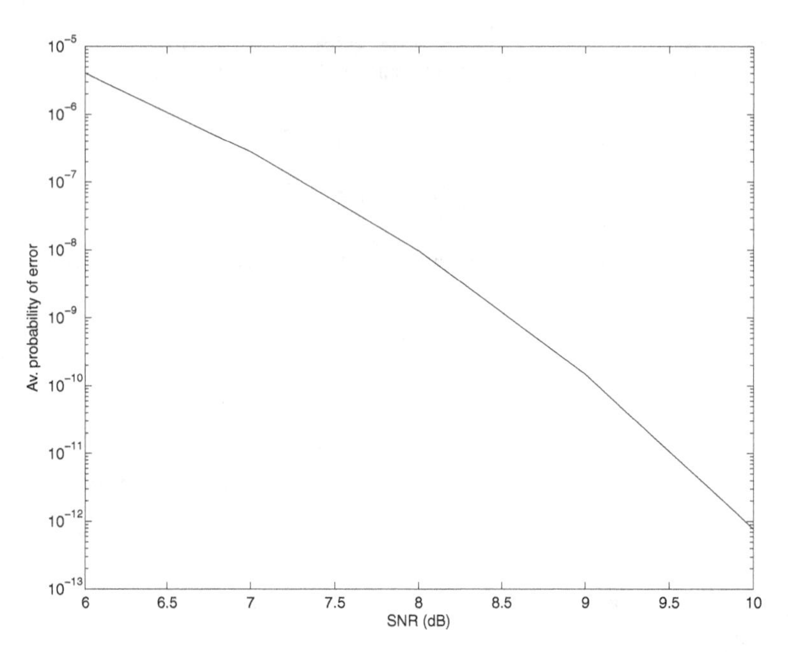

Figure 1.6 Plot of the average probability of error P_e vs. SNR(dB) for detection of $\mathbf{s}_0$ and $\mathbf{s}_1$ of (1.27) in WGN.

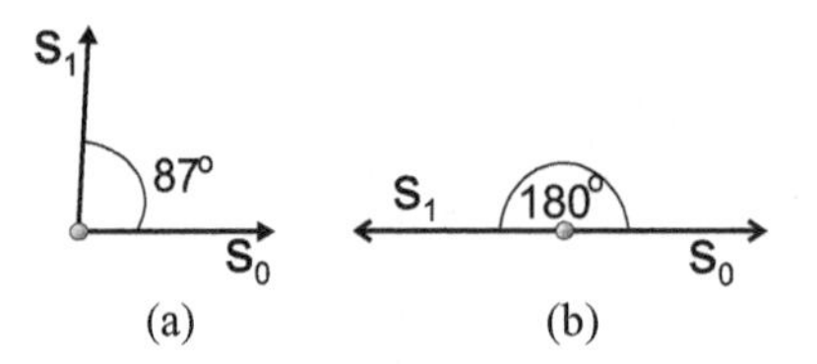

Figure 1.7 (a) $\mathbf{s}_0$ and $\mathbf{s}_1$ of (1.9) are separated by $87°$. (b) $\mathbf{s}_0$ and $\mathbf{s}_1$ of (1.26) are separated by $180°$.

for the two new signal vectors is significantly smaller compared to the two original signal vectors for the same SNR. Specifically, at SNR = 6 (dB), using the two vectors in (1.9) yields a $P_e = 1.73 \times 10^{-2}$, while using the two vectors in (1.26) yields a $P_e = 4.07 \times 10^{-6}$. As shown in (1.12), the two original signal vectors $\mathbf{s}_0$ and $\mathbf{s}_1$ of (1.9) have a correlation of $\mathbf{s}_1^T \mathbf{s}_0 = 1$, resulting in an angle between these two vectors given by $\cos^{-1}(\mathbf{s}_1^T \mathbf{s}_0 / E) = \cos^{-1}(1/19) = 87°$.

But the two new signal vectors $\mathbf{s}_0$ and $\mathbf{s}_1$ of (1.26) have a correlation of $\mathbf{s}_1^T \mathbf{s}_0 = -19$, resulting in an angle between these two vectors given by $\cos^{-1}(\mathbf{s}_1^T \mathbf{s}_0 / E) = \cos^{-1}(-19/19) = 180°$. The geometric relationships for these two sets of signal vectors are shown in Fig. 1.7. We note, from Fig. 1.7(a), the effective distance from the tip of the vectors $\mathbf{s}_0$ to $\mathbf{s}_1$ of (1.9) is much shorter than the effective distance from the tip of the vectors $\mathbf{s}_0$ to $\mathbf{s}_1$ of (1.26) as shown in Fig. 1.7(b). In Chapter 4, we will show that for two signal vectors of equal energy in WGN, the most favorable geometric relationship is for the two signal vectors to be "anti-podal" with a separating angle of $180°$ as shown for the two signal vectors of (1.26). Results in Chapter 4 will show that no two signal vectors of length 5 having equal energies for the AWGN channel binary detection problem can yield a smaller average probability of error for a specified SNR than that given by the expression in (1.33) and plotted in Fig. 1.6. The importance of the geometric relationship between signal vectors will be explored further in Chapter 5.

1.5 BPSK communication systems for different ranges

Now, we want to explore various factors that impact the receiver's detector SNR and the performance of the communication system. These factors depend on the transmitter power, the distance from a transmitter to the receiver, and various internal parameters of the receiver. In order to transmit data over the RF spectrum, we must use an authorized (agreed upon by international and national spectrum allocation agencies) carrier frequency f_0 suitable for the appropriate communication purposes. The carrier frequency f_0 determines the dimension of the transmit and receive antennas (i.e., lower frequencies need larger size antennas) and the RF propagation properties (i.e., multipath, scattering, propagation loss, etc. are highly frequency dependent) for a specific communication system requirement. For our purpose, we consider the simplest binary communication system based on the Binary Phase-Shift-Keyed (BPSK) modulation format, which is extensively used in microwave communication around the world. The two transmit signal waveforms $s_0(t)$ and $s_1(t)$ are two anti-podal sine waveforms defined by

$$\mathbf{s}_0(t) = A\sin(2\pi f_0 t),\ 0 \leq t \leq T,\ s_1(t) = -A\sin(2\pi f_0 t),\ 0 \leq t \leq T, \tag{1.34}$$

where each binary datum takes T seconds, and the information rate is given by $R = 1/T$ (bits/sec). The power P_{T} of each transmitter waveform is given by

$$P_{\mathrm{T}} = A^2/2. \tag{1.35}$$

Let P denote the power density on a sphere at a distance of r for an omni-directional transmit antenna. Then

$$P = P_{\mathrm{T}} A_{\mathrm{e}}/(4\pi r^2), \tag{1.36}$$

where A_{e} is the effective area of an antenna. Then the antenna gain G relative to an isotropic (i.e., omni-directional or omni) antenna is

$$G = 4\pi A_{\mathrm{e}}/\lambda^2. \tag{1.37}$$

The effective area of an omni-antenna is

$$A_{\mathrm{e}} = \lambda^2/4\pi. \tag{1.38}$$

Using (1.38) in (1.37) shows the gain G of an omni-antenna equals unity (consistent with its definition). Denote the transmit antenna gain as G_{T} and the receive antenna gain as G_{R}, then the received power P_{R} is given by

$$P_{\mathrm{R}} = P_{\mathrm{T}} G_{\mathrm{R}} G_{\mathrm{T}}/(4\pi r/\lambda)^2 = P_{\mathrm{T}} G_{\mathrm{R}} G_{\mathrm{T}}/L_{\mathrm{p}}, \tag{1.39}$$

where $L_{\rm p}$ denotes the path or space loss due to $1/r^2$ loss for EM waves traveling in free space given by

$$L_{\rm p} = (4\pi r/\lambda)^2, \tag{1.40}$$

where the wavelength $\lambda = c/f_0$, with $c = 2.9979 \times 10^8$ meter/sec, and f_0 is the frequency of the sinusoidal waveform. If we want to include other losses, we can denote it by L_0, and modify (1.39) to

$$P_{\rm R} = P_{\rm T}G_{\rm R}G_{\rm T}/(4\pi r/\lambda)^2 = P_{\rm R} = P_{\rm T}G_{\rm R}G_{\rm T}/(L_{\rm p}L_0). \tag{1.41}$$

The input noise to the receiver has the thermal noise $N_{\rm n}$ with a flat spectrum determined by the temperature T in kelvin facing the antenna given by

$$N_{\rm n} = kTW, \tag{1.42}$$

where k is the Boltzmann constant of 1.3807×10^{-23}, and W is the bandwidth of the RF receiver. At nominal open-air scenario, we can take $T = 290$ kelvin. Under various cooling conditions (e.g., for radio astronomy and high-performance communication applications), T can be reduced. On the other hand, if the antenna is directed toward the sun, then T can be as high as 5000 kelvin. Indeed, looking into the sun is not good for the human eyes nor good for a receiver antenna. However, various internal noises (e.g., amplifier noise, line loss, A/D noise) in the receiver can be modeled in various ways. One simplified system engineering approach is to find an equivalent temperature $T_{\rm e}$ modeling the amplifier and line losses. Furthermore, the total equivalent noise at the input of the receiver is modeled by an equivalent system temperature $T_{\rm S}$ consisting of two terms,

$$T_{\rm S} = T + T_{\rm e}. \tag{1.43}$$

Then the complications involving the gain and noise figure of the amplifier are all merged into the equivalent input temperature of $T_{\rm e}$. With T taken to be $T = 290$ kelvin, for a very high-quality receiver, $T_{\rm S}$ can be taken to be $T = 570$ kelvin, while for a noisy low-quality receiver with long line-feeds, etc., $T_{\rm S}$ can be as high as 4000 kelvin. We will neglect A/D noise at this point. An equivalent receiver input noise spectral density N_0 is then defined by

$$N_0 = kT_{\rm S}. \tag{1.44}$$

Now, a simplified equivalent transmission channel is modeled by a continuous-time AWGN channel given by

$$X(t) = \begin{cases} \tilde{s}_0(t) + N(t), \ 0 \leq t \leq T, \ H_0 \\ \tilde{s}_1(t) + N(t), \ 0 \leq t \leq T, \ H_1 \end{cases} \tag{1.45}$$

where $\tilde{s}_0(t)$ and $\tilde{s}_1(t)$ are the two resulting sinusoidal waveforms at the input of the receiver with power $P_{\rm R}$ due to the propagation of $s_0(t)$ and $s_1(t)$ of (1.34) and N is an equivalent continuous-time WGN process with zero-mean and a two-sided power spectral density of $N_0/2$ over all frequencies, where N_0 is the equivalent input noise spectral density given by (1.44). If the receiver has a gain of G_R, then the total receiver output noise at the detector is given by

$$N_{out} = G_R N_0 W = G_R k T_S W. \tag{1.46}$$

Let the receiver input power P_R be given by (1.41). We also assume the detector bandwidth R is set equal to W. For a binary BPSK system, in which each data bit has a T second duration, we can set $R = 1/T$. Then the receiver output signal power S_{out} at the detector is given by

$$S_{out} = G_R P_R. \tag{1.47}$$

We can define the signal-to-noise ratio (SNR) at the detector as the ratio of the detector signal power to the detector noise power by

$$\begin{aligned} \text{SNR} &= S_{out}/N_{out} = G_R P_R/(G_R k T_S W) = P_R/(k T_S W) \\ &= P_R T/(k T_S) = E/N_0, \end{aligned} \tag{1.48}$$

where the energy per bit E is defined by

$$E = P_R T. \tag{1.49}$$

For the BPSK system, it is known that at the detector, the average error probability satisfies

$$P_e = \mathrm{Q}(\sqrt{2E/N_0}) = \mathrm{Q}(\sqrt{2\text{SNR}}). \tag{1.50}$$

In order to achieve an average error probability of $P_e = 10^{-6} = \mathrm{Q}(4.753)$, which is a commonly used measure of the quality of a high-performance digital communication system, then we need to have

$$4.753 = \sqrt{2\text{SNR}}, \tag{1.51}$$

or

$$\text{SNR} = 11.296. \tag{1.52}$$

By using the value of the SNR of (1.52) in (1.48) and using P_R of (1.41), we can solve for the required transmitter power P_T as follows:

$$\begin{aligned} P_T &= 11.296 \times k \times T_S \times R \times L_p \times L_0/(G_T \times G_R) \\ &= 11.296 \times 1.3807 \times 10^{-23} \times (4\pi/(2.9979 \times 10^8))^2 \times T_S \times R \times f_0^2 \\ &\quad \times r^2 \times L_0/(G_T G_R) \\ &= 2.7404 \times 10^{-37} \times T_S \times R \times f_0^2 \times r^2 \times L_0/(G_T G_R). \end{aligned} \tag{1.53}$$

Now, consider four examples of interest.

Example 1.1 Consider a typical low-cost intra-room communication system with range $r = 3$ meters, rate $R = 1$M bits/sec, frequency $f_0 = 2.4$ GHz, and $T_S = 4000$ kelvin. Assume omni-antennas with $G_T = G_R = 1$ and $L_0 = 1$. Then $P_T = 5.68 \times 10^{-8}$ watts. □

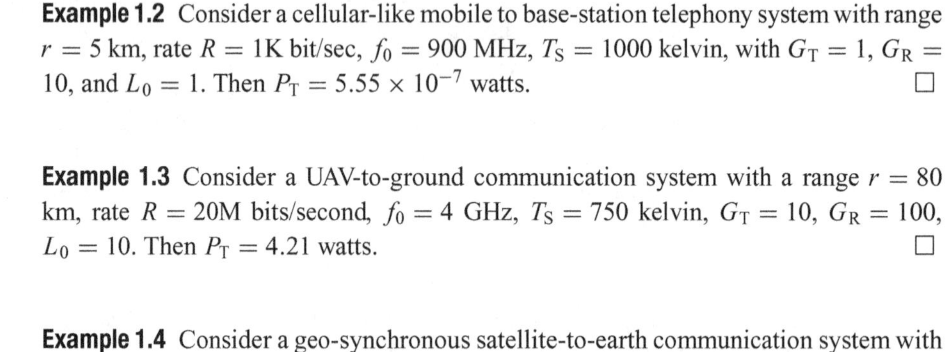

Example 1.2 Consider a cellular-like mobile to base-station telephony system with range $r = 5$ km, rate $R = 1\text{K}$ bit/sec, $f_0 = 900$ MHz, $T_S = 1000$ kelvin, with $G_T = 1$, $G_R = 10$, and $L_0 = 1$. Then $P_T = 5.55 \times 10^{-7}$ watts. □

Example 1.3 Consider a UAV-to-ground communication system with a range $r = 80$ km, rate $R = 20\text{M}$ bits/second, $f_0 = 4$ GHz, $T_S = 750$ kelvin, $G_T = 10$, $G_R = 100$, $L_0 = 10$. Then $P_T = 4.21$ watts. □

Example 1.4 Consider a geo-synchronous satellite-to-earth communication system with a slant range $r = 4 \times 10^7$ meters, rate $R = 500\text{M}$ bits/sec, $f_0 = 4$ GHz, $T_S = 500$ kelvin, $G_T = 3 \times 10^4$, $G_R = 1000$, and $L_0 = 10$. Then $P_T = 584.7$ watts. □

Several comments are worthy of attention.

1. The transmitter power P_T in the above model is the radiated RF power. The actual transmitter power needed to radiate P_T watts into space will be higher depending on the efficiencies of the high-power RF amplifiers in converting DC power to radiating RF power.
2. Of course, additional DC powers are needed to operate the other subsystems of the transmitter and receiver of a BPSK system. In a UAV or satellite systems, considerable additional powers are needed for other non-communication purposes.
3. The evaluated P_T values for the four cases represent the minimum transmitter radiated powers in order to achieve an average error probability error of $P_e = 10^{-6}$. All sorts of additional realistic system and propagation losses and degradations were not included in these calculations.
4. The transmission from a mobile to a base station in Example 1.2 assumed an idealistic line-of-sight propagation with no multipath and other disturbances.
5. For any one of the four cases, suppose we use that P_T value, but in practice suffer an additional loss modeled by $L_0 = 2$ (which is a very modest possible loss). Then from (1.41), the received power P_R will decrease by a factor of 2, and from (1.48), the SNR value will also decrease by a factor of 2. But from (1.50)–(1.52), this means the argument of the Q(·) function will decrease by a factor of square root of 2. Thus, now the new $P_e = \text{Q}(4.752/\sqrt{2}) = \text{Q}(3.361) = 3.611 \times 10^{-4}$, which is much higher than the original desired $P_e = 1 \times 10^{-6}$. Thus, under this scenario, the system performance may not be acceptable.
6. The bottom line is that since the Q(·) function is a very rapidly decreasing function, a small decrease of its argument can cause a large increase in its value. In practice, if one wants to design a robust BPSK system (to be able to operate even in various unexpected system and propagation degraded conditions), then one needs to include additional "margins" by increasing the actual operating transmit power possibly considerably above those evaluated using (1.53).

7. Modern communication systems may also incorporate more advanced modulation (more efficient than the simple BPSK modulation considered here) and coding techniques in order to achieve the required P_e using a lower SNR than that of the BPSK system.
8. Examples 1.2, 1.3, and 1.4 show that a desired high value of SNR at the receiver detector can be obtained not only from a high transmit power P_T value, but also by using high gain transmit and receive antennas with high values of G_T and G_R. These high-gain antennas can be obtained through various methods including using beam-forming techniques.
9. Since modern portable/mobile communication systems all use batteries, there are urgent needs to reduce battery power/energy usages but still meet the required system performances. We have shown that various methods can be utilized to meet these needs.

1.6 Estimation problems

There are many parameters of theoretical and practical interest in any complex communication, radar, and signal processing system. Various methods can be used to estimate these parameters. In this section, we consider some simple examples dealing with basic estimation methods that will motivate us to consider the more detailed study of parameter estimation in Chapter 7.

1.6.1 Two simple estimation problems

Example 1.5 Consider an everyday problem involving an estimation. Suppose we know a train leaves from station A at 3 p.m. and arrives at station C at 3:40 p.m. over a distance of 10 miles. Assuming the train travels at a constant speed, we want to estimate when the train will arrive at station B, which is 6.5 miles from station A. Clearly, this is an estimation problem that can be modeled by a polynomial of degree 1. Denote the traveled distance by x and the traveled time by y. Then we have two known pairs of data, with $(x_A, y_A) = (0, 0)$ and $(x_C, y_C) = (10, 40)$. Then a straight line given by a polynomial of degree 1, $y = ax + b$, can be found to satisfy these two pairs of data, yielding

$$y = (40/10)x, \tag{1.54}$$

where the two coefficients are given by $a = 40/10 = 4$ and $b = 0$. Once we have the model of (1.54), then to reach station B with a distance of $x_B = 6.5$ miles, we can find the travel time of $y_B = 26$ minutes. □

Example 1.6 Now, consider a problem in which the dependent scalar variable y is linearly related to the independent scalar variable x in the form of

$$y = a_0 x. \tag{1.55}$$

Since there is only one parameter a_0 in (1.55), then from any pair of known values of (x', y') satisfying (1.55), we can determine the parameter $a_0 = y'/x'$. However, in many practical situations, each measurement of y at time m, denoted by $y(m)$, is corrupted by an additive noise (AN) N at time m, denoted by $M(m)$, on the independent variable x at time m, denoted by $x(m)$ for $i = 1, \ldots n$. Then we have the AN channel model of

$$y(i) = a_0 x(i) + N(i),\ i = 1, \ldots, n. \tag{1.56}$$

Using a vector notation with $\mathbf{y} = [y(1), \ldots, y(n)]^T$, $\mathbf{x} = [x(1), \ldots, x(n)]^T$, and $\mathbf{N} = [N(1), \ldots, N(n)]^T$, (1.56) can be expressed as

$$\mathbf{y} = a_0 \mathbf{x} + \mathbf{N}. \tag{1.57}$$

In (1.57), while $\mathbf{y}$, $\mathbf{x}$, and $\mathbf{N}$ are all $n \times 1$ column vectors, the coefficient a_0 is a scalar parameter. Unlike the noise-free model of (1.55), where a single pair of known values of (x', y') can determine a_0 perfectly, we may want to use n pairs of known $(\mathbf{x}, \mathbf{y})$ to average over the random effects of the noise $N(i)$, $i = 1, \ldots, n$, to estimate a_0, possibly better than using only one specific pair of $(x(i), y(i))$.

Suppose

$$\mathbf{x} = [1,\ 2,\ 3,\ 4,\ 5,\ 6,\ 7,\ 8,\ 9,\ 10,\ 11,\ 12]^T, \tag{1.58}$$

and $a_0 = 0.5$. In the noise-free model of (1.55), then

$$\mathbf{y} = [0.5,\ 1,\ 1.5,\ 2,\ 2.5,\ 3,\ 3.5,\ 4,\ 4.5,\ 5,\ 5.5,\ 6]^T. \tag{1.59}$$

As stated earlier, any one of the 12 known pairs of $(x(i), y(i))$, such as $x(5) = 5$ and $y(5) = 2.5$, could yield the parameter $a_0 = 2.5/5 = 0.5$. Indeed, if we plot these 12 pairs of $(\mathbf{x}, \mathbf{y})$ points in Fig. 1.8(a), we notice a straight line of slope 0.5 goes over exactly all these 12 pairs of points. Thus, the estimate of $a_0 = 0.5$ is clearly optimum.

Now, consider an AN channel model of (1.57), where $\mathbf{x}$ is still given by (1.58), and assume all the components of $\mathbf{N}$ are specific realizations of independent zero mean Gaussian r.v.'s with unit variances, yielding

$$\begin{aligned}\mathbf{y}_1 = [&0.0951,\ 1.2007,\ 2.3324,\ 3.0888,\ 2.4007,\ 3.7403,\ 3.4688,\ 4.6312,\ 4.6326,\\ &4.8711, 5.9383,\ 6.6590]^T.\end{aligned} \tag{1.60}$$

A plot of these new 12 pairs of $(\mathbf{x}, \mathbf{y}_1)$ points are given in Fig. 1.8(b). Due to the effects of the noise $\mathbf{N}$, these 12 pairs of points do not lie on a straight line. However, intuitively, we want to draw a straight line so that as many of these points as possible lie close to the straight line. A straight line chosen in an ad hoc manner with a slope of $\tilde{a}_0 = 0.56$ was placed in Fig. 1.8(b). We note some of the points are above the straight line and some are below the straight line. An important question is, what is the "optimum" $\hat{a}_0$ that can be used in Fig. 1.8(b)? □

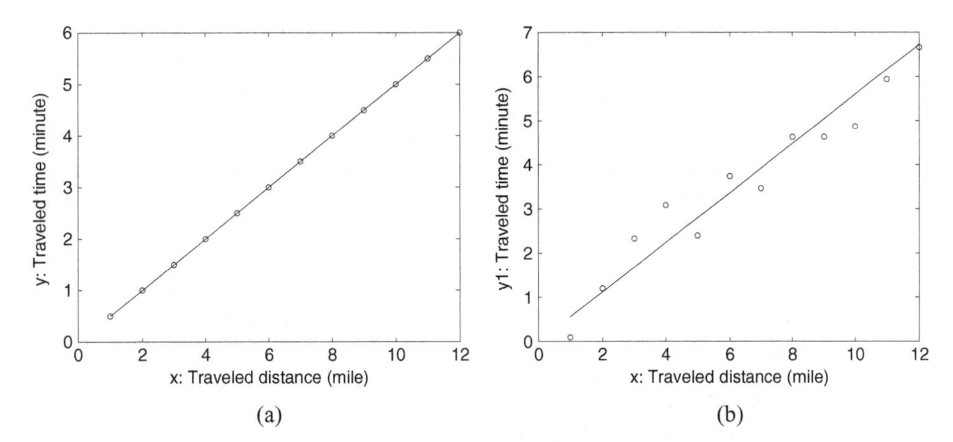

Figure 1.8 (a) Plot of $\mathbf{x}$ of (1.58) vs. noise-free $\mathbf{y}$ of (1.59) with a straight line of slope $a_0 = 0.5$. (b) Plot of $\mathbf{x}$ of (1.58) vs. noisy $\mathbf{y}_1$ of (1.60) with a straight line of slope $\tilde{a}_0 = 0.56$.

1.6.2 Least-absolute-error criterion

In order to talk about an "optimum" estimate of a_0, we must propose a criterion of optimality. Thus, in estimation theory, selecting a criterion of goodness is an important matter, as we will demonstrate in this section. Consider the error $e(i)$ between the observed $y(i)$ and what the noise-free model of (1.55) would predict from $a_0 x(i)$ for $i = 1, \ldots, n$, defined by

$$e(i) = y(i) - a_0 x(i), \; i = 1, \ldots, n. \tag{1.61}$$

Writing (1.61) in a vector notation, we have

$$\mathbf{e} = \mathbf{y} - a_0 \mathbf{x}, \tag{1.62}$$

where the error vector $\mathbf{e} = [e(1), \ldots, e(n)]^T$.

Back to our problem, with $\mathbf{x}$ given by (1.58) and $\mathbf{y}_1$ given by (1.60), how do we pick a_0 to minimize $\mathbf{e}$ of (1.62)? First, we can choose the absolute-error (AE) criterion defined by

$$\varepsilon_{\mathrm{AE}}(a_0) = \sum_{i=1}^{n} |y(i) - a_0 x(i)|, \tag{1.63}$$

where the AE $\varepsilon_{\mathrm{AE}}(a_0)$ expression is a function of a_0 for the given pair of vectors $\{\mathbf{x}, \mathbf{y}\}$. Then the least-absolute-error (LAE) criterion of the estimate of a_0 is defined by

$$\varepsilon_{\mathrm{AE}}(\hat{a}_0^{\mathrm{AE}}) = \min_{a_0 \in \mathbb{R}} \{\varepsilon_{\mathrm{AE}}(a_0)\} = \min_{a_0 \in \mathbb{R}} \left\{ \sum_{i=1}^{n} |y(i) - a_0 x(i)| \right\}. \tag{1.64}$$

In (1.64), since $\varepsilon_{\mathrm{AE}}(a_0)$ is a function of a_0 for the given pair of vectors $\{\mathbf{x}, \mathbf{y}\}$, we want to minimize $\varepsilon_{\mathrm{AE}}(a_0)$ with respect to (w.r.t.) a_0 among all possible real-valued numbers. The a_0 that actually attains the minimum of $\varepsilon_{\mathrm{AE}}(a_0)$ is denoted by $\hat{a}_0^{\mathrm{AE}}$ as the LAE criterion estimate of this problem.

The graphical interpretation of picking the optimum $\hat{a}_0^{\mathrm{AE}}$ under the LAE criterion in Fig. 1.8(b) is to minimize the sum of the magnitude of the vertical deviations from each

Salviati: Adunque, come questi osservatori sien tall, e che pur con tutto ciò abbianc
errato e peró convenga emendar loro errori, per poter dalle loro osservazioni
ritrar quel pio di notizia che sia possibile, conveniente cosa é che nol gli
applichiamo le minori e pio vicine emende e correzioni che si possa, purch' elle
bastino a ritrar l 'osservazioni dall' impossibilitá all possibilitá: si che,
verblgrazia, se si puo temperar un mainfesto errore ed un patente impossibile di
una loro osservazione con l' aggiunere o detrar 2 o ver 3 minuti, e con tale emenda
ridurlo al possibile, non si deva volerlo aggiustare con la giunta o suttrazione
del 15 o 20 o 50.

Therefore these observations being such, and that yet notwithstanding they did
err, and so consequently needed correction, that so one might from their
observations infer the best hints that may be; it is convenient that we apply unto
them the least and nearest emendations and corrections that may be; so that
they do but suffice to reduce the observations from impossibility to possibility;
so as, for instance if one may but correct a manifest error, and an apparent
impossibility of one of their observations by the addition or subtraction of
two or three minutes, and with that amendment to reduce it to possibility, a
man ought not to essay to adjust it by the addition or subtraction of 15, 20, or 50.

Figure 1.9 An extract of a paragraph (in Italian followed by an English translation) from "Dialogo Dei Massimi Sistemi" by Galileo Galilei, 1632.

$(x(i), y_1(i))$ point to the straight line. Intuitively, the AE criterion is reasonable, since a positive valued error of a given amount seems to be as undesirable as a negative-valued error of the same magnitude. The previously chosen straight line in an ad hoc manner with a slope of $\tilde{a}_0$ in Fig. 1.8(b) had points both above and below the straight line.

It is most interesting that the AE criterion was first hinted at by Galileo [8] in his astronomical data analysis of heavenly bodies in the seventeenth century, as shown in the text (first in Italian) followed by an English translation given in Fig. 1.9.

Unfortunately, the analytical solution of the LAE solution $\hat{a}_0^{\text{AE}}$ of (1.64) requires a complicated optimization procedure, which was not possible in Galileo's time. Today, for example, we can use Matlab's iterative complicated minimization "fminsearch" function (among many other optimization tools) to solve the problem of (1.64). In any case, the LAE solution for the given $\mathbf{x}$ of (1.58) and $\mathbf{y}_1$ of (1.60) yields the optimum coefficient $\hat{a}_0^{\text{AE}} = 0.5148$, while the minimum AE $\varepsilon_{\text{AE}}(\hat{a}_0^{\text{AE}}) = 4.889$, while using the arbitrarily chosen $a_0 = 0.56$ yields an $\varepsilon_{\text{AE}}(a_0 = 0.56) = 5.545$. In other words, there is no other coefficient a_0 such that $\varepsilon_{\text{AE}}(a_0)$ can be made smaller than $\varepsilon_{\text{AE}}(\hat{a}_0^{\text{AE}}) = 4.889$.

1.6.3 Least-square-error criterion

We have seen the LAE solution requires a sophisticated minimization operation, which was not practical before the days of modern digital computations. Thus, we may want to seek another error criterion so that the optimum solution based on that criterion can be solved more readily. It turns out Laplace and Gauss (both also in their astronomical data analysis work) formulated the least square-error (SE) criterion as follows:

$$\varepsilon_{\text{LS}}^2(a_0) = \sum_{i=1}^{n}(y(i) - a_0 x(i))^2 = \sum_{i=1}^{n}(y^2(i) - 2a_0 x(i)y(i) + a_0^2 x^2(i)). \tag{1.65}$$

Then the solution for a_0 based on the LSE criterion becomes

$$\varepsilon_{\mathrm{LS}}^2(\hat{a}_0^{\mathrm{LS}}) = \min_{a_0 \in \mathbb{R}}\{\varepsilon_{\mathrm{LS}}^2(a_0)\} = \min_{a_0 \in \mathbb{R}}\left\{\sum_{i=1}^{n}(y^2(i) - 2a_0x(i)y(i) + a_0^2x^2(i))\right\}. \tag{1.66}$$

In particular, the minimization of $\varepsilon_{\mathrm{LS}}^2(a_0)$ (i.e., the bracketed term on the r.h.s. of (1.66)) as a function of a_0 can be performed using a variational method. Thus, setting to zero the partial derivative of $\varepsilon_{\mathrm{LS}}^2(a_0)$ w.r.t. a_0 results in

$$\begin{aligned}\frac{\partial\varepsilon_{\mathrm{LS}}^2(a_0)}{\partial a_0} &= \frac{\partial\left\{\sum_{i=1}^{M}(y^2(i) - 2a_0x(i)y(i) + a_0^2x^2(i))\right\}}{\partial a_0}\\ &= \sum_{i=1}^{n} -2x(i)y(i) + 2a_0x^2(i) = 0.\end{aligned} \tag{1.67}$$

The solution of a_0 in (1.67) denoted by $\hat{a}_0^{\mathrm{LS}}$ is given by

$$\hat{a}_0^{\mathrm{LS}} = \frac{\sum_{i=1}^{n} x(i)y(i)}{\sum_{i=1}^{n} x^2(i)}. \tag{1.68}$$

If we take the second partial derivative of $\varepsilon_{\mathrm{LS}}^2(a_0)$ w.r.t. a_0 (by taking the first partial derivative of the expression in (1.67) w.r.t. to a_0 again), then we obtain

$$\frac{\partial^2\varepsilon_{\mathrm{LS}}^2(a_0)}{\partial a_0^2} = 2\sum_{i=1}^{n} x^2(i) > 0. \tag{1.69}$$

But since $\varepsilon_{\mathrm{LS}}^2(a_0)$ of (1.65) is a quadratic (i.e., a second-order polynomial) equation of a_0, then $\varepsilon_{\mathrm{LS}}^2(a_0)$ function must be either a $\cup$ shaped or a $\cap$ shaped function. Then an extremum solution (i.e., the solution obtained from $\partial\varepsilon_{\mathrm{LS}}^2(a_0)/\partial a_0 = 0$) yields a minimum solution if $\varepsilon_{\mathrm{LS}}(a_0)$ is $\cup$ shaped or yields a maximum solution if $\varepsilon_{\mathrm{LS}}^2(a_0)$ is $\cap$ shaped. However, from (1.67) and (1.69), the $\varepsilon_{\mathrm{LS}}^2(a_0)$ of (1.65) is $\cup$ shaped, since the second partial derivative of $\varepsilon_{\mathrm{LS}}^2(a_0)$ w.r.t. a_0 at the point of $\partial\varepsilon_{\mathrm{LS}}^2(a_0)/\partial a_0 = 0$ is always positive. However, for a quadratic equation, the only local minimum solution given by $\hat{a}_0^{\mathrm{LS}}$ is indeed a global minimum solution. This means the solution $\hat{a}_0^{\mathrm{LS}}$ given by (1.68) is not only optimum in the LSE sense, but is also computationally tractable involving only M sums in its numerator term, M sums in its denominator term, plus one division (without needing non-linear optimization computations). Thus, the LSE solution can be performed with hand calculations just as in the days of Laplace and Gauss, unlike the LAE solution, which requires solving a non-linear optimization problem.

For $\mathbf{x}$ of (1.58) and $\mathbf{y}_1$ of (1.60), the optimum LSE solution based on (1.68) yields $\hat{a}_0^{\mathrm{LS}} = 0.5312$ with the associated minimum SE $\varepsilon_{\mathrm{LS}}^2(\hat{a}_0^{\mathrm{LS}}) = 2.807$. If we use an arbitrary $a_0 = 0.56$, then $\varepsilon_{\mathrm{LS}}^2(a_0 = 0.56) = 3.347$. In other words, there is no a_0 such that $\varepsilon_{\mathrm{LS}}^2(a_0)$ can be smaller than $\varepsilon_{\mathrm{LS}}^2(\hat{a}_0^{\mathrm{LS}}) = 2.807$.

1.6.4 Estimation robustness

The computational simplicity of the LS method is a major factor in its near universal usage in data analysis and estimation. However, the LSE method does have a major

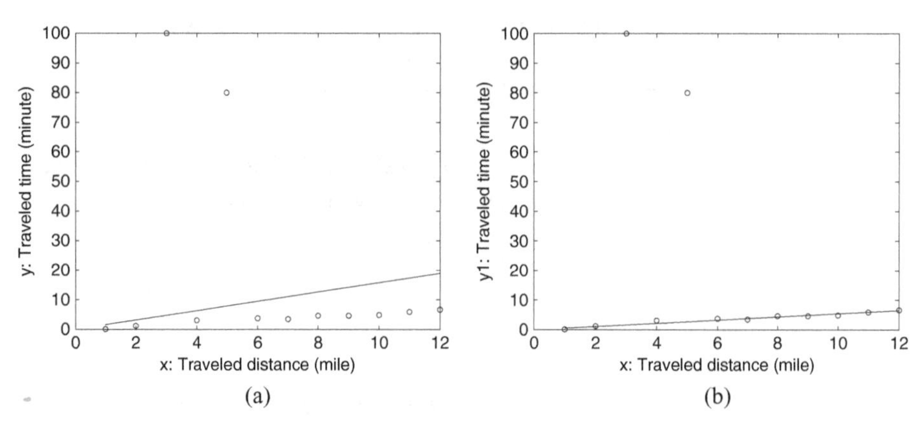

Figure 1.10 (a) Plot of $\mathbf{x}$ of (1.58) vs. $\mathbf{y}_2$ of (1.70) with 2 outliers and a straight line of slope $\hat{a}_0^{\mathrm{LS}} = 1.579$. (b) Plot of $\mathbf{x}$ of (1.58) vs. $\mathbf{y}_2$ of (1.70) with 2 outliers and a straight line of slope $\hat{a}_0^{\mathrm{AE}} = 0.5398$.

non-robust property, when there are some "outlier" (x, y) pairs of points, where y is very different from what $a_0 x$ predicts (due to some very large errors).

Example 1.7 Consider $\mathbf{x}$ of (1.58) and $\mathbf{y}_1$ of (1.60) of Example 1.6. Replace $\mathbf{y}_1$ by $\mathbf{y}_2$ defined by

$$\mathbf{y}_2 = [0.0951,\ 1.2007,\ 100,\ 3.0888,\ 80,\ 3.7403,\ 3.4688,\ 4.6312,\ 4.6326,\ 4.8711,\ 5.9383,\ 6.6590]^T, \tag{1.70}$$

where $y_2(3) = 100$ (instead of the previous $y_1(3) = 2.3324$) and $y_2(5) = 80$ (instead of the previous $y_1(5) = 2.4007$). We can still use (1.68) to find the LSE criterion solution, yielding $\hat{a}_0^{\mathrm{LS}} = 1.579$ with the associated minimum square-error $\varepsilon_{\mathrm{LS}}^2(\hat{a}_0^{\mathrm{LS}}) = 1.567 \times 10^4$. A plot of $\mathbf{x}$ versus $\mathbf{y}_2$ with a straight line of slope $\hat{a}_0^{\mathrm{LS}} = 1.579$ is shown in Fig. 1.10(a). However, using the LSE criterion, it seems most of the pairs of (x, y_2) points are not close to the straight line with a slope given by $\hat{a}_0^{\mathrm{LS}} = 1.579$. Indeed, using the slope of $\hat{a}_0^{\mathrm{LS}} = 1.579$, for $x = 4$, the predicted $y = 6.315$, is quite different from the noise free solution of $y = 2$. However, if we use the LAE criterion, its optimum solution $\hat{a}_0^{\mathrm{AE}} = 0.5398$ with the associated LAE $\varepsilon_{\mathrm{AE}}^2(\hat{a}_0^{\mathrm{AE}}) = 179$. A plot of $\mathbf{x}$ versus $\mathbf{y}_2$ with a straight line of slope $\hat{a}_0^{\mathrm{AE}} = 0.5398$ is shown in Fig. 1.10(b). Indeed, using the AE criterion, it seems most of the pairs of (x, y_2) points are much closer to the straight line with a slope given by $\hat{a}_0^{\mathrm{AE}} = 0.5398$. This means, with the exception of the two outlier pairs of points (3, 100) and (5, 80), the straight line with a slope given by $\hat{a}_0^{\mathrm{AE}} = 0.5398$ provides a quite good estimate of the y_2 values for a given x value. By using the slope of $\hat{a}_0^{\mathrm{AE}} = 0.5398$, for $x = 4$, the estimated $y = 2.159$, which is much closer to the noise-free solution of $y = 2$. Thus, when large errors occur (due to obvious human errors in data recordings or very large electronic noise disturbances in a few instances), the optimum AE criterion solution may provide a more robust solution than the optimum SE criterion solution. The result can be seen from the fact, due to the squaring effect in the SE criterion, that the squares of the large errors exert much more influence than the squares of small errors in the LS optimization operations. Thus, the estimated straight line tries to move closer to the outlier pair of points, neglecting most of the other pairs of points. Since it is most likely that these outliers probably do not represent the true behaviors of the x-y relationships, the optimum AE criterion pays less attention to the influence of these outliers, thus probably providing a better estimation of the behaviors of the x-y relationships (as demonstrated in Fig. 1.10(b)). □

1.6.5 Minimum mean-square-error criterion

So far, in using either the AE or square-error criteria, we have not exploited any statistical information on the noise **N** in the AN channel model. Now, we want to introduce the mean-square (MS) error criterion, in which by exploiting some statistical correlation information among different samples of the data, the mean-square-error (MSE) may be minimized.

Consider a real-valued random sequence

$$\{X(i),\ -\infty < i < \infty\}, \tag{1.71}$$

whose statistical moments are known. Denote its n-th moment M_n by

$$M_n = \mathrm{E}\left\{X^n(i)\right\},\ 1 \leq n, \tag{1.72}$$

which has the same value for all i. We assume its first moment $m_1 = \mathrm{E}\{X(i)\} = 0$. The second moment $m_2 = \mathrm{E}\{X^2(i)\} = \sigma^2$ is the variance. Denote its autocorrelation sequence $R(n)$ by

$$R(n) = \mathrm{E}\left\{X(i)X(i+n)\right\},\ -\infty < n < \infty. \tag{1.73}$$

Then $R(n)$ is an even-valued function of n and $R(0) = \sigma^2$. We assume all the moments and all the autocorrelation sequence values of this random sequence are known. At this point, we will not discuss how these moments can be evaluated from observations of this random sequence. There are various ways to perform these evaluations.

Example 1.8 Given a real-valued random sequence $\{X(i),\ -\infty < i < \infty\}$ satisfying all the conditions of (1.71)–(1.73), suppose we want to estimate the r.v. $X(2)$ using the r.v. $a_1X(1)$, where a_1 is a coefficient to be chosen appropriately. Then the estimation error is given by

$$e(a_1) = X(2) - a_1X(1). \tag{1.74}$$

Consider the use of the mean-square-error (MSE) criterion applied to (1.74) given by

$$\begin{aligned} \varepsilon_{\mathrm{MS}}(a_0) &= \mathrm{E}\left\{(X(2) - a_1X(1))^2\right\} = \mathrm{E}\left\{(X(2))^2 - 2a_1X(1)X(2) + a_1^2(X(1))^2\right\} \\ &= R(0) - 2a_1R(1) + a_1^2R(0). \end{aligned} \tag{1.75}$$

Then the minimum mean-square-error (MMSE) criterion solution for the optimum $\hat{a}_1^{\text{MS}}$ satisfies

$$\varepsilon_{\text{MS}}^2(\hat{a}_1^{\text{MS}}) = \min_{a_1 \in \mathbb{R}} \left\{ \varepsilon_{\text{MS}}^2(a_1) \right\} = \min_{a_1 \in \mathbb{R}} \left\{ R(0) - 2a_1 R(1) + a_1^2 R(0) \right\}. \tag{1.76}$$

Comparing the MSE $\varepsilon_{\text{MS}}^2(a_1)$ of (1.76) to the LSE $\varepsilon_{\text{LS}}^2(a_0)$ of (1.66), we notice $\varepsilon_{\text{MS}}^2(a_1)$ is again a quadratic function of a_1. Thus, we can use the variational method used in (1.67) to obtain

$$\frac{\partial \varepsilon_{\text{MS}}^2(a_1)}{\partial a_1} = -2R(1) + 2a_1 R(0) = 0. \tag{1.77}$$

Thus,

$$\hat{a}_1^{\text{MS}} = \frac{R(1)}{R(0)} \tag{1.78}$$

is the global minimum solution of $\varepsilon_{\text{MS}}^2(a_1)$, following the same arguments showing that the $\hat{a}_0^{\text{LS}}$ solution of (1.67) is the global minimum solution of $\varepsilon_{\text{LS}}^2(a_0)$. Furthermore, the associated $\varepsilon_{\text{LS}}(a_0)$ minimum is given by

$$\varepsilon_{\text{MS}}^2(\hat{a}_1^{\text{MS}}) = R(0) - \frac{R^2(1)}{R(0)} = R(0)\left(1 - \frac{R^2(1)}{R^2(0)}\right), \tag{1.79}$$

and the minimum normalized $\varepsilon_{\text{N}}^2(\hat{a}_1^{\text{MS}})$ is given by

$$\varepsilon_N^2(\hat{a}_1^{\text{MS}}) = \frac{\varepsilon_{\text{MS}}^2(\hat{a}_1^{\text{MS}})}{R(0)} = 1 - \frac{R^2(1)}{R^2(0)}. \tag{1.80}$$

In using MSE analysis, it is not the value of $\varepsilon_{\text{MS}}^2(a_1)$, but its normalized value that is important. For a fixed $R(1)/R(0)$ value, $\varepsilon_{\text{MS}}^2(a_1)$ in (1.79) increases linearly with $R(0)$, but $\varepsilon_{\text{N}}^2(\hat{a}_1^{\text{MS}})$ is dependent only on the $R(1)/R(0)$ value. Indeed, if $R(1) = \text{E}\{X(2)X(1)\} = 0$, it implies that on the average the r.v. $X(2)$ has no statistical correlation to the r.v. $X(1)$. Thus, trying to use $a_1 X(1)$ to estimate $X(2)$ should be statistically irrelevant. In this case, $\hat{a}_1^{\text{MS}} = 0/R(0) = 0$, then (1.79) shows the associated $\varepsilon_{\text{MS}}^2(\hat{a}_1^{\text{MS}}) = R(0)$ and (1.80) shows $\varepsilon_N^2(\hat{a}_1^{\text{MS}}) = 1$.

We can generalize (1.74) by trying to use $a_1 X(1) + a_2 X(2)$ to estimate $X(3)$, and obtain the equivalent version of (1.75) to (1.79) as

$$e(a_1, a_2) = X(3) - (a_1 X(1) + a_2 X(2)), \tag{1.81}$$

$$\varepsilon_{\text{MS}}(a_1, a_2) = R(0) - 2a_1 R(2) - 2a_2 R(1) + a_1^2 R(0) + a_2^2 R(0), \tag{1.82}$$

$$\begin{aligned} \varepsilon_{\text{MS}}^2(\hat{a}_1^{\text{MS}}, \hat{a}_2^{\text{MS}}) &= \min_{[a_1, a_2]^T \in \mathbb{R}^2} \{\varepsilon_{\text{MS}}(a_1, a_2)\} \\ &= \min_{[a_1, a_2]^T \in \mathbb{R}^2} \left\{ R(0) - 2a_1 R(2) - 2a_2 R(1) + a_1^2 R(0) + a_2^2 R(0) \right\}, \end{aligned} \tag{1.83}$$

$$\frac{\partial \varepsilon_{\text{MS}}^2(a_1, a_2)}{\partial a_1} = -2R(2) + 2a_1 R(0) + 2a_2 R(1) = 0, \tag{1.84}$$

$$\frac{\partial \varepsilon_{\text{MS}}^2(a_1, a_2)}{\partial a_2} = -2R(1) + 2a_2 R(0) + 2a_1 R(1) = 0. \tag{1.85}$$

Then the optimum solutions for $\hat{a}_1^{\mathrm{MS}}$ and $\hat{a}_2^{\mathrm{MS}}$ from (1.84) and (1.85) can be expressed as

$$\begin{bmatrix} R(0) & R(1) \\ R(1) & R(0) \end{bmatrix} \begin{bmatrix} \hat{a}_1^{\mathrm{MS}} \\ \hat{a}_2^{\mathrm{MS}} \end{bmatrix} = \begin{bmatrix} R(2) \\ R(1) \end{bmatrix}. \tag{1.86}$$

Equation (1.86) can be written as

$$\mathbf{R}_2 \hat{\mathbf{a}}_2 = \mathbf{r}_2, \tag{1.87}$$

where

$$\hat{\mathbf{a}}_2 = \begin{bmatrix} \hat{a}_1^{\mathrm{MS}} \\ \hat{a}_2^{\mathrm{MS}} \end{bmatrix}, \quad \mathbf{R}_2 = \begin{bmatrix} R(0) & R(1) \\ R(1) & R(0) \end{bmatrix}, \quad \hat{\mathbf{R}}_2 = \begin{bmatrix} R(2) \\ R(1) \end{bmatrix}. \tag{1.88}$$

If $\mathbf{R}_2$ is not singular, so its inverse $\mathbf{R}_2^{-1}$ exists, then the explicit solution for $\hat{\mathbf{a}}_2$ is given by

$$\hat{\mathbf{a}}_2 = \mathbf{R}_2^{-1} \mathbf{r}_2. \tag{1.89}$$

In practice, we can solve the linear system of two equations with two unknowns in (1.86) and (1.87) without needing to actually use the inverse of the $\mathbf{R}_2$ matrix. From a numerical computational point of view, using an inverse of a matrix is not recommended for actual computations. The associated $\varepsilon_{\mathrm{MS}}^2(\hat{\mathbf{a}}_2)$ becomes

$$\varepsilon_{\mathrm{MS}}^2(\hat{\mathbf{a}}_2) = R(0) - \mathbf{r}_2^T \mathbf{R}_2^{-1} \mathbf{r}_2. \tag{1.90}$$

Once we have formulated the optimum $\hat{a}_1^{\mathrm{MS}}$ and $\hat{a}_2^{\mathrm{MS}}$ as two components of $\hat{\mathbf{a}}_2$ with its solution given by the linear system of equations expressed by (1.87) and (1.89), we can generalize (1.74) to n r.v.'s. Specifically, denote the $n \times 1$ vector $\mathbf{X}_n$ of r.v.'s and the associated $n \times 1$ $\mathbf{a}_M$ coefficient vector by

$$\mathbf{X}_n = [X(1), \ldots, X(n)]^T, \quad \mathbf{a}_n = [a_1, \ldots, a_n]^T, \tag{1.91}$$

with the estimation error given by

$$e(\mathbf{a}_n) = (X(n+1) - \mathbf{a}_2^T \mathbf{X}_n), \tag{1.92}$$

and the MMSE solution $\hat{\mathbf{a}}_n^{\mathrm{MS}}$ obtained from

$$\varepsilon_{\mathrm{MS}}^2(\hat{\mathbf{a}}_n^{\mathrm{MS}})\} = \min_{\mathbf{a}_n \in \mathbb{R}^n} \left\{ R(0) - 2\mathbf{a}_n^T \mathbf{r}_n + \mathbf{a}_n^T \mathbf{R}_n \mathbf{a}_n \right\}. \tag{1.93}$$

Then

$$\hat{\mathbf{a}}_n^{\mathrm{MS}} = \begin{bmatrix} \hat{a}_1^{\mathrm{MS}} \\ \vdots \\ \hat{a}_n^{\mathrm{MS}} \end{bmatrix}, \quad \mathbf{R}_n = \begin{bmatrix} R(0) & R(1) & \cdots & R(n-1) \\ R(1) & R(0) & R(1) & \vdots \\ \vdots & \cdots & \ddots & R(1) \\ R(n-1) & \cdots & R(1) & R(0) \end{bmatrix}, \quad \mathbf{r}_n = \begin{bmatrix} R(n) \\ \vdots \\ R(1) \end{bmatrix}. \tag{1.94}$$

Thus the $\hat{\mathbf{a}}_n^{\mathrm{MS}}$ obtained from (1.93) also satisfies

$$\mathbf{R}_n \hat{\mathbf{a}}_n^{\mathrm{MS}} = \mathbf{r}_n, \tag{1.95}$$

and is given explicitly by

$$\hat{\mathbf{a}}_n^{\text{MS}} = \mathbf{R}_n^{-1}\mathbf{r}_n, \tag{1.96}$$

assuming $\mathbf{R}_n$ is non-singular. In practice, various computationally more stable methods can be used to solve for $\hat{\mathbf{a}}_n^{\text{MS}}$ in (1.95). The MMSE expression $\varepsilon_{\text{MS}}(\hat{\mathbf{a}}_n^{\text{MS}})$ is given by

$$\varepsilon_{\text{MS}}(\hat{\mathbf{a}}_n^{\text{MS}})\} = R(0) - \mathbf{r}_n^T\mathbf{R}_n^{-1}\mathbf{r}_n. \tag{1.97}$$

Example 1.9 Consider a far-field interfering random process source $S(t)$ with its wavefronts incident at an angle of θ relative to the normal of the x-axis with an auxiliary sensor denoted $X(t)$ and a main sensor denoted $Y(t)$ spaced a distance of L meters to the right of the auxiliary sensor as shown in Fig. 1.11 Assume both sensors are bandlimited to $(-B/2, B/2)$ and the far-field source $S(t)$ has a flat spectrum density of S_0 over $(-B/2, B/2)$. Then the autocorrelation function of the $S(t)$ random process is given by

$$\begin{aligned} R(t) &= \int_{-B/2}^{B/2} S_0 e^{j2\pi ft}df = S_0 B\frac{\sin(\pi Bt)}{\pi Bt} \\ &= R(0)\frac{\sin(\pi Bt)}{\pi Bt}, \quad -\infty < t < \infty. \end{aligned} \tag{1.98}$$

From Fig. 1.11, we note that a wavefront first impinges at $X(t)$, before it reaches $Y(t)$, needing to travel an additional propagation distance of d given by

$$d = L\sin(\theta) = c\tau, \tag{1.99}$$

$\sin(\theta) = d/L$, or $\theta = \sin^{-1}(d/L)$ is the incident angle. Given a propagation speed of c, this wavefront has a delay of τ seconds between its arrival at $x(t)$ and $y(t)$. Consider n samples from both sensors with a sampling interval of T are packed as $n \times 1$ vectors

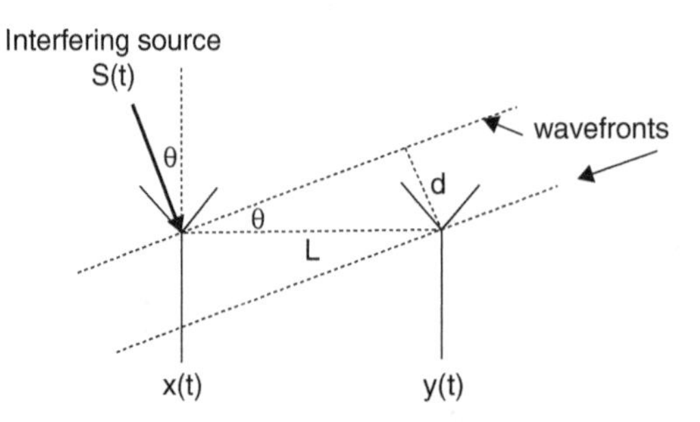

Figure 1.11 Interfering source $S(t)$ incident at angle θ to an auxiliary sensor $X(t)$ and a main sensor $Y(t)$.

given by

$$\mathbf{X}(i) = \begin{bmatrix} S(iT) \\ S((i+1)T) \\ \vdots \\ S((i+n-1)T) \end{bmatrix}, \ \mathbf{Y}(i) = \begin{bmatrix} S(iT+\tau) \\ S((i+1)T+\tau) \\ \vdots \\ S((i+n-1)T+\tau) \end{bmatrix}. \tag{1.100}$$

Now, we want to use $a_1\mathbf{X}(i)$ to estimate $\mathbf{Y}(i)$ in the MMSE sense. The error vector $\mathbf{e}(a_1)$ of this estimation is defined by

$$\mathbf{e}(a_1) = \mathbf{Y}(i) - a_1\mathbf{X}(i). \tag{1.101}$$

Then the MSE $\varepsilon^2(a_1)$ is given by

$$\varepsilon^2(a_1) = \mathrm{E}\left\{(\mathbf{Y}(i) - a_1\mathbf{X}(i))^T (\mathbf{Y}(i) - a_1\mathbf{X}(i))\right\} = nR(0) - 2a_1 nR(\tau) + a_1^2 nR(\tau), \tag{1.102}$$

and the MMSE solution $\hat{a}_1^{\mathrm{MS}}$ is given explicitly by

$$\hat{a}_1^{\mathrm{MS}} = \frac{R(\tau)}{R(0)}, \tag{1.103}$$

with its associated MMSE given by

$$\varepsilon^2(\hat{a}_1^{\mathrm{MS}}) = n\left(R(0) - \frac{R^2(\tau)}{R(0)}\right) = nR(0)\left(1 - \frac{R^2(\tau)}{R^2(0)}\right), \tag{1.104}$$

and its normalized MMSE (NMMSE) (normalized with respect to the total power $nR(0)$ of the $n \times 1$ vector $\mathbf{X}(i)$) is given by

$$\varepsilon_N^2\left(\hat{a}_1^{\mathrm{MS}}\right) = \frac{\varepsilon^2(\hat{a}_1^{\mathrm{MS}})}{nR(0)} = \left(1 - \frac{R^2(\tau)}{R^2(0)}\right). \tag{1.105}$$

If we use the autocorrelation function of this source given by (1.98), then (1.105) becomes

$$\varepsilon_N^2\left(\hat{a}_n^{\mathrm{MS}}\right) = 1 - \left(\frac{\sin(\pi B\tau)}{\pi B\tau}\right)^2. \tag{1.106}$$

For small values of $B\tau$, using the approximation of $\sin(x) \approx x - x^3/3!$, (1.106) becomes

$$\varepsilon_N^2(\hat{a}_n^{\mathrm{MS}}) \approx (\pi^2/3)(B\tau)^2. \tag{1.107}$$

Equation (1.107) has the interpretation that while the original interfering power of $S(t)$ collected by the main sensor has a total power of $nR(0)$, by subtracting the vector $a_1\mathbf{X}(i)$ from the vector $\mathbf{Y}(i)$, the resulting interfering noise has been reduced by a factor of $(\pi^2/3)(B\tau)^2$. For example, if $B\tau = 0.1$, then the interfering power in $\mathbf{Y}(i)$ has been reduced by a factor of 0.032. This reduction comes from the fact that the components of the random vector $\mathbf{X}(i)$ are statistically correlated to the components of the random vector $\mathbf{Y}(i)$, and by using $\hat{a}_1^{\mathrm{MS}}$ of (1.103) appropriately in the subtraction process in (1.101), this reduction was optimized. □

Example 1.10 The results in Example 1.9 can be used with some simplifications and assumptions to analyze the performance of a sidelobe canceller (SLC) used in many radar systems. Suppose a steerable antenna of the main sensor is steered toward a very weak desired target signal $S_t(t)$ (assumed to be uncorrelated to $S(t)$), resulting in a detector power of P_t. Unfortunately, the interfering power from $S(t)$ in Example 1.9 still is able to come into the main sensor with a power of $P_I = nR(0)$, which is assumed to be many times greater than P_t. The power of all the other noises in the main sensor receiver's detector, denoted by P_n, is also assumed to be much smaller than P_I. Then the original SNR at the detector is given by

$$\mathrm{SNR_{Orig}} = \frac{P_t}{P_s + P_n} \approx \frac{P_t}{P_s}. \tag{1.108}$$

However, upon using the configuration of Fig. 1.11 and the cancellation of (1.101), most of the interfering $S(t)$ has been subtracted from $\mathbf{Y}(i)$. The optimum error $\mathbf{e}(\hat{a}_1)$ is denoted by

$$\mathbf{e}(\hat{a}_1) = \mathbf{Y}(i) - \hat{a}_1\mathbf{X}(i) \approx \mathbf{S}_t(i) = [S_t(iT), \ldots, S_t((i+n-1)T)]^T. \tag{1.109}$$

Now, the SNR after using the SLC of (1.101) becomes

$$\begin{aligned}\mathrm{SNR_{SLC}} &= \frac{P_t}{nR(0)\left(1 - \dfrac{R^2(\tau)}{R^2(0)}\right)} = \left(\frac{P_t}{P_s}\right)\frac{1}{\left(1 - \dfrac{R^2(\tau)}{R^2(0)}\right)} = \frac{\mathrm{SNR_{Orig}}}{\left(1 - \dfrac{R^2(\tau)}{R^2(0)}\right)} \\ &\approx \frac{\mathrm{SNR_{Orig}}}{(\pi^2/3)(B\tau)^2}.\end{aligned} \tag{1.110}$$

Thus, $\mathrm{SNR_{SLC}}$ of the (SLC) can be increased significantly by cancelling out most of the statistically correlated interfering $S(t)$ existing in both $\mathbf{X}(i)$ and $\mathbf{Y}(i)$. In (1.109)–(1.110), we still assume other noises at the input to the receiver detector are much smaller than the reduced interfering noises. □

1.7 Conclusions

In Section 1.1, we provided some brief introductory comments on the historical backgrounds of detection and estimation and how they have been used in modern communication, radar, and signal processing systems. In Section 1.2, we considered a simple binary decision problem to motivate the need for using the ML decision rule. We showed the average probability of error using the optimum decision is lower than an ad hoc decision rule. In Section 1.3, we discussed the use of a correlation receiver to perform binary detection of two given deterministic signal vectors without noise and with noise. In Section 1.4, we introduced the concept of the importance of SNR in determining the performance of a binary correlation receiver. The geometry of the signal vectors is also shown to be important in determining the performance of the receiver. In Section 1.5, we wanted to study the performance of a BPSK communication system as a function of various parameters encountered in various possible realistic scenarios. The range equation indicates the received power in terms of the transmit power, losses due to free-space propagation and other factors, and gains from transmit and receive antennas. The SNR at the detector of a receiver is shown to depend on the range equation and equivalent receiver noise. The performance of the BPSK system depends on the detector SNR. The necessary transmit powers needed to satisfy the proper operation of four BPSK systems for a simple short-distance scenario, for a cellular mobile-to-base station scenario, for a UAV-to-ground scenario, and for a geosynchronous satellite-to-earth are analyzed in four examples. In Section 1.6.1, we studied two simple estimation problems in the presence of no noise and with noise. In Section 1.6.2, we consider the concept of a criterion of goodness in any estimation problem. The intuitive and reasonable absolute-error criterion was introduced. We mentioned that Galileo hinted at the use of the AE criterion in his astronomical data analysis work. An optimum estimator based on the least-absolute-error criterion was proposed. Unfortunately, the solution of the LAE optimum estimator needs to use a non-linear optimization program. In Section 1.6.3, the square-error criterion and the least-square-error estimator were introduced. Fortunately, the solution for the LSE estimator can be obtained by solving a linear system of equations. We also mentioned that Gauss and Laplace used the LSE estimator to tackle their astronomical analysis. In Section 1.6.4, we studied the robustness of the estimation results using the LAE and LSE methods. In Section 1.6.5, we considered estimation problems with random errors and introduced the mean-square-error criterion and the minimum mean-square-error estimation problem. Again, the MMSE solution can be found by solving a linear system of equations. Several examples show the use of linear mean-square estimation by exploiting the statistical correlations among the components of the random data. The concepts and examples in Chapter 1 are used to provide some intuitive motivations for the rest of the chapters in this book.

1.8 Comments

1. One of the earliest and extremely detailed treatments of "statistical communication theory" was the 1140-page comprehensive book by Middleton [9]. This book covered essentially all known works in detection and related topics up to about 1960. Another equally pioneering work in detection theory was the 1946 PhD thesis of the Soviet scientist Kotel'nikov that later appeared in an English translation in 1960 [10]. While experimental radar systems were built by many countries in the 1920s–1930s, these early radar efforts were mainly directed at the RF hardware and field testings. Due to the classified nature of the radar efforts in World War II, many of the details of who did what at what times are not completely clear. It appears that some of the early radar detection theory efforts were the work on SNR and matched filter (equivalent to correlation operation) of North [11] and the importance of the LR test of Davies [12] (also appeared in [3]). A comprehensive treatment of the U.S. radar efforts was collected

in the 28 volumes of the MIT Radiation Laboratory Series [13]. Volume 24 of this series is most related to radar detection [14]. Some additional background references relevant to various issues discussed in Chapter 1 include references [15]–[21].

2. We provide some Matlab computational details on the solution of the LAE estimator $\hat{a}_0^{\mathrm{AE}}$ in Section 1.6.2. Define the Matlab function res1(a) by:

```
function [m] = res1(a);
a0 = 0.5;
x = [1:12]';
y0 = a0*x;
randn('seed', 57);
y = y0 + 0.5 * randn(12, 1);
e = y - x*a;
m=norm(e,1);
```

Use the Matlab optimization function fminsearch to minimize res1(a) with initial starting points of $a = 0.4$ and $a = 0.6$. Then fminsearch('res1',0.4) yields the optimum $\hat{a}_0^{\mathrm{AE}}(1) = 0.5147656250000003$ and fminsearch('res1',0.6) yields an essentially identical $\hat{a}_0^{\mathrm{AE}}(2) = 0.5147460937499999$. Using res1($\hat{a}_0^{\mathrm{AE}}(1)$) yields $\varepsilon_{\mathrm{AE}}(\hat{a}_0^{\mathrm{AE}}(1)) = 4.889473661607013$ and res1($\hat{a}_0^{\mathrm{AE}}(2)$) yields $\varepsilon_{\mathrm{AE}}(\hat{a}_0^{\mathrm{AE}}(2)) = 4.889434599107013$. However, the arbitrarily chosen $a_0 = 0.56$ yields an $\varepsilon_{\mathrm{AE}}(a_0 = 0.56) = 5.545384981110282$.

3. Section 1.6.3 showed the LSE solution $\hat{a}_0^{\mathrm{LS}}$ has an analytically closed form given by (1.68). Nevertheless, it is interesting to solve for $\hat{a}_0^{\mathrm{LS}}$ using the Matlab fminsearch approach as follows. Define the Matlab function res(a) by:

```
function [n] = res(a);
a0 = 0.5;
x = [1:12]';
y0 = a0*x;
randn('seed', 57);
y = y0 + 0.5*randn(12, 1);
e = y - x*a;
n = (norm(e))^2;
```

The only difference between the function res1(a) and the function res(a) is that the AE criterion norm of "m=norm(e,1)" is now replaced by the SE criterion norm-square of "n = (norm(e))^2". Use the Matlab optimization function fminsearch to minimize res(a) with initial starting points of a = 0.4 and a = 0.6. Then fminsearch('res',0.4) yields the optimum $\hat{a}_0^{\mathrm{LS}}(1) = 0.5311718750000005$ and fminsearch('res',0.6) yields an essentially identical $\hat{a}_0^{\mathrm{LS}}(2) = 0.5311523437499996$. Using res($\hat{a}_0^{\mathrm{LS}}(1)$) yields $\varepsilon_{\mathrm{LS}}^2(\hat{a}_0^{\mathrm{LS}}(1)) = 2.806593940369510$ and res1($\hat{a}_0^{\mathrm{LS}}(2)$) yields $\varepsilon_{\mathrm{LS}}^2(\hat{a}_0^{\mathrm{LS}}(2)) = 2.806594136345985$. However, for an arbitrary $a_0 = 0.56$, $\varepsilon_{\mathrm{LS}}(a_0 = 0.56) = 3.346860175305835$. Of course, in practice, it is computationally simpler to find $\hat{a}_0^{\mathrm{LS}}$ using the analytically closed form given by (1.68) than using the optimization fminsearch program. Our purpose here is just to show that the optimization fminsearch program used for the minimization under the AE criterion can be easily used for the SE criterion as well.

References

[1] R. A. Fisher. Two new properties of mathematical likelihood. *Proc. Royal Soc., London, Ser. A*, 1934.

[2] J. Neyman and E. S. Pearson. Contributions to the theory of testing statistical hypothesis. *Stat. Res. Memoir*, vol. 1, 1936; vol. 2, 1938.

[3] P. M. Woodward. *Probability and Information Theory, with Applications to Radar*. Pergamon Press, 1953.

[4] A. N. Kolmogorov. Interpolation and extrapolation of stationary sequences. *Izvestia Acad. Nauk., USSR*, 1941.

[5] N. Wiener. *Extrapolation, Interpolation, and Smoothing of Stationary Time Series*. Wiley, 1949.

[6] C. F. Gauss. Theoria motus corporum coelestium in sectionibus conicis solem ambientium. Published in 1809. (LS work completed in 1795.)

[7] P. S. Laplace. Théorie analytique des probabilités (1812). http://en.wikipedia.org/wiki/Pierre-Simon_Laplace.

[8] Galileo Galilei. Dialogo dei massimi sistemi. 1632. English translation provided by T. Salusbury, 1961.

[9] D. Middleton. *An Introduction to Statistical Communication Theory*. McGraw-Hill, 1960.

[10] V. A. Kotel'nikov. *Theory of Optimum Noise Immunity*. Government Power Engineering Press, 1956; translated by R. A. Silverman, McGraw-Hill, 1960.

[11] D. O. North. An analysis of the factors which determine signal/noise discrimination in pulse-carrier system. Technical Report PTR-6C, RCA, June 1943.

[12] I. L. Davies. On determining the presence of signals in noise. *Proc. IEEE, Part III*, Mar. 1952.

[13] *MIT Radiation Laboratory Series*, 1950.

[14] J. L. Lawson and G. E. Uhlenbeck. Threshold signals. In *MIT Radiation Laboratory Series*. McGraw-Hill, 1950.

[15] T. K. Sarkar et al. *History of Wireless*. Wiley, 2006.

[16] P. Gralla. *How Wireless Works*. Que, 2002.

[17] G. W. Stimson. *Introduction to Airborne Radar*. SciTech, 2nd ed., 1998.

[18] http://en.wikipedia.org/wiki/Least_absolute_deviation.

[19] http://en.wikipedia.org/wiki/Least_squares.

[20] Y. Dodge, ed. *L_1-Statistical Procedures and Related Topics*. Institute of Statistical Studies, 1997.

[21] J. Aldrich. *R.A. Fisher and the Making of Maximum Likelihood 1912–1922*. Statistical Science, 1997.

Problems

1.1 In studying any subject, it is always illuminating to understand what some of the earlier pioneers in that subject have worked on. Some famous early researchers that made wireless communication possible include theoretical physicists like J. C. Maxwell and experimental physicists like H. R. Hertz, and experimental radio engineers like G. Marconi and N. Tesla, among many other people. You may want to use Wikipedia as well as the book, "History of Wireless," by T. K. Sarkar et al. [15], among many other sources to study this fascinating subject.

1.2 From Section 1.4, consider a fixed WGN variance of σ^2 in (4.1) and fixed equal energy signal vectors $\|\mathbf{s}_0\|^2 = \|\mathbf{s}_1\|^2 = E$ where the SNR $= E/\sigma^2$. Then the average error probability P_e varies greatly as a function of the normalized correlation $\rho(\mathbf{s}_0, \mathbf{s}_1)/E$ as defined in (4.25) of Chapter 4. Furthermore, denote the angle (in units of radians) separating these two vectors by $\cos^{-1}(\rho)$, or by $d = 180 \times \cos^{-1}(\rho)/\pi$ in units of degrees. From (4.28) of Chapter 4, setting $P_\mathrm{FA} = P_\mathrm{e}$, then $P_\mathrm{e} = \mathrm{Q}(\frac{-\sqrt{2E-2\rho}}{2\sigma}) = \mathrm{Q}(-\sqrt{\mathrm{SNR}(1-\rho)/2})$. Plot P_e as a function of the angle of separation d, for SNR $= 1, 5, 10$. Hint: Given the large dynamic range of the values of P_e, plot P_e using $semilogy(P_\mathrm{e})$ versus d. Are the values in your plot for SNR $= 10$ consistent with those values given by (1.33) and shown in Fig. 1.6 where $d = 180°$? Note: The P_e given by (1.26) and shown in Fig. 1.5 used a sub-optimum detector as given by (1.22), and thus should not be compared with those values of the P_e obtained by using (4.23) which used the optimum detector.

1.3 Consider the four examples in Section 1.5 with the stated assumptions. Now we allow some possible changes of parameters.

(a) In Example 1.1, suppose we increase the rate R to 1 Gbits/sec. What is the required P_T to support this rate?
(b) In Example 1.2, suppose we want to send some image data at $R = 100$ Kbits/sec. What is the required P_T to support this rate? If this new P_T is too demanding for a portable cell phone, what is a reasonable modification we can make on the cell phone?
(c) In Example 1.3, suppose we want the UAV to increase the range r to 160 km, operating in a possibly very rainy scenario with the propagation loss $L_0 = 1E3$. What is the required P_T to support this rate? Is this a reasonable requirement on a fairly small UAV? If we can use a higher gain antenna of $G_\mathrm{T} = 1E2$, what is the required P_T now to support this rate?
(d) In Example 1.4, instead of using a geo-synchronous satellite, we use a low-altitude satellite at a slant range $r = 1E4$. What is the required P_T to support this rate? If we now want to use lower gain $G_\mathrm{T} = 1E3$, what is the new P_T to support the same data rate?

1.4 In Section 1.5, we considered the BPSK communication system range and associated SNR as functions of antenna properties. Now, we want to consider the similar radar system range and associated SNR also as a function of antenna and scattering properties. While there are various similarities between the BPSK communication problem and the radar problem, the most basic difference is in the communication problem. Propagation is only one way from the transmitter antenna to the receiver antenna, while in the radar problem, propagations are two ways, from the transmitter antenna to the target and then scattering from the target back to the receiver antenna. From an isotropic radar transmitter antenna outputting a power of P_T, the power density at the range of

r is given by $P_T/(4\pi r^2)$. Denote G as the antenna gain toward the target; the target has a cross-section effective area of σ, and assume the target reflects all the intercepted power isotropically. Then the reflected power density becomes $(P_T G\sigma)/(4\pi r^2)$. Denote the radar receiver's antenna having an effective area of A, then the power at the receiver becomes $P_R = (P_T G A\sigma)/(4\pi r^2)^2$. The crucial difference between this P_R for the radar receiver and the communication receiver of (1.39) is that now the radar receiver power decreases as $1/r^4$, while that of the communication receiver power decreases as $1/r^2$. If we use the relationship between the effective area A and G and λ, then we have $A = G\lambda^2/(4\pi)$ (similar to (1.38)), then P_R becomes $P_R = (P_T G^2\lambda^2\sigma)/((4\pi)^3 r^4)$, which is often called the basic radar equation. We can denote the noise power density at the input of the radar receiver by $N_0 = kT_S$, where k is Boltzmann's constant and T_S is the equivalent system temperature (as in (1.44), and the input noise power at the radar receiver by $N = N_0 W$, where W is the equivalent (one-sided) bandwidth. Then the radar receiver's SNR is given by $SNR = P_R/N = (P_T G^2\lambda^2\sigma)/((4\pi)^3 r^4 N_0 W)$. In practice, since there may be additional radar system losses denoted by L, and $N_0 W$ may be replaced by $T_0 W F$, where T_0 is set to $290K$, and F denotes the receiver noise figure, then the operational SNR becomes $SNR_O = (P_T G^2\lambda^2\sigma)/((4\pi)^3 r^4 k T_0 W F)$.

Consider a land-based radar with a $P_T = 1E6$ watts, frequency $=$ 8 GHz, $G = 40$ dB, bandwidth $W = 5$ MHz, receiver noise figure $F = 6$ dB, loss $=$ 6 dB, and range $r = 30$ km to 150 km, in increments of 1 km. Plot SNR_O for $\sigma = 5\ \text{m}^2$, $0.5\ \text{m}^2$, $0.05\ \text{m}^2$. For a target at a range of $r = 40$ km, with a $\sigma = 0.5\ \text{m}^2$, can the radar receiver's SNR_O be above 15 dB?

1.5 In Section 1.6.1, (1.58) shows a 12×1 vector $\mathbf{x}$ and (1.60) shows a 12×1 noise perturbed vector $\mathbf{y}_1$. Now, take the same $\mathbf{x}$ vector as in (1.58), but consider a new noise perturbed vector $\mathbf{y}_1 = [0.092, 1.21, 2.4, 3.1, 2.5, 4.1, 3.5, 4.7, 4.8, 4.9, 6.1, 6.67]^T$. Consider the LAE criterion of Section 1.6.2 to find the new optimum coefficient $\hat{a}_0^{\text{AE}}$ using the Matlab fminsearch.m function. Find the associated minimum AE error of $\varepsilon_{\text{AE}}^2(\hat{a}_0^{\text{AE}})$. Consider $\tilde{a}_1 = 0.43$ and $\tilde{a}_2 = 0.57$. Find the associated AE error of $\varepsilon_{\text{AE}}^2(\tilde{a}_1)$ and $\varepsilon_{\text{AE}}^2(\tilde{a}_2)$. How do $\varepsilon_{\text{AE}}^2(\tilde{a}_1)$ and $\varepsilon_{\text{AE}}^2(\tilde{a}_2)$ compare to $\varepsilon_{\text{AE}}^2(\hat{a}_0^{\text{AE}})$?

1.6 Consider the same $\mathbf{x}$ vector as in (1.58) and a new noise-perturbed vector $\mathbf{y}_1 = [0.092, 1.21, 2.4, 3.1, 2.5, 4.1, 3.5, 4.7, 4.8, 4.9, 6.1, 6.67]^T$. In other words, we have the same $\mathbf{x}$ and $\mathbf{y}_1$ vectors as in Problem 1.5. However, under the LSE criterion of Section 1.6.3, use the explicit method of (1.68) to find the new optimum coefficient $\hat{a}_0^{\text{LS}}$ and the associated minimum LS error of $\varepsilon_{\text{LS}}^2(\hat{a}_0^{\text{LE}})$. In addition, also use the Matlab fminsearch method to minimize the residual of $\|\mathbf{y}_1 - a\mathbf{x}\|$, with an initial starting point of 0.4. Compare the optimum coefficient $\hat{a}_0^{\text{LS}}$ and the associated minimum LS error of $\varepsilon_{\text{LS}}^2(\hat{a}_0^{\text{LE}})$ found using the Matlab fminsearch method to that found using the explicit method of (1.68).

Consider $\tilde{a}_1 = 0.4$ and $\tilde{a}_2 = 0.57$. Find the associated LS error of $\varepsilon_{\text{LS}}^2(\tilde{a}_1)$ and $\varepsilon_{\text{LS}}^2(\tilde{a}_2)$. How do these two associated LS errors compare to that of the $\varepsilon_{\text{LS}}^2(\hat{a}_0^{\text{LE}})$?

1.7 Consider a new 12×1 noise perturbed vector $\mathrm{y}_2 = [0.092,\ 1.21,\ 100,\ 3.1,\ 80,\ 4.1,\ 3.5,\ 4.7,\ 4.8,\ 4.9,\ 6.1,\ 6.67]^T$, which is identical to the y_1 vector encountered in Problems 1.5 and 1.6, except for the two extreme large outliers with values of $\mathrm{y}_2(3) = 100$ and $\mathrm{y}_2(5) = 80$. First, use the LSE criterion and (1.68) to find the optimum coefficient $\hat{a}_0^{\mathrm{LS}}$. Second, use the LAE criterion (using the Matlab fminsearch method) to find the optimum coefficient $\hat{a}_0^{\mathrm{AE}}$. In both cases, use the same $\mathbf{x}$ vector of (1.58) used in Problems 1.5 and 1.6. Next, first plot $\mathbf{x}$ vs. $\mathbf{y}_2$ and a straight line of slope $\hat{a}_0^{\mathrm{LS}}$. Then plot $\mathbf{x}$ vs. $\mathbf{y}_2$ and a straight line of slope $\hat{a}_0^{\mathrm{AE}}$. Discuss the advantage of the robustness of this linear estimation problem using the LAE criterion over the LSE criterion.

1.8 Consider the MMSE estimation of $X(n+1)$ by $a_1 X(1) + a_2 X(2) + \ldots + a_n X(n)$ of Section 1.65. Assume $\{X(m),\ \infty < m < \infty\}$ is a zero mean random sequence with an autocorrelation sequence given by $R(n) = (0.5)^{|n|}$, $-\infty < n < \infty$. Find the optimum MMSE coefficient vector $\hat{\mathbf{a}}_n = [\hat{a}_1,\ \ldots,\ \hat{a}_n]^T$ and the associated $\varepsilon^2_{\mathrm{MS}}(\hat{\mathbf{a}}_n)$, for $n = 1, 2, 3, 4$. For this given $R(n)$, did the additional terms in $a_1 X(1) + \ldots + a_{n-1} X(n-1)$ help in reducing the $\varepsilon^2_{\mathrm{MS}}(\hat{\mathbf{a}}_n)$? Why?

1.9 Consider the MMSE estimation of $X(n+1)$ by $a_1\ X(1) + a_2\ X(2) + \ldots + a_n\ X(n)$ of Section 6.5. Assume $\{X(m),\ \infty < m < \infty\}$ is a zero mean random sequence with an autocorrelation sequence given by $R(n) = (0.5)^{|n|} + (0.3)^{|n|}$, $-\infty < n < \infty$. Find the optimum MMSE coefficient vector $\hat{\mathbf{a}}_n = [\hat{a}_1,\ \ldots,\ \hat{a}_n]^T$ and the associated $\varepsilon^2_{\mathrm{MS}}(\hat{\mathbf{a}}_n)$, for $n = 1, 2, 3, 4$. For this given $R(n)$, did the additional terms in $a_1 X(1) + \ldots + a_{n-1} X(n-1)$ help in reducing the $\varepsilon^2_{\mathrm{MS}}(\hat{\mathbf{a}}_n)$?

2 Review of probability and random processes

In order to understand fully the rest of the materials in this book, one needs to know some elementary concepts in probability and random processes. We assume the reader has already been exposed formally to some basic probability theory (e.g., in a one quarter/semester course). Thus, in Section 2.1, we essentially only list the probability concepts and provide some examples to describe each concept needed for the study of random processes. In Section 2.2, we introduce the importance of Gaussian random vector and associated n-dimensional, marginal, and conditional probability density functions. Section 2.3 introduces random processes, starting with definitions, then some motivations, various examples, and more details. In Section 2.4, the concepts of stationarity and particularly wide-sense stationarity are presented. The importance of autocorrelation and power spectral density functions is considered. Section 2.5 summarizes the Gaussian random process which constitutes the simplest and most important model for random noises in communication systems. In Section 2.6, different concepts of modeling and computing statistical parameters in probability and random processes based on ensemble and time averagings and ergodicity are treated. Section 2.7 considers discrete-time wide-sense stationary random sequences. The probability and random process materials covered in this chapter will be needed in the rest of this book.

2.1 Review of probability

A *sample space* $\mathbb{S}$ is the set of all outcomes (also called realizations) of a random experiment. An element of $\mathbb{S}$ is called a sample point and is denoted by s.

Example 2.1 The sample space of tossing a coin is $\mathbb{S} = \{H, T\}$. □

Example 2.2 The sample space of tossing two coins is $\mathbb{S} = \{HH, HT, TH, TT\}$. □

Example 2.3 The sample space of tossing a die is $\mathbb{S} = \{1, 2, 3, 4, 5, 6\}$. □

Example 2.4 The sample space of the time of failure of a device manufactured at time t_0 is $\mathbb{S} = [t_0, \infty)$. □

A *random event* is a set of certain outcomes of an experiment with some specified property.

Example 2.5 $\mathbb{A} = \{HH, HT, TH\}$ is a random event containing at least one head in the toss of two coins in the experiment of Example 2.2. □

Probability is a function P with its domain consisting of random events and with its range on [0, 1].

Example 2.6 If the two coins in Example 2.5 have equal probability of having a head or a tail, then $\mathrm{P}(\mathbb{A}) = 0.75$. □

A *random variable* (r.v.) X is a real-valued function with its domain of $\mathbb{S}$ and a range of real-valued numbers.

Example 2.7 In Example 2.2, define a r.v. X by $X(TT) = -2$, $X(TH) = -1.3$, $X(HT) = 1.2$, and $X(HH) = 2$. □

A *probability cumulative distribution function* (cdf) of a r.v. X, denoted by $F_X(x) = \mathrm{P}(X \leq x)$, is a function with its domain defined on the real line and its range on [0, 1].

Example 2.8 In Example 2.2, assume the two coins with equal probability of having a head or a tail and defined by the r.v. X in Example 2.7. Then this cdf $F_X(x)$ is given by

$$F_X(x) = \begin{cases} 0, & -\infty < x < -2, \\ 0.25, & -2 \leq x < 1.3, \\ 0.5, & -1.3 \leq x < 1.2, \\ 0.75, & 1.2 \leq x < 2, \\ 1, & 2 \leq x < \infty. \end{cases} \quad \square \tag{2.1}$$

A *probability density function* (pdf) $f_X(x) = dF_X(x)/dx$ is the derivative of the cdf of the r.v. X.

Example 2.9 The pdf $f_X(x)$ corresponding to the cdf of (2.1) is given by

$$\begin{aligned} f_X(x) = {} & 0.25\delta(x+2) + 0.25\delta(x+1.3) + 0.25\delta(x-1.2) \\ & + 0.25\delta(x-2), \quad -\infty < x < \infty. \quad \square \end{aligned} \tag{2.2}$$

A *discrete r.v.* is a r.v. in which X only takes on a finite or countably infinite number of real values.

Example 2.10 The r.v. X considered in Examples 2.7–2.9 is a discrete r.v. □

A *continuous r.v.* is a r.v. in which X takes on values on the real line.

Example 2.11 If we define the time of failure in Example 2.4 as the r.v. X for that problem, then X is a continuous r.v. □

Two events $\mathbb{A}$ and $\mathbb{B}$ are *independent* if $\mathrm{P}(\mathbb{A} \cap \mathbb{B}) = \mathrm{P}(\mathbb{A})\mathrm{P}(\mathbb{B})$.

Example 2.12 In Example 2.2, we assume all four outcomes have equal probability of 0.25. Let $\mathbb{A}$ be the event in which the first outcome is a head and $\mathbb{B}$ be the event in which the second outcome is a head. Then $\mathrm{P}(\mathbb{A}) = \mathrm{P}(\mathbb{B}) = 0.5$ and $\mathrm{P}(\mathbb{A} \cap \mathbb{B}) = 0.25$. Since $0.25 = \mathrm{P}(\mathbb{A} \cap \mathbb{B}) = \mathrm{P}(\mathbb{A})\mathrm{P}(\mathbb{B}) = 0.5 \times 0.5$, then the two events $\mathbb{A}$ and $\mathbb{B}$ are independent. □

The *conditional probability* $\mathrm{P}(\mathbb{A} \mid \mathbb{B})$ is defined as $\mathrm{P}(\mathbb{A} \mid \mathbb{B}) = \mathrm{P}(\mathbb{A} \cap \mathbb{B})/\mathrm{P}(\mathbb{B})$.

Example 2.13 In Example 2.12, $\mathrm{P}(\mathbb{A} \mid \mathbb{B}) = \mathrm{P}(\mathbb{A} \cap \mathbb{B})/\mathrm{P}(\mathbb{B}) = 0.25/0.5 = 0.25$. □

Thus far, we have only considered one-dimensional r.v.'s. There are many physical and human-made situations in which we need to use two or more random variables to model the random phenomena.

Example 2.14 Suppose our left hand tosses a die with its sample space defined by $\mathbb{S}_L = \{1, 2, 3, 4, 5, 6\}$ and the right hand tosses a coin with its sample space defined by $\mathbb{S}_R = \{H, T\}$.

Then, we can define a r.v. X with the sample space $\mathbb{S}_L$ and a r.v. Y with a sample space $\mathbb{S}_R$. □

The *joint probability cumulative distribution function* $F_{X,Y}(x, y)$ of the r.v.'s X and Y is defined by $F_{X,Y}(x, y) = \mathrm{P}(X \leq x, Y \leq y)$.

Example 2.15 Let $\mathrm{P}(X = 1, Y = -1) = 1/8$, $\mathrm{P}(X = 1, Y = 4) = 1/4$, $\mathrm{P}(X = 3, Y = -1) = 3/8$, and $\mathrm{P}(X = 3, Y = 4) = 1/8$. Then the joint probability cumulative distribution function $F_{X,Y}(x, y)$ is given by

$$F_X(x) = \begin{cases} 0, \; x < 1 \text{ or } y < -1, \\ 1/8, \; 1 \leq x < 3, \; -1 \leq y < 4, \\ 1/2, \; 3 \leq x, \; -1 \leq y < 4, \\ 3/8, \; 1 \leq x < 3, \; 4 \leq y, \\ 1, \; 3 \leq x < \infty, \; 4 \leq y < \infty. \end{cases} \quad \square \tag{2.3}$$

The *joint probability density function* $f_{X,Y}(x, y)$ of the r.v.'s X and Y is defined by $f_{X,Y}(x, y) = \partial^2 F_{X,Y}(x, y)/\partial x \partial y$.

Example 2.16 The joint pdf $f_{X,Y}(x, y)$ of Example 2.15 is given by

$$f_{X,Y}(x, y) = 0.125\delta(x-1)\delta(y+1) + 0.25\delta(x-1)\delta(y-4)$$
$$+ 0.375\delta(x-3)\delta(y+1) + 0.125\delta(x-3)\delta(y-4). \quad \square \tag{2.4}$$

The *conditional probability density function* $f_{X|Y}(x|y)$ is defined by $f_{X|Y}(x|y) = f_{X,Y}(x, y)/f_Y(y)$.

Example 2.17 Consider $f_{X,Y}(x, y) = 2xy$, $0 \le y \le 2x \le 2$. Then

$$f_Y(y) = \int_{x=-\infty}^{\infty} f_{X,Y}(x, y)dx = \int_{x=y/2}^{x=1} 2xydx = y - 0.25y^3, \; 0 \le y \le 2. \tag{2.5}$$

Thus,

$$f_{X|Y}(x|y) = f_{X,Y}(x, y)/f_Y(y) = 2x/(1 - 0.25y^2), 0 \le y \le 2, \; 0.5y \le x \le 1. \quad \square \tag{2.6}$$

The *independence of two r.v.'s* X and Y can be defined equivalently by

$$f_{X,Y}(x, y) = f_X(x)f_Y(y) \Leftrightarrow f_{X|Y}(x|y) = f_X(x) \Leftrightarrow f_{Y|X}(y|x) = f_Y(y). \tag{2.7}$$

In general, let $X_1, \ldots, X_N$ be N r.v.'s defined on a sample space $\mathbb{S}$ with a probability function P. Then the *multi-dimensional* (*multivariate*) *cdf* $F_{X_1,\ldots,X_N}(x_1, \ldots, x_N)$ of these N r.v.'s is defined by

$$F_{X_1,\ldots,X_N}(x_1, \ldots, x_N) = \mathrm{P}(s \in \mathbb{S} : X_1(s) \le x_1, \ldots, X_N(s) \le x_N). \tag{2.8}$$

Similarly, the pdf $f_{X_1,\ldots,X_N}(x_1, \ldots, x_N)$ is defined by

$$f_{X_1,\ldots,X_N}(x_1, \ldots, x_N) = \frac{\partial^N F_{X_1,\ldots,X_N}(x_1, \ldots, x_N)}{\partial x_1 \partial x_2 \ldots \partial x_N}. \tag{2.9}$$

For simplicity of notation, we define the *random vector* $\mathbf{X}$ as

$$\mathbf{X} = \begin{bmatrix} X_1 \\ X_2 \\ \vdots \\ X_N \end{bmatrix} = [X_1, \; X_2, \cdots, \; X_N]^T, \tag{2.10}$$

and its realization (which is a vector of deterministic numbers) as $\mathbf{x}$ given by

$$\mathbf{x} = \begin{bmatrix} x_1 \\ x_2 \\ \vdots \\ x_N \end{bmatrix} = [x_1, \; x_2, \cdots, \; x_N]^T. \tag{2.11}$$

Using the notations of (2.10) and (2.11), the cdf of (2.8) can be denoted by $F_{\mathbf{X}}(\mathbf{x})$ and the pdf of (2.9) can be denoted by $f_{\mathbf{X}}(\mathbf{x})$.

The *mean* μ of the r.v. X is the average of X defined by $\mu = \mathrm{E}\{X\}$, where $\mathrm{E}\{\cdot\}$ is the expectation (i.e., statistical (ensemble) averaging) operator.

Example 2.18 The mean of the r.v. in Example 2.9 is given by
$\mu = \mathrm{E}\{X\} = 0.25(-2 - 1.3 + 1.2 + 2) = -0.025.$ □

The *variance* σ^2 of the r.v. X is defined by $\sigma^2 = \mathrm{E}\{(X - \mu)^2\} = \mathrm{E}\{X^2\} - \mu^2$.

Example 2.19 The variance σ^2 of the r.v. in Example 2.9 is given by
$\sigma^2 = 0.25(4 + 1.69 + 1.44 + 4) - 0.025^2 = 2.7819.$ □

The *moment* m_n of the r.v. X is defined by $m_n = \mathrm{E}\{X^n\}$.

Example 2.20 The first moment $m_1 = \mu$. The variance $\sigma^2 = m_2 - \mu^2$. □

Central Limit Theorem states that if all the independent r.v.'s $X_1, \ldots, X_n$, have the same mean μ and same variance σ^2, denote $S_n = X_1 + \cdots + X_n$, then for large n, $\mathrm{P}\left(\frac{S_n - n\sigma}{\sigma\sqrt{n}} \leq t\right) \approx \Phi(t)$, where $\Phi(\cdot)$ is the cdf of a Gaussian r.v. of mean zero and unit variance.

Example 2.21 Suppose a fair (i.e., equally probable) die is rolled 30 times. Denote X_i the value of the i-th roll. Then the mean and variance of X_i are given by $\mu_i = 7/2$ and $\sigma_i^2 = 35/12$ and the mean and variance of S_n are given by $\mu_{S_{30}} = 105$ and $\sigma_{S_{30}}^2 = 175/2$. The approximate probability of the sum of all the values of the dice between 90 and 120 is then given by $\mathrm{P}(90 \leq S_{30} \leq 120) = \Phi\left(\frac{120-105}{\sqrt{175/2}}\right) - \Phi\left(\frac{90-105}{\sqrt{175/2}}\right) \approx 0.89.$ □

Some well-known one-dimensional real-valued r.v.'s with their domains, probabilities, and pdfs are:

Example 2.22

$$\text{Bernoulli r.v. } X : \mathbb{S}_X = \{0,\ 1\};\ \mathrm{P}(X = 0) = p,\ \mathrm{P}(X = 1) = q = 1 - p,$$
$$0 \leq p \leq 1.$$
$$\text{Poisson r.v. } X : \mathbb{S}_X = \{0, 1, \ldots\};\ \mathrm{P}(X = k) = (\lambda^k/k!)e^{-\lambda},\ k = 0,\ 1, \ldots.$$
$$\text{Uniform r.v. } X : \mathbb{S}_X = [a,\ b];\ f_X(x) = 1/(b - a),\ a \leq x \leq b.$$
$$\text{Exponential r.v. } X : \mathbb{S}_X = [0, \infty);\ f_X(x) = \lambda e^{-\lambda x},\ 0 \leq x < \infty.$$
$$\text{Laplacian r.v. } X : \mathbb{S}_X = (-\infty,\ \infty);\ f_X(x) = (\lambda/2)e^{-\lambda|x|},\ \infty < x < \infty.$$
$$\text{Rayleigh r.v. } X : \mathbb{S}_X = [0, \infty);\ f_X(x) = (x/\alpha^2)e^{-x^2/(2\alpha^2)},\ 0 \leq x < \infty.$$

Gaussian r.v. $X : \mathbb{S}_X = (-\infty, \infty)$ has a pdf defined by

$$f_X(x) = (1/\sqrt{2\pi\sigma^2})e^{-(x-\mu)^2/(2\sigma^2)}, \quad -\infty < x < \infty. \tag{2.12}$$

The r.v. X having the pdf of (2.12) is often denoted by $X \sim \mathcal{N}(\mu, \sigma^2)$. □

2.2 Gaussian random vectors

Gaussian random variables and random vectors are used extensively in the rest of this book. One reason is that from the Central Limit Theorem, the sum of a large number of independent random variables (of arbitrary probability distribution) with finite moments has a Gaussian distribution in the limit as the number of terms n goes to infinity. Even when the number of terms in the sum is not large, the central part of the resulting distribution is well approximated by a Gaussian distribution. Thus, the random noise, due to the excitations of a large number of electrons at the front end of a receiver in free space is well-approximated by a Gaussian distribution. Another reason is that Gaussian pdfs have various analytical properties leading to their popular usages. Furthermore, as we will show in this section, linear operations on jointly Gaussian random variables and random vectors remain Gaussian and thus also lead to simple explicit expressions.

The pdf $f_X(x)$ of a one-dimensional real-valued Gaussian r.v. has already been defined in (2.12). Now, consider a $N \times 1$ real-valued Gaussian random vector $\mathbf{X}$, denoted by (2.10), with a mean vector $\boldsymbol{\mu} = \mathrm{E}\{\mathbf{X}\}$, and a covariance matrix $\boldsymbol{\Lambda}$ defined by

$$\boldsymbol{\Lambda} = \mathrm{E}\{(\mathbf{X} - \boldsymbol{\mu})(\mathbf{X} - \boldsymbol{\mu})^T\} = \begin{bmatrix} \sigma_{11} & \cdots & \sigma_{1N} \\ \vdots & \ddots & \vdots \\ \sigma_{N1} & \cdots & \sigma_{NN} \end{bmatrix}, \tag{2.13}$$

where the $(i - j)$th element of $\boldsymbol{\Lambda}$ is given by

$$\sigma_{ij} = \mathrm{E}\{(X_i - \mu_i)(X_j - \mu_j)\}, \; i, \; j = 1, 2, \ldots, N, \tag{2.14}$$

and

$$\sigma_{ii} = \sigma_i^2 = \mathrm{E}\{(X_i - \mu_i)^2\} = \text{variance of } X_i, \; i = 1, 2, \ldots, N. \tag{2.15}$$

Then the pdf of the Gaussian random vector $\mathbf{X}$ with a non-singular covariance matrix $\boldsymbol{\Lambda}$ is given by

$$f_{\mathbf{X}}(\mathbf{x}) = \frac{1}{(2\pi)^{n/2} |\boldsymbol{\Lambda}|^{1/2}} e^{-(1/2)(\mathbf{x}-\boldsymbol{\mu})^T \boldsymbol{\Lambda}^{-1}(\mathbf{x}-\boldsymbol{\mu})}, \tag{2.16}$$

where $|\boldsymbol{\Lambda}|$ is the determinant of $\boldsymbol{\Lambda}$. This Gaussian random vector $\mathbf{X}$ with the pdf of (2.16) is often denoted by $\mathbf{X} \sim \mathcal{N}(\boldsymbol{\mu}, \boldsymbol{\Lambda})$.

Any two r.v.'s, X_i and X_j, $i \neq j$ are *uncorrelated* if $\sigma_{ij} = 0$. If these two r.v.'s are independent, then their joint pdf $f_{X_i,X_j}(x_i, x_j) = f_{X_i}(x_i) f_{X_j}(x_j)$ and

$$\begin{aligned} \sigma_{ij} &= \mathrm{E}_{X_i,X_j}\left\{(X_i - \mu_i)(X_j - \mu_j)\right\} \\ &= \mathrm{E}_{X_i}\{(X_i - \mu_i)\}\, \mathrm{E}_{X_j}\left\{(X_j - \mu_j)\right\}, \end{aligned} \tag{2.17}$$

showing they are uncorrelated. In other words, for any two r.v.'s, independence implies uncorrelatedness. However, if the N r.v.'s are mutually uncorrelated, satisfying (2.17) for all $i, j = 1, 2, \ldots, N$, the covariance matrix of (2.13) becomes a diagonal matrix

$$\boldsymbol{\Lambda} = \begin{bmatrix} \sigma_1^2 & \cdots & 0 \\ \vdots & \ddots & \vdots \\ 0 & \cdots & \sigma_n^2 \end{bmatrix}, \tag{2.18}$$

then

$$\boldsymbol{\Lambda}^{-1} = \text{diag}(1/\sigma_1^2, \ldots, 1/\sigma_N^2), \tag{2.19}$$

and

$$|\boldsymbol{\Lambda}| = \prod_{i=1}^{N} \sigma_i^2. \tag{2.20}$$

Thus, the pdf of the N uncorrelated Gaussian random vector $\mathbf{X}$ of (2.16) now becomes

$$\begin{aligned} f_{\mathbf{X}}(\mathbf{x}) &= \frac{1}{(2\pi)^{N/2} \prod_{i=1}^{N} \sigma_i} e^{-\sum_{i=1}^{N}(x_i-\mu_i)^2/(2\sigma_i^2)} \\ &= \prod_{i=1}^{N} \frac{1}{\sqrt{2\pi}\,\sigma_i} e^{-(x_i-\mu_i)^2/(2\sigma_i^2)} = \prod_{i=1}^{N} f_{X_i}(x_i), \end{aligned} \tag{2.21}$$

where each $f_{X_i}(x_i)$ is a one-dimensional Gaussian pdf given by (2.12). Equation (2.21) shows all the components of the N uncorrelated Gaussian random vector $\mathbf{X}$ are independent. In other words, for a Gaussian random vector, uncorrelatedness of all the components implies independence of all the components.

2.2.1 Marginal and conditional pdfs of Gaussian random vectors

When the components of an N-dimensional Gaussian random vector are mutually independent, the pdf of the N-dimensional pdf factors is given by the product of N one-dimensional Gaussian pdf's as shown in (2.10). This factorization property is crucially dependent on having the exponential functions in these Gaussian pdfs having the semi-group property (i.e., $\exp(A + B) = \exp(A) \times \exp(B)$). It turns out various Gaussian marginal and conditional pdfs needed in the remaining chapters of this book also have relatively simple forms due to the presence of the exponential functions in these pdfs.

Theorem 2.1 *Consider an N-dimensional real-valued Gaussian random column vector $\mathbf{X}_N$ having a pdf given by (2.16). The marginal pdf of the P-dimensional random vector $\mathbf{X}_P = [X_1, \ldots, X_P]^T$, taken from the first P components of the N-dimensional vector $\mathbf{X}_N = [\mathbf{X}_P^T, \mathbf{X}_{N-P}^T]^T$, with $1 \le P \le N-1$, is given by*

$$f_{\mathbf{X}_P}(\mathbf{x}_P) = \frac{1}{(2\pi)^{P/2} |\boldsymbol{\Lambda}_P|^{1/2}} e^{-(1/2)(\mathbf{x}_P - \boldsymbol{\mu}_P)^T \boldsymbol{\Lambda}_P^{-1} (\mathbf{x}_P - \boldsymbol{\mu}_P)}, \tag{2.22}$$

where its covariance matrix $\mathbf{\Lambda}_P$ *is the upper-left hand corner matrix of the originally given* $N \times N$ *partitioned covariance matrix* $\mathbf{\Lambda}_N$ *given by*

$$\mathbf{\Lambda}_N = \begin{bmatrix} \mathbf{\Lambda}_P & \mathbf{\Lambda}_{P(N-P)} \\ \mathbf{\Lambda}_{P(N-P)}^T & \mathbf{\Lambda}_{N-P} \end{bmatrix}, \tag{2.23}$$

and its mean vector $\boldsymbol{\mu}_P$ *is the first* P *components of the mean vector* $\boldsymbol{\mu}_N$ *denoted by*

$$\boldsymbol{\mu}_N = \mathrm{E}\left\{\left[\mathbf{X}_P^T, \mathbf{X}_{N-P}^T\right]^T\right\} = \left[\boldsymbol{\mu}_P^T, \boldsymbol{\mu}_{N-P}^T\right]^T. \tag{2.24}$$

Proof: The proof is given in Appendix 2.A. □

Now, consider the N-dimensional Gaussian pdf given in (2.13)–(2.16). Suppose we want to find the conditional pdf of the $(N-P)$-dimensional vector $\mathbf{X}_{N-P}$ conditioned on the P-dimensional vector $\mathbf{X}_P$ having the value of $\mathbf{a}$. Then the conditional pdf of $\mathbf{X}_{N-P}|\{\mathbf{X}_P = \mathbf{a}\}$ is also a Gaussian pdf with a $(N-P) \times 1$ mean vector $\boldsymbol{\mu}_{N-P|P}$ and a $(N-P) \times (N-P)$ covariance matrix $\mathbf{\Lambda}_{N-P|P}$ given by Theorem 2.2.

Theorem 2.2 *Consider an* N*-dimensional real-valued Gaussian random column vector* $\mathbf{X}_N = [\mathbf{X}_P^T, \mathbf{X}_{N-P}^T]^T \sim \mathcal{N}(\boldsymbol{\mu}_N, \mathbf{\Lambda}_N)$. *Let* $\mathbf{X}_P \sim \mathcal{N}(\boldsymbol{\mu}_P, \mathbf{\Lambda}_P)$, *where* $\boldsymbol{\mu}_P$ *is the first* P*-component of* $\boldsymbol{\mu}_N$ *and* $\mathbf{\Lambda}_P$ *is the upper-left* $(P \times P)$ *component of* $\mathbf{\Lambda}_N$, *and* $\mathbf{X}_{N-P} \sim \mathcal{N}(\boldsymbol{\mu}_{N-P}, \mathbf{\Lambda}_{N-P})$, *where* $\boldsymbol{\mu}_{N-P}$ *is the last* $(N-P)$ *component of* $\boldsymbol{\mu}_N$ *and* $\mathbf{\Lambda}_{N-P}$ *is the lower-right* $(N-P) \times (N-P)$ *component of* $\mathbf{\Lambda}_N$, *as denoted by (2.23)–(2.24). Then the conditional pdf of* $\mathbf{X}_{N-P}|\{\mathbf{X}_P = \mathbf{a}\}$ *is given by*

$$f_{\mathbf{X}_{N-P}|\mathbf{X}_P}(\mathbf{x}_{N-P}|\mathbf{x}_P = \mathbf{a})$$
$$= (2\pi)^{-(N-P)/2} \left|\mathbf{\Lambda}_{N-P|P}\right|^{-1/2} e^{(-1/2)(\mathbf{x}_{N-P}-\boldsymbol{\mu}_{N-P|P})^T \mathbf{\Lambda}_{N-P|P}^{-1}(\mathbf{x}_{N-P}-\boldsymbol{\mu}_{N-P|P})}, \tag{2.25}$$

where

$$\boldsymbol{\mu}_{N-P|P} = \boldsymbol{\mu}_{N-P} - \tilde{\mathbf{\Lambda}}_{N-P}^{-1}\tilde{\mathbf{\Lambda}}_{P(N-P)}^{T}(\mathbf{a} - \boldsymbol{\mu}_P), \tag{2.26}$$

and

$$\mathbf{\Lambda}_{N-P|P} = \tilde{\mathbf{\Lambda}}_{N-P}^{-1}. \tag{2.27}$$

Proof: The proof is given in Appendix 2.B.

Example 2.23 Consider a real-valued Gaussian pdf with $N = 3$ and $P = 1$, with a mean vector of $\boldsymbol{\mu}_3 = [0,\ 0,\ 0]^T$, a covariance matrix $\mathbf{\Lambda}_3$ and its inverse $\mathbf{\Lambda}_3^{-1} = \tilde{\mathbf{\Lambda}}_3$, and their sub-matrices $\mathbf{\Lambda}_1$, $\tilde{\mathbf{\Lambda}}_{12}^T$, and $\tilde{\mathbf{\Lambda}}_2$ given by

$$\mathbf{\Lambda}_3 = \begin{bmatrix} 5 & 1 & 1 \\ 1 & 4 & 1 \\ 1 & 1 & 3 \end{bmatrix}, \quad \tilde{\mathbf{\Lambda}}_3 = \begin{bmatrix} 0.2200 & -0.0400 & -0.0600 \\ -0.0400 & 0.2800 & -0.0800 \\ -0.0600 & -0.0800 & 0.3800 \end{bmatrix},$$

$$\mathbf{\Lambda}_1 = 5, \quad \tilde{\mathbf{\Lambda}}_{12}^T = \begin{bmatrix} -0.0400 \\ -0.0600 \end{bmatrix}, \quad \tilde{\mathbf{\Lambda}}_2 = \begin{bmatrix} 0.2800 & -0.0800 \\ -0.0800 & 0.3800 \end{bmatrix}. \tag{2.28}$$

By direct computations using the values of the sub-matrices in (2.28), we obtain

$$\boldsymbol{\Lambda}_1^{-1} + \tilde{\boldsymbol{\Lambda}}_{12}\tilde{\boldsymbol{\Lambda}}_2^{-1}\tilde{\boldsymbol{\Lambda}}_{12}^T$$

$$= 0.2000 + \begin{bmatrix} -0.0400 & -0.0600 \end{bmatrix} \begin{bmatrix} 0.2800 & -0.0800 \\ -0.0800 & 0.3800 \end{bmatrix}^{-1} \begin{bmatrix} -0.0400 & -0.0600 \end{bmatrix}^T,$$

$$= 0.200 + 0.020 = 0.2200 = \tilde{\boldsymbol{\Lambda}}_1,$$

which agrees with (2.28). Furthermore,

$$\left|\tilde{\boldsymbol{\Lambda}}_2^{-1}\right| |\boldsymbol{\Lambda}_1| = 10 \times 5 = 50 = |\boldsymbol{\Lambda}_3|,$$

which agrees with (2.97). □

Example 2.24 Consider the same Gaussian pdf with $N = 3$ and $P = 1$ as given in Example 2.23. Now, we want to find the conditional pdf of $\mathbf{X}_2 = [X_2, X_3]^T$ given that $X_1 = 4$. By using the definition of a conditional pdf, we have

$$f_{[X_2,X_3]|X_1}(x_2, x_3|x_1 = 4) = f_{[X_1,X_2,X_3]}(x_1 = 4, x_2, x_3)/f_{X_1}(x_1 = 4)$$

$$= \frac{(2\pi)^{1/2}\,|\boldsymbol{\Lambda}_1|^{1/2}\, e^{-(1/2)[4,\, x_2,\, x_3]\boldsymbol{\Lambda}_3^{-1}[4,\, x_2,\, x_3]^T}}{(2\pi)^{3/2}\,|\boldsymbol{\Lambda}_3|^{1/2}\, e^{-(1/2)4^2/5}}$$

$$= \frac{e^{-0.16+0.16x_2-0.14x_2^2+0.08x_2x_3+0.24x_3-0.19x_3^2}}{(2\pi)|10|^{1/2}}. \tag{2.29}$$

However, by using (2.25)–(2.28), we have

$$\boldsymbol{\Lambda}_{2|1} = \tilde{\boldsymbol{\Lambda}}_2^{-1} = \begin{bmatrix} 3.8 & 0.8 \\ 0.8 & 2.8 \end{bmatrix}, \quad \left|\boldsymbol{\Lambda}_{2|1}\right| = 10, \quad \boldsymbol{\mu}_{2|1} = -\begin{bmatrix} 3.8 & 0.8 \\ 0.8 & 2.8 \end{bmatrix} \begin{bmatrix} -0.04 \\ -0.06 \end{bmatrix} \times 4. \tag{2.30}$$

Thus, (2.104) yields

$$f_{[X_2,X_3]|X_1}(x_2, x_3|x_1 = 4) = \frac{e^{-(1/2)[[x_2,\, x_3]-\boldsymbol{\mu}_{2|1}]\tilde{\boldsymbol{\Lambda}}_2[[x_2,\, x_3]-\boldsymbol{\mu}_{2|1}]^T}}{(2\pi)|10|^{1/2}}$$

$$= \frac{e^{-0.16+0.16x_2-0.14x_2^2+0.08x_2x_3+0.24x_3-0.19x_3^2}}{(2\pi)|10|^{1/2}}. \tag{2.31}$$

This shows the conditional pdf of (2.31) obtained using (2.25)–(2.27) of Theorem 2.2 in (2.16) is identical to the definition of a conditional pdf as given by (2.29). □

2.3 Random processes (stochastic processes)

There are two fundamentally different approaches to viewing a random process (r.p.). A r.p. is sometimes also called a stochastic process.

Table 2.1 Electric consumptions at 5 p.m. and 6 p.m.

x_2(6 p.m.) ↑			
100	1	5	3
90	4	5	4
80	3	3	3
	90	100	110 → x_1(5 p.m.)

The first approach is stated below:

I. A *random process* $\{X(s,t),\ t \in \mathbf{T},\ s \in \mathbb{S}\}$ is a family of random variables where the index t belongs to some real-valued parameter set $\mathbf{T}$ and s is an element of the sample space $\mathbb{S}$.

Then in this approach, the r.p. $\{X(s,t),\ t \in \mathbf{T},\ s \in \mathbb{S}\}$ for any fixed $\{t_1, \ldots, t_n\}$ is characterized by the r.v.'s $X_1 = X(t_1), \ldots, X_n = X(t_n)$. We can denote this set of r.v.'s as a random vector $\mathbf{X} = [X_1, \ldots, X_n]^T$. Then $\{X(s,t),\ t \in \mathbf{T},\ s \in \mathbb{S}\}$ is characterized by the cdf $F_{\mathbf{X}}(\mathbf{x})$ or the pdf $f_{\mathbf{X}}(\mathbf{x})$ of $\mathbf{X}$. Often for simplicity of notation, we denote a r.p. by $\{X(s,t),\ t \in T\}$ after suppressing the s dependency. Of course, either notation means the same r.p. Furthermore, in many applications (including all of our applications in communication), $\mathbf{T}$ is taken to be a real-valued discrete set or a continuous interval denoting time instants.

Example 2.25 Denote the consumption of electricity in kilowatts in a particular district of Los Angeles for the month of July at $t_1 = 5$ p.m. by the r.v. $X_1 = X(t_1)$, at $t_2 =$ 6 p.m. by the r.v. $X_2 = X(t_2)$, etc. Then this r.p. $\{X(t),\ t \in \mathbf{T}\}$, showing the hourly electric consumption from 5 p.m. to midnight, is represented by the random vector $\mathbf{X} = [X_1, \cdots, X_8]^T$. Then its statistical behavior is characterized by either $f_{\mathbf{x}}(\mathbf{x})$ or $F_{\mathbf{x}}(\mathbf{x})$. □

Example 2.26 Using the notation of Example 2.25, suppose among the 31 days of July, the observed number of days jointly having x_1 and x_2 consumptions (in units of kilowatts) is given in Table 2.1. For example, suppose there are 3 days at 5 p.m. using 100 kilowatts and at 6 p.m. using 80 kilowatts (as shown in the (3, 2) position of Table 2.1. We note that the sum of all the days in Table 2.1 equals 31. From Table 2.1, we can construct the two-dimensional pdf $f_{X_1,X_2}(x_1, x_2)$, given by

$$\begin{aligned} f_{X_1,X_2}(x_1,x_2) = (1/31)[&1\delta(x_1-90)\delta(x_2-100) + 5\delta(x_1-100)\delta(x_2-100) \\ &+ 3\delta(x_1-110)\delta(x_2-100) + 4\delta(x_1-90)\delta(x_2-90) \\ &+ 5\delta(x_1-100)\delta(x_2-90) + 4\delta(x_1-110)\delta(x_2-90) \\ &+ 3\delta(x_1-80)\delta(x_2-100) + 3\delta(x_1-90)\delta(x_2-80) \\ &+ 3\delta(x_1-90)\delta(x_2-80)],\ -\infty < x_1, x_2 < \infty. \end{aligned} \tag{2.32}$$

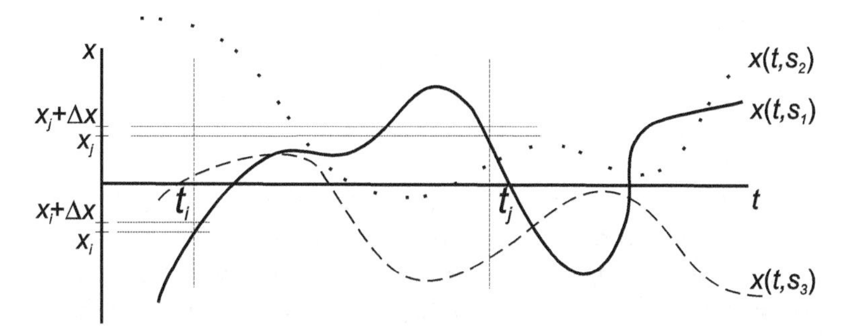

Figure 2.1 A sketch of a random process illustrating three realizations (curves) versus time.

The pdf $f_{X_1,X_2}(x_1, x_2)$ of (2.32) consists of the sum of 9 delta functions. The data in Table 2.1 are based only upon the statistic of 31 days over one particular month of July. Suppose we have the statistic for the month of July over five years. Since each year generates one such table, then we can take the average of these five tables to obtain a more statistical reliable single table. Furthermore, in Table 2.1, both x_1 and x_2 only take on three values, resulting in 9 entries in the table. Suppose we want finer resolution, with x_1 having N_1 values and x_2 having N_2 values. Then Table 2.1 needs to have $N_1 \times N_2$ entries and the corresponding pdf $f_{X_1,X_2}(x_1, x_2)$ will consist of the sum of $N_1 \times N_2$ delta functions. Next, suppose we want to construct the three-dimensional pdf $f_{X_1,X_2,X_3}(x_1, x_2, x_3)$, where x_1 can take N_1 values, x_2 can take N_2 values, x_3 can take N_3 values. Then we can construct a table with $N_1 \times N_2 \times N_3$ entries and the corresponding pdf $f_{X_1,X_2,X_3}(x_1, x_2, x_3)$ will consist of the sum of $N_1 \times N_2 \times N_3$ delta functions. We can continue in this manner, to construct the n-dimensional pdf $f_{\mathbf{X}}(\mathbf{x})$, where $\mathbf{x} = [x_1, x_2, \cdots, x_n]^T$, with a table of $N_1 \times N_2 \times \cdots \times N_n$ entries, and a corresponding pdf $f_{\mathbf{X}}(\mathbf{x})$ consists of the sum of $N_1 \times N_2 \times \cdots \times N_n$ delta functions. Clearly, the construction of a high-dimensional pdf from raw observed data can be quite involved. □

The second and alternative approach in defining a r.p. is stated below.

II. A *random process* $\{X(s, t),\ t \in \mathbf{T},\ s \in \mathbb{S}\}$ can also be characterized as a collection of curves (also called realizations) with the variable $t \in \mathbf{T}$ for each outcome of the realization $s \in \mathbb{S}$.

Figure 2.1 shows symbolically a collection of three curves with t as the variable (without loss of generality can be presumed to indicate time), and curve 1 denotes the realization of the sample point s_1 as a function of t, while curve 2 denotes the realization of the sample point s_2 as a function of t, etc. If the r.p. $X(t, s)$ takes on real values, then there are an uncountably infinite number of such possible realizations (or curves). The assignment of a probability on these curves becomes delicate, and more advanced mathematical measure-theoretic tools must be used in this formulation. Fortunately, for what we need to do in this course, we do not have to obtain the probability of these curves. It is also fortunate that A. N. Kolmogorov [1] showed that the two approaches (denoted as I and II in our previous notations) of defining a random process are theoretically equivalent. Thus, for the remainder of this course, we only need to consider the use of

n-dimensional cdf $F_{\mathbf{X}}(\mathbf{x})$ or pdf $f_{\mathbf{X}}(\mathbf{x})$ in characterizing a r.p. As we note in Fig. 2.1, at any given instant t_i, consider all the realizations (curves) of $s \in \mathbb{S}$ of the process $x(t_i, s)$ having values in $(x_i, x_i + \Delta x)$. In Fig. 2.1, we showed only three realizations $\{x(s_1, t),\ t \in \mathbf{T}\}$, $\{x(s_2, t),\ t \in \mathbf{T}\}$, and $\{x(s_3, t),\ t \in \mathbf{T}\}$. Take a vertical line located at t_i, and consider all the realizations intersecting this line with values in $(x_i, x_i + \Delta x)$. By forming this histogram for different values of x_i, we can obtain the pdf $f_{X(t_i)}(x_i)\Delta x$. In the limit as Δx goes to zero, we can conceptually obtain $f_{X(t_i)}(x_i)$. Similarly, by repeating at instant t_j for all values in $(x_j, x_j + \Delta x)$, we can conceptually obtain $f_{X(t_j)}(x_j)$. By considering those realizations having values $(x_i, x_i + \Delta x)$ at t_i as well as those having values $(x_j, x_j + \Delta x)$ at instant t_j, we can conceptually obtain the bivariate pdf of $f_{X(t_i),X(t_j)}(x_i, x_j)$. Similarly, by using n time instants, we can conceptually obtain the n-dimensional pdf of $f_{\mathbf{X}}(\mathbf{x})$. Again, the actual efforts involved in evaluating the n-dimensional pdf of $f_{\mathbf{X}}(\mathbf{x})$ for large n are non-trivial. □

Example 2.27 Consider a r.p. represented by a sinusoid with a random phase $\Theta(s)$, with a pdf $f_\Theta(\theta)$ and a sample space $\mathbb{S}$, and a random amplitude $A(s)$, independent of $\Theta(s)$, with a pdf $f_A(a)$ and a sample space $\mathbb{S}'$, given by

$$X(t, s, s') = A(s)\cos(2\pi f t + \theta(s')),\ -\infty < t < \infty,\ s \in \mathbb{S},\ s' \in \mathbb{S}'. \tag{2.33}$$

In the first approach in defining this r.p., we can take n sampling instants, $\{t_1,\ t_2, \ldots, t_n\}$, and find the n-dimensional pdf $f_{\mathbf{X}}(\mathbf{x})$, of the random vector $\mathbf{X} = [X_1, \ldots, X_n]^T$, where $X_1 = X(t_1), \ldots, X_n = X(t_n)$. In the second approach in defining this r.p., we have a collection of realizations $\{x(s, s'),\ -\infty < t < \infty,\ s \in \mathbb{S},\ s' \in \mathbb{S}'\}$. Then the actual evaluation of the n-dimensional pdf $f_{\mathbf{X}}(\mathbf{x})$ by taking the limit of the n-dimensional histogram of these realizations as mentioned just before Example 2.27 is non-trivial. □

2.4 Stationarity

A random process $\{X(t),\ t \in \mathbf{T}\}$ is *stationary of order n* if

$$f_{X(t_1),\ldots,X(t_n)}(x_1, \ldots, x_n) = f_{X(t_1+\varepsilon),\ldots,X(t_n+\varepsilon)}(x_1, \ldots, x_n), \tag{2.34}$$

for every $t_1, \ldots, t_n$ in $\mathbf{T}$, for every ε (such that $t_1 + \varepsilon, \ldots, t_n + \varepsilon$ are still in $\mathbf{T}$), and for any given positive integer n. To verify whether a random process is stationary of order n for some given n (e.g., $n = 5$) from an actual measured or estimated n-dimensional pdf is probably very difficult if not impossible.

Example 2.28 If all the r.v.'s in the r.p. are *independent and identically distributed* (i.i.d.), that is, all the r.v.'s have the identical distribution (i.e., $f_X(t_i)(x) = f_X(x),\ i = 1, \ldots, n$), then clearly

$$\begin{aligned} f_{X(t_1),\ldots,X(t_n)}(x_1,\ldots,x_n) &= f_{X(t_1)}(x_1)\ldots f_{X(t_n)}(x_n) \\ &= f_X(x_1)\ldots f_X(x_n) \\ &= f_{X(t_1+\varepsilon)}(x_1)\ldots f_{X(t_n+\varepsilon)}(x_n) \\ &= f_{X(t_1+\varepsilon),\ldots,X(t_n+\varepsilon)}(x_1,\ldots,x_n) \end{aligned} \tag{2.35}$$

and the r.p. is stationary of order n. □

A random process $\{X(t),\ t \in \mathbf{T}\}$ is *strictly stationary* if it is stationary of order n for every positive integer n. From any measured or estimated pdfs, we clearly cannot verify the validity of (2.35) for every positive integer n.

Example 2.29 A r.p. with i.i.d. r.v.'s is also strictly stationary since (2.35) is valid for every positive n. □

A random process $\{X(t), t \in \mathbf{T}\}$ is *wide-sense stationary* (WSS) if it satisfies

(1) $\mathrm{E}\{|X(t)|^2\} < \infty$; (2.36)

(2) $\mathrm{E}\{X(t)\} = \mu =$ constant for all $t \in \mathbf{T}$; (2.37)

(3) The autocorrelation function (sometimes also called the correlation function) $R(t_1 - t_2)$ of $\{X(t),\ t \in \mathbf{T}\}$ satisfies

$$\mathrm{E}\{X(t_1)X(t_2)\} = \int_{-\infty}^{\infty}\int_{-\infty}^{\infty} x_1 x_2 f_{x(t_1)x(t_2)}(x_1, x_2) dx_1 dx_2 = R(t_1 - t_2). \tag{2.38}$$

In general, the autocorrelation function $R_0(t_1, t_2)$ of any r.p. $\{X(t),\ t \in \mathbf{T}\}$ is defined by

$$\mathrm{E}\{X(t_1)X(t_2)\} = \int_{-\infty}^{\infty}\int_{-\infty}^{\infty} x_1 x_2 f_{x(t_1)x(t_2)}(x_1, x_2) dx_1 dx_2 = R_0(t_1, t_2). \tag{2.39}$$

We note in general the autocorrelation function $R_0(t_1, t_2)$ in (2.39) is a function of two variables depending on the two time instants t_1 and t_2, while the autocorrelation function $R(t_1 - t_2)$ in (2.38) of a WSS r.p. is a function of one variable given by $(t_1 - t_2)$. For a WSS r.p., $R(t_1 - t_2) = \mathrm{E}\{X(t_1)X(t_2)\} = \mathrm{E}\{X(t_1 + \varepsilon)X(t_2 + \varepsilon)\} = R((t_1 + \varepsilon) - (t_2 + \varepsilon)) = R(t_1 - t_2)$. Thus, for a WSS r.p., the autocorrelation function only depends on the time difference $(t_1 - t_2)$ of the two time instants $\{t_1, t_2\}$; it does not depend on the actual location of the two time instants $\{t_1, t_2\}$. However, for an arbitrary r.p. the autocorrelation function of (2.39) depends on the actual values of t_1 and t_2. For an arbitrary r.p., $R_0(t_1, t_2)$ need not be equal to $R_0(t_1 + \varepsilon, t_2 + \varepsilon)$ for all possible ε. In general, once a r.p. is verified to be WSS, it is easier to evaluate the autocorrelation function of the form of $R(t_1 - t_2)$, than to evaluate the autocorrelation function of the form $R_0(t_1, t_2)$.

Example 2.30 Consider the r.p. in Example 2.27, where $A(s)$ is a zero-mean Gaussian r.v. of variance σ^2 and $\theta(s)$ is set to zero. Thus, we have

$$X(t,s) = A(s)\cos(2\pi ft), \ -\infty < t < \infty, \ s \in \mathbb{S}. \tag{2.40}$$

Now, we want to check whether the r.p. of (2.40) is a WSS process. By direct evaluations using property 1 of (2.36), $\mathrm{E}\{|X(t,s)|^2\} \leq \mathrm{E}\{|A(s)|^2\} = \sigma^2 < \infty$, and property 2 of (2.37), $\mathrm{E}\{X(t,s)\} = 0$, and thus both properties are satisfied. However, the autocorrelation function

$$\begin{aligned} \mathrm{E}\{X(t_1)X(t_2)\} &= \mathrm{E}\{A(s)\cos(2\pi f t_1)A(s)\cos(2\pi f t_2)\} \\ &= \sigma^2 \cos(2\pi f t_1)\cos(2\pi f t_2) \\ &= (\sigma^2/2)\left[\cos(2\pi f(t_1 - t_2)) + \cos(2\pi f(t_1 + t_2))\right] \end{aligned} \tag{2.41}$$

is not a function of $(t_1 - t_2)$, since there is a dependency of $(t_1 + t_2)$ in the last term in (2.41). Thus, the r.p. of (2.41) is not a WSS r.p. □

Example 2.31 Consider the r.p. given by

$$X(t,s) = A(s)\cos(2\pi ft) + B(s)\sin(2\pi ft), \ -\infty < t < \infty, \ s \in \mathbb{S}, \tag{2.42}$$

where $A(s)$ and $B(s)$ are two independent r.v.'s with zero-means and the same variance σ^2. Direct evaluations show that the r.p. of (2.42) satisfies properties 1 and 2 of (2.36) and (2.37). Now, the autocorrelation function

$$\begin{aligned} \mathrm{E}\{X(t_1)X(t_2)\} &= \sigma^2 \cos(2\pi f t_1)\cos(2\pi f t_2) + \sigma^2 \sin(2\pi f t_1)\sin(2\pi f t_2) \\ &= \sigma^2 \cos(2\pi f(t_1 - t_2)) \end{aligned} \tag{2.43}$$

is only a function of $(t_1 - t_2)$, and thus property 3 of (2.38) is also satisfied. The r.p. of (2.42) is a WSS r.p. □

Consider a continuous-time WSS r.p. $\{X(t),\ t \in T\}$. The *power spectral density* (psd) $S(\omega)$ of a WSS r.p. is defined by

$$S(\omega) = \mathscr{F}\{R(t)\} = \int_{-\infty}^{\infty} R(t)e^{-i\omega t}dt, \ -\infty < \omega = 2\pi f < \infty, \tag{2.44}$$

where $\mathscr{F}\{\cdot\}$ denotes the Fourier transform operator, since the integral in (2.44) is just the Fourier transform of $R(t)$. Since $\{X(t),\ t \in T\}$ is a WSS r.p., its autocorrelation $R(t)$ is a function of a single variable t. The $(t_1 - t_2)$ variable encountered before in the WSS r.p. discussion is set to t in (2.44). In other words, we do not care about the value of t_1 or t_2, but care only about their difference, which is a single variable and denoted by t. If a r.p. is not WSS, its autocorrelation $R_0(\cdot,\cdot)$ is a function of two variables. Then the one-dimensional Fourier transform in (2.44) has to be generalized to a two-dimensional Fourier transform. Unfortunately, while a two-dimensional Fourier transform exists, the

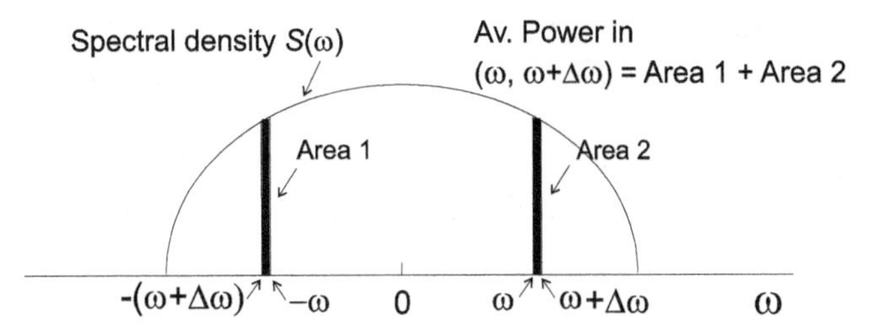

Figure 2.2 Sketch of the average power in the frequency band $(\omega, \omega + \Delta\omega)$ in terms of the integral of $S(\omega)$ over both $(-\omega - \Delta\omega, -\omega)$ and $(\omega, \omega + \Delta\omega)$.

resulting Fourier transform of $R_0(\cdot, \cdot)$ does not have all the desirable properties of the psd function $S(\omega)$ of (2.44). Thus, for all of our applications, we will consider a psd function only for a WSS r.p.

The psd function $S(\omega)$ has the interpretation as the average power density at radian frequency ω of the r.p., while $S(\omega)$ in (2.44) is called the "two-sided psd," since it is formally defined over the entire line $(-\infty, \infty)$. However, in engineering, we often use the "one-sided frequency" of $f = \frac{\omega}{2\pi}$ defined only over the positive values of $[0, \infty)$. Then the average power over the frequency band of $(\omega, \omega + \Delta\omega)$ is defined as the integral of $S(\omega)$ over both $(-\omega - \Delta\omega, -\omega)$ and $(\omega, \omega + \Delta\omega)$ as shown in (2.45) and Fig. 2.2.

$$\text{Average power at } (\omega, \omega + \Delta\omega) = \int_{-\omega-\Delta\omega}^{-\omega} S(\omega')d\omega' + \int_{\omega}^{\omega+\Delta\omega} S(\omega')d\omega'. \tag{2.45}$$

Furthermore, since the psd function $S(\omega)$ is the Fourier transform of the autocorrelation function $R(t)$, then the autocorrelation is the inverse transform of the psd function $S(\omega)$ defined by

$$R(t) = \mathscr{F}^{-1}\{S(\omega)\} = \frac{1}{2\pi}\int_{-\infty}^{\infty} S(\omega)e^{i\omega t}d\omega, \ -\infty < t < \infty. \tag{2.46}$$

The Fourier transform pair relationship between the psd function $S(\omega)$ and its autocorrelation function $R(t)$ is known as the Bochner–Khinchin property of a WSS r.p. We note that by setting $t = 0$ in (2.46),

$$R(0) = \mathrm{E}\left\{X^2(t)\right\} = \frac{1}{2\pi}\int_{-\infty}^{\infty} S(\omega)d\omega = \text{Average Power of the WSS r.p. } X(t). \tag{2.47}$$

Example 2.32 Consider a WSS r.p. with an autocorrelation function $R(t)$ given by

$$R(t) = e^{-\alpha|t|}, \ -\infty < t < \infty, \tag{2.48}$$

then its psd function $S(\omega)$, obtained by applying the Fourier transform to $R(t)$, is given by

$$S(\omega) = \frac{2\alpha}{\alpha^2 + \omega^2}, \ -\infty < t < \infty. \tag{2.49}$$

The Fourier transformation and the inverse Fourier transformation form a unique one-to-one relationship. That is, if a given $S(\omega)$ is the Fourier transform of a $R(t)$, then the inverse Fourier transform of $S(\omega)$ yields a unique function given by the original $R(t)$ function. Similarly, if $R(t)$ is the inverse Fourier transform of $S(\omega)$, then the Fourier transform of $S(\omega)$ yields a unique function given by the original $S(\omega)$. □

Example 2.33 A WSS r.p. $\{X(t),\ t \in \mathbf{T}\}$ is called a *white process* (w.p.) if its autocorrelation function is defined by

$$R(t) = C_0\delta(t),\ -\infty < t < \infty,\ 0 < C_0, \tag{2.50}$$

and its psd function $S(\omega)$ is given by

$$S(\omega) = C_0,\ -\infty < \omega < \infty. \quad \square \tag{2.51}$$

The "paradox" of a w.p. is that its total power is infinite since

$$\frac{1}{2\pi}\int_{-\infty}^{\infty} S(\omega)d\omega = \infty = \text{Power of } \{X(t)\}. \tag{2.52}$$

For a physical w.p., the psd function $S(\omega)$ in (2.52) is valid only for some region $[-\omega_0,\ \omega_0]$, with some large value of ω_0. Outside of this region $S(\omega)$ is indeed very small and approaches zero for very large values of ω. Thus, in this case, the power of the WSS w.p. can be made finite. Of course, for this case, the associated autocorrelation of (2.50) is not quite a delta function but can still be a very highly spiked function having a very small but not zero width centered about the origin.

Consider a WSS r.p. $\{X(t),\ t \in \mathbf{T}\}$ with a psd function $S(\omega)$ as the input to a linear time-invariant system with an impulse response function $h(t)$ and a transfer function $H(\omega) = \mathscr{F}\{h(t)\}$. Denote $\{Y(t),\ t \in \mathbf{T}\}$ as the r.p. at the output of this system. Then $\{Y(t),\ t \in \mathbf{T}\}$ is also a WSS r.p. and its psd function $S_Y(\omega)$ is given by

$$S_Y(\omega) = |H(\omega)|^2 S(\omega),\ -\infty < \omega < \infty. \tag{2.53}$$

Example 2.34 Consider the "white process" $\{X(t),\ t \in \mathbf{T}\}$ of Example 2.33 with its autocorrelation function $R(t)$ given by (2.50) and its psd function $S(\omega)$ given by (2.51). Consider a linear time-invariant system modeled by a RC low-pass filter with a transfer function given by

$$H(\omega) = \frac{1/(i\omega C)}{R + 1/(i\omega C)} = \frac{1}{1 + i\omega RC},\ -\infty < \omega < \infty, \tag{2.54}$$

and the square of the magnitude of the transfer function given by

$$|H(\omega)|^2 = \frac{1}{1 + \omega^2 R^2 C^2},\ -\infty < \omega < \infty. \tag{2.55}$$

Let the w.p. be the input to the RC low-pass filter with the transfer function of (2.54) and denote the output WSS r.p. by $\{Y(t),\ t \in \mathbf{T}\}$. Then the psd function of $S_Y(\omega)$ is given by

$$S_Y(\omega) = S(\omega)|H(\omega)|^2 = \frac{C_0}{1+\omega^2 R^2 C^2} = \frac{C_0/(R^2C^2)}{(1/\left(R^2C^2\right)) + \omega^2}, \quad -\infty < \omega < \infty. \tag{2.56}$$

By comparing (2.56) to (2.49), denoting $\alpha = 1/(RC)$, and using the unique one-to-one Fourier-inverse Fourier transformation property, then

$$R_Y(t) = \frac{C_0}{2RC} e^{-\frac{|t|}{RC}}, \quad -\infty < \omega < \infty. \quad \square \tag{2.57}$$

2.5 Gaussian random process

A random process $\{X(t),\ t \in \mathbf{T}\}$ is a *Gaussian r.p.* if for any $\{t_1, \ldots, t_n\}$ in $(-\infty, \infty)$, the associated random vector $\mathbf{X}$ taken from the r.p. at these n time instants has an n-dimensional Gaussian pdf $f_{\mathbf{X}}(\mathbf{x})$ of the form

$$f_{\mathbf{X}}(\mathbf{x}) = \frac{1}{(2\pi)^{n/2}|\mathbf{\Lambda}|^{1/2}} e^{-(1/2)(\mathbf{x}-\boldsymbol{\mu})^T \mathbf{\Lambda}^{-1}(\mathbf{x}-\boldsymbol{\mu})}, \tag{2.58}$$

where the $n \times 1$ mean vector $\boldsymbol{\mu}$ is given by

$$\boldsymbol{\mu} = \mathrm{E}\{\mathbf{X}\}, \tag{2.59}$$

and the $n \times n$ covariance matrix $\mathbf{\Lambda}$ is given by

$$\mathbf{\Lambda} = \mathrm{E}\{(\mathbf{X}-\boldsymbol{\mu})(\mathbf{X}-\boldsymbol{\mu})^T\} = \begin{bmatrix} \sigma_{11} & \cdots & \sigma_{1n} \\ \vdots & \ddots & \vdots \\ \sigma_{n1} & \cdots & \sigma_{nn} \end{bmatrix}, \tag{2.60}$$

with the $(i-j)$th element of $\mathbf{\Lambda}$ defined by $\sigma_{ij} = \mathrm{E}\{(X_i - \mu_i)(X_j - \mu_j)\}$, $i, j = 1, 2, \ldots, n$, $\sigma_{ii} = \sigma_i^2 = \mathrm{E}\{(X_i - \mu_i)^2\} =$ variance of X_i, and $|\mathbf{\Lambda}|$ is the determinant of the $\mathbf{\Lambda}$ matrix. We note the pdf in (2.58) depends only on the first and second moments of the random vector $\mathbf{X}$. The n-dimensional Gaussian pdf $f_{\mathbf{X}}(\mathbf{x})$ and the associated mean vector and covariance matrix given in (2.58)–(2.60) above are identical to those defined in (2.13)–(2.16).

Consider a linear time-invariant operation on a Gaussian r.p. $\{X(t),\ t \in \mathbf{T}\}$. Let $\mathbf{X}$ denote the $n \times 1$ Gaussian random vector with a mean vector $\boldsymbol{\mu}$ and a covariance matrix $\mathbf{\Lambda}$. Let $\mathbf{A}$ be a deterministic $n \times n$ non-singular matrix and $\mathbf{B}$ be a deterministic $n \times 1$ vector, and denote $\mathbf{Y}$ as the linear operation on $\mathbf{X}$ given by

$$\mathbf{Y} = \mathbf{AX} + \mathbf{B}. \tag{2.61}$$

For the linearly transformed $\mathbf{Y}$ vector, its $n \times 1$ mean vector $\boldsymbol{\mu}_{\mathbf{Y}}$ is given by

$$\boldsymbol{\mu}_{\mathbf{Y}} = \mathrm{E}\{\mathbf{Y}\} = \mathbf{A}\boldsymbol{\mu} + \mathbf{B}, \tag{2.62}$$

its $n \times n$ covariance matrix $\mathbf{\Lambda_Y}$ is given by

$$\mathbf{\Lambda_Y} = \mathrm{E}\{(\mathbf{Y} - \boldsymbol{\mu}_\mathbf{Y})(\mathbf{Y} - \boldsymbol{\mu}_\mathbf{Y})^T\} = \mathbf{A \Lambda A}^T, \tag{2.63}$$

and its pdf $f_\mathbf{Y}(\mathbf{y})$ is given by

$$f_\mathbf{Y}(\mathbf{y}) = \frac{1}{(2\pi)^{n/2}|\mathbf{\Lambda_Y}|^{1/2}} e^{-(1/2)(\mathbf{y}-\boldsymbol{\mu}_\mathbf{Y})^T \mathbf{\Lambda}_\mathbf{Y}^{-1}(\mathbf{y}-\boldsymbol{\mu}_\mathbf{Y})}. \tag{2.64}$$

In (2.64), the inverse of the covariance matrix $\mathbf{\Lambda_Y}$ is given by

$$\mathbf{\Lambda}_\mathbf{Y}^{-1} = \mathbf{A}^{-T}\mathbf{\Lambda}^{-1}\mathbf{A}^{-1}, \tag{2.65}$$

where $\mathbf{A}^{-T}$ denotes the inverse of the transpose of $\mathbf{A}$. Thus, a linear operation on the Gaussian r.p. $\{X(t),\ t \in \mathbf{T}\}$ yields another Gaussian r.p. $\{Y(t),\ t \in \mathbf{T}\}$, whose mean vector and covariance matrix can be found readily from (2.62) and (2.63).

Example 2.35 Let $\mathbf{X} = [X_1,\ X_2,\ X_3]^T$ be a real-valued random vector with a Gaussian pdf given by

$$f_\mathbf{X}(\mathbf{x}) = Ce^{(-1/2)(2x_1^2 - x_1x_2 + x_2^2 - 2x_1x_3 + 4x_3^2)}. \tag{2.66}$$

Since $[x_1,\ x_2,\ x_3]\mathbf{\Lambda}^{-1}[x_1,\ x_2,\ x_3]^T = 2x_1^2 - x_1x_2 + x_2^2 - 2x_1x_3 + 4x_3^2$, and we denote $\mathbf{\Lambda}^{-1} = [\sigma_{ij}],\ i, j = 1, 2, 3,$ then direct evaluation shows $\sigma_{11} = 2,\ \sigma_{22} = 1,\ \sigma_{33} = 4,\ \sigma_{23} = \sigma_{32} = 0,\ \sigma_{12} = \sigma_{21} = -1/2$, and $\sigma_{13} = \sigma_{31} = -1$. Thus,

$$\mathbf{\Lambda}^{-1} = \begin{bmatrix} 2 & -1/2 & -1 \\ -1/2 & 1 & 0 \\ -1 & 0 & 4 \end{bmatrix},\ \mathbf{\Lambda} = \begin{bmatrix} 2/3 & 1/3 & 1/6 \\ 1/3 & 7/6 & 1/12 \\ 1/6 & 1/12 & 7/24 \end{bmatrix},\ \boldsymbol{\mu} = \begin{bmatrix} 0 \\ 0 \\ 0 \end{bmatrix}. \tag{2.67}$$

Direct evaluation shows $|\mathbf{\Lambda}| = 1/6$. Thus, $C = \left((2\pi)^{3/2}|\mathbf{\Lambda}|^{1/2}\right)^{-1}$. Consider the linear operation on $\mathbf{X}$ given by $\mathbf{Y} = \mathbf{AX}$, where $\mathbf{A}$ is given by

$$\mathbf{A} = \begin{bmatrix} 1 & -1/4 & -1/2 \\ 0 & 1 & -2/7 \\ 0 & 0 & 1 \end{bmatrix}. \tag{2.68}$$

Direct evaluation using (2.62) and (2.63) with $\mathbf{B} = \mathbf{0}$ yields the pdf $f_\mathbf{Y}(\mathbf{y})$ given by

$$f_\mathbf{Y}(\mathbf{y}) = \left(6/(2\pi)^3\right)^{1/2} e^{-(1/2)(2y_1^2 + (7/8)y_2^2 + (24/7)y_3^2)}. \quad \square \tag{2.69}$$

Consider a WSS Gaussian r.p. $\{X(t),\ t \in \mathbf{T}\}$. Consider taking n sampling instants taken at $\{t_1, \cdots, t_n\}$ and an ε translating the n sampling instants to $\{t_1 + \varepsilon, \cdots, t_n + \varepsilon\}$. Denote the two sampled random vectors by

$$\mathbf{X} = [X(t_1), \ldots, X(t_n)]^T = [X_1, \ldots, X_n]^T, \tag{2.70}$$

$$\tilde{\mathbf{X}} = [X(t_1 + \varepsilon), \ldots, X(t_n + \varepsilon)]^T = [\tilde{X}_1, \ldots, \tilde{X}_n]^T. \tag{2.71}$$

Since the r.p. is WSS, all the r.v.'s have the constant value of μ, and thus the mean vectors of $\mathbf{X}$ and $\tilde{\mathbf{X}}$ are identical with

$$\boldsymbol{\mu} = \mathrm{E}\{\mathbf{X}\} = \mathrm{E}\{\tilde{\mathbf{X}}\} = [\mu, \ldots, \mu]^T. \tag{2.72}$$

The covariance matrices of $\mathbf{X}$ and $\tilde{\mathbf{X}}$ are also identical as shown by comparing (2.73) to (2.74):

$$\begin{aligned}\Lambda_{\mathbf{X}} &= \mathrm{E}\{(\mathbf{X} - \boldsymbol{\mu})(\mathbf{X} - \boldsymbol{\mu})^T\} = \left[\mathrm{E}\left\{(X(t_i) - \mu)\left(X(t_j) - \mu\right)\right\}\right] \\ &= \left[R(t_i - t_j) - \mu^2\right], \; i, \; j = 1, 2, \ldots, n,\end{aligned} \tag{2.73}$$

$$\begin{aligned}\Lambda_{\tilde{\mathbf{X}}} &= \mathrm{E}\{(\tilde{\mathbf{X}} - \boldsymbol{\mu})(\tilde{\mathbf{X}} - \boldsymbol{\mu})^T\} = \left[\mathrm{E}\left\{(X(t_i + \varepsilon) - \mu)(X(t_j + \varepsilon) - \mu)\right\}\right] \\ &= \left[R((t_i + \varepsilon) - (t_j + \varepsilon)) - \mu^2\right] = \left[R(t_i - t_j) - \mu^2\right], \; i, \; j = 1, 2, \ldots, n.\end{aligned} \tag{2.74}$$

Then, we can denote

$$\Lambda = \Lambda_{\mathbf{X}} = \Lambda_{\tilde{\mathbf{X}}}. \tag{2.75}$$

From (2.72) and (2.75), since the mean vectors and covariance matrices of $\mathbf{X}$ and $\tilde{\mathbf{X}}$ are identical, the corresponding n-dimensional pdfs $f_{\mathbf{X}}(\mathbf{x})$ and $f_{\tilde{\mathbf{X}}}(\mathbf{x})$ are also identical

$$f_{X(t_1),\ldots,X(t_n)}(x_1, \ldots, x_n) = f_{\mathbf{X}}(\mathbf{x}) = f_{\tilde{\mathbf{X}}}(\mathbf{x}) = f_{X(t_1+\varepsilon),\ldots,X(t_n+\varepsilon)}(x_1, \ldots, x_n). \tag{2.76}$$

But the l.h.s and r.h.s. of (2.76) satisfy the property of stationarity of order n of (2.35) for all positive integers n. From the above arguments, we have shown a WSS Gaussian process is a strictly stationary process.

2.6 Ensemble averaging, time averaging, and ergodicity

Consider a r.p. $\{X(t), \; t \in \mathbf{T}\}$. When we perform an expectation of $X(t)$, denoted by $\mathrm{E}\{X(t)\}$, we mean

$$\mathrm{E}\{X(t)\} = \int_{-\infty}^{\infty} x f_{X(t)}(x) dx, \tag{2.77}$$

which is denoted as an *ensemble averaging*, since it is an averaging in the statistical sense of the r.v. $X(t)$ with respect to the pdf $f_{X(t)}(x)$. The ensemble averaging of the r.v. $X(t)$ yields a deterministic number. On the other hand, a *time-averaging* of the realization of a r.p. $\{X(t), \; t \in \mathbf{T}\}$ is denoted by

$$\langle x(t) \rangle = \lim_{T \to \infty} \left\{ \frac{1}{T} \int_{-T/2}^{T/2} x(t) dt \right\}. \tag{2.78}$$

The time averaging operator is denoted by the symbol $< \cdot >$ as shown in (2.78). The time-averaged expression in (2.78) for each realization also yields a deterministic number. However, different realizations of the r.p. yield different numbers. The r.p. $\{X(t), \; t \in \mathbf{T}\}$ is said to be *ergodic with respect to its mean* if (2.77) equals (2.78) with probability one with respect to all the realizations.

For the more general case, we can denote the expectation of $h(X(t))$ as

$$\mathrm{E}\{h(X(t))\} = \int_{-\infty}^{\infty} h(x) f_{X(t)}(x) dx. \tag{2.79}$$

Similarly, the time averaging of the realization of $h(x(t))$ is denoted by

$$\langle h(x(t)) \rangle = \lim_{T\to\infty} \left\{ \frac{1}{T} \int_{-T/2}^{T/2} h(x(t)) dt \right\}. \tag{2.80}$$

The r.p. $\{X(t),\ t \in \mathbf{T}\}$ is said to be *ergodic with respect to* $h(X(t))$ if (2.79) equals (2.80) with probability one with respect to all the realizations.

Intuitively, a r.p. $\{X(t),\ t \in \mathbf{T}\}$ is ergodic if the statistical properties of the process for each r.v. $X(t)$ at time t are exhibited in the realization over the entire time duration of $(-\infty, \infty)$. If a r.p. is specified in an analytical manner, then there may be analytical ways to verify an ergodic property of the process. However, for any physically observed process, it is impossible to verify the equivalency of (2.77) to (2.78) or (2.79) to (2.80). Indeed, in physical measurement of a process, we can only observe a realization of the process for as long a duration as is feasible (but certainly not from $-\infty$ to ∞ in time). In practice, we may need to replace the region of integration in (2.78) or (2.80) by replacing $-T/2$ by a, $T/2$ by b, and $1/T$ by $1/(b-a)$, and take $(b-a)$ as long as possible. Then by assuming ergodicity, we use the time-averaged value as an approximation to the ensemble-averaged value of the r.p.

Example 2.36 Consider the r.p. defined by

$$X(t) = \cos(2\pi f t + \Theta),\ -\infty < t < \infty, \tag{2.81}$$

where the frequency f is a fixed real number and Θ is a uniformly distributed r.v. on $(0, 2\pi]$. Then, using (2.77), the ensemble-average $\mathrm{E}\{X(t)\} = 0$. Similarly, using (2.78), the time averaging of $x(t)$ for a realization of $\Theta = \theta$, for an $\theta \in (0, 2\pi]$, also yields $\langle x(t) \rangle = 0$, for any $\theta \in (0, 2\pi]$. Thus, we can conclude the r.p. of (2.81) is ergodic with respect to the mean of the r.p. □

2.7 WSS random sequence

A discrete-time random process $\{X(t),\ t \in \mathbf{T}\}$, where $\mathbf{T}$ is a discrete set, is also called a *random sequence* (r.s.). In particular, we consider a WSS r.s., where $\mathbf{T}$ is taken to be the set of all integers denoted by $\mathbb{Z}$. Thus, a r.s. is often denoted by $\{X(n),\ n \in \mathbb{Z}\}$. Assume the r.s. is real-valued, then its autocorrelation sequence $R(n)$ of the WSS r.s. is given by

$$R(n) = \mathrm{E}\{X(n+m)X(m)\},\ n \in \mathbb{Z}. \tag{2.82}$$

Its psd function $S(\omega)$, now defined on the interval $-\pi < \omega \leq \pi$, is given by the Fourier series expansion of

$$S(\omega) = \mathscr{F}\{R(n)\} = \sum_{n=-\infty}^{\infty} R(n) e^{-i\omega n},\ -\pi < \omega \leq \pi. \tag{2.83}$$

We can consider (2.83) as the discrete version corresponding to $S(\omega)$, the Fourier transform of the autocorrelation function $R(t)$ of a continuous-time WSS r.p. $X(t)$ given by (2.44). The Fourier coefficients $\{R(n), -\infty < n < \infty\}$ of $S(\omega)$ in (2.83) are given by

$$R(n) = \mathscr{F}^{-1}\{S(\omega)\} = \frac{1}{2\pi}\int_{-\pi}^{\pi} S(\omega)e^{i\omega n}d\omega, \ -\infty < n < \infty. \tag{2.84}$$

Thus, $R(n)$ of (2.84) corresponds to the inverse Fourier transform, $R(t)$ of the continuous-time psd of a WSS r.p. $X(t)$ of (2.46). We note, for a WSS r.s., its psd $S(\omega)$ of (2.83) and inside the integral of (2.84) are defined only over the interval of $-\pi < \omega \leq \pi$, while for the WSS r.p. $X(t)$, its psd $S(\omega)$ of (2.46) is defined over the interval $-\infty < \omega < \infty$. Furthermore, for a WSS r.s., its autocorrelation sequence $R(n)$ is defined over all integers, while for a WSS r.p., its autocorrelation function $R(t)$ is defined over the interval $-\infty < t < \infty$.

For a WSS r.s., its psd function $S(\omega)$ and the autocorrelation sequence $R(n)$ also form a one-to-one unique discrete Fourier transform pair. From (2.83), let $-n + m = p$, then $R(-n) = \mathrm{E}\{X(-n+m)X(m)\} = \mathrm{E}\{X(p)X(n+p)\} = \mathrm{E}\{X(n+p)X(p)\} = R(n)$, and thus the autocorrelation sequence $R(n)$ is an even function of n. From (2.84), we note the psd function $S(\omega)$ is also an even function of ω over the interval of $-\pi < \omega \leq \pi$.

Example 2.37 Consider a WSS r.s. with an autocorrelation sequence

$$R(n) = C_0 a^{|n|}, \ -\infty < n < \infty, \ 0 < C_0, \ 0 < |a| < 1. \tag{2.85}$$

Its psd function is given by

$$S(\omega) = \frac{C_0(1 - |a|^2)}{|e^{i\omega} - a|^2}, \ -\pi < \omega \leq \pi. \quad \square \tag{2.86}$$

Consider a WSS r.s. $\{X(n), \ n \in \mathbb{Z}\}$ with an autocorrelation sequence $R(n)$ and a psd function $S(\omega)$, input to a linear time-invariant discrete-time system with a discrete-time impulse response sequence given by $\mathbf{h} = \{\ldots, h_{-1}, h_0, h_1, \ldots\}$ and a discrete-time transfer function $H(\omega)$ given by

$$H(\omega) = \sum_{n=-\infty}^{\infty} h_n e^{-i\omega n}, \ -\pi < \omega \leq \pi. \tag{2.87}$$

Then the output r.s. $\{Y(n), \ n \in \mathbb{Z}\}$ is also WSS and its psd function $S_{Y(n)}(\omega)$ is given by

$$S_{Y(n)}(\omega) = S(\omega)|H(\omega)|^2, \ -\pi < \omega \leq \pi. \tag{2.88}$$

We note the psd function $S_{Y(n)}(\omega)$ of (2.88) at the output of a discrete-time transfer function is analogous to the psd function $S_Y(\omega)$ of (2.53) at the output of a continuous-time transfer function.

Example 2.38 Consider a discrete-time linear time-invariant transfer function $H(\omega)$ whose magnitude response is given by

$$|H(\omega)| = \frac{\sqrt{(1-|a|^2)}}{|e^{i\omega} - a|}, \ -\pi < \omega \leq \pi, \ 0 < |a| < 1. \tag{2.89}$$

Consider a WSS "white sequence" (w.s.) $\{X(n),\ n \in \mathbb{Z}\}$, whose autocorrelation sequence $R(n)$ is given by

$$R(n) = \begin{cases} C_0\ ,\ n = 0,\ 0 < C_0, \\ 0\ ,\ n \neq 0. \end{cases} \tag{2.90}$$

By direct evaluation, its corresponding psd $S(\omega)$ is given by

$$S(\omega) = C_0, \ -\pi < \omega \leq \pi. \tag{2.91}$$

Then from (2.88), the psd function $S_{Y(n)}(\omega)$ of the r.s. $\{Y(n),\ n \in \mathbb{Z}\}$ at the output of the linear time-invariant discrete-time system with a magnitude transfer function given by (2.89) driven by an input w.s. $\{X(n),\ n \in \mathbb{Z}\}$ is given by

$$S_{Y(n)}(\omega) = S(\omega)|H(\omega)|^2 = \frac{C_0(1-|a|^2)}{|e^{i\omega} - a|^2}, \ -\pi < \omega \leq \pi. \tag{2.92}$$

Thus, the autocorrelation of the output sequence, from (2.85)–(2.86), is given by

$$R_{Y(n)}(n) = C_0 a^{|n|}, \ -\infty < n < \infty, \ 0 < C_0, \ 0 < |a| < 1. \quad \square \tag{2.93}$$

2.8 Conclusions

In Section 2.1, basic concepts of probability, including sample space, event, probability, random variable, cdf, pdf, conditional probability, joint probability cdf and pdf, random vector, multi-dimensional pdf, and the Central Limit Theorem were reviewed. A list of commonly encountered scalar r.v.'s and associated probabilities was also given. Section 2.2 introduced the n-dimensional Gaussian pdf as characterized by its mean vector and covariance matrix. Some detailed properties of marginal and conditional Gaussian pdfs were given. In Section 2.3, two equivalent definitions of a random process over continuous time were presented. Section 2.4 considered stationary random processes in which some parameters of the processes are time-invariant. In particular, the concept of wide-sense stationarity of the autocorrelation function depending only on their time difference and not on absolute time instants was presented. This process led to the important concept of the power spectral density function specifying the frequency characteristics of its power. In Section 2.5, random processes with Gaussian pdfs led to a simple closure property that linear time-invariant operations on Gaussian processes remained Gaussian processes. Section 2.6 discussed the difference of ensemble (statistical) averaging when the pdfs of the random variables were available versus time averaging when only realizations (observations) of the process were given. Ergodicity provided the link that showed the equivalence of these two kinds of averaging. In Section 2.7, some basic concepts of wide-sense stationary random sequences analogous to those of wide-sense stationary random processes over continuous time were discussed.

2.9 Comments

The study of communication and estimation theories is based on the usage of probability, statistics, and random processes as applied to signals, noises, and processing systems. Until recently, the earliest known documentation of attempts to formulate a probabilistic model to understand the relative frequency concept in games of chance appeared to be correspondences between P. Fermat and B. Pascal in the sixteenth century [2]. However, in 1998, the "Archimedes Codex" that vanished for many years appeared and was sold to a new owner for $2 million [3]. Codex is essentially a process of binding sheets of sheep skins to form a book in ancient times. Archimedes was the famous Greek mathematician and scientist (287 BC–212 BC), who lived in the town of Syracuse in Sicily. His works were recorded in Codices A, B, and C. Codices A and B were known to medieval scholars (who told the world of his theory of buoyancy and specific gravity, the Archimedes screw, and many other geometric theories), but are believed to be totally lost now. The "Archimedes Codex," or Codex C, also known to medieval scholars, was nominally a priest's praying book written on top of Archimedes's writings. Scientists used the latest X-ray techniques to retrieve Archimedes' original writings. In Codex C, Archimedes solved the ancient "Stomachion" problem, which consisted of 14 specific polygons originally cut up from a square. The challenge was to find all the possible different ways these polygons could reconstruct the square. Modern mathematical group theorists were not able to solve this problem analytically; but using brute-force computer enumerations, that number was found only in 2003 to have 536 basic solutions, with 32 rotations, leading to a final number of $17\,152 = 536 \times 32$ possible ways. In Codex C, Archimedes obtained the same number analytically, treating it as a combinatorial problem. Thus, Archimedes can be considered to be the "father of combinatorics," which is a branch of mathematics, including the theory of counting such as those of combinations and permutations encountered in discrete probability theory. While Archimedes did not originate the concept of probability, he introduced the logical concept of counting, which is relevant in discrete probability theory. Incidentally, in his concept of counting, his understanding of "infinity" was fully consistent with modern understanding of "infinity" [3]. In the seventeenth and eighteenth centuries, both P. S. Laplace [4] and F. Gauss [5] made major contributions in probability and estimation theories. However, with the usage of measure theory due to E. Borel, H. Lebesgue, J. Radon, and M. Fréchet, and the concept of the sample space by R. von Mises in the early twentieth century, A. N. Kolmogorov [1] in 1933 was able to establish probability theory on an axiomatic basis, and then made probability a branch of mathematics. Some early well-known books on probability theory include [6–9]. Some early well-known books on random processes include [10–12]. Books dealing with statistical theory are not listed here.

The 1958 work of W. B. Davenport and W. L. Root [13] was probably one of the earliest textbooks dealing with probability and random processes as applied to noises and signals in communication and radar systems. The 1965 work of A. Papoulis [14] was probably one of the earliest textbooks introducing probability and random processes to electrical engineering students. Since then there have been several dozens of introductory probability and random process books, of varying degrees of rigor and intentions (as indicated by their titles), written mainly for electrical engineering students. A small sample of them are listed in [15–27].

The materials on probability and random processes in this chapter can be found in [13–27]. Details on the proofs in Appendices 2.A and 2.B can be found in [28].

2.A Proof of Theorem 2.1 in Section 2.2.1

By definition of the marginal pdf $f_{\mathbf{X}_p}(\mathbf{x}_p)$, it is obtained by integrating over the $N-P$ components of $\mathbf{x}_{N-P}$ in the integral

$$f_{\mathbf{X}_p}(\mathbf{x}_p) = \int_{\mathbf{x}_{N-P}} \frac{1}{(2\pi)^{N/2}|\mathbf{\Lambda}_N|^{1/2}} e^{-(1/2)(\mathbf{x}_N-\boldsymbol{\mu}_N)^T \mathbf{\Lambda}_N^{-1}(\mathbf{x}_N-\boldsymbol{\mu}_N)} d\mathbf{x}_{N-P}. \tag{2.94}$$

For the non-singular covariance matrix $\mathbf{\Lambda}_N$ of (2.23), its inverse is given by

$$\mathbf{\Lambda}_N^{-1} = \tilde{\mathbf{\Lambda}}_N = \begin{bmatrix} \tilde{\mathbf{\Lambda}}_P & \tilde{\mathbf{\Lambda}}_{P(N-P)} \\ \tilde{\mathbf{\Lambda}}_{P(N-P)}^T & \tilde{\mathbf{\Lambda}}_{N-P} \end{bmatrix}, \tag{2.95}$$

where from Schur's Identity ([28], p. 6), $\tilde{\mathbf{\Lambda}}_P$ is given by

$$\tilde{\mathbf{\Lambda}}_P = \mathbf{\Lambda}_P^{-1} + \tilde{\mathbf{\Lambda}}_{P(N-P)} \tilde{\mathbf{\Lambda}}_{N-P}^{-1} \tilde{\mathbf{\Lambda}}_{P(N-P)}^T, \tag{2.96}$$

and from Jacobi's Theorem ([28], p. 7), $|\mathbf{\Lambda}_N|$ is given by

$$|\mathbf{\Lambda}_N| = \left|\tilde{\mathbf{\Lambda}}_{N-P}^{-1}\right| |\mathbf{\Lambda}_P|. \tag{2.97}$$

In many probability and statistical proofs, if a r.v. or random vector does not have a zero mean, define a new r.v. or random vector after subtracting its mean. After performing the proof assuming a zero mean, then in the final expression replace the variable or vector by the variable or vector minus its mean, respectively. In the following part of the proof from (2.98)–(2.102), we assume the mean vector is the zeroth vector. By using (2.95), the quadratic form in the exponential term in (2.94) can be expressed as

$$\begin{aligned} \mathbf{x}_N^T \mathbf{\Lambda}_N^{-1} \mathbf{x}_N = \mathbf{x}_P^T \mathbf{\Lambda}_P^{-1} \mathbf{x}_P + \mathbf{x}_{N-P}^T \tilde{\mathbf{\Lambda}}_{P(N-P)}^T \mathbf{x}_P + \mathbf{x}_P^T \tilde{\mathbf{\Lambda}}_{P(N-P)} \mathbf{x}_{N-P} \\ + \mathbf{x}_{N-P}^T \tilde{\mathbf{\Lambda}}_{N-P} \mathbf{x}_{N-P}. \end{aligned} \tag{2.98}$$

Upon completing the square in the $\mathbf{x}_{N-P}$ term by adding and subtracting $\mathbf{x}_P^T \tilde{\mathbf{\Lambda}}_{P(N-P)} \tilde{\mathbf{\Lambda}}_{(N-P)}^{-1} \tilde{\mathbf{\Lambda}}_{P(N-P)}^T \mathbf{x}_P$ in (2.98), we obtain

$$\begin{aligned} \mathbf{x}_N^T \mathbf{\Lambda}_N^{-1} \mathbf{x}_N = \left(\mathbf{x}_{N-P} + \tilde{\mathbf{\Lambda}}_{(N-P)}^{-1} \tilde{\mathbf{\Lambda}}_{P(N-P)}^T \mathbf{x}_P\right)^T \tilde{\mathbf{\Lambda}}_{N-P} \left(\mathbf{x}_{N-P} + \tilde{\mathbf{\Lambda}}_{(N-P)}^{-1} \tilde{\mathbf{\Lambda}}_{P(N-P)}^T \mathbf{x}_P\right) \\ + \mathbf{x}_P^T \left(\tilde{\mathbf{\Lambda}}_P - \tilde{\mathbf{\Lambda}}_{P(N-P)} \tilde{\mathbf{\Lambda}}_{(N-P)}^{-1} \tilde{\mathbf{\Lambda}}_{P(N-P)}^T\right) \mathbf{x}_P. \end{aligned} \tag{2.99}$$

By using (2.96), we obtain

$$\mathbf{x}_P^T\left(\tilde{\mathbf{\Lambda}}_P - \tilde{\mathbf{\Lambda}}_{P(N-P)}\tilde{\mathbf{\Lambda}}_{(N-P)}^{-1}\tilde{\mathbf{\Lambda}}_{P(N-P)}^T\right)\mathbf{x}_P = \mathbf{x}_P^T\mathbf{\Lambda}_P^{-1}\mathbf{x}_P. \tag{2.100}$$

Then using (2.100) in (2.99), we have

$$\begin{aligned}\mathbf{x}_N^T\mathbf{\Lambda}_N^{-1}\mathbf{x}_N &= \left(\mathbf{x}_{N-P} + \tilde{\mathbf{\Lambda}}_{(N-P)}^{-1}\tilde{\mathbf{\Lambda}}_{P(N-P)}^T\mathbf{x}_P\right)^T\tilde{\mathbf{\Lambda}}_{N-P}\left(\mathbf{x}_{N-P} + \tilde{\mathbf{\Lambda}}_{(N-P)}^{-1}\tilde{\mathbf{\Lambda}}_{P(N-P)}^T\mathbf{x}_P\right)\\ &\quad + \mathbf{x}_P^T\mathbf{\Lambda}_P^{-1}\mathbf{x}_P.\end{aligned} \tag{2.101}$$

Now, use (2.101) in the integral of (2.94) to yield

$$\begin{aligned}
&f_{\mathbf{X}_p}(\mathbf{x}_P)\\
&= \int_{\mathbf{x}_{N-P}} \frac{1}{(2\pi)^{\frac{N}{2}}|\mathbf{\Lambda}_N|^{\frac{1}{2}}} e^{-\frac{1}{2}\mathbf{x}_P^T\mathbf{\Lambda}_P^{-1}\mathbf{x}_P}\\
&\quad \cdot e^{-\frac{1}{2}\left(\mathbf{x}_{N-P}+\tilde{\mathbf{\Lambda}}_{(N-P)}^{-1}\tilde{\mathbf{\Lambda}}_{P(N-P)}^T\mathbf{x}_P\right)^T\tilde{\mathbf{\Lambda}}_{N-P}\left(\mathbf{x}_{N-P}+\tilde{\mathbf{\Lambda}}_{(N-P)}^{-1}\tilde{\mathbf{\Lambda}}_{P(N-P)}^T\mathbf{x}_P\right)} d\mathbf{x}_{N-P}\\
&= e^{-\frac{1}{2}\mathbf{x}_P^T\mathbf{\Lambda}_P^{-1}\mathbf{x}_P}\frac{1}{(2\pi)^{\frac{N}{2}}|\mathbf{\Lambda}_N|^{\frac{1}{2}}}\\
&\quad \cdot \int_{\mathbf{x}_{N-P}} e^{\frac{-\left(\mathbf{x}_{N-P}+\tilde{\mathbf{\Lambda}}_{(N-P)}^{-1}\tilde{\mathbf{\Lambda}}_{P(N-P)}^T\mathbf{x}_P\right)^T\tilde{\mathbf{\Lambda}}_{N-P}\left(\mathbf{x}_{N-P}+\tilde{\mathbf{\Lambda}}_{(N-P)}^{-1}\tilde{\mathbf{\Lambda}}_{P(N-P)}^T\mathbf{x}_P\right)}{2}} d\mathbf{x}_{N-P}\\
&= e^{\frac{-\mathbf{x}_P^T\mathbf{\Lambda}_P^{-1}\mathbf{x}_P}{2}}\int_{\mathbf{x}_{N-P}}\frac{1}{(2\pi)^{\frac{N}{2}}|\mathbf{\Lambda}_N|^{\frac{1}{2}}} e^{\frac{-\mathbf{x}_{N-P}^T\tilde{\mathbf{\Lambda}}_{N-P}\mathbf{x}_{N_P}}{2}} d\mathbf{x}_{N-P}\\
&= \frac{1}{(2\pi)^{\frac{P}{2}}|\mathbf{\Lambda}_P|^{\frac{1}{2}}} e^{\frac{-\mathbf{x}_P^T\mathbf{\Lambda}_P^{-1}\mathbf{x}_P}{2}}\int_{\mathbf{x}_{N-P}}\frac{1}{(2\pi)^{\frac{N-P}{2}}|\tilde{\mathbf{\Lambda}}_{N-P}^{-1}|^{\frac{1}{2}}} e^{\frac{-\mathbf{x}_{N-P}^T\tilde{\mathbf{\Lambda}}_{N-P}\mathbf{x}_{N_P}}{2}} d\mathbf{x}_{N-P}\\
&= \frac{1}{(2\pi)^{\frac{P}{2}}|\mathbf{\Lambda}_P|^{\frac{1}{2}}} e^{\frac{-\mathbf{x}_P^T\mathbf{\Lambda}_P^{-1}\mathbf{x}_P}{2}}.
\end{aligned} \tag{2.102}$$

In (2.102), the second equation follows from the first equation, since $\mathbf{x}_P^T\Lambda_P^{-1}\mathbf{x}_P$ is independent of $d\mathbf{x}_{N-P}$; the third equation follows from the second equation since $\tilde{\mathbf{\Lambda}}_{N-P}^{-1}\tilde{\mathbf{\Lambda}}_{P(N-P)}^T\mathbf{x}_P$ is a constant independent of $d\mathbf{x}_{N-P}$ in its integral and thus the integral on the second equation is the same as the integral in the third equation; the fourth equation follows from the third equation by using (2.97); and finally the fifth equation follows from the fourth equation since the integral in the fifth equation equals one. Equation (2.102) shows the P-dimensional marginal pdf of the original N-dimensional Gaussian pdf is still Gaussian for any P from 1 to $(N-1)$.

When the random vector $\mathbf{x}_N$ has the mean vector $\boldsymbol{\mu}_N$, then the expression given by (2.101) takes the form of

$$\begin{aligned}(\mathbf{x}_N - \boldsymbol{\mu}_N)^T\mathbf{\Lambda}_N^{-1}(\mathbf{x}_N - \boldsymbol{\mu}_N) &= \left(\mathbf{x}_{N-P} - \boldsymbol{\mu}_{N-P} + \tilde{\mathbf{\Lambda}}_{N-P}^{-1}\tilde{\mathbf{\Lambda}}_{P(N-P)}^T(\mathbf{x}_P - \boldsymbol{\mu}_P)\right)^T\\ &\quad \cdot\tilde{\mathbf{\Lambda}}_{N-P}\left(\mathbf{x}_{N-P} - \boldsymbol{\mu}_{N-P} + \tilde{\mathbf{\Lambda}}_{N-P}^{-1}\tilde{\mathbf{\Lambda}}_{P(N-P)}^T(\mathbf{x}_P - \boldsymbol{\mu}_P)\right)\\ &\quad + (\mathbf{x}_P - \boldsymbol{\mu}_P)^T\mathbf{\Lambda}_P^{-1}(\mathbf{x}_P - \boldsymbol{\mu}_P).\end{aligned} \tag{2.103}$$

Then the last line of (2.102) takes the form of (2.103). In summary, if $\mathbf{X}_N = [\mathbf{X}_P^T, \mathbf{X}_{N-P}^T]^T \sim \mathcal{N}(\boldsymbol{\mu}_N, \boldsymbol{\Lambda}_N)$, then $\mathbf{X}_P \sim \mathcal{N}(\boldsymbol{\mu}_P, \boldsymbol{\Lambda}_P)$, where $\boldsymbol{\mu}_P$ is the first P-components of $\boldsymbol{\mu}_N$ and $\boldsymbol{\Lambda}_P$ is the upper-left $P \times P$ components of $\boldsymbol{\Lambda}_N$. □

2.B Proof of Theorem 2.2 in Section 2.2.1

The conditional pdf of $\mathbf{X}_{N-P}|\mathbf{X}_P$ is defined by $f_{\mathbf{X}_{N-P}|\mathbf{X}_P}(\mathbf{x}_{N-P}|\mathbf{x}_P) = f_{\mathbf{X}_N}(\mathbf{x}_N)/f_{\mathbf{X}_P}(\mathbf{x}_P)$. However, when $\mathbf{X}_N$ is Gaussian, then

$$\begin{aligned} f_{\mathbf{X}_N}(\mathbf{x}_N) &= f_{\mathbf{X}_{N-P}|\mathbf{X}_P}(\mathbf{x}_{N-P}|\mathbf{x}_P = \mathbf{a}) f_{\mathbf{X}_P}(\mathbf{x}_P = \mathbf{a}) \\ &= (2\pi)^{-N/2}|\boldsymbol{\Lambda}_N|^{-1/2} e^{-\frac{1}{2}(\mathbf{x}_N - \boldsymbol{\mu}_N)^T \boldsymbol{\Lambda}_N^{-1}(\mathbf{x}_N - \boldsymbol{\mu}_N)}. \end{aligned} \tag{2.104}$$

The quadratic form in (2.104) can be decomposed as follows. From (2.103), we associate the first term on the r.h.s. of (2.103) to yield the quadratic form in $f_{\mathbf{X}_{N-P}|\mathbf{X}_P}(\mathbf{x}_{N-P}|\mathbf{x}_P = \mathbf{a})$ given by (2.25) and the second term on the r.h.s. of (2.103) to yield the quadratic form in $f_{\mathbf{X}_P}(\mathbf{x}_P = \mathbf{a})$. Of course, the property $|\boldsymbol{\Lambda}_N|^{-1/2} = \left|\tilde{\boldsymbol{\Lambda}}_{N-P}^{-1}\right|^{-1/2} |\boldsymbol{\Lambda}_P|^{-1/2}$ as given by (2.97) is also used.

References

[1] A. N. Kolmogorov. Foundations of the theory of probability (*translated from "Grundbegriffe dur Wahrscheinlichkeitrechnung"), Ergebnisse der Mathematik II*, 1933.

[2] Correspondences between P. Fermat and B. Pascal. http://en.wikipedia.org/wiki/History_of_probability.

[3] R. Netz and W. Noel. *The Archimedes Codex*. Weidenfeld & Nicolson, 2007.

[4] P. S. Laplace. Théorie analytique des probabilités (1812). http://en.wikipedia.org/wiki/Pierre-Simon_Laplace.

[5] F. Gauss. Theoria combinationis observationum erroribus minimis obnoxiae (1823). http://www-groups.dcs.st-and.ac.uk/~history/Biographies/Gauss.html.

[6] P. Levy. *Calcul des probabilities*. Gauthiers-Villars, 1925.

[7] H. Cramér. *Random Variables and Probability Distributions*. Cambridge University Press, 1937.

[8] W. Feller. *An Introduction to Probability Theory and Its Applications*. Wiley, vol. 1, 1957; vol. 2, 1966.

[9] M. Loève. *Probability Theory*. Van Nostrand, 1960.

[10] J. L. Doob. *Stochastic Processes*. Wiley, 1953.

[11] A. Blanc-Lapierre and R. Fortet. *Théorie des fonctions aléatoires*. Masson et Cie, 1953.

[12] U. Grenander and M. Rosenblatt. *Statistical Analysis of Stationary Time Series*. Wiley, 1957.

[13] W. B. Davenport Jr. and W. L. Root. *An Introduction to the Theory of Random Signals and Noise*. McGraw-Hill, 1958.

[14] A. Papoulis. *Probability, Random Variables and Stochastic Processes*. McGraw-Hill, 1st ed., 1965; 4th ed., 2002.

[15] R. B. Ash. *Basic Probability Theory*. Wiley, 1970.

[16] E. Wong. *Stochastic Processes in Information and Dynamical Systems*. McGraw-Hill, 1970.
[17] A. Leon-Garcia. *Probability and Random Processes for Electrical Engineering*. A. Wesley, 1st ed., 1970; 3rd ed., 2008.
[18] C. W. Helstrom. *Probability and Stochastic Processes for Engineers*. Macmillan, 1984.
[19] J. B. Thomas. *Introduction to Probability*. Springer-Verlag, 1986.
[20] A. V. Balakrishnan. *Introduction to Random Processes in Engineering*. Wiley, 1995.
[21] Y. Viniotis. *Probability and Random Processes for Electrical Engineers*. McGraw-Hill, 1998.
[22] D. Stirzaker. *Probability and Random Variables – A Beginner's Guide*. Cambridge University Press, 1999.
[23] M. B. Pursley. *Random Processes in Linear Systems*. Prentice-Hall, 2002.
[24] P. Olofsson. *Probability, Statistics, and Stochastic Processes*. Wiley, 2005.
[25] T. L. Fine. *Probability and Probabilistic Reasoning for Electrical Engineering*. Pearson Prentice-Hall, 2006.
[26] J. A. Gubner. *Probability and Random Processes for Electrical Engineers*. Cambridge University Press, 2006.
[27] S. Kay. *Intuitive Probability and Random Processes using Matlab*. Springer, 2006.
[28] K. S. Miller. *Multidimensional Gaussian Distributions*. Wiley, 1964.

Problems

2.1 Write down the probability density and probability distribution functions as well as the mean and variance for the following random variables: Bernoulli; binomial; Poisson; Cauchy; exponential; Gaussian (one-dimensional); Laplace; Rayleigh. (No derivation is needed; but you should be able to derive them for yourselves.)

2.2

(a) Consider $Y = 2X + 3$ where X is a uniform r.v. on $[0, 1]$. Find $F_Y(y)$ and $f_Y(y)$ in terms of $F_X(x)$ and $f_X(x)$.
(b) Let the r.v. X with $\mathcal{N}(0, 1)$ pass through a half-wave rectifier, with a transfer function $y = g(x) = xu(x)$, where $u(x)$ is a unit step function equal to one for $x \geq 0$ and zero elsewhere. Find $F_Y(y)$ and $f_Y(y)$ in terms of $F_X(x)$ and $f_X(x)$.
(c) Let the r.v. X be $\mathcal{N}(0, \sigma^2)$ and the r.v. $Y = X^2$. Find $E[Y]$.

2.3 The random vector $[X, Y]^T$ has the joint pdf of

$$f_{X,Y}(x, y) = c(x + y), \qquad 0 < x < 1,\ 0 < y < 1.$$

(a) Find c.
(b) Find the joint cdf of $[X, Y]^T$.
(c) Find the marginal cdfs and the marginal pdfs of X and Y.

2.4 Let

$$f_{X,Y}(x, y) = \begin{cases} (1/9)e^{-y/3}, & 0 \leq x \leq y < \infty, \\ 0, & \text{elsewhere.} \end{cases}$$

Find the covariance of X and Y and the correlation coefficient of X and Y.

Gaussian random variable/vector and process

2.5 Let $\mathbf{X} = [X_1, X_2, X_3]^T$ be a real-valued Gaussian random vector with $f_{\mathbf{X}}(\mathbf{x}) = C \exp[-\frac{1}{2}(2x_1^2 - x_1x_2 + x_2^2 - 2x_1x_3 + 4x_3^2)]$.

(a) Find the covariance matrix $\mathbf{\Lambda}$ of $\mathbf{X}$.
(b) Find the constant C.
(c) Let $\mathbf{Y} = \mathbf{BX}$ where

$$\mathbf{B} = \begin{bmatrix} 1 & -1/4 & -1/2 \\ 0 & 1 & -2/7 \\ 0 & 0 & 1 \end{bmatrix}.$$

Find Y_1, Y_2, and Y_3 explicitly in terms of X_1, X_2, and X_3.
(d) Find X_1, X_2, and X_3 explicitly in terms of Y_1, Y_2, and Y_3.
(e) Find the pdf of $f_{\mathbf{Y}}(\mathbf{y})$ explicitly in terms of y_1, y_2, and y_3.
(f) Show the r.v.'s Y_1, Y_2, and Y_3 are mutually independent and find their means and variances.

2.6 Let $Z = aX + bY$, where a and b are real-valued numbers, $X \sim \mathcal{N}(\mu_X, \sigma_X^2)$, $Y \sim \mathcal{N}(\mu_Y, \sigma_Y^2)$, and X and Y are independent r.v.'s.

(a) Find μ_Z.
(b) Find σ_Z^2.
(c) Show Z is a Gaussian r.v. using the characteristic function method. What are the μ_Z and σ_Z^2 you found using the characteristic function method? Are they the same as those you found in part (a) and part (b)? Hint: The characteristic function of the continuous r.v. Z is defined by $\phi_Z(\omega) = \mathrm{E}\left\{e^{i\omega Z}\right\}$. If Z is a Gaussian r.v. with a mean μ_Z and variance σ_Z^2, then $\phi_Z(\omega) = e^{i\omega\mu_Z} e^{-\omega^2\sigma_Z^2/2} \Leftrightarrow Z \sim \mathcal{N}(\mu_Z, \sigma_Z^2)$.

2.7 Let $Z = X + Y$, where X is a Poisson r.v. with a parameter λ_X, Y is also a Poisson r.v. with a parameter λ_Y, and X and Y are independent random variables.

(a) Find μ_Z.
(b) Find σ_Z^2.
(c) Show Z is a Poisson r.v. using the moment generation method. What are the μ_Z and σ_Z^2 you found using the moment generation method? Are they the same as those you found in part (a) and part (b)? Hint: The moment generating function of a discrete r.v. Z is defined by $\phi_Z(t) = \mathrm{E}\{e^{tZ}\}$. If Z is a Poisson r.v. with the parameter λ, then its moment generating function $\phi_Z(t) = \sum_{k=0}^{\infty} (e^{kt}e^{-\lambda}\lambda^k)/k! = e^{-\lambda} \sum_{k=0}^{\infty} (e^t\lambda)^k/k! = e^{-\lambda}e^{e^t\lambda}$.
(d) Show $Z = X + Y$ is a Poisson r.v. using only the independence property of the two r.v.'s and the Poisson property of X and Y.
(e) Suppose $Z = 2X + 3Y$, where X is a Poisson r.v. with a parameter λ_X, Y is also a Poisson r.v. with a parameter λ_Y, and X and Y are independent random variables. Is Z still a Poisson r.v.?

Random process

2.8 Consider the random process

$$X(t) = A \sin(2\pi f t + \Phi), \quad -\infty < t < \infty. \qquad (\Delta)$$

(a) Suppose f and Φ are positive-valued constants, and A is a positive-valued r.v. with a pdf $f_A(a)$, $0 < a < \infty$. Is $X(t)$ a stationary random process of order 1?
(b) Suppose A and f are positive-valued constants, and Φ is a uniformly distributed r.v. on $[0, 2\pi)$. Is $X(t)$ a stationary random process of order 1?
(c) Let the assumptions in (b) still be valid. Is $X(t)$ a w.s. stationary random process?

Wide-sense stationary random process

2.9 Let $\{X(n), n \in z\}$ be a real-valued w.s. stationary random sequence defined on the set of all integers z. Then $R(n) = \mathrm{E}\{X(m+n)X(m)\}$ and the power spectral density is defined by

$$S(\omega) = \sum_{n=-\infty}^{\infty} R(n)e^{-\omega n}, \quad -\pi \le \omega \le \pi$$

$$R(n) = \frac{1}{2\pi}\int_{-\pi}^{\pi} S(\omega)e^{\omega n}\, d\omega, \quad n \in z. \qquad (*)$$

For $R(0) = 1$, $R(\pm 1) = 1/2$, $R(\pm 2) = 1/4$, $R(m) = 0$, $|m| \ge 3$, find its power spectral density $S(\omega)$, $-\pi \le \omega \le \pi$. Show your $S(\omega)$ does yield the originally specified $R(n)$ by using (*) explicitly.

2.10 Consider a discrete-time linear time-invariant system with an input w.s. stationary sequence $\{X(n), n \in z\}$ having a power spectral density $S_X(\omega)$, $-\pi \le \omega < \pi$. The output sequence $\{Y(n), n \in z\}$ is given by

$$Y(n) = (X(n-1) + X(n) + X(n+1))/3, \qquad n \in z.$$

Find the output power spectral density $S_Y(\omega)$ explicitly in terms of $S_X(\omega)$, $-\pi \le \omega < \pi$.

2.11 Consider a continuous-time random process defined by $X(t, \Theta) = A\cos(\omega t + \Theta)$, $-\infty < t < \infty$, where A and ω are constants and Θ is a uniformly distributed r.v. on $[0, 2\pi)$.

(a) Find the ensemble-averaged mean $\mathrm{E}\{X(t, \Theta)\}$, where $\mathrm{E}\{\cdot\}$ is averaging with respect to the r.v. Θ.
(b) Find the ensemble-averaged autocorrelation function $R_X(t, t+\tau) = \mathrm{E}\{X(t,\Theta)X(t+\tau,\Theta)\}$. Use the identity $\cos(u)\cos(v) = (1/2)\cos(u+v) + (1/2)\cos(u-v)$. Is $R_X(t, t+\tau)$ a function of t?
(c) Find the time-averaged mean $< X(t,\theta) >= \lim_{T\to\infty} \frac{1}{2T}\int_{-T}^{T} X(t,\theta)dt$.
(d) Find the time-averaged autocorrelation $< X(t,\theta)X(t+\tau,\theta) >= \lim_{T\to\infty} \frac{1}{2T}\int_{-T}^{T} X(t,\theta)X(t+\tau,\theta)dt$.
(e) Is the ensemble-averaged mean equal to the time-averaged mean for this process? Is the ensemble-averaged autocorrelation equal to the time-averaged autocorrelation

for this process? A random process is said to be ergodic if all ensemble averages are equal to the time averages.

2.12 Let $\{X(t), t \in \mathbb{R}\}$ be a real-valued w.s. stationary Gaussian random process on the real line. Then it is straightforward to show that $\{X(t), t \in \mathbb{R}\}$ is a strictly stationary random process. Denote a complex-valued random process by $\{Z(t) = X(t) + iY(t), t \in \mathbb{R}\}$, where $\{X(t), t \in \mathbb{R}\}$ and $\{Y(t), t \in \mathbb{R}\}$ are both real-valued. If $\{Z(t) = X(t) + iY(t), t \in \mathbb{R}\}$ is a complex-valued w.s. stationary Gaussian random process, is it a strictly stationary random process? If not, what additional assumption must be imposed on $\{Z(t) = X(t) + iY(t), t \in \mathbb{R}\}$?

3 Hypothesis testing

Hypothesis testing is a concept originated in statistics by Fisher [1] and Neyman–Pearson [2] and forms the basis of detection of signals in noises in communication and radar systems.

3.1 Simple hypothesis testing

Suppose we measure the outcome of a real-valued r.v. X. This r.v. can come from two pdf's associated with the hypotheses, H_0 or H_1. Under H_0, the conditional probability of X is denoted by $p_0(x) = p(x|H_0)$, $-\infty < x < \infty$, and under H_1, the conditional probability of X is denoted by $p_1(x) = p(x|H_1)$, $-\infty < x < \infty$. This hypothesis is called "simple" if the two conditional pdf's are fully known (i.e., there are no unknown parameters in these two functions). From the observed x value (which is a realization of the r.v. X), we want to find a strategy to decide on H_0 or H_1 in some optimum statistical manner.

Example 3.1 The binary hypothesis problem in deciding between H_0 or H_1 is ideally suited to model the radar problem in which the hypothesis H_0 is associated with the absence of a target and the hypothesis H_1 is associated with the presence of a target. In a binary hypothesis problem, there are four possible states, whether H_0 or H_1 is true and whether the decision is to declare H_0 or to declare H_1. Table 3.1 summarizes these four states and the associated names and probabilities. Suppose H_0 is true and the decision is to declare H_0, assuming a target is absent. Then this event is "uneventful" and our decision is good. Suppose H_0 is true and the decision is to declare H_1, then this is a "false alarm" and our decision is bad. Suppose H_1 is true and the decision is to declare H_0, then this is a "miss" and our decision can be very bad (assuming the target has some serious ill intention toward us). Finally, suppose H_1 is true and the decision is to declare H_1, then this is a correct "detection" and our decision is good. The probability associated with each state is also given in Table 3.1 and will be defined in (3.1)–(3.4). □

We denote the decision region $\mathbb{R}_1$ as the set of all observed real values that we will declare H_1, and $\mathbb{R}_0$ as the set of all observed real values that we will declare H_0. The two sets $\mathbb{R}_1$ and $\mathbb{R}_0$ are disjoint (i.e., having no points in common) and their union is the entire real line (i.e., $\mathbb{R}_1 \cup \mathbb{R}_0 = (-\infty, \infty)$).

Table 3.1 Four states in a binary detection problem.

Decision	Truth H_0	Truth H_1
H_0	Uneventful (Good) $P_U = 1 - P_{FA}$	Miss (Bad) $P_M = 1 - P_D$
H_1	False alarm (Bad) P_{FA}	Detection (Good) P_D

We want to evaluate the probability associated with each of these four states in Table 3.1. The probability of a false alarm is denoted by P_{FA} and is defined by

$$P_{\text{False alarm}} = P_{\mathrm{FA}} = \int_{\mathbb{R}_1} p_0(x)dx. \tag{3.1}$$

This false alarm event occurs when the hypothesis is H_0 (thus the $p_0(x)$ pdf is applicable) and its probability is obtained by integration over all the elements in $\mathbb{R}_1$ (which results in declaring H_1) in the r.h.s. of (3.1). The probability of a detection is denoted by P_{D} and is defined by

$$P_{\text{Detection}} = P_{\mathrm{D}} = \int_{\mathbb{R}_1} p_1(x)dx. \tag{3.2}$$

This detection event occurs when the hypothesis is H_1 (thus the $p_1(x)$ pdf is applicable) and its probability is obtained by integration over all the elements in $\mathbb{R}_1$ (which results in declaring H_1) in the r.h.s. of (3.2). The probability of an uneventful event is the probability when H_0 is true, and we declare H_0. Thus, it is the complement of P_{FA} as defined by

$$P_{\text{Uneventful event}} = P_{\mathrm{U}} = 1 - P_{\mathrm{FA}}. \tag{3.3}$$

Similarly, the probability of a miss event occurs when H_1 is true, but we declare H_0. Thus, it is the complement of P_D as defined by

$$P_{\text{Miss}} = P_{\mathrm{M}} = 1 - P_{\mathrm{D}}. \tag{3.4}$$

Ideally, we want P_{D} as close to unity as possible and simultaneously want P_{FA} as close to zero as possible. In general, it is not possible to make P_{D} close to one without letting P_{FA} also become large. Similarly, it is not possible to make P_{FA} close to zero without letting P_{D} also become small. Statistical decision theory offers several different statistical optimization methods to address this problem.

3.2 Bayes criterion

In the Bayes criterion method, we define a cost function c_{ij} given by

$$c_{ij} = \text{Cost of choosing hypothesis } H_1 \text{ when } H_j \text{ is true}, \ i, j = 0, 1. \tag{3.5}$$

Equivalently, we can define the cost matrix $\mathbf{C}$ in terms of c_{ij} given by

$$\mathbf{C} = \begin{bmatrix} c_{00} & c_{01} \\ c_{10} & c_{11} \end{bmatrix}. \tag{3.6}$$

In the Bayes criterion method, we also assume the prior probability (sometimes also called the a priori probability) of the hypothesis H_0 is known and can be denoted by

$$\mathrm{P}(H_0) = \pi_0, \ \mathrm{P}(H_1) = \pi_1 = 1 - \pi_0. \tag{3.7}$$

Then the prior probability of the hypothesis H_1 is also known and is the complement of the prior probability of the hypothesis H_0.

Next, we define the average cost $\bar{c}$ per decision by

$$\begin{aligned} \bar{c} &= \pi_0[c_{00}(1 - P_{\mathrm{FA}}) + c_{10} P_{\mathrm{FA}}] + (1 - \pi_0)[c_{01}(1 - P_{\mathrm{D}}) + c_{11} P_{\mathrm{D}}] \\ &= \pi_0 \left[c_{00} \int_{\mathbb{R}_0} p_0(x)dx + c_{10} \int_{\mathbb{R}_1} p_0(x)dx \right] \\ &\quad + (1 - \pi_0) \left[c_{01} \int_{\mathbb{R}_0} p_1(x)dx + c_{11} \int_{\mathbb{R}_1} p_1(x)dx \right]. \end{aligned} \tag{3.8}$$

The Bayes criterion decision rule is the rule chosen to minimize the average cost.

Theorem 3.1 *The solution for the optimum decision regions $\mathbb{R}_0$ and $\mathbb{R}_1$ under the Bayes criterion decision rule is determined by using the likelihood ratio (LR) function*

$$\Lambda(x) = \frac{p_1(x)}{p_0(x)}, \tag{3.9}$$

where $p_1(x)$ and $p_0(x)$ are the conditional pdfs of the observed r.v. X under H_1 and H_0, respectively. Then

$$\mathbb{R}_0(x) = \{x : \Lambda(x) \leq \Lambda_0\} = \{x : p_1(x) \leq \Lambda_0 p_0(x)\} \tag{3.10a}$$

$$\mathbb{R}_1(x) = \{x : \Lambda(x) \geq \Lambda_0\} = \{x : p_1(x) \geq \Lambda_0 p_0(x)\} \tag{3.10b}$$

where the LR constant is defined by

$$\Lambda_0 = \frac{\pi_0(c_{10} - c_{00})}{\pi_1(c_{01} - c_{11})}, \tag{3.11}$$

where π_1 and $\pi_0 = 1 - \pi_1$ are the prior probabilities of H_1 and H_0 respectively, and c_{ij} is the cost of choosing H_i when H_j is true for $i, j = 0, 1$. We assume by definition of the cost, $c_{01} \geq c_{11}$ and $c_{10} \geq c_{00}$. That is, it is more costly to make an error than not to make an error given either H_1 or H_0.

Proof: Prove that the $\mathbb{R}_0$ and $\mathbb{R}_1$ given by (3.10) using the LR function in (3.9) yield the optimum solution under the Bayes criterion of minimizing the average cost $\bar{c}$ of

$$\begin{aligned} \bar{c} &= \pi_0 \left[c_{00} \int_{\mathbb{R}_0} p_0(x)dx + c_{10} \int_{\mathbb{R}_1} p_0(x)dx \right] \\ &\quad + \pi_1 \left[c_{01} \int_{\mathbb{R}_0} p_1(x)dx + c_{11} \int_{\mathbb{R}_1} p_1(x)dx \right]. \end{aligned} \tag{3.12}$$

Let $\mathbb{R}'_0$ and $\mathbb{R}'_1$ be any other decision regions for deciding H_0 and H_1 based on any other decision strategy.

Denote $\check{\mathbb{R}}_1 = \mathbb{R}'_0 \cap \mathbb{R}_1$ and $\check{\mathbb{R}}_2 = \mathbb{R}_0 \cap \mathbb{R}'_1$. Then, using any other decision strategy, the associated average cost $\bar{c}'$ is given by

$$\bar{c}' = \pi_0 \left[c_{00} \int_{\mathbb{R}'_0} p_0(x)dx + c_{10} \int_{\mathbb{R}'_1} p_0(x)dx \right] + \pi_1 \left[c_{01} \int_{\mathbb{R}'_0} p_1(x)dx + c_{11} \int_{\mathbb{R}'_1} p_1(x)dx \right]. \tag{3.13}$$

We want to show $\bar{c}' \geq \bar{c}$ or $\bar{c}' - \bar{c} \geq 0$. Thus, start with

$$\begin{aligned} \bar{c}' - \bar{c} &= \pi_0 \left\{ c_{00} \left[\int_{\mathbb{R}'_0} p_0(x)dx - \int_{\mathbb{R}_0} p_0(x)dx \right] + c_{10} \left[\int_{\mathbb{R}'_1} p_0(x)dx - \int_{\mathbb{R}_1} p_0(x)dx \right] \right\} \\ &\quad + \pi_1 \left\{ c_{01} \left[\int_{\mathbb{R}'_0} p_1(x)dx - \int_{\mathbb{R}_0} p_1(x)dx \right] + c_{11} \left[\int_{\mathbb{R}'_1} p_1(x)dx - \int_{\mathbb{R}_1} p_1(x)dx \right] \right\} \end{aligned} \tag{3.14}$$

Rewrite

$$\mathbb{R}'_0 = (\mathbb{R}'_0 \cap \mathbb{R}_0) \cup (\mathbb{R}'_0 \cap \mathbb{R}_1) = (\mathbb{R}'_0 \cap \mathbb{R}_0) \cup \check{\mathbb{R}}_1 \tag{3.15}$$

$$\mathbb{R}_0 = (\mathbb{R}_0 \cap \mathbb{R}'_0) \cup (\mathbb{R}_0 \cap \mathbb{R}'_1) = (\mathbb{R}_0 \cap \mathbb{R}'_0) \cup \check{\mathbb{R}}_2 \tag{3.16}$$

$$\mathbb{R}'_1 = (\mathbb{R}'_1 \cap \mathbb{R}_1) \cup (\mathbb{R}'_1 \cap \mathbb{R}_0) = (\mathbb{R}'_1 \cap \mathbb{R}_1) \cup \check{\mathbb{R}}_2 \tag{3.17}$$

$$\mathbb{R}_1 = (\mathbb{R}_1 \cap \mathbb{R}'_1) \cup (\mathbb{R}_1 \cap \mathbb{R}'_0) = (\mathbb{R}_1 \cap \mathbb{R}'_1) \cup \check{\mathbb{R}}_1. \tag{3.18}$$

Then for any integrand (omitting the integrands for simplicity of presentation), we have from (3.15) and (3.16)

$$\int_{\mathbb{R}'_0} - \int_{\mathbb{R}_0} = \int_{\check{\mathbb{R}}_1} - \int_{\check{\mathbb{R}}_2} \tag{3.19}$$

and from (3.17) and (3.18)

$$\int_{\mathbb{R}'_1} - \int_{\mathbb{R}_1} = \int_{\check{\mathbb{R}}_2} - \int_{\check{\mathbb{R}}_1}. \tag{3.20}$$

By using (3.19) and (3.20) in (3.14), we now have

$$\begin{aligned} \bar{c}' - \bar{c} &= \pi_0 \left\{ c_{00} \left[\int_{\check{\mathbb{R}}_1} p_0(x)dx - \int_{\check{\mathbb{R}}_2} p_0(x)dx \right] + c_{10} \left[\int_{\check{\mathbb{R}}_2} p_0(x)dx - \int_{\check{\mathbb{R}}_1} p_0(x)dx \right] \right\} \\ &\quad + \pi_1 \left\{ c_{01} \left[\int_{\check{\mathbb{R}}_1} p_1(x)dx - \int_{\check{\mathbb{R}}_2} p_1(x)dx \right] + c_{11} \left[\int_{\check{\mathbb{R}}_2} p_1(x)dx - \int_{\check{\mathbb{R}}_1} p_1(x)dx \right] \right\} \\ &= \pi_0 (c_{10} - c_{00}) \left\{ \left[\int_{\check{\mathbb{R}}_2} - \int_{\check{\mathbb{R}}_1} \right] p_0(x)dx \right\} - \pi_1 (c_{01} - c_{11}) \left\{ \left[\int_{\check{\mathbb{R}}_2} - \int_{\check{\mathbb{R}}_1} \right] p_1(x)dx \right\}. \end{aligned} \tag{3.21}$$

But from (3.11), we have

$$\pi_1(c_{01} - c_{11})\Lambda_0 = \pi_0(c_{10} - c_{00}). \tag{3.22}$$

Use the r.h.s. of (3.22) in the first term of the r.h.s. of (3.21) to obtain

$$\bar{c}' - c = \pi_1(c_{01} - c_{11})\left\{\int_{\breve{\mathbb{R}}_2} [\Lambda_0 p_0(x) - p_1(x)]\, dx + \int_{\breve{\mathbb{R}}_1} [p_1(x) - \Lambda_0 p_0(x)]\, dx\right\}. \tag{3.23}$$

But note, $\breve{\mathbb{R}}_2$ is a subset of $\mathbb{R}_0$ and thus (3.10a) is valid, then the first integral on the r.h.s. of (3.23) is non-negative. Similarly, note $\breve{\mathbb{R}}_1$ is a subset of $\mathbb{R}_1$ and thus (3.10b) is valid, then the second integral on the r.h.s. of (3.23) is also non-negative. But since we assume $(c_{01} - c_{11})$ is non-negative, then under H_1, the cost of an error is greater than the cost with no error. Since π_1 is non-negative, then

$$\bar{c}' - \bar{c} \geq 0. \tag{3.24}$$

Thus, the average cost of any other decision strategy can not be lower than the average cost obtained by using the Bayes criterion with decision regions given by (3.10a) and (3.10b). □

Example 3.2 Consider $p_0(x)$ and $p_1(x)$ as two Gaussian pdfs with their means given by $\mu_0 = 3$ and $\mu_1 = 7$, with the same variance $\sigma^2 = 1$, as shown in Fig. 3.1(a). Since the LR function $\Lambda(x)$ is the ratio of two exponentials, which yields another exponential, then $\ln(\Lambda(x))$ is linear in x as can be seen in Fig. 3.1(b). Indeed, from Fig. 3.1(a), at $x = 5$, we note $p_0(5) = p_1(5)$ and thus $\ln(\Lambda(5)) = \ln(1) = 0$, as shown in Fig. 3.1(b). □

Example 3.3 Consider a scalar additive Gaussian channel shown in Fig. 3.1 and modeled by

$$X = \begin{cases} x_0 + N, & H_0, \\ x_1 + N, & H_1, \end{cases} \tag{3.25}$$

where N is a zero mean Gaussian noise of variance σ^2. Under H_0 the constant x_0 is transmitted and under H_1 the constant x_1 is transmitted. Then under H_0, the received r.v. $X = x_0 + N$ has a Gaussian pdf $p_0(x)$ with a mean of $\mu_0 = x_0$ and variance σ^2, while under H_1, $X = x_1 + N$ has a Gaussian pdf $p_1(x)$ with a mean of $\mu_1 = x_1$ and the same variance σ^2. Without loss of generality, we can assume $x_0 < x_1$. We note that, for the zero mean Gaussian noise N of variance σ^2 of Example 3.3, when biased by the constants of x_0 and x_1 under H_0 and H_1, the additive Gaussian channel yields the two pdfs of Example 3.2. From the observed x at the receiver, the LR function is given by

$$\begin{aligned} \Lambda = \frac{p_1(x)}{p_0(x)} &= \frac{(1/\sqrt{2\pi\sigma^2})e^{-\frac{(x-x_1)^2}{2\sigma^2}}}{(1/\sqrt{2\pi\sigma^2})e^{-\frac{(x-x_0)^2}{2\sigma^2}}} \\ &= \exp\left\{\frac{1}{2\sigma^2}\left[-(x^2 - 2xx_1 + x_1^2) + (x^2 - 2xx_0 + x_0^2)\right]\right\}. \end{aligned} \tag{3.26}$$

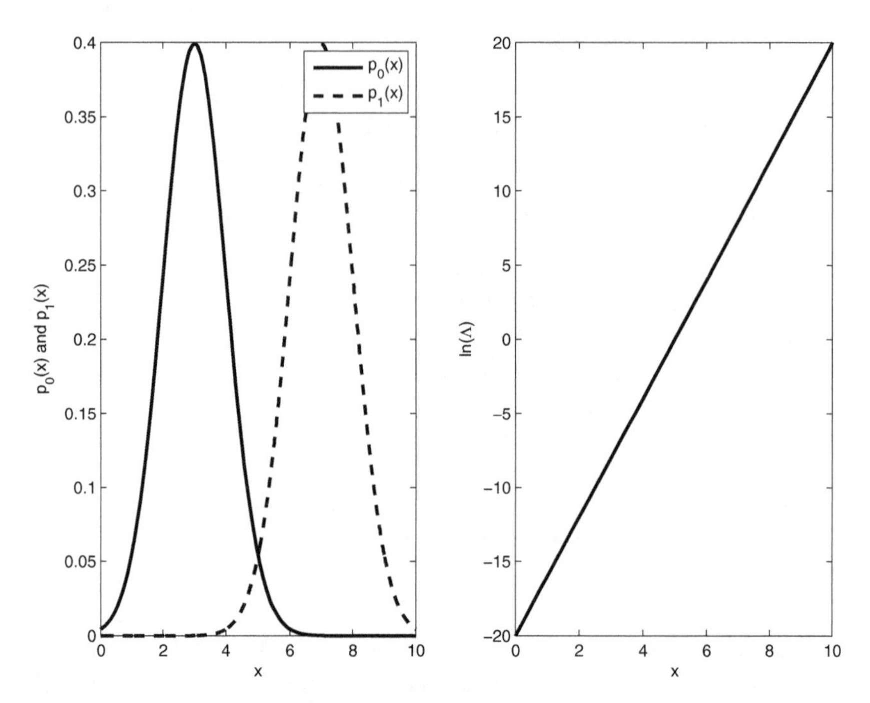

Figure 3.1 (a) Plot of $p_0(x)$ and $p_1(x)$ versus x of Example 3.2; (b) Plot of $\ln(\Lambda(x))$ versus x of Example 3.2.

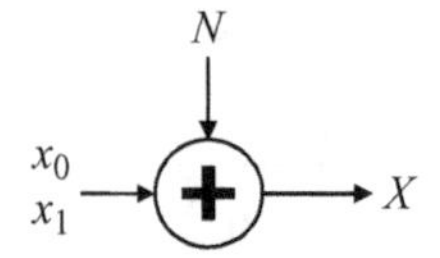

Figure 3.2 Block diagram of an additive Gaussian channel.

Then from (3.10b), the Bayes criterion decision region $\mathbb{R}_1$ is given by

$$\begin{aligned}
\mathbb{R}_1 &= \{x : \Lambda(x) \geq \Lambda_0\} = \{x : \ln(\Lambda(x)) \geq \ln(\Lambda_0)\} \\
&= \left\{x : \frac{1}{2\sigma^2}\left\{-(x^2 - 2xx_1 + x_1^2) + (x^2 - 2xx_0 + x_0^2)\right\} \geq \ln(\Lambda_0)\right\} \\
&= \left\{x : \left\{2xx_1 - x_1^2 - 2xx_0 + x_0^2\right\} \geq 2\sigma^2 \ln(\Lambda_0)\right\} \\
&= \left\{x : x(2x_1 - 2x_0) \geq x_1^2 - x_0^2 + 2\sigma^2 \ln(\Lambda_0)\right\} \\
&= \left\{x : x \geq \frac{x_1^2 - x_0^2 + 2\sigma^2 \ln(\Lambda_0)}{2(x_1 - x_0)}\right\} = \{x : x \geq x_c\}.
\end{aligned} \tag{3.27}$$

Note, x_c is a constant that depends on x_0, x_1, Λ_0, and σ^2. Similarly, from (3.10a) the decision region $\mathbb{R}_0$ is given by

$$\mathbb{R}_0 = \{x : x \leq x_c\}. \tag{3.28}$$

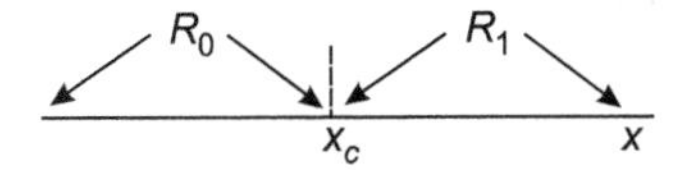

Figure 3.3 Bayes criterion decision region $\mathbb{R}_1$ and decision region $\mathbb{R}_0$ for Example 3.3.

Equations (3.27)–(3.28) show the Bayes criterion decision region $\mathbb{R}_1$ for Example 3.3 is given by the half-line to the right of x_c and the decision region $\mathbb{R}_0$ is given by the half-line to the left of x_c. If $\pi_0 = \pi_1 = 1/2$ and the two relative costs are equal (i.e., $(c_{10} - c_{00}) = (c_{01} - c_{11})$), then from (3.11) the LR constant $\Lambda_0 = 1$. Thus,

$$\mathbb{R}_1 = \left\{ x : x \geq \frac{x_1^2 - x_0^2}{2(x_1 - x_0)} \right\} = \left\{ x : x \geq \frac{x_1 + x_0}{2} = x_c \right\}. \tag{3.29}$$

Note that $x_c = (x_1 + x_0)/2$ is just the midpoint between the peak value of $p_1(x)$ and the peak value of $p_0(x)$. □

Example 3.4 Let the Bayes costs satisfy

$$c_{00} = c_{11} = 0, \; c_{01} = c_{10} = 1. \tag{3.30}$$

Then the average cost of (3.11) becomes

$$\bar{c} = \pi_0 P_{\mathrm{FA}} + \pi_1 P_{\mathrm{M}} = \text{average probability of error} = P_{\mathrm{e}}, \tag{3.31}$$

where P_{FA} is the error under H_0 and P_{M} is the error under H_1. When these two errors are averaged with respect to their prior probabilities of π_0 and π_1, then (3.31) shows that the average probability of error P_{e} is equal to the average cost. Thus, the decision regions associated with the Bayes criterion, when the cost conditions of (3.30) are satisfied, are equivalent to the decision regions under the minimum average probability of error criterion. For a binary digital communication system, the minimum average probability of error criterion may be the most relevant criterion to determine the performance of the system. □

3.3 Maximum a posteriori probability criterion

In the Bayes criterion decision rule, suppose the two relative costs are equal, as given by

$$c_{01} - c_{11} = c_{10} - c_{00}, \tag{3.32}$$

but the two prior probabilities need not be equal. Then the Bayes LR constant Λ_0 of (3.11) becomes

$$\Lambda_0 = \frac{\pi_0(c_{10} - c_{00})}{\pi_1(c_{01} - c_{11})} = \frac{\pi_0}{\pi_1}. \tag{3.33}$$

Then the Bayes criterion decision rule for $\mathbb{R}_1$ becomes

$$\mathbb{R}_1 = \left\{ x : \Lambda(x) = \frac{p_1(x)}{p_0(x)} \geq \Lambda_0 = \frac{\pi_0}{\pi_1} \right\}$$
$$= \{x : \pi_1 p_1(x) \geq \pi_0 p_0(x)\} = \{x : \tilde{p}(x, H_1) \geq \tilde{p}(x, H_0)\}$$
$$= \{x : p(H_1|x)p(x) \geq p(H_0|x)p(x)\} = \{x : p(H_1|x) \geq p(H_0|x)\}. \quad (3.34)$$

The expression $p(H_1|x)$ or $p(H_0|x)$ is called an a posteriori probability, since it relates to the fact that the value of the r.v. X has been observed with the value of x. Thus, the decision region $\mathbb{R}_1$ obtained based on the last term in (3.34) is called the Maximum A Posteriori (MAP) Criterion. Under the MAP criterion, $\mathbb{R}_0$ is given by

$$\mathbb{R}_0 = \{x : p(H_1|x) \leq p(H_0|x)\}. \quad (3.35)$$

3.4 Minimax criterion

Under the Bayes criterion, we need to know:

(1) The prior probability of H_0 denoted by $\pi_0 = \mathrm{P}(H_0)$, and the prior probability of H_1, denoted by $\pi_1 = \mathrm{P}(H_1) = 1 - \pi_0$;
(2) The cost function c_{ij}, $i, j = 0, 1$;
(3) The condition probabilities $p_0(x) = p(x|H_0)$ and $p_1(x) = p(x|H_1)$.

However, in many situations, we may not have sufficient statistics on the occurrence of the events for H_0 and H_1 to obtain meaningful π_0 and π_1. Suppose we believe $\mathrm{P}(H_0)$ is π_0 and proceed to perform the Bayes criterion decision rule, but in reality $\mathrm{P}(H_0)$ is π. Under this condition, what is the true average cost and how does one minimize this expression? The Minimax criterion provides a quantitative approach to attack this problem.

In order to study this problem, let us review and introduce some new notations. From our earlier discussion, we denote π_0 as the assumed prior probability of hypothesis H_0. Now, we denote π as the true prior probability of hypothesis H_0. The likelihood ratio (LR) function is denoted by $\Lambda(x) = p_1(x)/p_0(x)$, $-\infty < x < \infty$. The LR constant Λ_0 (previously used under the Bayes criterion) based on π_0, expressed to show its dependency on π_0, can be denoted by $\Lambda_0 = \Lambda_0(\pi_0) = [\pi_0(c_{10} - c_{00})]/[(1 - \pi_0)(c_{01} - c_{11})]$. The associated decision region for hypothesis H_1 is denoted by $\mathbb{R}_1(\pi_0) = \{x : \Lambda(x) \geq \Lambda_0(x_0)\}$ and the decision region for hypothesis H_0 is denoted by $\mathbb{R}_0(\pi_0) = \{x : \Lambda(x) \leq \Lambda_0(x_0)\}$. Now, the LR constant based on π is denoted by $\Lambda_0(\pi) = \pi(c_{10} - c_{00})/(1 - \pi)(c_{01} - c_{11})$. Then decision region $\mathbb{R}_1$ for hypothesis H_1 based on the LR test using $\Lambda_0(\pi)$ (when the true prior probability for $\mathrm{P}(H_0)$ is π) is denoted by $\mathbb{R}_1(\pi) = \{x : \Lambda(x) \geq \Lambda_0(x)\}$. Similarly, the decision region $\mathbb{R}_0$ for hypothesis H_0 based on the LR test using $\Lambda_0(\pi)$ (when the true prior probability $\mathrm{P}(H_0)$ is π) is denoted by $\mathbb{R}_0(\pi) = \{x : \Lambda(x) \leq \Lambda_0(x_0)\}$.

Using the above notations, the true average cost based on using the LR test assuming the prior probability is π_0 [and using decision regions $\mathbb{R}_1(\pi_0)$ and $\mathbb{R}_0(\pi_0)$] when in reality the true prior probability is π and is given by

$$\begin{aligned}
\bar{c}(\pi, \pi_0) &= \pi \left[c_{10} \int_{\mathbb{R}_1(\pi_0)} p_0(x)dx + c_{00} \int_{\mathbb{R}_0(\pi_0)} p_0(x)dx \right] \\
&\quad + (1-\pi) \left[c_{01} \int_{\mathbb{R}_0(\pi_0)} p_1(x)dx + c_{11} \int_{\mathbb{R}_1(\pi_0)} p_1(x)dx \right] \\
&= \pi \left[c_{10} P_{\mathrm{FA}}(\pi_0) + c_{00} P_{\mathrm{U}}(\pi_0) \right] + (1-\pi) \left[c_{01} P_{\mathrm{M}}(\pi_0) + c_{11} P_{\mathrm{D}}(\pi_0) \right]. \quad (3.36)
\end{aligned}$$

In terms of (3.36) the previously considered Bayes criterion average cost $\bar{c}$ can be rewritten as

$$\begin{aligned}
\bar{c}(\pi, \pi) &= \bar{c}(\pi) \\
&= \pi \left[c_{10} \int_{\mathbb{R}_1(\pi)} p_0(x)dx + c_{00} \int_{\mathbb{R}_0(\pi)} p_0(x)dx \right] \\
&\quad + (1-\pi) \left[c_{01} \int_{\mathbb{R}_0(\pi)} p_1(x)dx + c_{11} \int_{\mathbb{R}_1(\pi)} p_1(x)dx \right] \\
&= \pi \left[c_{10} P_{\mathrm{FA}}(\pi) + c_{00} P_{\mathrm{U}}(\pi) \right] + (1-\pi) \left[c_{01} P_{\mathrm{M}}(\pi) + c_{11} P_{\mathrm{D}}(\pi) \right]. \quad (3.37)
\end{aligned}$$

We note, under the Bayes criterion, when the assumed prior probability and the true prior probability are the same, it does not matter whether we denote this probability by either π_0 or π. Thus,

$$\bar{c}(\pi) = \bar{c}(\pi, \pi) = \bar{c}(\pi_0) = \bar{c}(\pi_0, \pi_0). \quad (3.38)$$

From the optimality of the Bayes criterion in minimizing the average cost, we know

$$\bar{c}(\pi, \pi) = \min_{\bar{c} = \mathbb{C}_\pi} \bar{c}, \quad (3.39)$$

where

$$\mathbb{C}_\pi = \left\{ \begin{array}{l} \text{Space of average cost associated with all decision rules} \\ \text{based on using } \mathrm{P}(H_0) = \pi \text{ and } c_{ij},\ i,\ j = 0,\ 1 \end{array} \right\}.$$

Clearly,

$$\bar{c}(\pi, \pi_0) \in \mathbb{C}_\pi, \quad (3.40)$$

and from (3.39) and (3.41), we have

$$\bar{c}(\pi, \pi_0) \geq \bar{c}(\pi),\ 0 \leq \pi_0 \leq 1,\ 0 \leq \pi \leq 1. \quad (3.41)$$

Consider $\bar{c}(\pi, \pi_0)$ as a function of π for any fixed π_0. Then it satisfies:

(1) $\bar{c}(\pi, \pi_0)$ is a linear function of π in $[0,\ 1]$ as can be seen from (3.36);
(2) $\bar{c}(\pi_0, \pi_0) = \bar{c}(\pi_0)$ as can be seen from (3.36)–(3.38);
(3) $\bar{c}(\pi)$ is a concave function of π in $[0,\ 1]$ as defined below.

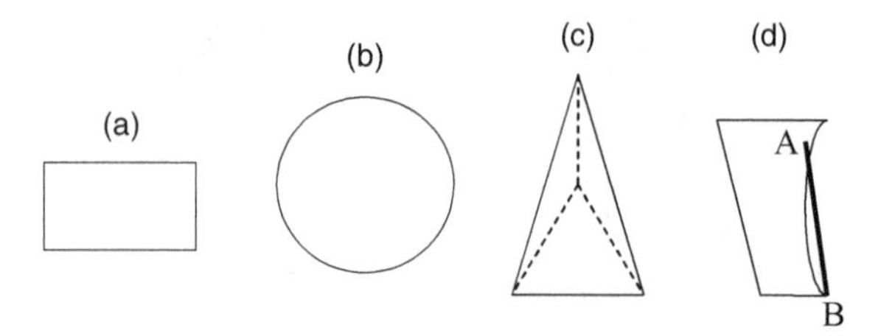

Figure 3.4 (a) Convex set; (b) Convex set; (c) Convex set; (d) Not a convex set.

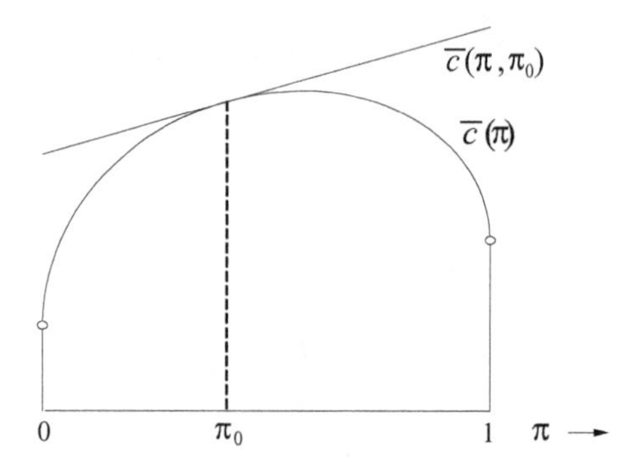

Figure 3.5 $\bar{c}(\pi)$ is a concave function for $\pi \in [0, 1]$.

In order to demonstrate the concavity of $\bar{c}(\pi)$, first consider the definition of a convex set. A set is called convex if any point on the chord generated from any two points in the set is also in the set. For example, a rectangle in 2D is a convex set (Fig. 3.4a), a disk in 2D is a convex set (Fig. 3.4b), a pyramid with a triangular base in 3D is a convex set (Fig. 3.4c), and a sphere in 3D is a sphere. The figure in Fig. 3.4d is not a convex set, since there are points on the chord from point A to point B not in the set. A non-negative valued function $f(x)$ is said to be a concave function over $[a, b]$, if $f(x) \leq$ tangent of $f(x)$, for any $x \in [a, b]$. Furthermore, the body enclosed by the concave function $y = f(x)$ over $[a, b]$ and the chords $\{x = a, y \in [0, f(a)]\}$, $\{x \in [a, b], y = 0\}$, $\{x \in [a, b], y \in [f(a), f(b)]\}$, and $\{x = b, y \in [0, f(b)]\}$ is a convex set in 2D. $\bar{c}(\pi)$ is a concave function for $\pi \in [0, 1]$ as shown in Fig. 3.5.

Under the Minimax criterion, we want to pick π_0 such that the maximum of $\bar{c}(\pi, \pi_0)$ is minimized for all $\pi \in [0, 1]$. Denote this optimum π_0 by $\hat{\pi}_0$.That is

$$\max_{\pi \in [0, 1]} \bar{c}(\pi, \hat{\pi}_0) = \min_{\pi_0 \in [0, 1]} \max_{\pi \in [0, 1]} \bar{c}(\pi, \pi_0). \tag{3.42}$$

We note that if $\hat{\hat{\pi}}_0$ is such that

$$\bar{c}\left(\hat{\hat{\pi}}_0\right) = \max_{\pi_0} \bar{c}(\pi_0) \tag{3.43}$$

is in the interior of $[0, 1]$, then $\hat{\pi}_0 = \hat{\hat{\pi}}_0$ is the solution under the Minimax criterion (see Fig. 3.6a). Otherwise, we choose $\hat{\pi}_0$ to be the boundary point (i.e., either $\hat{\pi}_0 = 0$ or $\hat{\pi}_0 = 1$) that yields the maximum of $\bar{c}(\pi_0)$ (see Fig. 3.6b and Fig. 3.6c).

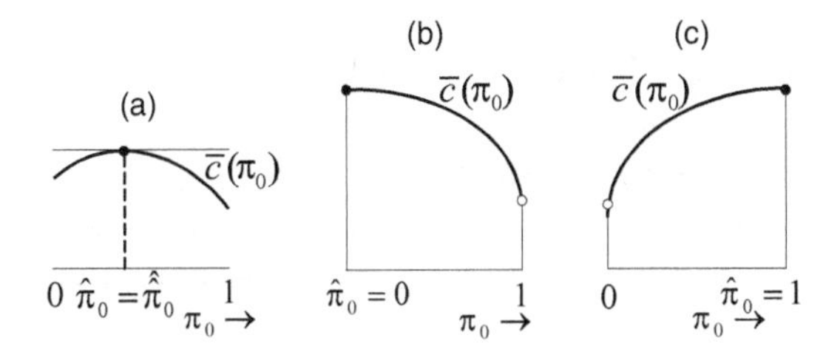

Figure 3.6 (a) $\hat{\pi}_0$ is an interior point; (b) $\hat{\pi}_0 = 0$ is a boundary point; (c) $\hat{\pi}_0 = 1$ is a boundary point.

When $\hat{\hat{\pi}}_0$ is an interior point of $[0,\ 1]$, then we can find $\hat{\pi}_0 = \hat{\hat{\pi}}_0$ by solving either of the following two equations in (3.44) or (3.45).

$$\left.\frac{\partial \bar{c}(\pi_0)}{\partial \pi_0}\right|_{\pi_0=\hat{\hat{\pi}}_0} = 0 = \left[c_{10} P_{\mathrm{FA}}\left(\hat{\hat{\pi}}_0\right) + c_{00} P_{\mathrm{U}}\left(\hat{\hat{\pi}}_0\right)\right] - \left[c_{01} P_{\mathrm{M}}\left(\hat{\hat{\pi}}_0\right) + c_{11} P_{\mathrm{D}}\left(\hat{\hat{\pi}}_0\right)\right] \tag{3.44}$$

or

$$\frac{\partial \bar{c}\left(\pi, \hat{\hat{\pi}}_0\right)}{\partial \pi} = 0 = \left[c_{10} P_{\mathrm{FA}}\left(\hat{\hat{\pi}}_0\right) + c_{00} P_{\mathrm{U}}\left(\hat{\hat{\pi}}_0\right)\right] - \left[c_{01} P_{\mathrm{M}}\left(\hat{\hat{\pi}}_0\right) + c_{11} P_{\mathrm{D}}\left(\hat{\hat{\pi}}_0\right)\right]. \tag{3.45}$$

Then the Minimax average cost is given by

$$\bar{c}\left(\hat{\hat{\pi}}_0\right) = \left[c_{10} P_{\mathrm{FA}}\left(\hat{\hat{\pi}}_0\right) + c_{00} P_{\mathrm{U}}\left(\hat{\hat{\pi}}_0\right)\right] = \left[c_{01} P_{\mathrm{M}}\left(\hat{\hat{\pi}}_0\right) + c_{11} P_{\mathrm{D}}\left(\hat{\hat{\pi}}_0\right)\right]. \tag{3.46}$$

3.5 Neyman–Pearson criterion

In many situations, not only are the prior probabilities of H_0 and H_1 not available, but the costs c_{ij}, $i,\ j = 0,\ 1$, may also be difficult to define. There are situations in which H_1 occurs so rarely and its consequences are so great (e.g., what happened in New York City on 9/11), we may not want to use the minimum average cost criterion (either the standard Bayes criterion or the Minimax criterion) to design the receiver. Instead, we may want to insure that P_{FA} is below some acceptable small number and then want to maximize P_{D} (or minimize $P_{\mathrm{M}} = 1 - P_{\mathrm{D}}$).

The Neyman–Pearson (NP) criterion is the decision rule that chooses P_{FA} to be fixed at or below some value and then maximizes P_{D}. The NP criterion results in a LR test. (The proof of this claim is omitted, since it is essentially the same as the proof of the Bayes criterion decision rule resulting in a LR test.) Under the NP criterion with a given LR function $\Lambda(x) = p_1(x)/p_0(x)$, choose the LR constant Λ_0 (with details to be specified shortly), so that the two decision regions are given by

$$\mathbb{R}_1 = \{x : \Lambda(x) \geq \Lambda_0\}, \tag{3.47}$$

$$\mathbb{R}_0 = \{x : \Lambda(x) \leq \Lambda_0\}. \tag{3.48}$$

In order to find Λ_0 easily, consider the following brief discussion on the concept of the transformation of a random variable. Let X be a r.v. and let $g(\cdot)$ be a given function. Then $Y = g(X)$ is another r.v. In particular, take $g(\cdot)$ to be the function $\Lambda(\cdot) = p_1(\cdot)/p_0(\cdot)$. Then $\Lambda(X) = p_1(X)/p_0(X)$ is another r.v. Since X has the pdf $p_1(x)$ under H_1 and the pdf $p_0(x)$ under H_0, then denote the pdfs of $\Lambda(X)$ under H_1 by $\tilde{p}_1(\Lambda)$ and denote under H_0 by $\tilde{p}_0(\Lambda)$. However, $\tilde{p}_1(\Lambda)$ and $\tilde{p}_0(\Lambda)$ are related to $p_1(x)$ and $p_0(x)$ simply by

$$\tilde{p}_1(\Lambda) = \frac{p_1(x)dx}{d\Lambda}, \tag{3.49}$$

$$\tilde{p}_0(\Lambda) = \frac{p_0(x)dx}{d\Lambda}. \tag{3.50}$$

Under the LR test, from (3.47)–(3.50), we have

$$\alpha = P_{\text{FA}} = \int_{\mathbb{R}_1} p_0(x)dx = \int_{\Lambda_0}^{\infty} \tilde{p}_0(\Lambda)d\Lambda, \tag{3.51}$$

$$\beta = P_{\text{D}} = \int_{\mathbb{R}_1} p_1(x)dx = \int_{\Lambda_0}^{\infty} \tilde{p}_1(\Lambda)d\Lambda, \tag{3.52}$$

where we also introduced the notation of $\alpha = P_{\text{FA}}$ and $\beta = P_{\text{D}}$. Thus, given $p_1(x)$ and $p_0(x)$, we can obtain $\tilde{p}_1(\Lambda)$ and $\tilde{p}_0(\Lambda)$ from (3.49)–(3.50). Then for a specified α, we can pick Λ_0 such that the integral of $\tilde{p}_0(\Lambda)$ from Λ_0 to ∞ in (3.51) yields the desired α.

In general, we can also find the LR constant Λ_0 from the "receiving operational curve" (ROC) defined as a plot of β versus α with both the domain and the range of [0, 1]. Using (3.51)–(3.52), upon differentiation with respect to Λ_0, we obtain

$$\frac{d\alpha}{d\Lambda_0} = -\tilde{p}_0(\Lambda_0), \tag{3.53}$$

$$\frac{d\beta}{d\Lambda_0} = -\tilde{p}_1(\Lambda_0). \tag{3.54}$$

By dividing (3.54) by (3.53), we have

$$\frac{d\beta}{d\alpha} = \frac{\dfrac{d\beta}{d\Lambda_0}}{\dfrac{d\alpha}{d\Lambda_0}} = \frac{\tilde{p}_0(\Lambda_0)}{\tilde{p}_1(\Lambda_0)} = \frac{p_1(x_0)}{p_0(x_0)} = \Lambda_0. \tag{3.55}$$

Figure 3.7 shows a ROC given by the plot of β versus α as a light dark curve. Equation (3.55) tells us that the slope of the tangent of β versus α yields the LR constant Λ_0 shown as a heavy dark straight line in Fig. 3.7. The average cost (3.46) under the Minimax criterion can be restated as

$$[c_{10}\alpha + c_{00}(1-\alpha)] = [c_{01}(1-\beta) + c_{11}\beta]. \tag{3.56}$$

From (3.56), we can solve for β in terms of α, which yields a straight line

$$\beta = ((c_{00} - c_{01}) + \alpha(c_{10} - c_{00}))\,/(c_{11} - c_{01}). \tag{3.57}$$

Thus, the Minimax operating curve under the Minimax criterion is given by the straight line of β versus α defined by (3.56) that intersects the ROC at the (α, β) point in

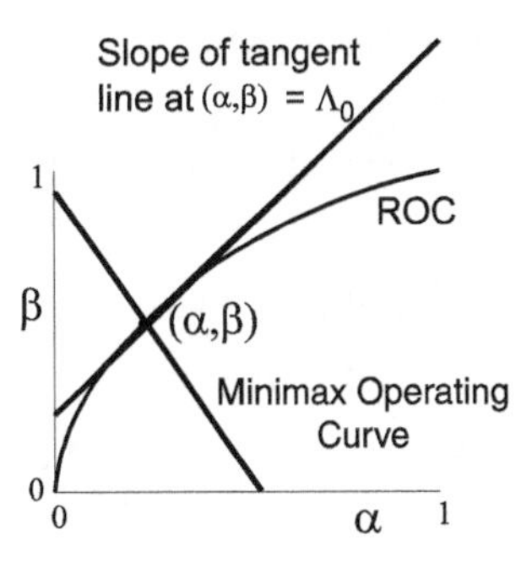

Figure 3.7 Receiving operating curve, LR constant Λ_0, and Minimax operating curve.

Fig. 3.7. Thus, Fig. 3.7 provides all the information for the operating parameters under the Bayes, Minimax, and NP criteria. Operationally, under all three criteria, from the given conditional pdfs $p_1(x)$ and $p_0(x)$, we obtain the LR function $\Lambda(x)$. Furthermore, the optimum decision rules under all three criteria are all given by the LR test. Since each value of the LR constant Λ_0 yields a pair of $(\alpha, \ \beta)$ points on the ROC, by varying Λ_0, all the ROC points can be generated (at least conceptually). Furthermore, the slope of the tangent of the ROC at any point given by $(\alpha, \ \beta)$ is the LR constant Λ_0 associated with that point. Under the Bayes criterion, the LR constant Λ_0 is given. Thus, the $(\alpha, \ \beta)$ values under the Bayes criterion can be determined (at least conceptually) from that point on the ROC such that the tangent of the slope at that point has the value of Λ_0. Under the NP criterion, for a specified $P_{\mathrm{FA}} = \alpha$, the maximum P_{D} is that β given by the $(\alpha, \ \beta)$ point on the ROC. Under the Minimax criterion, the straight line of β versus α of (3.57) intersects the ROC and shows the operating point at (α, β).

3.6 Simple hypothesis test for vector measurements

Now, consider the received data in deciding the two hypotheses as given by the realization $\mathbf{x} = [x_1, \ldots, x_n]^T$ from a column random vector. All the theory we have developed so far for the realization x from a scalar random variable X generalizes to the n-dimensional realization vector $\mathbf{x}$ in an almost trivial manner. The prior pdfs under H_1 and H_0 are denoted by the n-dimensional pdfs $p_1(\mathbf{x})$ and $p_0(\mathbf{x})$ respectively. The LR function is now defined by

$$\Lambda(\mathbf{x}) = \frac{p_1(\mathbf{x})}{p_0(\mathbf{x})}. \tag{3.58}$$

We note, since $p_1(\mathbf{x})$ and $p_0(\mathbf{x})$ are both scalar-valued functions of a vector variable, their ratio, $\Lambda(\mathbf{x})$, is also a scalar-valued function of a vector variable. Then the LR constant Λ_0 is also a scalar just like the previously considered scalar case. The LR tests for decisions $\mathbb{R}_1$ and $\mathbb{R}_0$ are given by

$$\mathbb{R}_1 = \{\mathbf{x} : \Lambda(\mathbf{x}) \geq \Lambda_0\}, \tag{3.59}$$

$$\mathbb{R}_0 = \{\mathbf{x} : \Lambda(\mathbf{x}) \leq \Lambda_0\}, \tag{3.60}$$

where the actual value of Λ_0 depends on whether the Bayes, Minimax, or NP criterion is used.

Consider a strictly monotonic increasing scalar-valued function $h(\cdot)$ applied to $\Lambda(\mathbf{x})$, then (3.59)–(3.60) now become

$$\mathbb{R}_1 = \{\mathbf{x} : \Lambda(\mathbf{x}) \geq \Lambda_0\} = \{\mathbf{x} : h(\Lambda(\mathbf{x})) \geq h(\Lambda_0)\}, \tag{3.61}$$

$$\mathbb{R}_0 = \{\mathbf{x} : \Lambda(\mathbf{x}) \leq \Lambda_0\} = \{\mathbf{x} : h(\Lambda(\mathbf{x})) \leq h(\Lambda_0)\}. \tag{3.62}$$

Then P_{D} and P_{FA} can be expressed in terms of the new pdf's of $\tilde{\tilde{p}}_1(h)$ and $\tilde{\tilde{p}}_0(h)$ in terms of the original pdf's of $p_1(\mathbf{x})$ and $p_0(\mathbf{x})$ given by

$$\tilde{\tilde{p}}_1(h') = \left. \frac{p_1(\mathbf{x})}{\left| \dfrac{\partial^n h(\Lambda(\mathbf{x}))}{\partial x_1 \cdots \partial x_n} \right|} \right|_{\mathbf{x}=\Lambda^{-1}(h^{-1}(h'))}, \tag{3.63}$$

$$\tilde{\tilde{p}}_0(h') = \left. \frac{p_0(\mathbf{x})}{\left| \dfrac{\partial^n h(\Lambda(\mathbf{x}))}{\partial x_1 \cdots \partial x_n} \right|} \right|_{\mathbf{x}=\Lambda^{-1}(h^{-1}(h'))}, \tag{3.64}$$

$$P_{\mathrm{D}} = \int_{\mathbb{R}_1} p_1(\mathbf{x}) d\mathbf{x} = \int_{\Lambda_0}^{\infty} \tilde{p}_1(\Lambda) d\Lambda = \int_{h(\Lambda_0)}^{\infty} \tilde{\tilde{p}}_1(h) dh, \tag{3.65}$$

$$P_{\mathrm{FA}} = \int_{\mathbb{R}_1} p_0(\mathbf{x}) d\mathbf{x} = \int_{\Lambda_0}^{\infty} \tilde{p}_0(\Lambda) d\Lambda = \int_{h(\Lambda_0)}^{\infty} \tilde{\tilde{p}}_0(h) dh. \tag{3.66}$$

In the expressions in the denominator of (3.63) and (3.64), we assumed the $h(\Lambda(\cdot))$ function is n-th order differential with respect to $x_1, \ldots, x_n$. The $|\cdot|$ expression denotes the absolute value and the expression for the value of $\mathbf{x}$ to be used in the r.h.s. of (3.63), and (3.64) assumes the two inverse functions $\Lambda^{-1}(\cdot)$ and $h^{-1}(\cdot)$ also exist. □

Example 3.5 Take $h(\Lambda(\mathbf{x})) = \ln(\Lambda(\mathbf{x}))$. □

Similarly, consider a strictly monotonic decreasing scalar-valued function $h(\cdot)$ applied to $\Lambda(\mathbf{x})$, then (3.59)–(3.60) now become

$$\mathbb{R}_1 = \{\mathbf{x} : \Lambda(\mathbf{x}) \geq \Lambda_0\} = \{\mathbf{x} : h(\Lambda(\mathbf{x})) \leq h(\Lambda_0)\}, \tag{3.67}$$

$$\mathbb{R}_0 = \{\mathbf{x} : \Lambda(\mathbf{x}) \leq \Lambda_0\} = \{\mathbf{x} : h(\Lambda(\mathbf{x})) \geq h(\Lambda_0)\}. \tag{3.68}$$

Example 3.6 Take $h(\Lambda(\mathbf{x})) = e^{-\Lambda(\mathbf{x})}$. □

For the decision regions given in (3.67) and (3.68), their P_{D} and P_{FA} can be expressed by

$$P_{\mathrm{D}} = \int_{\mathbb{R}_1} p_1(\mathbf{x}) d\mathbf{x} = \int_{\Lambda_0}^{\infty} \tilde{p}_1(\Lambda) d\Lambda = \int_{-\infty}^{h(\Lambda_0)} \tilde{\tilde{p}}_1(h) dh, \tag{3.69}$$

$$P_{\mathrm{FA}} = \int_{\mathbb{R}_1} p_0(\mathbf{x}) d\mathbf{x} = \int_{\Lambda_0}^{\infty} \tilde{p}_0(\Lambda) d\Lambda = \int_{-\infty}^{h(\Lambda_0)} \tilde{\tilde{p}}_0(h) dh. \tag{3.70}$$

Any function $g(\mathbf{x})$ that replaces and summarizes the data vector $\mathbf{x}$ is called a statistic. Furthermore, the statistic $g(\mathbf{x})$ is called a sufficient statistic for a parameter θ (which for the binary hypothesis problem is either H_0 or H_1) if

$$p(\mathbf{x}|g(\mathbf{x}) = g_1) = p(\mathbf{x}|g(\mathbf{x}) = g_1, \theta). \tag{3.71}$$

That is, $g(\mathbf{x})$ contains as much of the information as $\mathbf{x}$ about θ. In practice, we want to find a $g(\mathbf{x})$ that is much simpler (e.g., needs fewer bits for storage or transmission as compared to $\mathbf{x}$), yet contains as much information as the original raw data vector $\mathbf{x}$ about θ.

Theorem 3.2 ***Factorization Theorem***
$g(\mathbf{x})$ is a sufficient statistic for θ if

$$p(\mathbf{x}|\theta) = k(g(\mathbf{x})|\theta)l(\mathbf{x}). \tag{3.72}$$

That is, $p(\mathbf{x}|\theta)$ factors as the product of a part, $l(\mathbf{x})$, not dependent of θ, and a part, $k(g(\mathbf{x})|\theta)$, that depends on θ only through $g(\mathbf{x})$. The proof of this theorem is complicated and will not be given here.

Example 3.7 The LR function $\Lambda(\mathbf{x})$ is a sufficient statistic for $\theta = H_1$ or H_0.

Proof: Equation (3.73) follows from the definition of $\Lambda(\mathbf{x})$ and (3.74) is an identity

$$p(\mathbf{x}|H_1) = \Lambda(\mathbf{x})p(\mathbf{x}|H_0), \tag{3.73}$$

$$p(\mathbf{x}|H_0) = 1 \cdot p(\mathbf{x}|H_0). \tag{3.74}$$

Then, in the factorization theorem of (3.72),

$$p(\mathbf{x}|\theta) = k(g(\mathbf{x})|\theta)l(\mathbf{x}),\ \theta = H_1,\ H_0, \tag{3.75}$$

use

$$k(g(\mathbf{x})|\theta) = \begin{cases} \Lambda(\mathbf{x}), & \theta = H_1, \\ 1, & \theta = H_0, \end{cases} \tag{3.76}$$

$$l(\mathbf{x}) = p(\mathbf{x}|H_0). \tag{3.77}$$

Thus, by using $g(\mathbf{x}) = \Lambda(\mathbf{x})$ in (3.73)–(3.77), this shows it satisfies the factorization theorem and thus $\Lambda(\mathbf{x})$ is a sufficient statistic for $\theta = H_1$ or H_0. □

Example 3.7 shows why the LR test is so important and basic in binary hypothesis testing.

Example 3.8 Let $g(\mathbf{x})$ be a sufficient statistic for $\theta = H_1$ or H_0. Then use the factorization theorem to obtain

$$\Lambda(\mathbf{x}) = \frac{p(\mathbf{x}|H_1)}{p(\mathbf{x}|H_0)} = \frac{k(g(\mathbf{x})|H_1)l(\mathbf{x})}{k(g(\mathbf{x})|H_0)l(\mathbf{x})} = \frac{k(g(\mathbf{x})|H_1)}{k(g(\mathbf{x})|H_0)} = \tilde{\Lambda}(g(\mathbf{x})). \tag{3.78}$$

Thus, the LR function of the observed data vector $\mathbf{x}$ equals the LR function of the sufficient statistic $g(\mathbf{x})$. □

Example 3.9 Consider the observed vector $\mathbf{x} = [x_1, \ldots, x_n]^T$ to be an element of the n-dimensional Euclidean space $\mathbb{R}^n$. Consider the use of a sufficient statistic as that of the proper selection of coordinates in $\mathbb{R}^n$ for "theoretical data compression." Specifically, let the sufficient statistic $g(\mathbf{x})$ be one coordinate of $\mathbb{R}^n$ and the remaining $(n-1)$ coordinates be denoted $\mathbf{y}$. Then

$$\Lambda(\mathbf{x}) = \Lambda(g(\mathbf{x}), \mathbf{y}) = \frac{p(g(\mathbf{x}), \mathbf{y}|H_1)}{p(g(\mathbf{x}), \mathbf{y}|H_0)} = \frac{p_1(g(\mathbf{x})|H_1)p_2(\mathbf{y}|g(\mathbf{x}), H_1)}{p_1(g(\mathbf{x})|H_0)p_2(\mathbf{y}|g(\mathbf{x}), H_0)}. \tag{3.79}$$

From the definition of $g(\mathbf{x})$ as a sufficient statistic for $\theta = H_1$ or H_0 in (3.71), then

$$p_2(\mathbf{y}|g(\mathbf{x}), H_1) = p_2(\mathbf{y}|g(\mathbf{x})), \tag{3.80}$$

$$p_2(\mathbf{y}|g(\mathbf{x}), H_0) = p_2(\mathbf{y}|g(\mathbf{x})). \tag{3.81}$$

Then use (3.80)–(3.81) on the r.h.s. of (3.79) to obtain

$$\Lambda(\mathbf{x}) = \frac{p_1(g(\mathbf{x})|H_1)}{p_1(g(\mathbf{x})|H_0)} = \tilde{\Lambda}(g(\mathbf{x})). \tag{3.82}$$

Thus, (3.82) shows the LR function of the one-dimensional $g(\mathbf{x})$ is equal to the LR function of the n-dimensional data vector $\mathbf{x}$. Thus, we do not need to use the $(n-1)$ coordinates data vector $\mathbf{y}$. The sufficient statistic $g(\mathbf{x})$ can be considered to be a data compression of $\mathbf{x}$ with regard to hypothesis testing of H_1 versus H_0. □

3.7 Additional topics in hypothesis testing (*)

In this section, we consider several additional topics in hypothesis testing. In Section 3.7.1, we consider the sequential likelihood ratio test (SLRT), where the LR is compared to two thresholds, and can be considered to be a generalization of the Neyman–Pearson method where the LR is compared to a single threshold. SLRT permits termination of the detection test when a prescribed probability of false alarm is satisfied. In Section 3.7.2, we introduce the concept of a composite hypothesis and the uniformly most powerful test. In Section 3.7.3, we consider one of the simplest non-parametric hypothesis methods based on the sign test.

3.7.1 Sequential likelihood ratio test (SLRT)

The hypothesis testing problems we discussed in Sections 3.1–3.6 were all based on a common assumption, i.e., the number of observations is predetermined to be a constant. This is not the case in the sequential likelihood ratio test (SLRT), where the number of observations required by the test is dependent on the outcomes of the observations. The

SLRT repeats the test, taking the most recent observation into account, and proceeds until a prescribed P_{FA} and P_{D} are achieved.

Consider the case where the observations $x_1, x_2, \ldots$ are obtained sequentially. Each observation is assumed statistically independent, and is generated from either of the following hypotheses:

$$H_0 : x \sim p_0(x) = p(x \,|\, H_0), \tag{3.83}$$

$$H_1 : x \sim p_1(x) = p(x \,|\, H_1). \tag{3.84}$$

At each step n, one of the following three decisions will be made based on a probability test on the observations $\mathbf{x}_n = [x_1, x_2, \ldots, x_n]$:

$$\text{(i) } \hat{H} = H_1. \tag{3.85}$$

$$\text{(ii) } \hat{H} = H_0. \tag{3.86}$$

$$\text{(iii) Take one more observation } x_{n+1} \text{ and then go to step } n+1. \tag{3.87}$$

The whole process is continued until either the decision H_1 or H_0 is made. This is called the *sequential test* for a simple binary hypothesis. Note that the number of required observations to make the final decision (H_1 or H_0) will be dependent on the realizations of the observations, and therefore the number of required observations in a sequential test is a random variable. This is very different from the conventional binary hypothesis testing, where the number of observations is a fixed number.

If we denote the set of all possible observation vectors $\mathbf{x}_n$ leading to the n-th testing step by $\mathbb{R}^{(n)}$, then the above testing procedure essentially divides the set $\mathbb{R}^{(n)}$ into three decision regions: $\mathbb{R}_1^{(n)}$, $\mathbb{R}_0^{(n)}$, and $\mathbb{R}_\times^{(n)}$, and each corresponds to decision (i), (ii), and (iii), respectively. In the sequential likelihood ratio test, these decision regions are defined as

$$\mathbb{R}_1^{(n)} = \{\mathbf{x}_n : \{B < L_m(\mathbf{x}_m) < A,\ m = 1, 2, \ldots, n-1\} \cap \{L_n(\mathbf{x}_n) \geq A\}\}, \tag{3.88}$$

$$\mathbb{R}_0^{(n)} = \{\mathbf{x}_n : \{B < L_m(\mathbf{x}_m) < A,\ m = 1, 2, \ldots, n-1\} \cap \{L_n(\mathbf{x}_n) \leq B\}\}, \tag{3.89}$$

$$\mathbb{R}_\times^{(n)} = \{\mathbf{x}_n : B < L_m(\mathbf{x}_m) < A,\ m = 1, 2, \ \ldots,\ n\}, \tag{3.90}$$

where A and B are two constants and

$$L_m(\mathbf{x}_m) = \frac{p_1(\mathbf{x}_m)}{p_0(\mathbf{x}_m)} = \frac{\prod_{i=1}^m p_1(x_i)}{\prod_{i=1}^m p_0(x_i)} \tag{3.91}$$

is the likelihood ratio at step m. It can be shown that the above sequential test is guaranteed to end up in H_1 or H_0 in probability 1 as n approaches infinity, i.e.,

$$\lim_{n\to\infty} \mathrm{P}\left(\mathbf{x}_n \in \mathbb{R}_\times^{(n)}\right) = 0, \text{ or } \lim_{n\to\infty} \mathrm{P}\left(\mathbf{x}_n \in \{\mathbb{R}_1^{(n)} \cup \mathbb{R}_0^{(n)}\}\right) = 1. \tag{3.92}$$

From (3.88)–(3.90), it is clear that the performance of the sequential test is completely determined by the constants A and B. A fundamental relation among A, B, and the probability of false alarm $P_{\text{FA}} = \alpha$ and the probability of detection $P_{\text{D}} = \beta$ can be derived as follows.

We first express the following four common performance measures of the test in terms of $\mathbb{R}_0^{(n)}$ and $\mathbb{R}_1^{(n)}$:

$$\alpha = P_{\mathrm{FA}} = \mathrm{P}(\hat{H} = H_1 | H_0) = \sum_{n=1}^{\infty} \int_{\mathbb{R}_1^{(n)}} p_0(\mathbf{x}_n) d\mathbf{x}_n, \tag{3.93}$$

$$1 - \alpha = P_{\mathrm{U}} = \mathrm{P}(\hat{H} = H_0 | H_0) = \sum_{n=1}^{\infty} \int_{\mathbb{R}_0^{(n)}} p_0(\mathbf{x}_n) d\mathbf{x}_n, \tag{3.94}$$

$$1 - \beta = P_{\mathrm{M}} = \mathrm{P}(\hat{H} = H_0 | H_1) = \sum_{n=1}^{\infty} \int_{\mathbb{R}_0^{(n)}} p_1(\mathbf{x}_n) d\mathbf{x}_n, \tag{3.95}$$

$$\beta = P_{\mathrm{D}} = \mathrm{P}(\hat{H} = H_1 | H_1) = \sum_{n=1}^{\infty} \int_{\mathbb{R}_1^{(n)}} p_1(\mathbf{x}_n) d\mathbf{x}_n. \tag{3.96}$$

From (3.88) and (3.89), we have

$$p_1(\mathbf{x}_n) \geq A p_0(\mathbf{x}_n), \text{ for } \mathbf{x}_n \in \mathbb{R}_1^{(n)}, \tag{3.97}$$

$$p_1(\mathbf{x}_n) \leq B p_0(\mathbf{x}_n), \text{ for } \mathbf{x}_n \in \mathbb{R}_0^{(n)}. \tag{3.98}$$

Combining (3.93), (3.96), and (3.97), we have

$$\beta = \sum_{n=1}^{\infty} \int_{\mathbb{R}_1^{(n)}} p_1(\mathbf{x}_n) d\mathbf{x}_n \geq A \sum_{n=1}^{\infty} \int_{\mathbb{R}_1^{(n)}} p_0(\mathbf{x}_n) d\mathbf{x}_n = A\alpha. \tag{3.99}$$

Thus, an upper bound on A is obtained as $A \leq \beta/\alpha$. A lower bound on B can be similarly obtained by using (3.94), (3.95), and (3.98):

$$1 - \beta = \sum_{n=1}^{\infty} \int_{\mathbb{R}_0^{(n)}} p_1(\mathbf{x}_n) d\mathbf{x}_n \leq B \sum_{n=1}^{\infty} \int_{\mathbb{R}_0^{(n)}} p_0(\mathbf{x}_n) d\mathbf{x}_n = B(1 - \alpha), \tag{3.100}$$

which gives the bound $B \geq (1 - \beta)/(1 - \alpha)$. If we assume the excess of the likelihood ratio crossing the boundary is negligible, then the values of A and B can be approximated by

$$A \approx \frac{\beta}{\alpha}, \quad B \approx \frac{1 - \beta}{1 - \alpha}. \tag{3.101}$$

Equation (3.101) is known as Wald's approximation. It has also been shown by Wald [3] that the use of (3.101) cannot result in any appreciable increase in the value of either P_{FA} or P_{M} and therefore the threshold (3.101) is commonly used in the SLRT.

One of the major advantages of the SLRT is that the average number of observations required to achieve the same α and β is smaller than that of the standard likelihood ratio test. To compute the average number of observations $\bar{n}$ required in SLRT, we first denote the log-likelihood ratio of x_i as

$$z_i = \ln \left\{ \frac{p_1(x_i)}{p_0(x_i)} \right\}, \tag{3.102}$$

then the log-likelihood ratio of $[x_1, x_2, \dots, x_n]$ can be expressed as

$$Z_n = \ln\left\{\frac{\prod_{i=1}^{n} p_1(x_i)}{\prod_{i=1}^{n} p_0(x_i)}\right\} = \sum_{i=1}^{n} z_i. \tag{3.103}$$

The average of Z_n at the termination under hypotheses H_0 and H_1 can be computed as

$$\begin{aligned} \mathrm{E}\{Z_n|H_0\} &= \mathrm{E}\left\{Z_n|\hat{H} = H_0, H_0\right\}\mathrm{P}\left(\hat{H} = H_0|H_0\right) \\ &\quad + \mathrm{E}\left\{Z_n|\hat{H} = H_1, H_0\right\}\mathrm{P}\left(\hat{H} = H_1|H_0\right) \\ &\approx (\ln B)(1-\alpha) + (\ln A)\alpha, \end{aligned} \tag{3.104}$$

$$\begin{aligned} \mathrm{E}\{Z_n|H_1\} &= \mathrm{E}\left\{Z_n|\hat{H} = H_0, H_1\right\}\mathrm{P}\left(\hat{H} = H_0|H_1\right) \\ &\quad + \mathrm{E}\left\{Z_n|\hat{H} = H_1, H_1\right\}\mathrm{P}\left(\hat{H} = H_1|H_1\right) \\ &\approx (\ln B)(1-\beta) + (\ln A)\beta. \end{aligned} \tag{3.105}$$

From the independent and identically distributed property of the observations, given the underlying hypothesis, we have

$$\mathrm{E}\{Z_n|H_k\} = \mathrm{E}\left\{\sum_{i=1}^{n} z_i|H_k\right\} = \mathrm{E}\{n|H_k\}\,\mathrm{E}\{z|H_k\},\ i = 0, 1. \tag{3.106}$$

Combining (3.104), (3.105), and (3.106), we can then approximate the average number of observations in the SLRT under H_0 and H_1 as

$$\bar{n}_{H_0} = \mathrm{E}\{n|H_0\} \approx \frac{(\ln B)(1-\alpha) + (\ln A)\alpha}{\mathrm{E}\{z|H_0\}}, \tag{3.107}$$

$$\bar{n}_{H_1} = \mathrm{E}\{n|H_1\} \approx \frac{(\ln B)(1-\beta) + (\ln A)\beta}{\mathrm{E}\{z|H_1\}}. \tag{3.108}$$

In the following, we show the advantage of reducing the average number of observations using the SLRT compared to that of the standard log-likelihood ratio test through a practical example.

Example 3.10 Consider the case where the observations $x_1, x_2, \dots$ are generated independently from either of the following two hypotheses:

$$H_0 : x \sim \mathcal{N}(0, \sigma^2), \tag{3.109}$$

$$H_1 : x \sim \mathcal{N}(v, \sigma^2). \tag{3.110}$$

It follows that

$$p_0(x) = \frac{1}{\sqrt{2\pi\sigma^2}} e^{-\frac{x^2}{2\sigma^2}}, \tag{3.111}$$

$$p_1(x) = \frac{1}{\sqrt{2\pi\sigma^2}} e^{-\frac{(x-v)^2}{2\sigma^2}}. \tag{3.112}$$

Sequential likelihood ratio test

From (3.102), the likelihood function z_i can then be obtained as

$$z_i = \frac{x_i v}{\sigma^2} - \frac{v^2}{2\sigma^2}. \tag{3.113}$$

The expected value of z_i under H_0 and H_1 can also be easily computed as

$$\mathrm{E}\{z_i | H_0\} = -\frac{v^2}{2\sigma^2}, \tag{3.114}$$

$$\mathrm{E}\{z_i | H_0\} = \frac{v^2}{\sigma^2} - \frac{v^2}{2\sigma^2} = \frac{v^2}{2\sigma^2}. \tag{3.115}$$

From (3.107) and (3.108), we then have

$$\bar{n}_{H_0} \approx \frac{(\ln B)(1-\alpha) + (\ln A)\alpha}{\left(-\frac{v}{2\sigma^2}\right)}, \tag{3.116}$$

$$\bar{n}_{H_1} \approx \frac{(\ln B)(1-\beta) + (\ln A)\beta}{\left(\frac{v^2}{2\sigma^2}\right)}. \tag{3.117}$$

Standard (fixed sample size) likelihood ratio test

It can be easily shown that a sufficient statistic of n observations is

$$\Gamma = \sum_{i=1}^{n} X_i \gtrless_{H_0}^{H_1} \gamma_0. \tag{3.118}$$

Under H_0 and H_1, the sufficient statistic Γ has the following distribution:

$$p_0(\gamma) = \frac{1}{\sqrt{2\pi n\sigma^2}} \exp\left\{-\frac{\gamma^2}{2n\sigma^2}\right\}, \tag{3.119}$$

$$p_1(\gamma) = \frac{1}{\sqrt{2\pi n\sigma^2}} \exp\left\{-\frac{(\gamma - nv)^2}{2n\sigma^2}\right\}. \tag{3.120}$$

It follows that

$$P_{\mathrm{FA}} = \alpha = \int_{\gamma_0}^{\infty} p_0(\gamma) d\gamma = \frac{1}{\sqrt{2\pi}} \int_{\frac{\gamma_0}{\sigma\sqrt{n}}}^{\infty} e^{-t^2/2} dt = \mathrm{Q}\left(\frac{\gamma_0}{\sigma\sqrt{n}}\right), \tag{3.121}$$

$$P_{\mathrm{D}} = \beta = \int_{\gamma_0}^{\infty} p_1(\gamma) d\gamma = \frac{1}{\sqrt{2\pi}} \int_{\frac{\gamma_0 - nv}{\sigma\sqrt{n}}}^{\infty} e^{-t^2/2} = \mathrm{Q}\left(\frac{\gamma_0 - nv}{\sigma\sqrt{n}}\right). \tag{3.122}$$

From (3.121) and (3.122), we have

$$\gamma_0 = \sigma\sqrt{n}\mathrm{Q}^{-1}(\alpha) = \sigma\sqrt{n}\mathrm{Q}^{-1}(\beta) + nv. \tag{3.123}$$

Then the fixed sample size n for a standard likelihood ratio test is obtained as

$$n = \frac{\left(\mathrm{Q}^{-1}(\alpha) - \mathrm{Q}^{-1}(\beta)\right)^2}{\left(\frac{v^2}{\sigma^2}\right)}. \tag{3.124}$$

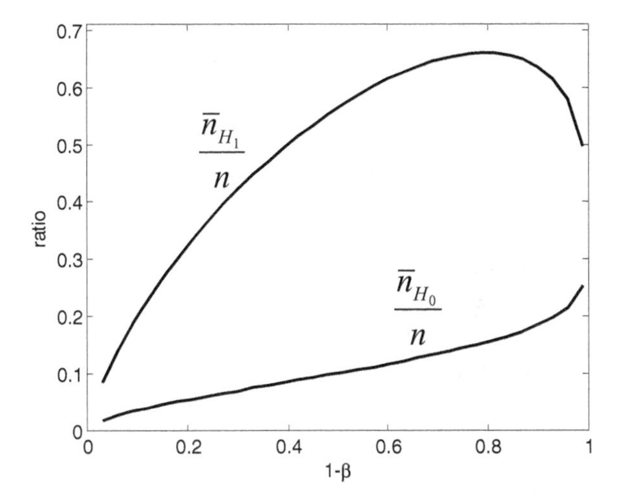

Figure 3.8 $\bar{n}_{H_0}/n$ and $\bar{n}_{H_1}/n$ versus β in Example 3.10 ($\alpha = 10^{-4}$).

Figure 3.8 shows the ratio $\bar{n}_{H_0}/n$ and $\bar{n}_{H_1}/n$ versus β, simulated under $\alpha = 10^{-4}$. It is observed from the figure that the number of observations required by the SLRT is significantly smaller than that required by the standard likelihood ratio test. At a probability of detection $P_D = \beta$ around 0.5, $\bar{n}_{H_1}$ is approximately $0.6n$, and $\bar{n}_{H_0}$ is approximately $0.1n$, which amounts to 40% and 90% saving in using the SLRT compared to the standard, fixed-sample size likelihood ratio test. □

3.7.2 Uniformly most powerful test

In Section 3.1, we considered *simple hypothesis testing* when both the pdf $p_0(\mathbf{x})$ under H_0 and the pdf $p_1(\mathbf{x})$ under H_1 are fully known. However, there are many practical problems in which one or both of these pdf's may have some unknown parameters $\boldsymbol{\Theta} = [\theta_1, \ldots, \theta_k]$. Then, we denote these as *composite hypothesis testing* problems. In this section, we want to study optimum decision rules that can handle these issues. For simplicity of presentation, we restrict H_0 to be a simple hypothesis but allow H_1 to be a composite hypothesis. We will introduce some well-known statistical hypothesis testing notations and concepts that are more general than the simple hypothesis testing detection problems considered earlier.

For the binary hypothesis testing problem in statistical literatures, the hypothesis H_0 is called the *null hypothesis* and the hypothesis H_1 is called the *alternative hypothesis*. Let the realization or observed random vector be in the n-dimensional Euclidean vector space denoted by $\mathbb{R}^n$, whose subset $\mathbb{R}_1$ is used to declare H_1 and whose subset $\mathbb{R}_0$ is used to declare H_0. Let H_0 be a simple hypothesis, then what we have been calling the probability of false alarm P_{FA} in detection theory is now called the *significance level* of the test α and is given by

$$\alpha = P_{FA} = \int_{\mathbb{R}_1} p_0(\mathbf{x})d\mathbf{x}. \tag{3.125}$$

Let H_1 be a composite hypothesis with a pdf denoted by $p_1(\mathbf{x}; \boldsymbol{\Theta})$. Then what we have been calling the probability of detection P_D in detection theory is now called the *power* of the test β and is given by

$$\beta = P_D = \int_{\mathbb{R}_1} p_1(\mathbf{x}; \boldsymbol{\Theta}) d\mathbf{x}. \tag{3.126}$$

Of course, if H_1 is a simple hypothesis, then the power of the test β in (3.126) uses the pdf $p_1(\mathbf{x})$. In statistical literatures, $\mathbb{R}_1$ is called the critical region. The problem of finding the critical region $\mathbb{R}_1$ for the *most powerful test* (i.e., finding the largest power β) of a simple H_0 hypothesis and a simple H_1 hypothesis is that of constraining α to satisfy (3.125) and maximize β of (3.126) using the pdf $p_1(\mathbf{x})$. Thus, the most powerful test solution for $\mathbb{R}_1$ is given by the Neyman–Pearson test of (3.53) using the LR test of

$$\mathbb{R}_1 = \left\{ \mathbf{x} : \Lambda(\mathbf{x}) = \frac{p_1(\mathbf{x})}{p_0(\mathbf{x})} \geq \Lambda_0 \right\}, \tag{3.127}$$

where Λ_0 is a real-valued constant. Then $\mathbb{R}_0$ is the complement of $\mathbb{R}_1$ in $\mathbb{R}^n$, which can be expressed symbolically as $\mathbb{R}_0 = \mathbb{R}^n - \mathbb{R}_1$, and given explicitly as in (3.48) by

$$\mathbb{R}_0 = \left\{ \mathbf{x} : \Lambda(\mathbf{x}) = \frac{p_1(\mathbf{x})}{p_0(\mathbf{x})} \leq \Lambda_0 \right\}. \tag{3.128}$$

If $\mathbf{X}$ is a discrete random vector, there may not be a critical region $\mathbb{R}_1$ that attains the equality of (3.125). Then the most powerful test is to maximize (3.127) among the largest $\mathbb{R}_1$ region to achieve the largest value in (3.125) constrained to be less than α.

Now, let H_1 be a composite hypothesis with a pdf denoted by $p_1(\mathbf{x}; \boldsymbol{\Theta})$. Then the above LR function

$$\Lambda(\mathbf{x}; \boldsymbol{\Theta}) = \frac{p_1(\mathbf{x}; \boldsymbol{\Theta})}{p_0(\mathbf{x})} \tag{3.129}$$

is a function of the parameter $\boldsymbol{\Theta}$. However, if the critical region $\mathbb{R}_1$ of (3.127) is not a function of the parameter $\boldsymbol{\Theta}$, then this test is called *uniformly most powerful* (UMP) with respect to all alternative hypothesis pdfs $p_1(\mathbf{x}; \boldsymbol{\Theta})$.

Example 3.11 Under hypothesis H_0, let the pdf $p_0(\mathbf{x})$ represent n independent Gaussian random variables of zero mean and known variance σ^2 given by

$$p_0(\mathbf{x}) = (2\pi\sigma^2)^{-n/2} \exp\left(-\sum_{m=1}^{n} \frac{x_m^2}{2\sigma^2} \right). \tag{3.130}$$

Under hypothesis H_1, let the pdf $p_1(\mathbf{x})$ represent n independent Gaussian random variables of unknown mean $\mu > 0$ and known variance σ^2 given by

$$p_1(\mathbf{x}; \mu) = (2\pi\sigma^2)^{-n/2} \exp\left(-\sum_{m=1}^{n} \frac{(x_m - \mu)^2}{2\sigma^2} \right). \tag{3.131}$$

Clearly, H_0 is a simple hypothesis and H_1 is a composite hypothesis with its pdf depending on the scalar parameter $\Theta = \mu > 0$. From the LR test of (3.127), we have

$$\mathbb{R}_1 = \left\{ \mathbf{x} : \Lambda(\mathbf{x}) = \exp\left(-\frac{1}{2\sigma^2} \left\{ \sum_{m=1}^{n} \left[x_m^2 - 2\mu x_m + \mu^2 - x_m^2 \right] \right\} \right) \geq \Lambda_0 \right\}, \tag{3.132}$$

or equivalently,

$$\begin{aligned} \mathbb{R}_1 &= \left\{ \mathbf{x} : \left(-\frac{1}{2\sigma^2} \left\{ \sum_{m=1}^{n} \left[-2\mu x_m + \mu^2 \right] \right\} \right) \geq \ln(\Lambda_0) \right\} \\ &= \left\{ \mathbf{x} : 2\mu \sum_{m=1}^{n} x_m \geq 2\sigma^2 \ln(\Lambda_0) + n\mu^2 \right\}. \end{aligned} \tag{3.133}$$

Since $\mu > 0$, then (3.133) can be expressed as

$$\mathbb{R}_1 = \left\{ \mathbf{x} : \bar{x} = \left(\frac{1}{n} \right) \sum_{m=1}^{n} x_m \geq \left(\frac{\sigma^2}{n\mu} \right) \ln(\Lambda_0) + \left(\frac{1}{2} \right) \mu = \gamma_0 \right\}, \tag{3.134}$$

where $\bar{x}$ denotes the sample mean of the n observed samples and γ_0 is sometimes called the *critical value*. Under H_0, $\bar{x}$ is the realization of a zero mean Gaussian r.v. with a variance of σ^2/n; under H_1, $\bar{x}$ is the realization of a Gaussian r.v. with a mean of μ and a variance of σ^2/n. Then the significance level α from (3.125) is given by

$$\alpha = \int_{\mathbb{R}_1} p_0(\mathbf{x}) d\mathbf{x} = \int_{\gamma_0}^{\infty} \frac{1}{\sqrt{2\pi\sigma^2/n}} \exp\left(-\bar{x}^2 n / 2\sigma^2 \right) d\bar{x} = \mathrm{Q}(\gamma_0 \sqrt{n}/\sigma), \tag{3.135}$$

or $\gamma_0 = \sigma \mathrm{Q}^{-1}(\alpha)/\sqrt{n}$. Then the critical region from (3.134) shows

$$\mathbb{R}_1 = \left\{ \mathbf{x} : \bar{x} \geq \gamma_0 = \sigma \mathrm{Q}^{-1}(\alpha)/\sqrt{n} \right\}, \tag{3.136}$$

which is not a function of the parameter μ. Thus, the LR test given by (3.132)–(3.134) is UMP. However, the power β of (3.126) can be evaluated explicitly by

$$\begin{aligned} \beta = \int_{\mathbb{R}_1} p_1(\mathbf{x}; \mu) d\mathbf{x} &= \int_{\gamma_0}^{\infty} \frac{1}{\sqrt{2\pi\sigma^2/n}} \exp\left(-(\bar{x} - \mu)^2 n / 2\sigma^2 \right) d\bar{x} \\ &= \mathrm{Q}\left(\frac{\sigma \mathrm{Q}^{-1}(\alpha) - \mu\sqrt{n}}{\sigma} \right), \end{aligned} \tag{3.137}$$

which is a function of the parameter μ. □

Example 3.12 Consider the two hypotheses H_0 and H_1 of Example 3.11, except now the mean μ in the pdf of H_1 is allowed to take values over the entire real line of $-\infty < \mu < \infty$. Since μ can be positive-valued or negative-valued, the step going from (3.133) to (3.134) need not be valid. The direction of the inequality in (3.134) may change depending on the signum of the parameter μ, and thus the critical region is now a function of μ. This shows the LR test for this example is not UMP. □

3.7.3 Non-parametric sign test

Thus far, in all the binary hypothesis tests, pdfs under H_0 and H_1 were assumed to be available. In some practical problems, it may not be realistic to model these pdfs. Thus, various kinds of non-parametric binary hypothesis tests are possible. One of the simplest non-parametric binary detection methods is based on the sign test. Under H_0, the received data $x_m,\ m = 1, \ldots, n$ consists of the realization of n independent noise samples modeled by

$$x_m = v_m,\ m = 1 \ldots, n, \tag{3.138}$$

where v_m is the realization of the noise V_m assumed to be a continuous r.v. whose pdf $f_V(v)$ is not known, but has a median value of 0 (i.e., $\mathrm{P}(V_m \le 0) = \mathrm{P}(V_m \ge 0) = 1/2$). Under H_1, the received data are modeled by

$$x_m = s + v_m,\ m = 1,\ \ldots,\ n, \tag{3.139}$$

where the value of the deterministic signal s is unknown but assumed to be positive-valued. Consider the sign test given by

$$T_n = \sum_{m=1}^{n} \mathrm{U}(x_m) \begin{matrix} \ge \gamma_0, \text{ declare } H_1, \\ \le \gamma_0, \text{ declare } H_0, \end{matrix} \tag{3.140}$$

where $\mathrm{U}(x)$ is the unit step function taking a unity value when $x > 0$ and taking a value of 0 when $x \le 0$ and the threshold γ_0 is determined by the value of the P_{FA}. Intuitively, under H_0, since each x_m is as likely to be positive as well as negative-valued, each $\mathrm{U}(x_m)$ is as likely to be a one as a zero, and thus the statistic T_n is most likely to have an integral value near $ns/2$. Under H_1, since each x_m is more likely to be positive-valued, then each $\mathrm{U}(x_m)$ is more likely to be a one than a zero, and thus the statistic T_n is more likely to have an integral value between $ns/2$ and ns. The conditional probability of $\mathrm{U}(X_m)$ under H_0 is given by

$$\mathrm{P}\left(\mathrm{U}(X_m) = 1 \mid H_0\right) = \mathrm{P}\left(\mathrm{U}(X_m) = 0 \mid H_0\right) = 1/2. \tag{3.141}$$

The conditional probability of $\mathrm{U}(X_m)$ under H_1 is given by

$$\mathrm{P}\left(\mathrm{U}(X_m) = 1 \mid H_1\right) = p,\ \mathrm{P}\left(\mathrm{U}(X_m) = 0 \mid H_1\right) = q = 1 - p, \tag{3.142}$$

where

$$p = \int_0^\infty p_V(x_m - s)dx_m = \int_{-s}^\infty p_V(v)dv > 0. \tag{3.143}$$

Under H_0, T_n has a binomial distribution with its pmf given by

$$p_t^{(0)} = \mathrm{P}(T_n = t \mid H_0) = \mathrm{C}_t^n(1/2)^t(1/2)^{n-t} = \mathrm{C}_t^n(1/2)^n,\ t = 0, \ldots, n\ , \tag{3.144}$$

where C_t^n is the combination of n items taken t at a time. Then the P_{FA} (i.e., the significance level α of the test) is given by

$$\alpha = P_{\mathrm{FA}} = \sum_{t=\lceil \gamma_0 \rceil}^{n} p_t^{(0)}, \tag{3.145}$$

Table 3.2 Probability of detection $P_D = \beta$ for $n = 10$ and 20 and desired $P_{FA} = \alpha_D = 0.20$ and 0.020 and actual $P_{FA} = \alpha_A$ with probability $p = 0.7$ and $p = 0.9$ and associated threshold γ_0.

			Probability of detection $P_D = \beta$	
Threshold γ_0	Desired α_D	Actual α_A	$p = 0.7$	$p = 0.9$
$n = 10$				
$\gamma_0 = 7$	$\alpha_D = 0.20$	$\alpha_A = 0.17188$	$\beta = 0.64961$	$\beta = 0.98720$
$\gamma_0 = 9$	$\alpha_D = 0.020$	$\alpha_A = 0.010741$	$\beta = 0.14931$	$\beta = 0.73610$
$n = 20$				
$\gamma_0 = 13$	$\alpha_D = 0.20$	$\alpha_A = 0.13159$	$\beta = 0.77227$	$\beta = 0.99969$
$\gamma_0 = 16$	$\alpha_D = 0.020$	$\alpha_A = 0.005909$	$\beta = 0.23751$	$\beta = 0.95683$

where $\lceil \gamma_0 \rceil$ is the ceiling of γ_0 (i.e., smallest integer larger than or equal to γ_0). We note, P_{FA} of (3.145), as well as $\lceil \gamma_0 \rceil$ (and thus the critical region $\mathbb{R}_1$) are not functions of the parameter s. Under H_1, T_n also has a binomial distribution with its pmf given by

$$p_t^{(1)} = \mathrm{P}(T_n = t \mid H_1) = \mathrm{C}_t^n p^t q^{n-t}, \tag{3.146}$$

where $p = (1 - q)$ is defined by (3.143). Then the P_D (i.e., the power of the test) is given by

$$\beta = P_D = \sum_{t=\lceil \gamma_0 \rceil}^{n} p_t^{(1)}. \tag{3.147}$$

Example 3.13 Consider a received datum of length $n = 10$ and 20, $p = 0.7$ and 0.9. By using the model based on (3.138)–(3.146), we obtain the following results, tabulated in Table 3.2.

In Table 3.2 with $n = 10$, suppose the desired P_{FA} is taken to be $\alpha_D = 0.20$. Then using (3.145), $\gamma_0 = 6$ yields an actual $\alpha_A = 0.37695 > \alpha_D = 0.20$, which is not acceptable. Thus, we have to take a larger $\gamma_0 = 7$, yielding an actual $\alpha_A = 0.17188 < \alpha_D = 0.20$, which is acceptable. Then for $p = 0.7$, using (3.146)–(3.147), $\beta = 0.64961$. For $p = 0.9$, $\beta = 0.98720$. For $\alpha_D = 0.020$, $\gamma_0 = 8$ yields an actual $\alpha_A = 0.054688 > \alpha_D = 0.020$, which is not acceptable. However, a larger $\gamma_0 = 9$, yielding an actual $\alpha_A = 0.010741 < \alpha_D = 0.020$, which is acceptable. Then for $p = 0.7$, $\beta = 0.14931$. For $p = 0.9$, $\beta = 0.73610$. Similar operations are used for the $n = 20$ case. As expected, for the same fixed desired $P_{FA} = \alpha_D = 0.20$ and 0.020, the corresponding probability of detections $P_D = \beta$ are larger for $n = 20$ compared to $n = 10$ in Table 3.2. □

Example 3.14 In Example 3.13, the sign test of (3.140) was applied to the received signal of (3.139) where the noise V_m was arbitrary except its median was constrained to have the value of zero (i.e., $\mathrm{P}(V_m \leq 0) = \mathrm{P}(V_m \geq 0) = 1/2$). Now, in (3.139), we assume the noises V_m, $m = 1, \ldots, n$, are i.i.d. Gaussian with zero mean

Table 3.3 Comparisons of probability of detection $P_D(\text{Gaussian}) = \beta_G$ to $P_D(\text{SignTest}) = \beta_{ST}$ for $n = 10$ and 20, $\alpha_A = 0.17188$ and 0.010742, and $p = 0.7$ and 0.9.

		β_G	
n	α_A	$p = 0.7$	$p = 0.9$
$n = 10$			
	$\alpha_A = 0.17188$	$\beta_G = 0.76162 > \beta_{ST} = 0.64961$	$\beta_G = 0.99905 > \beta_{ST} = 0.98720$
	$\alpha_A = 0.010742$	$\beta_G = 0.26073 > \beta_{ST} = 0.14931$	$\beta_G = 0.96022 > \beta_{ST} = 0.73610$
$n = 20$			
	$\alpha_A = 0.17188$	$\beta_G = 0.88995 > \beta_{ST} = 0.77227$	$\beta_G = 0.999998 > \beta_{ST} = 0.99968$
	$\alpha_A = 0.010742$	$\beta_G = 0.43158 > \beta_{ST} = 0.23751$	$\beta_G = 0.99934 > \beta_{ST} = 0.95683$

and unit variance. Consequently, we want to compare the probability of detection $P_D = \beta$ in Table 3.2 of Example 3.13, with the present case when we know each V_m has a Gaussian distribution. Now, under H_0, removing the U($\cdot$) operation in T_n of (3.140) yields $T_n = V_1 + \ldots + V_n$, which is Gaussian with $T_n \sim \mathcal{N}(0, n)$. Thus, we have $\alpha = \Phi(-\gamma_0/\sqrt{n})$. For $n = 10$, with $\alpha_A = 0.17188$, then $\gamma_0 = 2.993924$, and with $\alpha_A = 0.010742$, then $\gamma_0 = 7.2712429$. Under H_1, removing the U($\cdot$) operation in T_n of (3.140) yields $T_n = (s + V_1) + \cdots + (s + Vn)$, which is also Gaussian with $T_n \sim \mathcal{N}(ns, n)$. Since the noise pdf $p_V(v)$ is Gaussian, (3.143) yields $s = \Phi^{-1}(p)$. For $p = 0.7$, $s_1 = 0.52440$, and for $p = 0.9$, $s_2 = 1.28155$. Now, for V_m, $m = 1, \ldots, n$, being i.i.d. Gaussian, denote $P_D(\text{Gaussian}) = \beta_G = \Phi(-(\gamma - ns)/n^{0.5})$. In Table 3.3, we compare $P_D(\text{Gaussian}) = \beta_G$ to the corresponding $P_D(\text{Sign Test}) = \beta_{ST}$ of Table 3.2 for $n = 10$ and 20, $\alpha_A = 0.17188$ and 0.010742, and $p = 0.7$ and 0.9.

As shown in Table 3.3, all corresponding $P_D(\text{Gaussian}) = \beta_G$ values are greater than $P_D(\text{Sign Test}) = \beta_{ST}$ values, since the probability of detection results given by $P_D(\text{Gaussian}) = \beta_G$ are based on the Neyman–Pearson criterion, while those probability of detection results given by $P_D(\text{Sign Test}) = \beta_{ST}$ are based on an ad hoc sub-optimum detection scheme. □

3.8 Conclusions

Thus far in all the above discussions, pdfs under H_0 and H_1 are assumed to be known. Various non-parametric hypothesis tests are available when these pdfs are not available. Section 3.7.3 considers the binary hypothesis sign test. As expected, for a specified system, its performances (i.e., probability of detection and probability of false alarm) associated with a non-parametric test are inferior compared to those when the pdfs of H_0 and H_1 are known.

3.9 Comments

Detection theory used in digital communication and radar systems is based on statistical hypothesis testing theory. The first person to study this problem was Bayes [4], who originated the concept of the Bayes rule in the eighteenth century. Fisher originated the concepts of the likelihood and maximum likelihood in the 1920–1930s [1]. Neyman and Pearson originated the basic principles of statistical hypothesis testing, including the Bayes criterion, the Minimax criterion, the Neyman–Pearson criterion [2], and the uniformly most powerful test [5]. Wald made significant contributions on the mathematical aspects of these problems [6]. Some interesting discussions on various interpretations of likelihood were provided in [7]. Well-known statistics books dealing with hypothesis testing include [8–10]. Early use of LR tests for radar detection was discussed in [11, 12]. Early communication theory books dealing with hypothesis testing included [13–19]. More recent books dealing with hypothesis testing included [20–23]. All digital communication system books have a section on binary detection, but will not be referenced here. Reference [24] provided a detailed radar system oriented book on hypothesis testing. The sequential hypothesis testing method was advocated by Wald in [25] and [26]. Chapters on the basic sequential method appeared in chapter 24 of [8] and sequential detection for communications appeared in [17], [20], and [23]. An introduction to non-parametric statistical inference method was provided in [27], and with applications to communications was given in [28]. A collection of tutorial and detailed technical papers on non-parametric methods for detection appeared in [29].

References

[1] R. A. Fisher. Two new properties of mathematical likelihood. *Proc. Royal Soc., London, Ser. A*, 1934.
[2] J. Neyman and E. Pearson. *On the Problem of the Most Efficient Tests of Statistical Hypotheses*, volume 231 of *A. Phil. Trans. Royal Soc.*, 1933.
[3] A. Wald. Sequential tests of statistical hypotheses. *Ann. Math. Statist.*, pages 117–186, 1945.
[4] T. P. Bayes. An essay towards solving a problem in the doctrine of chances. *Phil. Trans. Royal Soc., London*, 1763.
[5] J. Neyman and E. S. Pearson. Sufficient statistics and uniformly most powerful tests of statistical hypotheses. *Stat. Res. Memoir*, vol. 1, 1936.
[6] A. Wald. Contributions to the theory of statistical hypotheses and tests of statistical hypotheses. *Ann. Math. Statist.*, 10, 1939.
[7] A. W. F. Edwards. *Likelihood.* Cambridge University Press, 1972.
[8] M. G. Kendall and A. Stewart. *The Advanced Theory of Statistics*, vols. 1–2. Griffin, 1958.
[9] A. Wald. *Statistical Decision Theory.* Wiley, 1950.
[10] E. L. Lehmann. *Testing Statistical Hypothesis.* Wiley, 1950.
[11] I. L. Davies. On determining the presence of signals in noise. *Proc. IEE, pt. III*, Mar. 1952.
[12] P. M. Woodward. *Probability and Information Theory, with Applications to Radar*. Pergamon Press, 1953.
[13] D. Middleton. *An Introduction to Statistical Communication Theory.* McGraw-Hill, 1960.

[14] V. A. Kotel'nikov. *Theory of Optimum Noise Immunity*. Government Power Engineering Press, 1956, translated by R. A. Silverman, McGraw-Hill, 1960.
[15] J. M. Wozencraft and I. M. Jacobs. *Principles of Communication Engineering*. Wiley, 1965.
[16] H. L. Van Trees. *Detection, Estimation, and Modulation, Part I*. Wiley, 1968.
[17] C. W. Helstrom. *Statistical Theory of Signal Detection*. Pergamon, 2nd ed., 1969.
[18] A. D. Whalen. *Detection of Signals in Noise*. Academic Press, 1st ed., 1971. R. N. McDonough and A. D. Whalen, 2d ed., 1995.
[19] V. H. Poor. *An Introduction to Detection and Estimation*. Springer-Verlag, 1988.
[20] C. W. Helstrom. *Statistical Theory of Signal Detection*. Pergamon, 2nd ed., 1960. An expanded version appeared as *Elements of Signal Detection and Estimation*, Prentice-Hall, 1995.
[21] S. M. Kay. *Fundamentals of Statistical Signal Processing, Volume 2: Detection Theory*. Prentice Hall, 1998.
[22] R. D. Hippenstiel. *Detection Theory*. CRC Press, 2002.
[23] T. A. Schonhoff and A. A. Giordano. *Detection and Estimation Theory*. Pearson, 2006.
[24] J. V. DiFranco and W. L. Rubin. *Radar Detection*. Artech House, 1980.
[25] A. Wald. *Sequential Analysis*. Dover, 1947.
[26] A. Wald and J. Wolfowitz. Optimum character of the sequential probability ratio test. *Ann. Math. Statist.*, 19, 1948.
[27] J. D. Gibbons. *Nonparametric Statistical Inference*. McGraw-Hill, 1971.
[28] J. D. Gibson and J. L. Melsa. *Introduction to Nonparametric Detection and Applications*. Academic Press, 1975.
[29] S. A. Kassam and J. B. Thomas, eds. *Nonparametric Detection Theory and Applications*. Dowden, Hutchinson and Ross, 1980.

Problems

3.1 In detection problems dealing with Gaussian r.v. and random process, we always need to evaluate the complementary Gaussian distribution function:

$$\mathrm{Q}(x) = (1/\sqrt{2\pi}) \int_x^{\infty} e^{-t^2/2} dt. \tag{3.148}$$

Since $\mathrm{Q}(x)$ does not have a closed form expression, we can approximate by various means.

(a) Most numerical routines (including Matlab) define $\mathrm{erf}(x) = \frac{2}{\sqrt{\pi}} \int_0^x e^{-t^2} dt$ and $\mathrm{erfc}(x) = 1 - \mathrm{erf}(x)$. Find $\mathrm{Q}(\cdot)$ in terms of $\mathrm{erfc}(\cdot)$.
(b) Let $\mathrm{Q}_1(x) = (1/x\sqrt{2\pi})\, e^{-x^2/2}, \quad x > 0$, be a simple approximation to $\mathrm{Q}(x)$.
(c) Let $\mathrm{Q}_2(x) = e^{-x^2/2}(b_1 t + b_2 t^2 + b_3 t^3 + b_4 t^4 + b_5 t^5)/\sqrt{2\pi}$, $x \geq 0$, be a better approximation to $\mathrm{Q}(x)$, where

$$t = \frac{1}{1+px}, \quad p = 0.2316419,$$
$$b_1 = 0.319381530, \; b_2 = -0.356563782,$$
$$b_3 = 1.781477937, \; b_4 = -1.821255978,$$
$$b_5 = 1.330274429. \tag{3.149}$$

Tabulate Q(x), Q$_1$(x), and Q$_2$(x) for $x = 0.5, 1, 1.5, 2, 2.5, 3, 3.5, 4, 4.5, 5$. Plot (or sketch) $\log_{10} \mathrm{Q}_1(x)$, $\log_{10} \mathrm{Q}_1(x)$, and $\log_{10} \mathrm{Q}_2(x)$ versus x values.

3.2 Consider the binary detection problem where

$$X[n] = \begin{cases} W[n], & n = 0, \ldots, N-1, \quad H_0, \\ A + w[n], & n = 0, \ldots, N-1, \quad H_1, \end{cases}$$

where $A > 0$ and $W[n]$ is a WGN with zero mean and variance σ^2. Assume

$$c_{ij} = \begin{cases} 0, & i = j, \\ 1, & i \neq j, \end{cases}, \quad \forall i, \ j = 0, \ 1. \tag{3.150}$$

(a) Let $\mathrm{P}(H_0) = \mathrm{P}(H_1) = 1/2$. For the minimum average error probability criterion receiver, find $\mathbb{R}_1$ and $\mathbb{R}_0$, and the minimum average error probability.
(b) Repeat part (a), except use $\mathrm{P}(H_0) = 1/4; \mathrm{P}(H_1) = 3/4$.
(c) Repeat parts (a) and (b), except use the MAP criterion.
(d) Sketch the decision regions $\mathbb{R}_1$ and $\mathbb{R}_0$ and the thresholds, for parts (a) and (b), using $N = A = \sigma^2 = 1$.

3.3 Let $X = X_1^2 + \cdots X_n^2$ with $X_i \sim \mathcal{N}(0, 1)$. Then X is a chi-square r.v. of degree n with a pdf given by $p_X(x) = 2^{-n/2}(1/\Gamma(n/2))x^{-1+n/2}e^{-x/2}, x \geq 0$, where the gamma function is defined by $\Gamma(a) = \int_0^\infty e^{-t}t^{a-1}dt$. In particular, if a is an integer, then $\Gamma(a + 1) = a!$ is the factorial of a. Consider the binary hypothesis test where under H_0, X is a chi-square r.v. of degree $n = 2$, and under H_1, X is a chi-square r.v. of degree $n = 6$. Find the two LR decision regions $\mathbb{R}_1$ and $\mathbb{R}_0$, when the LR constant $\Lambda_0 = 1$. Find the associated P_{FA} and P_{D}.

3.4 Consider the conditional pdfs under H_0 and H_1 given by

$$p_0(x) = \begin{cases} \frac{3}{2}(1-x)^2, & 0 \leq x < 2, \\ 0, & \text{elsewhere}, \end{cases}$$

$$p_1(x) = \begin{cases} \frac{3}{4}x(2-x), & 0 \leq x < 2, \\ 0, & \text{elsewhere}. \end{cases}$$

(a) Find the LR function.
(b) Under the Bayes criterion (with a given Λ_0), sketch the decision regions $\mathbb{R}_0$ and $\mathbb{R}_1$. Hint: For any real number a, $x(2a - x) = a^2 - (a - x)^2$.
(c) Let $P_{\mathrm{FA}} = 0.001$. Find $\mathbb{R}_0$ and $\mathbb{R}_1$ explicitly.

3.5 Consider a NP criterion detection problem where under H_0 we have $X = N$, with the pdf of N given by

$$p_N(n) = \begin{cases} e^{-n}, & 0 \leq n < \infty, \\ 0, & n < 0, \end{cases}$$

and under H_1, $X = 1 + N$. Let $P_{\mathrm{FA}} = 0.1$.

(a) Show the NP criterion decision regions $\mathbb{R}_0$ and $\mathbb{R}_1$ are not unique.
(b) Suppose $\mathbb{R}_1 = \{x : x \geq x_1\}$. Find x_1, $\mathbb{R}_0$, and P_{D}.

(c) Suppose $\mathbb{R}_1 = \{x : 1 \leq x \leq x_2\}$. Find x_2, $\mathbb{R}_0$, and P_{D}.
(d) For any valid NP criterion decision regions (with $P_{\text{FA}} = 0.1$), find P_{D}.

3.6 A random variable X is distributed according to a Cauchy distribution

$$p(x) = \frac{m}{\pi(m^2 + x^2)}.$$

The parameter m can take on either of two values, $m_0 = 1$ and $m_1 = 2$.

(a) Find the LR test to decide on the basis of a single measurement of X between the two hypotheses H_0 (i.e., $m = m_0$) and H_1 (i.e., $m = m_1$).
(b) Under the Neyman–Pearson criterion, find P_{D} explicitly in terms of P_{FA}. (We assume both P_{D} and P_{FA} are non-zero valued.) Hint 1: Find P_{FA} in terms of some threshold γ_0 (obtained from the LR test), then find γ_0 in terms of P_{FA}. Finally, find P_{D} in terms of γ_0, which is a function of P_{FA}. Hint 2: $\int \frac{dx}{a^2+x^2} = (1/a)\tan^{-1}(x/a)$.

3.7 The binary hypothesis testing method can be used to model the following simple industrial engineering problem. Consider a manufacturing process of a device, denoted as H_0, with a probability of failure of $p = 0.05$. Consider a new manufacturing process method, denoted as H_1, with a probability of failure of $p = 0.02$. Suppose we inspect 200 devices but do not know which process was used in the manufacturing. Denote X as the number of devices in the lot of 200 that failed. Assume all failures are independent. Declare H_1 if $X \leq 5$ and declare H_0 if $X > 5$. Find the probability of declaring H_1 given H_0 is true. In statistics, this is called the probability of error of the first kind, $P_{\text{I}}(\cdot)$. For us, we call it the probability of a false alarm, $P_{\text{FA}}(\cdot)$. Consider also finding the probability of declaring H_0 given H_1 is true. In statistics, this is called the probability of error of the second kind, $P_{\text{II}}(\cdot)$. For us, we call it the probability of a miss, $P_{\text{M}}(\cdot)$. Hint: Use the Poisson approximation to the binomial approximation when the number of terms n is large, the probability p is small, and np has a medium value. Suppose you are an analytical industrial engineer. Having evaluated the above two probabilities, how do you interpret this information (i.e., use a few sentences with English words and some numbers) to impress upon your manufacturing supervisor that this new manufacturing process is good for the plant?

3.8 Denote $\mathbf{X} = [X_1, X_2, \ldots, X_n]^T$, where the X_i's are n independent Bernoulli r.v.'s, each with a probability function $p(x) = \theta^x(1-\theta)^{1-x}$, $x = 0, 1$. Since the parameter θ is given, the above probability can be written in any one of the form of $p(x) = p(x;\theta) = p(x|\theta)$. Consider the statistic $g(\mathbf{x}) = \sum_{i=1}^{n} x_i$, where x_i is the value of a realization of the r.v. X_i. Show $g(\cdot)$ is a sufficient statistic. Hint: Use the factorization theorem and find what is a possible $k(g(\mathbf{x}|\theta))$ and what is a possible $l(\mathbf{x})$.

3.9 Consider the optimum decision rule under the Minimax criterion where under H_0 we receive only a zero mean unit variance Gaussian noise r.v. and under H_1 we receive one-volt in the presence of the zero mean unit variance Gaussian noise r.v. Let $c_{00} = c_{11} = 0$ and $c_{01} = c_{10} = 1$.

(a) Find $\bar{C}(\pi, \pi) = \bar{C}(\pi)$ analytically in terms of Q(·) with π as a variable.
(b) Evaluate $\bar{C}(\pi)$ for $\pi = 0.001, 0.01, 0.1, 0.2, 0.3, 0.4, 0.5, 0.6, 0.7, 0.8, 0.9, 0.99, 0.999$.
(c) Find $\bar{C}(\pi, \pi_0)$ analytically.
(d) Evaluate $\bar{C}(\pi, \pi_0)$ for $\pi = 0.001, 0.01, 0.1, 0.2, 0.3, 0.4, 0.5, 0.6, 0.7, 0.8, 0.9, 0.99, 0.999$ with $\pi_0 = 0.1, 0.5, 0.9$.
(e) Plot $\bar{C}(\pi)$ versus π as well as $\bar{C}(\pi, \pi_0)$ versus π for $\pi_0 = 0.1, 0.5, 0.9$. Does $\bar{C}(\pi)$ appear to be a concave function? For this problem, you can use $Q_2(\cdot)$ of Problem 3.1(c) as an excellent approximation to Q(·) for $x \geq 0$. However, the above $Q_2(\cdot)$ approximation is only valid for $x \geq 0$. Since Q(·) is known to have symmetry about $x = 0$, $y = 0.5$ (i.e., $Q(-x) = 1 - Q(x)$, $x \geq 0$), you can generate a new approximation for Q(·) based on $Q_2(\cdot)$ valid for $-\infty < x < \infty$.

3.10 Consider a slight generalization of Example 3.10 of Section 3.7.1. Under hypothesis H_0, let the pdf $p_0(\mathbf{x})$ represent n independent Gaussian random variables of known mean μ_0 and known variance σ^2. Under hypothesis H_1, let the pdf $p_1(\mathbf{x})$ represent n independent Gaussian random variables of unknown mean μ_1, but known to be $\mu_1 > \mu_0$ and have known variance σ^2. Show the P_{FA} is not a function of the unknown parameter μ_1, and thus this LR test is UMP.

3.11 Consider problem 3.6, in which $m_0 = 0$ and the parameter $m_1 \neq 0$. Show this P_{FA} is a function of the unknown parameter m_1, and thus this LR test is not UMP.

3.12 Consider Example 3.10 in Section 3.7.1 based on Sequential Detection. Under H_0, all i.i.d. Gaussian r.v.'s X_i, $i = 1, \ldots, n$, have zero mean and unit variance. Under H_1, all i.i.d. Gaussian r.v.'s X_i, $i = 1, \ldots, n$, have unit mean and unit variance. Take $P_{\text{FA}} = \alpha = 0.05$ and $P_{\text{M}} = 1 - \beta = 0.05$. Show in the SPRT, $A = 19$, $B = 0.0526$, $\bar{n}_{H_1} = 5.299$, $\bar{n}_{H_0} = 5.3229$, and theoretical number of terms needed in conventional non-sequential detection $n = 8.398$. Perform a simulation study of this problem using Matlab for 10 realizations of data under H_0 and 10 realizations of data under H_1. In both cases, use randn('seed', y) with $y = 2000 : 2009$. Show all 10 L_n values and averaged number of terms used in these 10 realizations for both the H_0 and the H_1 cases.

3.13 Consider the use of the SPRT method applied to the decision of the probability of a "head" in the coin tossing problem. Under H_1, we assume the probability of a "head" is θ_1, under H_0, we assume the probability of a "head" is θ_0, with $\theta_1 > \theta_0$. Let the outcome of each coin tossing be denoted by $Z_i = 1$ for a "head" and $Z_i = 0$ for a "tail." We assume all the Z_i r.v.'s are i.i.d. Denote $X_m = \sum_{i=1}^{m} Z_i$ as the number of successes in a trail of length m.

(a) Find the LR function $L_m(x_m)$ corresponding to (3.91) for this problem.
(b) Take the ln of all the terms in $B < L_m(x_m) < A$ to yield

$$\ln(B) < \ln(L_m(x_m)) < \ln(A) \,. \qquad (*)$$

Rewrite $(*)$ as $b + c\,m < x_m < a + c\,m$. Find a, b, and c in terms of A, B, θ_1, and θ_0.

(c) For $\alpha = 1 - \beta = 0.05$, and $\theta_1 = 0.6$ and $\theta_0 = 0.2$, find A and B. Find $\bar{n}_{H_0}$ and $\bar{n}_{H_1}$.
(d) Perform a simulation study of this problem using Matlab with 10 realizations for H_0 and then 10 realizations for H_1 with $\theta_1 = 0.6$ and $\theta_0 = 0.2$. From the simulation results show $L_m(x_m)$ and also the corresponding n data under these two simulations (similar to using the same technique as in problem 3.12).

3.14 Duplicate the results in Example 3.13 of Section 3.7.3 and show Table 3.2 with $p = 0.8$.

3.15 Duplicate the results in Example 3.14 of Section 3.7.3 and show Table 3.3 with $p = 0.8$.

4 Detection of known binary deterministic signals in Gaussian noises

In Chapter 3, we have shown that for a binary hypothesis test, under the Bayes, the Minimax, and the Neyman–Pearson criteria, all of these optimum decision rules result in LR tests only with different LR constants. In this chapter, we will elaborate on these LR tests to obtain various receiver structures for the detection of known binary deterministic signal vectors and waveforms in the presence of white and colored Gaussian noises. While many possible channel noise models have been proposed, the Gaussian noise model is justified from the result of the central limit theorem and represents the simplest and also most commonly used noise model in communication, radar, and signal processing.

4.1 Detection of known binary signal vectors in WGN

Consider the data vector $\mathbf{X}$ given by

$$\mathbf{X} = \begin{cases} \mathbf{s}_0 + \mathbf{N}, & H_0, \\ \mathbf{s}_1 + \mathbf{N}, & H_1, \end{cases} \tag{4.1}$$

where

$$\mathbf{X} = [X_1, \ldots, X_n]^T, \; \mathbf{s}_0 = [s_{01}, \ldots, s_{0n}]^T, \\ \mathbf{s}_1 = [s_{11}, \ldots, s_{1n}]^T, \; \mathbf{N} = [N_1, \ldots, N_n]^T, \tag{4.2}$$

with $\mathbf{s}_0$ and $\mathbf{s}_1$ being two known $n \times 1$ deterministic signal vectors and $\mathbf{N}$ is a $n \times 1$ Gaussian noise vector with the mean vector given by

$$\mathrm{E}\{\mathbf{N}\} = \mathbf{0}, \tag{4.3}$$

the $n \times n$ covariance matrix $\mathbf{\Lambda}$ given by

$$\mathbf{\Lambda} = \mathrm{E}\left\{\mathbf{N}\mathbf{N}^T\right\} = \sigma^2 \mathbf{I}_n = \begin{bmatrix} \sigma^2 & 0 & 0 & 0 \\ 0 & \sigma^2 & 0 & 0 \\ 0 & 0 & \ddots & 0 \\ 0 & 0 & 0 & \sigma^2 \end{bmatrix}, \tag{4.4}$$

where the determinant of $\mathbf{\Lambda}$ is denoted by $|\mathbf{\Lambda}| = \sigma^{2n}$, the inverse of $\mathbf{\Lambda}$ is denoted by $\mathbf{\Lambda}^{-1} = (1/\sigma^2)\mathbf{I}_n$, and the pdf of $\mathbf{N}$ is given by

$$p_{\mathbf{N}}(\mathbf{n}) = \frac{1}{(2\pi)^{n/2}|\mathbf{\Lambda}|^{1/2}} \exp\left(-\frac{\mathbf{n}^T\mathbf{n}}{2\sigma^2}\right) = \frac{1}{(2\pi)^{n/2}\sigma^n} \exp\left(-\frac{\mathbf{n}^T\mathbf{n}}{2\sigma^2}\right). \tag{4.5}$$

A noise vector $\mathbf{N}$ with a covariance given by (4.4), where all the off-diagonal components are zero, is called a *white noise* (WN) vector. Since the noise $\mathbf{N}$ is a Gaussian vector satisfying (4.4)–(4.5), it is called a *white Gaussian noise* (WGN) vector.

In light of the signal model of (4.1) and that the noise $\mathbf{N}$ is a WGN vector, the pdf of $\mathbf{X}$ under hypothesis H_0 is given by

$$\begin{aligned} p_0(\mathbf{x}) &= \frac{1}{(2\pi)^{n/2}|\mathbf{\Lambda}|^{1/2}} \exp\left(-\frac{(\mathbf{x}-\mathbf{s}_0)^T(\mathbf{x}-\mathbf{s}_0)}{2\sigma^2}\right) \\ &= \frac{1}{(2\pi)^{n/2}\sigma^n} \exp\left(-\frac{1}{2\sigma^2}\sum_{i=1}^{n}(x_i - s_{0i})^2\right), \end{aligned} \tag{4.6}$$

and the pdf of $\mathbf{X}$ under hypothesis H_1 is given by

$$\begin{aligned} p_1(\mathbf{x}) &= \frac{1}{(2\pi)^{n/2}|\mathbf{\Lambda}|^{1/2}} \exp\left(-\frac{(\mathbf{x}-\mathbf{s}_1)^T(\mathbf{x}-\mathbf{s}_1)}{2\sigma^2}\right) \\ &= \frac{1}{(2\pi)^{n/2}\sigma^n} \exp\left(-\frac{1}{2\sigma^2}\sum_{i=1}^{n}(x_i - s_{1i})^2\right). \end{aligned} \tag{4.7}$$

Now, consider a LR test for H_0 and H_1. As considered in Sections 3.2–3.5, the value of the LR constant Λ_0 varies depending on the decision criterion (e.g., Bayes, Minimax, or NP).

The LR function

$$\begin{aligned} \Lambda(\mathbf{x}) = \frac{p_1(\mathbf{x})}{p_0(\mathbf{x})} &= \exp\left\{\frac{-1}{2\sigma^2}\sum_{i=1}^{n}\left[(x_i - s_{1i})^2 - (x_i - s_{0i})^2\right]\right\} \\ &= \exp\left\{\frac{-1}{2\sigma^2}\left[\mathbf{x}^T\mathbf{x} - \mathbf{x}^T\mathbf{s}_1 - \mathbf{s}_1^T\mathbf{x} + \mathbf{s}_1^T\mathbf{s}_1 - \mathbf{x}^T\mathbf{x} + \mathbf{x}^T\mathbf{s}_0 + \mathbf{s}_0^T\mathbf{x} - \mathbf{s}_0^T\mathbf{s}_0\right]\right\} \\ &= \exp\left\{\frac{-1}{2\sigma^2}\left[2\mathbf{x}^T(\mathbf{s}_0 - \mathbf{s}_1) + \mathbf{s}_1^T\mathbf{s}_1 - \mathbf{s}_0^T\mathbf{s}_0\right]\right\}. \end{aligned} \tag{4.8}$$

Then the decision regions $\mathbf{R}_1$ and $\mathbf{R}_0$ are given by

$$\begin{aligned} \mathbf{R}_1 &= \{\mathbf{x} : \Lambda(\mathbf{x}) \geq \Lambda_0\} = \{\mathbf{x} : \ln(\Lambda(\mathbf{x})) \geq \ln(\Lambda_0)\} \\ &= \left\{\mathbf{x} : \frac{-1}{2\sigma^2}\left\{2\mathbf{x}^T(-\mathbf{s}_1 + \mathbf{s}_0) + \mathbf{s}_1^T\mathbf{s}_1 - \mathbf{s}_0^T\mathbf{s}_0\right\} \geq \ln(\Lambda_0)\right\} \\ &= \left\{\mathbf{x} : \left\{2\mathbf{x}^T(\mathbf{s}_1 - \mathbf{s}_0)\right\} \geq 2\sigma^2\ln(\Lambda_0) + \mathbf{s}_1^T\mathbf{s}_1 - \mathbf{s}_0^T\mathbf{s}_0\right\}, \\ &= \left\{\mathbf{x} : \left\{\mathbf{x}^T(\mathbf{s}_1 - \mathbf{s}_0)\right\} \geq \sigma^2\ln(\Lambda_0) + \frac{\mathbf{s}_1^T\mathbf{s}_1}{2} - \frac{\mathbf{s}_0^T\mathbf{s}_0}{2}\right\} \\ &= \left\{\mathbf{x} : \left\{\mathbf{x}^T(\mathbf{s}_1 - \mathbf{s}_0)\right\} \geq \gamma_0\right\} = \{\mathbf{x} : \gamma \geq \gamma_0\}, \end{aligned} \tag{4.9}$$

$$\begin{aligned} \mathbf{R}_0 &= \{\mathbf{x} : \Lambda(\mathbf{x}) \leq \Lambda_0\} = \{\mathbf{x} : \ln(\Lambda(\mathbf{x})) \leq \ln(\Lambda_0)\} \\ &= \left\{\mathbf{x} : \left\{\mathbf{x}^T(\mathbf{s}_1 - \mathbf{s}_0)\right\} \leq \gamma_0\right\} = \{\mathbf{x} : \gamma \leq \gamma_0\}, \end{aligned} \tag{4.10}$$

where the LR constant Λ_0 can be chosen from (3.11), (3.42)–(3.46), or (3.55). From (4.9) and (4.10), we define

$$\gamma = \mathbf{x}^T(\mathbf{s}_1 - \mathbf{s}_0) = \sum_{i=1}^{n} x_i(s_{1i} - s_{oi}) \tag{4.11}$$

as a *sufficient statistic* for the detection of H_0 or H_1. Thus, the scalar γ of (4.11) is the realization of the scalar random variable

$$\Gamma = \mathbf{X}^T(\mathbf{s}_1 - \mathbf{s}_0) = \sum_{i=1}^{n} X_i(s_{1i} - s_{0i}), \tag{4.12}$$

which is as good (i.e., sufficient) as the raw observed data vector $\mathbf{x}$ (which is a $n \times 1$ dimensional vector), or the LR function $\Lambda(\mathbf{x})$ (in the definition of $\mathbf{R}_1$ and $\mathbf{R}_0$), for the detection of H_1 or H_0.

Now, consider the performance of the LR receiver based on the use of $\mathbf{R}_1$ and $\mathbf{R}_0$ in (4.9)–(4.10). From (4.12), since each X_i is a Gaussian random variable, then Γ is a linear combination of scalar Gaussian random variables and is still a scalar Gaussian random variable. But for a scalar Gaussian random variable, it is fully characterized by just its mean and its variance values. The statistics of the data vector $\mathbf{X}$ change depending on H_0 or H_1. Specifically, the conditional means of Γ under H_0 or H_1 are given by

$$\mu_0 = \mathrm{E}\{\Gamma | H_0\} = \mathrm{E}\left\{(\mathbf{s}_0 + \mathbf{N})^T(\mathbf{s}_1 - \mathbf{s}_0)\right\} = \mathbf{s}_0^T(\mathbf{s}_1 - \mathbf{s}_0) = \sum_{i=1}^{n} s_{0i}(s_{1i} - s_{0i}), \tag{4.13}$$

$$\mu_1 = \mathrm{E}\{\Gamma | H_1\} = \mathrm{E}\left\{(\mathbf{s}_1 + \mathbf{N})^T(\mathbf{s}_1 - \mathbf{s}_0)\right\} = \mathbf{s}_1^T(\mathbf{s}_1 - \mathbf{s}_0) = \sum_{i=1}^{n} s_{1i}(s_{1i} - s_{0i}). \tag{4.14}$$

The conditional variance of Γ under H_0 is given by

$$\begin{aligned}
\sigma_0^2 &= \mathrm{E}\left\{[\Gamma - \mu_0]^2 | H_0\right\} = \mathrm{E}\left\{\left[\mathbf{X}^T(\mathbf{s}_1 - \mathbf{s}_0) - \left(\mathbf{s}_0^T(\mathbf{s}_1 - \mathbf{s}_0)\right)\right]^2 | H_0\right\} \\
&= \mathrm{E}\left\{\left[(\mathbf{s}_0 + \mathbf{N})^T(\mathbf{s}_1 - \mathbf{s}_0) - \left(\mathbf{s}_0^T(\mathbf{s}_1 - \mathbf{s}_0)\right)\right]^2\right\} \\
&= \mathrm{E}\left\{\left[(\mathbf{s}_0 + \mathbf{N} - \mathbf{s}_0)^T(\mathbf{s}_1 - \mathbf{s}_0)\right]^2\right\} \\
&= \mathrm{E}\left\{\left[\mathbf{N}^T(\mathbf{s}_1 - \mathbf{s}_0)\right]^2\right\} = \mathrm{E}\left\{[(\mathbf{N}, (\mathbf{s}_1 - \mathbf{s}_0))]^2\right\} \\
&= \mathrm{E}\left\{(\mathbf{N}, (\mathbf{s}_1 - \mathbf{s}_0))(\mathbf{N}, (\mathbf{s}_1 - \mathbf{s}_0))\right\} \\
&= \mathrm{E}\left\{((\mathbf{s}_1 - \mathbf{s}_0), \mathbf{N})(\mathbf{N}, (\mathbf{s}_1 - \mathbf{s}_0))\right\} = \mathrm{E}\left\{(\mathbf{s}_1 - \mathbf{s}_0)^T\mathbf{N}\mathbf{N}^T(\mathbf{s}_1 - \mathbf{s}_0)\right\} \\
&= \left[(\mathbf{s}_1 - \mathbf{s}_0)^T\sigma^2\mathbf{I}_n(\mathbf{s}_1 - \mathbf{s}_0)\right] = \sigma^2\left[(\mathbf{s}_1 - \mathbf{s}_0)^T(\mathbf{s}_1 - \mathbf{s}_0)\right] = \sigma^2\|(\mathbf{s}_1 - \mathbf{s}_0)\|^2.
\end{aligned} \tag{4.15}$$

The fifth, sixth, and seventh lines of (4.15) use the notations and properties of the inner product and the square of the norm $\|\cdot\|$ of a vector. Let $\mathbf{x} = [x_1, \ldots, x_n]^T$ and $\mathbf{y} = [y_1, \ldots, y_n]^T$ be two arbitrary $n \times 1$ real-valued vectors, then their *inner product* is scalar-valued and is defined by

$$(\mathbf{x}, \mathbf{y}) = \mathbf{x}^T \mathbf{y} = \sum_{i=1}^{n} x_i y_i = \sum_{i=1}^{n} y_i x_i = \mathbf{y}^T \mathbf{x} = (\mathbf{y}, \mathbf{x}). \tag{4.16}$$

Equation (4.16) shows the inner product of two vectors is commutative (i.e., the order of the two vectors in the inner product can be interchanged) and is equivalent to the definition of the correlation of the two vectors given in (1.10). Furthermore, when the two vectors are identical, then

$$(\mathbf{x}, \mathbf{x}) = \mathbf{x}^T \mathbf{x} = \sum_{i=1}^{n} x_i x_i = \|\mathbf{x}\|^2, \tag{4.17}$$

and $\|\mathbf{x}\|$ is called the *norm* of the vector $\mathbf{x}$ and has the interpretation as the length of the vector $\mathbf{x}$ from the origin to the point $\mathbf{x} = [x_1, \ldots, x_n]^T$ in the n-dimensional $\mathbb{R}^n$ space. If $\mathbf{x}$ is a vector with its components $\{x_1, \ldots, x_n\}$ representing a voltage source sampled at n time instants (over $(n-1)$ time intervals), then each x_i^2 has the unit of power, and sum of n such terms in (4.17) over $(n-1)$ time intervals yields $\|\mathbf{x}\|^2$ with the interpretation as the *energy* of the vector $\mathbf{x}$.

The conditional variance of Γ under H_1 is identical to that under H_0 and is given by

$$\sigma_1^2 = \sigma^2 \|(\mathbf{s}_1 - \mathbf{s}_0)\|^2 = \sigma_0^2. \tag{4.18}$$

From (4.9), the probability of a false alarm P_{FA} (i.e., the probability of declaring hypothesis H_1 when H_0 is true) is given by

$$\begin{aligned}
P_{\mathrm{FA}} &= \int \cdots \int_{R_1} p_0(\mathbf{x}) d\mathbf{x} = \int_{\Lambda_0}^{\infty} \tilde{p}_0(\Lambda) d\Lambda = \mathrm{P}(\Gamma \geq \gamma_0 | H_0) \\
&= \frac{1}{\sqrt{2\pi}\sigma_0} \int_{\gamma_0}^{\infty} \exp\left(\frac{(\gamma - \mu_0)^2}{2\sigma_0^2}\right) d\gamma = \frac{1}{\sqrt{2\pi}} \int_{\frac{\gamma_0 - \mu_0}{\sigma_0}}^{\infty} \exp\left(\frac{-t^2}{2}\right) dt \\
&= \mathrm{Q}\left(\frac{\gamma_0 - \mu_0}{\sigma_0}\right) = \mathrm{Q}\left(\frac{\sigma^2 \ln(\Lambda_0) + \frac{\|\mathbf{s}_1\|^2}{2} - \frac{\|\mathbf{s}_0\|^2}{2} - (\mathbf{s}_0, \mathbf{s}_1) + \|\mathbf{s}_0\|^2}{\sigma \|(\mathbf{s}_1 - \mathbf{s}_0)\|}\right) \\
&= \mathrm{Q}\left(\frac{\sigma^2 \ln(\Lambda_0) + \frac{\|\mathbf{s}_1\|^2}{2} - (\mathbf{s}_0, \mathbf{s}_1) + \frac{\|\mathbf{s}_0\|^2}{2}}{\sigma \|(\mathbf{s}_1 - \mathbf{s}_0)\|}\right) \\
&= \mathrm{Q}\left(\frac{\sigma^2 \ln(\Lambda_0) + \frac{\|\mathbf{s}_1 - \mathbf{s}_0\|^2}{2}}{\sigma \|(\mathbf{s}_1 - \mathbf{s}_0)\|}\right),
\end{aligned} \tag{4.19}$$

where $\mathrm{Q}(\cdot)$ is the complementary Gaussian distribution function of a zero mean and unit variance random variable. It is interesting to note that the first integral to the right of the equal sign of P_{FA} in (4.19) following from the definition of P_{FA} is an n-dimensional integral over the region of $\mathbb{R}_1$. However, by using the scalar-valued sufficient statistic Γ, all the subsequent integrals in (4.19) are one-dimensional integrals, leading to the integral defined by the Q(·) function. The fact that we can reduce a problem needing to perform an n-dimensional integral to an equivalent problem needing to perform only a one-dimensional integral is a very significant attribute of the usefulness of the concept of the sufficient statistic of Γ. If the LR constant $\Lambda_0 = 1$, then the above P_{FA} in (4.19) reduces to

$$P_{\text{FA}} = \text{Q}\left(\frac{\|(\mathbf{s}_1 - \mathbf{s}_0)\|}{2\sigma}\right). \tag{4.20}$$

Similarly, from (4.9) the probability of detection P_{D} (i.e., the probability of declaring hypothesis H_1 when H_1 is true) is given by

$$\begin{aligned}
P_{\text{D}} &= \int \cdots \int_{\mathbb{R}_1} p_1(\mathbf{x}) d\mathbf{x} = \int_{\Lambda_0}^{\infty} \tilde{p}_1(\Lambda) d\Lambda = \text{P}(\Gamma \geq \gamma_0 | H_1) \\
&= \frac{1}{\sqrt{2\pi}\sigma_1} \int_{\gamma_0}^{\infty} \exp\left(\frac{(\gamma - \mu_1)^2}{2\sigma_1^2}\right) d\gamma = \frac{1}{\sqrt{2\pi}} \int_{\frac{\gamma_0 - \mu_1}{\sigma_1}}^{\infty} \exp\left(\frac{-t^2}{2}\right) dt \\
&= \text{Q}\left(\frac{\gamma_0 - \mu_1}{\sigma_1}\right) = \text{Q}\left(\frac{\sigma^2 \ln(\Lambda_0) + \frac{\|\mathbf{s}_1\|^2}{2} - \frac{\|\mathbf{s}_0\|^2}{2} + (\mathbf{s}_0, \mathbf{s}_1) - \|\mathbf{s}_1\|^2}{\sigma \|(\mathbf{s}_1 - \mathbf{s}_0)\|}\right) \\
&= \text{Q}\left(\frac{\sigma^2 \ln(\Lambda_0) - \frac{\|\mathbf{s}_1\|^2}{2} + (\mathbf{s}_0, \mathbf{s}_1) - \frac{\|\mathbf{s}_0\|^2}{2}}{\sigma \|(\mathbf{s}_1 - \mathbf{s}_0)\|}\right) \\
&= \text{Q}\left(\frac{\sigma^2 \ln(\Lambda_0) - \frac{\|\mathbf{s}_1 - \mathbf{s}_0\|^2}{2}}{\sigma \|(\mathbf{s}_1 - \mathbf{s}_0)\|}\right).
\end{aligned} \tag{4.21}$$

Analogous to the P_{FA} case, the first integral to the right of the first equal sign of P_{D} in (4.21) is an n-dimensional integral, while all the subsequent integrals are all one-dimensional integrals using the sufficient statistic of Γ. If the LR constant $\Lambda_0 = 1$, then the above P_{D} in (4.21) becomes

$$P_{\text{D}} = \text{Q}\left(\frac{-\|\mathbf{s}_1 - \mathbf{s}_0\|}{2\sigma}\right) = 1 - P_{\text{FA}}. \tag{4.22}$$

In general, the average probability of error P_{e} defined as the average over the two kinds of errors obtained by weighting the probability of occurrence of each kind of error is given by

$$P_{\text{e}} = \pi_0 P_{\text{FA}} + (1 - \pi_0)(1 - P_{\text{D}}). \tag{4.23}$$

If the LR constant $\Lambda_0 = 1$, then the above P_{e} in (4.23) becomes

$$P_{\text{e}} = \pi_0 P_{\text{FA}} + (1 - \pi_0) P_{\text{FA}} = P_{\text{FA}} = 1 - P_{\text{D}}. \tag{4.24}$$

The P_{FA}, P_e, and P_D expressions in (4.19)–(4.24) are valid for any two arbitrary signal vectors $\mathbf{s}_1$ and $\mathbf{s}_0$ with any LR constant Λ_0. Equation (1.10) defined $\mathbf{s}_1^T\mathbf{s}_0 = (\mathbf{s}_1, \mathbf{s}_0)$ as the correlation between $\mathbf{s}_1$ and $\mathbf{s}_0$. From (4.16) then we can define the *normalized correlation* between $\mathbf{s}_1$ and $\mathbf{s}_0$ as

$$\rho = \frac{(\mathbf{s}_1, \mathbf{s}_0)}{\sqrt{E_1 E_0}}, \tag{4.25}$$

where the energy of $\mathbf{s}_0$ is denoted by $E_0 = \|\mathbf{s}_0\|^2$ and the energy of $\mathbf{s}_1$ is denoted by $E_1 = \|\mathbf{s}_1\|^2$. If the two vectors are orthogonal, then $(\mathbf{s}_1, \mathbf{s}_0) = 0$ and $\rho = 0$. If the two vectors are anti-podal, then $\mathbf{s}_1 = -\mathbf{s}_0$, then $\|\mathbf{s}_1\|^2 = \|\mathbf{s}_0\|^2 = E$ and (4.25) shows $\rho = -E/E = -1$. In general,

$$\begin{aligned}\|\mathbf{s}_1 - \mathbf{s}_0\|^2 &= (\mathbf{s}_1 - \mathbf{s}_0, \mathbf{s}_1 - \mathbf{s}_0) = \|\mathbf{s}_1\|^2 - 2(\mathbf{s}_1, \mathbf{s}_0) + \|\mathbf{s}_0\|^2 \\ &= E_1 - 2\rho\sqrt{E_1 E_0} + E_0. \end{aligned} \tag{4.26}$$

Then (4.19) becomes

$$P_{FA} = \mathrm{Q}\left(\frac{\sigma^2 \ln(\Lambda_0) + (E_1 - 2\rho\sqrt{E_1 E_0} + E_0)/2}{\sigma\sqrt{E_1 - 2\rho\sqrt{E_1 E_0} + E_0}}\right) \tag{4.27}$$

and with $\Lambda_0 = 1$, (4.20) becomes

$$P_{FA} = \mathrm{Q}\left(\frac{\sqrt{E_1 - 2\rho\sqrt{E_1 E_0} + E_0}}{2\sigma}\right). \tag{4.28}$$

Similarly, (4.21) now becomes

$$P_D = \mathrm{Q}\left(\frac{\sigma^2 \ln(\Lambda_0) - (E_1 - 2\rho\sqrt{E_1 E_0} + E_0)/2}{\sigma\sqrt{E_1 - 2\rho\sqrt{E_1 E_0} + E_0}}\right) \tag{4.29}$$

and with $\Lambda_0 = 1$, (4.22) becomes

$$P_D = \mathrm{Q}\left(\frac{-\sqrt{E_1 - 2\rho\sqrt{E_1 E_0} + E_0}}{2\sigma}\right). \tag{4.30}$$

Example 4.1 Now, relate the results on P_e of (4.28) to the P_e considered in Section 1.4, where we assumed there were full symmetries in the probabilities associated with H_0 and H_1, with $\mathrm{P}(H_1) = \mathrm{P}(H_0) = 1/2$ and LR constant of $\Lambda_0 = 1$. For two anti-podal signals with $\mathbf{s}_1 = -\mathbf{s}_0$, then $\|\mathbf{s}_1\|^2 = \|\mathbf{s}_0\|^2 = E$ and $\rho = -1$. We still define the SNR as given before in (1.18) with the expression of

$$\mathrm{SNR} = \frac{E}{n\sigma^2}. \tag{4.31}$$

Thus, for the two anti-podal signal vectors, from (4.28), we have

$$P_e = P_{FA} = \mathrm{Q}\left(\sqrt{\frac{4E}{(2\sigma)^2}}\right) = \mathrm{Q}\left(\sqrt{\frac{E}{\sigma^2}}\right) = \mathrm{Q}(\sqrt{n} \times \sqrt{\mathrm{SNR}}). \tag{4.32}$$

Indeed, since the two signal vectors $\mathbf{s}_1$ and $\mathbf{s}_0$ defined in (1.27) are anti-podal, it is not surprising that the P_e of (1.33) with $n = 5$ is identical to the P_e of (4.32) with $\sqrt{5} = 2.236$. □

Example 4.2 Consider again the use of the signal vectors $\mathbf{s}_1$ and $\mathbf{s}_0$ of (1.9), then (1.11) showed $\|\mathbf{s}_1\|^2 = \|\mathbf{s}_0\|^2 = E = 19$, while (1.12) showed $(\mathbf{s}_1, \mathbf{s}_0) = 1$ and now (4.25) shows a normalized correlation of $\rho = 1/19$. We still assume $\mathrm{P}(H_1) = \mathrm{P}(H_0) = 1/2$ with the LR constant of $\Lambda_0 = 1$. However, now we use the sufficient statistic of (4.12) to obtain an average probability of error from (4.28) to yield

$$P_e = P_{FA} = \mathrm{Q}\left(\sqrt{\frac{2(1-\rho)E}{(2\sigma)^2}}\right) = \mathrm{Q}\big(\sqrt{(9/19)n \times \mathrm{SNR}}\big) = \mathrm{Q}\big(1.539\sqrt{\mathrm{SNR}}\big). \tag{4.33}$$

However, the average probability of error obtained in (1.26), denoted now by P'_e, yields

$$P'_e = \mathrm{Q}(1.059\sqrt{\mathrm{SNR}}). \tag{4.34}$$

Since the $\mathrm{Q}(\cdot)$ is a monotonically decreasing function of its argument, thus clearly

$$P'_e = \mathrm{Q}\big(1.059\sqrt{\mathrm{SNR}}\big) > \mathrm{Q}\big(1.539\sqrt{\mathrm{SNR}}\big) = P_e. \tag{4.35}$$

But how is this possible? We are using the same signal vectors $\mathbf{s}_1$ and $\mathbf{s}_0$ of (1.9) with the same geometric property of having equal energies of $E = 19$ and the same normalized correlation of $\rho = 1/19$. Except now we use the sufficient statistic Γ of (4.12), obtained from the LR test, given by

$$\Gamma = \mathbf{X}^T(\mathbf{s}_1 - \mathbf{s}_0) = (\mathbf{s}_1 - \mathbf{s}_0)^T\mathbf{X}, \tag{4.36}$$

while back in (1.22), we used an ad hoc chosen statistic Γ', denoted by

$$\Gamma' = \mathbf{X}^T\mathbf{s}_1 = \mathbf{s}_1^T\mathbf{X}. \tag{4.37}$$

Since the sufficient statistic Γ obtained from the LR test is optimum in the Bayes criterion sense (which includes the minimum average error of probability criterion), it is not surprising that the performance (i.e., the average error of probability) obtained using the statistic Γ' chosen arbitrarily may be inferior (i.e., higher average error of probability). □

4.2 Detection of known binary signal waveforms in WGN

Consider the observed data waveform is given by

$$X(t) = \begin{cases} s_0(t) + N(t), & H_0, \\ s_1(t) + N(t), & H_1, \end{cases} \quad 0 \le t \le T, \tag{4.38}$$

where $\mathbf{s}_0$ and $\mathbf{s}_1$ are two real-valued known deterministic signals and $N(t)$ is a real-valued wide-sense stationary Gaussian random process with its mean given by

$$\mu(t) = \mathrm{E}\{N(t)\} = 0, \tag{4.39}$$

and its autocorrelation function $R(\tau)$ and power spectral density function $S(f)$ are given by

$$R(\tau) = \mathrm{E}\{N(t)N(t+\tau)\} = \frac{N_0}{2}\delta(\tau), \; -\infty < \tau < \infty, \tag{4.40}$$

$$S(f) = \frac{N_0}{2}, \; -\infty < f < \infty. \tag{4.41}$$

A noise process $N(t)$ satisfying (4.39)–(4.41) is said to be a continuous-time white noise process. Since $N(t)$ is also Gaussian distributed, it is called a *white Gaussian noise* (WGN) process.

Now, consider an approach based on the limit of sampled waveforms. While a WGN $N(t)$ process has a constant power spectral density function $S(f)$ satisfying (4.41), let us first consider a bandlimited WGN $X_{\mathrm{BL}}(t)$ process whose power spectral density $S_{\mathrm{BL}}(t)$ is bandlimited to the frequency band $[-f_0, \; f_0]$ satisfying

$$S_{\mathrm{BL}}(f) = \begin{cases} \frac{N_0}{2}, & -f_0 < f < f_0, \\ 0, & \text{elsewhere}. \end{cases} \tag{4.42}$$

Then its autocorrelation function $R_{\mathrm{BL}}(\tau)$ is given by

$$\begin{aligned} R_{\mathrm{BL}}(\tau) &= \frac{N_0}{2}\int_{-f_0}^{f_0} \exp(i2\pi f\tau)df = \frac{N_0[\exp(i2\pi f_0\tau) - \exp(-i2\pi f_0\tau)]}{i4\pi\tau} \\ &= \frac{N_0 \sin(2\pi f_0 \tau)}{2\pi\tau}, \; -\infty < \tau < \infty. \end{aligned} \tag{4.43}$$

We note in (4.43), for

$$\tau = m/(2f_0), \; m = \pm 1, \; \pm 2, \; \ldots, \tag{4.44}$$

$$\sin(2\pi f_0(m/(2f_0))) = \sin(m\pi) = 0, \tag{4.45}$$

$$R_{\mathrm{BL}}(m/(2f_0)) = 0. \tag{4.46}$$

$\Delta\tau = 1/(2f_0)$ is called the *Nyquist sampling interval* and $2f_0$ is called the *Nyquist sampling rate* for a bandlimited waveform bandlimited to $[-f_0, \; f_0]$. We also note that

$$R_{\mathrm{BL}}(0) = \left.\frac{N_0 \sin(2\pi f_0\tau)}{2\pi\tau}\right|_{\tau=0} = \frac{N_0 2\pi f_0}{2\pi} = N_0 f_0 = \sigma^2 \tag{4.47}$$

is the variance of the wide-sense stationary bandlimited WGN $X_{\mathrm{BL}}(t)$ process denoted by σ^2.

Over the observation interval $[0, \; T]$, consider partitioning it into n uniform subintervals satisfying

$$n\Delta\tau = n/(2f_0) = T. \tag{4.48}$$

Now, let the bandwidth $f_0 \to \infty$ and the number of partitions $n \to \infty$ in (4.48), subject to the constraint of

$$\lim_{f_0\to\infty,\, n\to\infty} \{n/(2f_0)\} = T. \tag{4.49}$$

Then the integral on the r.h.s. of (4.50) is defined by the Riemann integration as the limit of the finite sum on the l.h.s. of (4.50).

$$\lim_{f_0\to\infty,\, n\to\infty} \left\{ \sum_{i=1}^{n} x_i(s_{1i} - s_{0i})\Delta\tau \right\} = \int_0^T x(\tau)(s_1(\tau) - s_0(\tau))d\tau. \tag{4.50}$$

But from (4.48), we defined the sufficient statistic γ by

$$\gamma = \mathbf{x}^T(\mathbf{s}_1 - \mathbf{s}_0) = \sum_{i=1}^{n} x_i(s_{1i} - s_{0i}). \tag{4.51}$$

Then using (4.51), (4.50) can be rewritten as

$$\lim_{f_0\to\infty,\, n\to\infty} \{\gamma\Delta\tau\} = \int_0^T x(\tau)(s_1(\tau) - s_0(\tau))d\tau. \tag{4.52}$$

But from (4.9), we defined γ_0 by

$$\gamma_0 = \sigma^2 \ln(\Lambda_0) + \frac{\mathbf{s}_1^T\mathbf{s}_1}{2} - \frac{\mathbf{s}_0^T\mathbf{s}_0}{2} = N_0 f_0 \ln(\Lambda_0) + \frac{\sum_{i=1}^{n} s_{1i}^2}{2} - \frac{\sum_{i=1}^{n} s_{0i}^2}{2}. \tag{4.53}$$

Then taking the limit of $\{\gamma_0\Delta\tau\}$ yields

$$\begin{aligned} \lim_{f_0\to\infty,\, n\to\infty} \{\gamma_0\Delta\tau\} &= \lim_{f_0\to\infty,\, n\to\infty} \left\{ N_0 f_0 \ln(\Lambda_0)\frac{1}{2f_0} + \left[\frac{\sum_{i=1}^{n}\left(s_{1i}^2 - s_{0i}^2\right)}{2}\right]\Delta\tau \right\} \\ &= \frac{N_0 \ln(\Lambda_0)}{2} + \frac{\int_0^T \left(s_1^2(\tau) - s_0^2(\tau)\right) d\tau}{2}. \end{aligned} \tag{4.54}$$

By taking the limit on the decision region $R_1 = \{\mathbf{x} : \gamma \geq \gamma_0\}$ given by the last expression in (4.9), we now have

$$\begin{aligned} \mathbb{R}_1 &= \left\{ x(t) : \lim_{f_0\to\infty,\, n\to\infty} \{\gamma\Delta\tau\} \geq \lim_{f_0\to\infty,\, n\to\infty} \{\gamma_0\Delta\tau\} \right\} \\ &= \left\{ x(t) : \int_0^T x(\tau)(s_1(\tau) - s_0(\tau))d\tau \geq \frac{N_0 \ln\Lambda_0 + \int_0^T \left(s_1^2(\tau) - s_0^2(\tau)\right) d\tau}{2} \right\}. \end{aligned} \tag{4.55}$$

From (4.55) we can define the sufficient statistic γ^c (corresponding to γ of the vector data case of (4.11)) for the continuous-time waveform binary hypothesis testing by

$$\gamma^c = \int_0^T x(\tau)(s_1(\tau) - s_0(\tau))d\tau, \tag{4.56}$$

and the threshold γ_0^c (corresponding to γ_0 of the vector data case in (4.53)) given by

$$\gamma_0^c = \frac{N_0 \ln(\Lambda_0)}{2} + \frac{\int_0^T s_1^2(\tau)d\tau}{2} - \frac{\int_0^T s_0^2(\tau)d\tau}{2}. \tag{4.57}$$

Then the decision region $\mathbb{R}_1$ of (4.55) can be expressed by

$$\mathbb{R}_1 = \{x(t) : \gamma^c \geq \gamma_0^c\}, \tag{4.58}$$

and the decision region $\mathbb{R}_0$ is given by

$$\mathbb{R}_0 = \{x(t) : \gamma^c \leq \gamma_0^c\}. \tag{4.59}$$

Similar to the vector case, we can define the *inner product* of two real-valued continuous-time waveforms $s_0(t)$ and $s_1(t)$ by

$$(s_1(t), s_0(t)) = \int_0^T s_1(t)s_0(t)dt = \int_0^T s_0(t)s_1(t)dt = (s_0(t), s_1(t)). \tag{4.60}$$

These waveforms are considered to be vectors in the infinite-dimensional space of finite square integral functions over $[0, T]$. We note that this inner product of the two functions is still commutative. Furthermore, if the two functions $s_1(t)$ are identical, then the square of the *norm* of $s_1(t)$ is defined by

$$(s_1(t), s_1(t)) = \int_0^T s_1^2(t)dt = \|s_1(t)\|^2. \tag{4.61}$$

If $s_1(t)$ is in the unit of voltage, then $s_i^2(t)$ has a unit of power, and its integral over $[0, T]$ is the square of the norm $\|s_i(t)\|^2$ and can be interpreted as the energy of $s_1(t)$ over $[0, T]$. Using the notations of inner product and norm, (4.56) and (4.57) can be expressed as

$$\gamma^c = (x(t), s_1(t) - s_0(t)), \tag{4.62}$$

$$\gamma_0^c = \frac{N_0 \ln(\Lambda_0)}{2} + \frac{\int_0^T s_1^2(\tau)d\tau}{2} - \frac{\int_0^T s_0^2(\tau)d\tau}{2}. \tag{4.63}$$

The performance of the continuous-time waveform binary hypothesis receiver can be obtained from the P_{FA} and the P_{D}. We note that since $X(t)$ is a Gaussian random process and the inner product (i.e., an integral) is a linear operation of $X(t)$ with $s_1(t) - s_0(t)$ in (4.62), then Γ^c is a scalar Gaussian random variable. Thus, similar to the vector data case, we only need to find the mean and the variance of Γ^c. Specifically, the means of Γ^c under H_0 and H_1 are given by

$$\begin{aligned}\mu_0 &= \mathrm{E}\{\Gamma^c | H_0\} = \mathrm{E}\{((s_0(t) + N(t)), (s_1(t) - s_0(t)))\} \\ &= (s_0(t), (s_1(t) - s_0(t))),\end{aligned} \tag{4.64}$$

$$\mu_1 = (s_1(t), (s_1(t) - s_0(t))). \tag{4.65}$$

The variances of Γ^c under H_0 and H_1 are given by

$$\begin{aligned}\sigma_0^2 &= \mathrm{E}\{[\Gamma^c - \mu_0]^2 | H_0\} = \mathrm{E}\{(N(t), (s_1(t) - s_0(t)))^2\} \\ &= \int_0^T \int_0^T \mathrm{E}\{N(t)N(\tau)\} [s_1(t) - s_0(t)] [s_1(\tau) - s_0(\tau)]\, dt d\tau \\ &= \frac{N_0}{2} \int_0^T \int_0^T \delta(t - \tau) [s_1(t) - s_0(t)] [s_1(\tau) - s_0(\tau)]\, dt d\tau = \sigma_1^2.\end{aligned} \tag{4.66}$$

The probability of false alarm P_{FA} is given by

$$P_{\text{FA}} = \frac{1}{\sqrt{2\pi}\sigma_0}\int_{\gamma_0^c}^{\infty} \exp\left(\frac{(\gamma^c-\mu_0)^2}{2\sigma_0^2}\right) d\gamma^c = \frac{1}{\sqrt{2\pi}}\int_{\frac{\gamma_0^c-\mu_0}{\sigma_0}}^{\infty} \exp\left(\frac{-t^2}{2}\right) dt$$

$$= \mathrm{Q}\left(\frac{\gamma_0^c-\mu_0}{\sigma_0}\right) = \mathrm{Q}\left(\frac{\frac{N_0}{2}\ln(\Lambda_0) + \frac{\|s_1(t)-s_0(t)\|^2}{2}}{\sqrt{\frac{N_0}{2}}\|s_1(t)-s_0(t)\|}\right). \tag{4.67}$$

If $\Lambda_0 = 1$, then (4.67) reduces to

$$P_{\text{FA}} = \mathrm{Q}\left(\frac{\|s_1(t)-s_0(t)\|}{\sqrt{2N_0}}\right). \tag{4.68}$$

The probability of detection P_{D} is given by

$$P_{\text{D}} = \frac{1}{\sqrt{2\pi}\sigma_1}\int_{\gamma_0^c}^{\infty} \exp\left(\frac{(\gamma^c-\mu_1)^2}{2\sigma_1^2}\right) d\gamma^c = \frac{1}{\sqrt{2\pi}}\int_{\frac{\gamma_0^c-\mu_1}{\sigma_1}}^{\infty} \exp\left(\frac{-t^2}{2}\right) dt$$

$$= \mathrm{Q}\left(\frac{\gamma_0^c-\mu_1}{\sigma_1}\right) = \mathrm{Q}\left(\frac{\frac{N_0}{2}\ln(\Lambda_0) - \frac{\|s_1(t)-s_0(t)\|^2}{2}}{\sqrt{\frac{N_0}{2}}\|s_1(t)-s_0(t)\|}\right). \tag{4.69}$$

If $\Lambda_0 = 1$, then (4.69) reduces to

$$P_{\text{D}} = \mathrm{Q}\left(\frac{-\|s_1(t)-s_0(t)\|}{\sqrt{2N_0}}\right). \tag{4.70}$$

In general, the average probability of an error P_{e} is given by

$$P_{\text{e}} = \pi_0 P_{\text{FA}} + (1-\pi_0)(1-P_{\text{D}}). \tag{4.71}$$

If $\Lambda_0 = 1$, then (4.71) reduces to

$$P_{\text{e}} = \pi_0 P_{\text{FA}} + (1-\pi_0)P_{\text{FA}} = P_{\text{FA}} = 1 - P_{\text{D}}. \tag{4.72}$$

Analogous to (4.25) and (4.26), we can define the normalized correlation ρ by

$$\rho = \frac{(s_1(t), s_0(t))}{\sqrt{E_1 E_0}}, \tag{4.73}$$

where the two energies are denoted by $E_0 = \|s_0(t)\|^2$ and $E_1 = \|s_1(t)\|^2$, and

$$\begin{aligned}\|s_1(t)-s_0(t)\|^2 &= (s_1(t)-s_0(t), s_1(t)-s_0(t)) \\ &= \|s_1(t)\|^2 - 2(s_1(t),\ s_0(t)) + \|s_0(t)\|^2 \\ &= E_1 - 2\rho\sqrt{E_1E_0} + E_0.\end{aligned} \tag{4.74}$$

Then (4.67) becomes

$$P_{\text{FA}} = \mathrm{Q}\left(\frac{\frac{N_0}{2}\ln(\Lambda_0) + (E_1 - 2\rho\sqrt{E_1E_0} + E_0)/2}{\sqrt{\frac{N_0}{2}(E_1 - 2\rho\sqrt{E_1E_0} + E_0)}}\right) \tag{4.75}$$

and with $\Lambda_0 = 1$, (4.68) becomes

$$P_{\text{FA}} = \text{Q}\left(\sqrt{\frac{E_1 - 2\rho\sqrt{E_1 E_0} + E_0}{2N_0}}\right). \tag{4.76}$$

Similarly, (4.69) becomes

$$P_{\text{D}} = \text{Q}\left(\frac{\frac{N_0}{2}\ln(\Lambda_0) - (E_1 - 2\rho\sqrt{E_1 E_0} + E_0)/2}{\sqrt{\frac{N_0}{2}(E_1 - 2\rho\sqrt{E_1 E_0} + E_0)}}\right), \tag{4.77}$$

and with $\Lambda_0 = 1$, (4.70) becomes

$$P_{\text{D}} = \text{Q}\left(-\sqrt{\frac{E_1 - 2\rho\sqrt{E_1 E_0} + E_0}{2N_0}}\right). \tag{4.78}$$

Example 4.3 Consider the two anti-podal sinusoidal signal waveforms of

$$s_0(t) = A\sin(2\pi f_0 t),\ 0 \le t \le T, \tag{4.79}$$

$$s_1(t) = -A\sin(2\pi f_0 t),\ 0 \le t \le T, \tag{4.80}$$

with equal energy E of both signals given by

$$\begin{aligned} E &= A^2 \int_0^T \sin^2(2\pi f_0 t)dt = \frac{A^2}{2}\int_0^T (1 - \cos(4\pi f_0 t))dt \\ &= \frac{A^2}{2}\left(T - \frac{\sin(4\pi f_0 T)}{4\pi f_0}\right) \approx \frac{A^2 T}{2}. \end{aligned} \tag{4.81}$$

The second term in the last bracketed expression of (4.81) is dropped under the condition of a large "time-bandwidth product" of $\text{f}_0\text{T} \gg 1$, which is equivalent to requiring that there are many periods of the sinusoids in the interval T. This condition is normally satisfied in almost all practical applications. From (4.73), the normalized correlation of the two signals yields $\rho = -1$. Assuming $\text{P}(H_0) = \text{P}(H_1) = 1/2$ and $\Lambda_0 = 1$, the average error of probability from (4.76) yields

$$P_{\text{e}} = \text{Q}\left(\sqrt{\frac{2E}{N_0}}\right) = \text{Q}(\sqrt{2\text{SNR}}), \tag{4.82}$$

and the probability of detection from (4.78) yields

$$P_{\text{D}} = \text{Q}\left(-\sqrt{\frac{2E}{N_0}}\right) = \text{Q}(-\sqrt{2\text{SNR}}), \tag{4.83}$$

where the SNR of a binary deterministic waveform of equal energy E in an AWGN channel with a two-sided noise spectral density of $N_0/2$ is denoted by

$$\text{SNR} = \frac{E}{N_0}. \tag{4.84}$$

This modulation system is called the binary-phase-shift-keyed (BPSK) system. □

Example 4.4 Consider the two orthogonal sinusoidal signal waveforms of

$$s_0(t) = A\sin(2\pi f_0 t),\ 0 \le t \le T, \tag{4.85}$$

$$s_1(t) = A\cos(2\pi f_0 t),\ 0 \le t \le T, \tag{4.86}$$

with equal energies E given by

$$E = \int_0^T s_0^2(t)dt = \int_0^T s_1^2(t)dt \approx \frac{A^2 T}{2}. \tag{4.87}$$

Then the normalized correlation function of these two waveforms is given by

$$\begin{aligned}\rho &= \frac{A^2 \int_0^T \sin(2\pi f_0 t)\cos(2\pi f_0 t)dt}{A^2 T/2} \\ &= \frac{2}{T}\int_0^T (1/2)\left(\sin(2\pi f_0 t + 2\pi f_0 t) + \sin(2\pi f_0 t - 2\pi f_0 t)\right)dt \\ &= \frac{1}{T}\int_0^T (\sin(4\pi f_0 t) + \sin(0))dt = \frac{1-\cos(4\pi f_0 T)}{4\pi f_0 T}.\end{aligned} \tag{4.88}$$

Under the large "time-bandwidth product" condition of $f_0 T \gg 1$, we have $\rho \approx 0$. Assuming $\mathrm{P}(H_0) = \mathrm{P}(H_1) = 1/2$ and $\Lambda_0 = 1$, the average error of probability from (4.76) yields

$$P_e = \mathrm{Q}\left(\sqrt{\frac{E}{N_0}}\right) = \mathrm{Q}(\sqrt{\mathrm{SNR}}), \tag{4.89}$$

and the probability of detection from (4.78) yields

$$P_\mathrm{D} = \mathrm{Q}\left(-\sqrt{\frac{E}{N_0}}\right) = \mathrm{Q}(-\sqrt{\mathrm{SNR}}). \tag{4.90}$$

The modulation system with the two orthogonal signal waveforms of (4.85) and (4.86) is called a binary frequency-shift-keyed (BFSK) system. By comparing (4.89) with (4.82) and (4.90) with (4.83), we see that a BFSK system requires having twice the SNR value to achieve the same P_e and P_D performances as that of a BPSK system. □

4.3 Detection of known deterministic binary signal vectors in colored Gaussian noise

Let $\{N_i,\ -\infty < i < \infty\}$ be a real-valued zero mean Gaussian wide-sense stationary random noise sequence. Let $\mathbf{N} = [N_1, \ldots, N_n]^T$ be a $n \times 1$ random vector formed from this sequence. Let its autocorrelation matrix be denoted by

$$\mathbf{R} = \mathrm{E}\{\mathbf{N}\mathbf{N}^T\}. \tag{4.91}$$

$\mathbf{N}$ is said to be a *colored Gaussian noise* (CGN) vector if $\mathbf{R}$ is an arbitrary non-negative-definite symmetric matrix. $\mathbf{N}$ is said to be a *white Gaussian noise* (WGN) vector if $\mathbf{R} = \sigma^2 \mathbf{I}_n$. Clearly, a WGN vector is a special case of the CGN vector.

A $n \times n$ symmetric matrix $\mathbf{R}$ is defined to be *positive semi-definite* if and only if for any arbitrary deterministic $n \times 1$ vector $\mathbf{y}$, the quadratic form satisfies

$$0 \leq \mathbf{y}^T \mathbf{R} \mathbf{y}, \text{ except } \mathbf{y} = \mathbf{0}. \tag{4.92}$$

The above matrix $\mathbf{R}$ is said to be *positive-definite*, if the $\leq$ term in (4.92) is replaced by the $<$ term.

Lemma 4.1 *Consider an arbitrary $n \times 1$ random vector $\mathbf{X} = [X_1, \ldots, X_n]^T$. Its autocorrelation matrix*

$$\mathbf{R} = \mathrm{E}\{\mathbf{X}\mathbf{X}^T\} \tag{4.93}$$

is positive semi-definite.

Proof: *Consider an arbitrary deterministic $n \times 1$ vector $\mathbf{y}$ and form the scalar random variable $S = \mathbf{y}^T \mathbf{X}$. Then the random variable S satisfies*

$$0 \leq \mathrm{E}\{S^2\} = \mathrm{E}\{SS^T\} = \mathrm{E}\{\mathbf{y}^T \mathbf{X}\mathbf{X}^T \mathbf{y}\} = \mathbf{y}^T \mathrm{E}\{\mathbf{X}\mathbf{X}^T\}\mathbf{y} = \mathbf{y}^T \mathbf{R} \mathbf{y} \tag{4.94}$$

and meets the condition of (4.92). This shows $\mathbf{R}$ is positive semi-definite. □

In practice, if $0 < \mathrm{E}\{S^2\}$, then the $n \times n$ autocorrelation matrix is positive-definite. If $\mathrm{X} = [X_1, \ldots, X_n]^T$ has a mean vector $\boldsymbol{\mu} = [\mu_1, \ldots, \mu_n]^T$, its covariance matrix $\boldsymbol{\Lambda} = \mathrm{E}\{(\mathbf{X} - \boldsymbol{\mu})(\mathbf{X} - \boldsymbol{\mu})^T\}$ is also positive semi-definite. Thus, for a zero mean random vector, its autocorrelation matrix, denoted by $\mathbf{R}$, is equivalent to its covariance matrix, denoted by $\boldsymbol{\Lambda}$. The proof of Lemma 4.1 can also be applied to the positive semi-definiteness and positive-definiteness of the covariance matrix. In practice, most real-life measured noise processes can be assumed to have zero means and certainly can be assumed to be measured not at zero degrees kelvin (see Section 1.5). Thus, their covariance matrices can be modeled as positive-definite.

Consider the observed data vector $\mathbf{X}$ given by

$$\mathbf{X} = \begin{cases} \mathbf{s}_0 + \mathbf{N}, & H_0, \\ \mathbf{s}_1 + \mathbf{N}, & H_1, \end{cases} \tag{4.95}$$

where

$$\mathbf{X} = [X_1, \ldots, X_n]^T, \ \mathbf{s}_0 = [s_{01}, \ldots, s_{0n}]^T,$$
$$\mathbf{s}_1 = [s_{11}, \ldots, s_{1n}]^T, \ \mathbf{N} = [N_1, \ldots, N_n]^T, \tag{4.96}$$

with $\mathbf{s}_0$ and $\mathbf{s}_1$ being two known $n \times 1$ deterministic signal vectors and $\mathbf{N}$ is a $n \times 1$ Gaussian noise vector with the mean vector given by

$$\mathrm{E}\{\mathbf{N}\} = \mathbf{0}, \tag{4.97}$$

and an arbitrary $n \times n$ positive-definite covariance matrix $\boldsymbol{\Lambda} = \mathrm{E}\{\mathbf{N}\mathbf{N}^T\}$. Since the pdf of $\mathbf{N}$ is given by

$$p_{\mathbf{N}}(\mathbf{n}) = \frac{1}{(2\pi)^{n/2} |\boldsymbol{\Lambda}|^{1/2}} \exp\left(-\frac{\mathbf{n}^T \boldsymbol{\Lambda}^{-1} \mathbf{n}}{2} \right), \tag{4.98}$$

then the pdf of $\mathbf{X}$ under H_0 is given by

$$p_0(\mathbf{x}) = p_{\mathbf{N}}(\mathbf{x} - \mathbf{s}_0) = \frac{1}{(2\pi)^{n/2}|\boldsymbol{\Lambda}|^{1/2}} \exp\left(-\frac{(\mathbf{x} - \mathbf{s}_0)^T \boldsymbol{\Lambda}^{-1} (\mathbf{x} - \mathbf{s}_0)}{2} \right), \tag{4.99}$$

and the pdf of $\mathbf{X}$ under H_1 is given by

$$p_1(\mathbf{x}) = p_{\mathbf{N}}(\mathbf{x} - \mathbf{s}_1) = \frac{1}{(2\pi)^{n/2}|\boldsymbol{\Lambda}|^{1/2}} \exp\left(-\frac{(\mathbf{x} - \mathbf{s}_1)^T \boldsymbol{\Lambda}^{-1} (\mathbf{x} - \mathbf{s}_1)}{2} \right). \tag{4.100}$$

The LR function is given by

$$\Lambda(\mathbf{x}) = \frac{p_1(\mathbf{x})}{p_0(\mathbf{x})} = \exp\left(-\frac{1}{2} \left[(\mathbf{x} - \mathbf{s}_1)^T \boldsymbol{\Lambda}^{-1} (\mathbf{x} - \mathbf{s}_1) - (\mathbf{x} - \mathbf{s}_0)^T \boldsymbol{\Lambda}^{-1} (\mathbf{x} - \mathbf{s}_0) \right] \right) \tag{4.101}$$

and

$$\ln(\Lambda(\mathbf{x})) = \mathbf{x}^T \boldsymbol{\Lambda}^{-1} (\mathbf{s}_1 - \mathbf{s}_0) + \frac{1}{2} \left(\mathbf{s}_0^T \boldsymbol{\Lambda}^{-1} \mathbf{s}_0 - \mathbf{s}_1^T \boldsymbol{\Lambda}^{-1} \mathbf{s}_1 \right). \tag{4.102}$$

Then decision region $\mathbb{R}_1$ is given by

$$\mathbb{R}_1 = \{\mathbf{x} : \ \Lambda(\mathbf{x}) \geq \Lambda_0\} = \{\mathbf{x} : \ \ln(\Lambda(\mathbf{x})) \geq \ln(\Lambda_0)\} = \{\mathbf{x} : \ \gamma \geq \gamma_0\}, \tag{4.103}$$

where the sufficient statistic γ for the CGN binary detection is now given by

$$\gamma = \mathbf{x}^T \boldsymbol{\Lambda}^{-1} (\mathbf{s}_1 - \mathbf{s}_0) \tag{4.104}$$

and the threshold is given by

$$\gamma_0 = \ln(\Lambda_0) + \frac{1}{2} \left(\mathbf{s}_1^T \boldsymbol{\Lambda}^{-1} \mathbf{s}_1 - \mathbf{s}_0^T \boldsymbol{\Lambda}^{-1} \mathbf{s}_0 \right). \tag{4.105}$$

Decision region $\mathbb{R}_0$ is then given by

$$\mathbb{R}_0 = \{\mathbf{x} : \ \Lambda(\mathbf{x}) \leq \Lambda_0\} = \{\mathbf{x} : \ \ln(\Lambda(\mathbf{x})) \leq \ln(\Lambda_0)\} = \{\mathbf{x} : \ \gamma \leq \gamma_0\}. \tag{4.106}$$

From (4.104), we know

$$\Gamma = \mathbf{X}^T \boldsymbol{\Lambda}^{-1} (\mathbf{s}_1 - \mathbf{s}_0) \tag{4.107}$$

is a Gaussian random variable with different means under H_0 and H_1. The conditional means of Γ under H_0 and H_1 are given by

$$\mu_0 = \mathrm{E}\{\Gamma | H_0\} = \mathrm{E}\{\mathbf{X}^T \boldsymbol{\Lambda}^{-1} (\mathbf{s}_1 - \mathbf{s}_0) | H_0\} = \mathbf{s}_0^T \boldsymbol{\Lambda}^{-1} (\mathbf{s}_1 - \mathbf{s}_0), \tag{4.108}$$

$$\mu_1 = \mathrm{E}\{\Gamma | H_1\} = \mathrm{E}\{\mathbf{X}^T \boldsymbol{\Lambda}^{-1} (\mathbf{s}_1 - \mathbf{s}_0) | H_1\} = \mathbf{s}_1^T \boldsymbol{\Lambda}^{-1} (\mathbf{s}_1 - \mathbf{s}_0). \tag{4.109}$$

The variances of Γ under H_0 and H_1 are identical and are given by

$$\begin{aligned} \sigma_0^2 &= \mathrm{E}\{[\Gamma - \mu_0]^2 | H_0\} = \mathrm{E}\{(\mathbf{s}_1 - \mathbf{s}_0)^T \boldsymbol{\Lambda}^{-1} \mathbf{N}\mathbf{N}^T \boldsymbol{\Lambda}^{-1} (\mathbf{s}_1 - \mathbf{s}_0)\} \\ &= (\mathbf{s}_1 - \mathbf{s}_0)^T \boldsymbol{\Lambda}^{-1} \boldsymbol{\Lambda} \boldsymbol{\Lambda}^{-1} (\mathbf{s}_1 - \mathbf{s}_0) = (\mathbf{s}_1 - \mathbf{s}_0)^T \boldsymbol{\Lambda}^{-1} (\mathbf{s}_1 - \mathbf{s}_0) = \sigma_1^2. \end{aligned} \tag{4.110}$$

Then the probability of a false alarm is given by

$$P_{\mathrm{FA}} = \mathrm{P}(\Gamma \geq \gamma_0 | H_0) = \frac{1}{\sqrt{2\pi}\sigma_0} \int_{\gamma_0}^{\infty} \exp\left(\frac{(\gamma - \mu_0)^2}{2\sigma_0^2}\right) d\gamma = \mathrm{Q}\left(\frac{\gamma_0 - \mu_0}{\sigma_0}\right)$$

$$= \mathrm{Q}\left(\frac{\ln(\Lambda_0) + \frac{1}{2}(\mathbf{s}_1 - \mathbf{s}_0)^T \mathbf{\Lambda}^{-1}(\mathbf{s}_1 - \mathbf{s}_0)}{\sqrt{(\mathbf{s}_1 - \mathbf{s}_0)^T \mathbf{\Lambda}^{-1}(\mathbf{s}_1 - \mathbf{s}_0)}}\right) \quad (4.111)$$

and the probability of detection P_{D} is given by

$$P_{\mathrm{D}} = \mathrm{P}(\Gamma \geq \gamma_0 | H_1) = \frac{1}{\sqrt{2\pi}\sigma_1} \int_{\gamma_0}^{\infty} \exp\left(\frac{(\gamma - \mu_1)^2}{2\sigma_1^2}\right) d\gamma = \mathrm{Q}\left(\frac{\gamma_0 - \mu_1}{\sigma_1}\right)$$

$$= \mathrm{Q}\left(\frac{\ln(\Lambda_0) - \frac{1}{2}(\mathbf{s}_1 - \mathbf{s}_0)^T \mathbf{\Lambda}^{-1}(\mathbf{s}_1 - \mathbf{s}_0)}{\sqrt{(\mathbf{s}_1 - \mathbf{s}_0)\mathbf{\Lambda}^{-1}(\mathbf{s}_1 - \mathbf{s}_0)}}\right), \quad (4.112)$$

and the average probability of error is given by

$$P_{\mathrm{e}} = \pi_0 P_{\mathrm{FA}} + (1 - \pi_0)(1 - P_{\mathrm{D}}). \quad (4.113)$$

In Section 4.4, we consider a whitening filter interpretation of the detection of two deterministic signal vectors in CGN.

4.4 Whitening filter interpretation of the CGN detector

In the detection of binary known deterministic signal vectors in CGN as modeled by (4.95)

$$\mathbf{X} = \begin{cases} \mathbf{s}_0 + \mathbf{N}, & H_0, \\ \mathbf{s}_1 + \mathbf{N}, & H_1, \end{cases}$$

if $\mathbf{N}$ is a WGN vector, then we know from (4.11) the sufficient statistic for binary detection is given by

$$\gamma_{\mathrm{WGN}} = \mathbf{x}^T(\mathbf{s}_1 - \mathbf{s}_0) = (\mathbf{x}, \mathbf{s}_1 - \mathbf{s}_0). \quad (4.114)$$

If $\mathbf{N}$ is a CGN vector then we know from (4.104) the sufficient statistic for binary detection is given by

$$\gamma_{\mathrm{CGN}} = \mathbf{x}^T \mathbf{\Lambda}^{-1}(\mathbf{s}_1 - \mathbf{s}_0). \quad (4.115)$$

Indeed, if $\mathbf{\Lambda} = \sigma^2 \mathbf{I}_n$, then

$$\gamma_{\mathrm{CGN}}|_{\mathbf{\Lambda}^{-1}=\mathbf{I}_n/\sigma^2} = \mathbf{x}^T \mathbf{\Lambda}^{-1}(\mathbf{s}_1 - \mathbf{s}_0)\big|_{\mathbf{\Lambda}^{-1}=\mathbf{I}_n/\sigma^2} = \gamma_{\mathrm{WGN}}/\sigma^2.$$

Thus, (4.114) is clearly a special case of (4.115). We note, γ_{WGN} normalized by the positive constant σ^2 does not affect its sufficient statistic property for binary detection of signals in WGN.

In (4.115), we can rewrite it as

$$\gamma_{\text{CGN}} = \mathbf{x}^T \boldsymbol{\Lambda}^{-1}(\mathbf{s}_1 - \mathbf{s}_0) = \mathbf{x}^T \mathbf{r} = (\mathbf{x}, \mathbf{r}) = \sum_{i=1}^{n} x_i r_i$$
$$= \mathbf{x}^T \boldsymbol{\Lambda}^{-1/2} \boldsymbol{\Lambda}^{-1/2}(\mathbf{s}_1 - \mathbf{s}_0) = (\mathbf{z}, \mathbf{u}) = \mathbf{z}^T \mathbf{u}, \tag{4.116}$$

where

$$\mathbf{r} = [r_1, \ldots, r_n]^T = \boldsymbol{\Lambda}^{-1}(\mathbf{s}_1 - \mathbf{s}_0), \tag{4.117}$$

$$\mathbf{z} = \boldsymbol{\Lambda}^{-1/2}\mathbf{x} \ \ (\text{or } \mathbf{z}^T = \mathbf{x}^T \boldsymbol{\Lambda}^{-1/2}), \tag{4.118}$$

$$\mathbf{u} = \boldsymbol{\Lambda}^{-1/2}(\mathbf{s}_1 - \mathbf{s}_0). \tag{4.119}$$

Under WGN, we use the received data vector $\mathbf{x}$ to perform an inner product with the known $(\mathbf{s}_1 - \mathbf{s}_0)$ vector. But under CGN, as shown in the last equation of (4.116), we do not use $\mathbf{x}$ but instead use a transformed vector $\mathbf{z}$, to perform an inner product with the vector $\mathbf{u}$. From (4.118) and (4.119), we note we need to transform both the $\mathbf{x}$ vector and the $(\mathbf{s}_1 - \mathbf{s}_0)$ vector by the matrix $\boldsymbol{\Lambda}^{-1/2}$. $\boldsymbol{\Lambda}^{-1/2}$ is called the *whitening filter* (or whitening matrix) for the detection of binary known deterministic signal vectors in CGN. But what is $\boldsymbol{\Lambda}^{-1/2}$?

First, let us briefly review the concept of the square root of a real number. Given a positive real number, say 4, then $4^{1/2} = \pm 2$ and the square of $4^{1/2}$ yields $(4^{1/2})(4^{1/2}) = (2)(2) = (-2)(-2) = 4$. Given a negative real number, say -3, then $(-3)^{1/2} = \pm i\sqrt{3}$ and the square of $(-3)^{1/2}$ yields $(-3)^{1/2}(-3)^{1/2} = (i\sqrt{3})(i\sqrt{3}) = (-i\sqrt{3})(-i\sqrt{3}) = -3$. Thus, the square root of a real number is clear. Given a complex number $z = r\exp(i\theta)$, $r \geq 0$, $0 \leq \theta < 2\pi$, then its square root $z^{1/2} = r^{1/2}\exp(i\theta/2)$ satisfies $z^{1/2}z^{1/2} = (r^{1/2}\exp(i\theta/2))(r^{1/2}\exp(i\theta/2)) = r\exp(i\theta) = z$. Thus, the square root of a complex number is also clear. Now, given an $n \times n$ matrix $\mathbf{A}$ of complex numbers, we want its square root $\mathbf{A}^{1/2}$ also to satisfy $(\mathbf{A}^{1/2})(\mathbf{A}^{1/2}) = \mathbf{A}$. But what is this $\mathbf{A}^{1/2}$? However, if we have a real-valued $n \times n$ diagonal matrix $\mathbf{D} = \text{diag}[d_1, \ldots, d_n]$, then its square root $\mathbf{D}^{1/2} = \text{diag}[d_1^{1/2}, \ldots, d_n^{1/2}]$ is well defined and satisfies $(\mathbf{D}^{1/2})(\mathbf{D}^{1/2}) = (\text{diag}[d_1^{1/2}, \ldots, d_n^{1/2}])(\text{diag}[d_1^{1/2}, \ldots, d_n^{1/2}]) = \text{diag}[d_1, \ldots, d_n] = \mathbf{D}$. So the essence of the issue is that given any real-valued $n \times n$ matrix $\mathbf{A}$, if we can transform it to have a factor that is a diagonal matrix, then we know how to take the square root of that diagonal matrix. If the $n \times n$ matrix is our covariance matrix $\boldsymbol{\Lambda}$, then the orthogonal transformation $\mathbf{T}$ of $\boldsymbol{\Lambda}$ will be able to find the square root $\boldsymbol{\Lambda}^{1/2}$ of $\boldsymbol{\Lambda}$.

Consider a $n \times n$ symmetric non-negative definite matrix $\boldsymbol{\Lambda}$. The *eigenvalue problem* of $\boldsymbol{\Lambda}$ is defined by

$$\boldsymbol{\Lambda}\boldsymbol{\theta}_i = \lambda_i \boldsymbol{\theta}_i, \ 1 \leq i \leq n, \tag{4.120}$$

where $\boldsymbol{\theta}_i = [\theta_{i1}, \ldots, \theta_{in}]^T$, $1 \leq i \leq n$, is an $n \times 1$ vector and is called the i-th *eigenvector* and λ_i, $1 \leq i \leq n$, is a non-negative valued real number called the i-th *eigenvalue* of the matrix $\boldsymbol{\Lambda}$. In particular, the eigenvectors obtained from (4.120) must be *orthogonal* in the sense that

$$(\boldsymbol{\theta}_i, \boldsymbol{\theta}_j) = 0, \ i \neq j, \ 1 \leq i, j \leq n. \tag{4.121}$$

Furthermore, we can normalize each eigenvector $\boldsymbol{\theta}_i$ to have unit norm by forming $\tilde{\boldsymbol{\theta}}_i = \boldsymbol{\theta}_i/\|\boldsymbol{\theta}_i\|$, $1 \le i \le n$, and then each $\tilde{\boldsymbol{\theta}}_i$, $1 \le i \le n$ satisfies the *normality* condition of

$$(\tilde{\boldsymbol{\theta}}_i, \tilde{\boldsymbol{\theta}}_i) = \|\tilde{\boldsymbol{\theta}}_i\|^2 = 1,\ 1 \le i \le n. \tag{4.122}$$

Since the eigenvalue problem equation of (4.120) is a linear equation (i.e., if $\boldsymbol{\theta}_i$ is a solution of (4.120), then $c\boldsymbol{\theta}_i$ for any constant c is a solution of (4.120)). If a $\boldsymbol{\theta}_i$ does not have a unit norm, then we can always normalize it by multiplying by the constant $c = 1/\|\boldsymbol{\theta}_i\|$. Thus, without loss of generality, we can always assume the set of eigenvectors $\{\boldsymbol{\theta}_i,\ 1 \le i \le n\}$ satisfies (4.121) and (4.122). Equations (4.121) and (4.122) define the *orthonormality* condition of

$$(\boldsymbol{\theta}_i, \boldsymbol{\theta}_j) = \boldsymbol{\theta}_i^T \boldsymbol{\theta}_j = \delta_{ij} = \begin{cases} 0,\ i \ne j, \\ 1,\ i = j, \end{cases} \quad 1 \le i, j \le n. \tag{4.123}$$

From the orthonormality properties of the eigenvectors $\{\boldsymbol{\theta}_i,\ 1 \le i \le n\}$ of $\boldsymbol{\Lambda}$, we can form a new $n \times n$ matrix $\mathbf{T}$ denoted by

$$\mathbf{T} = [\boldsymbol{\theta}_1, \ldots, \boldsymbol{\theta}_n]. \tag{4.124}$$

But the eigenvalue problem of (4.120) for each i can be rewritten as a matrix-matrix multiplication problem of the form

$$\boldsymbol{\Lambda}[\boldsymbol{\theta}_1, \ldots, \boldsymbol{\theta}_n] = [\lambda_1\boldsymbol{\theta}_1, \ldots, \lambda_n\boldsymbol{\theta}_n] = [\boldsymbol{\theta}_1, \ldots, \boldsymbol{\theta}_n] \begin{bmatrix} \lambda_1 & 0 & \cdots & 0 \\ 0 & \lambda_2 & \cdots & 0 \\ \vdots & \vdots & \ddots & \vdots \\ 0 & \cdots & 0 & \lambda_n \end{bmatrix}. \tag{4.125}$$

By using the notation of the $n \times n$ matrix $\mathbf{T}$ of (4.124), then (4.125) can be expressed as

$$\boldsymbol{\Lambda}\mathbf{T} = \mathbf{T}\mathbf{D}, \tag{4.126}$$

where the $n \times n$ diagonal matrix $\mathbf{D}$ is defined by

$$\mathbf{D} = \mathrm{diag}[\lambda_1,\ \lambda_2, \ldots, \lambda_n] = \begin{bmatrix} \lambda_1 & 0 & \cdots & 0 \\ 0 & \lambda_2 & \cdots & 0 \\ \vdots & \vdots & \ddots & \vdots \\ 0 & \cdots & 0 & \lambda_n \end{bmatrix}. \tag{4.127}$$

Since the columns of $\mathbf{T}$ in (4.124) are orthonormal vectors, then using the property of (4.123), we have

$$\mathbf{T}^T\mathbf{T} = \begin{bmatrix} \boldsymbol{\theta}_1^T \\ \boldsymbol{\theta}_2^T \\ \vdots \\ \boldsymbol{\theta}_n^T \end{bmatrix} [\boldsymbol{\theta}_1,\ \ldots,\ \boldsymbol{\theta}_n] = \mathbf{I}_n. \tag{4.128}$$

But if $\mathbf{T}^T\mathbf{T} = \mathbf{I}_n$ from (4.128), then

$$\mathbf{T}^T = \mathbf{T}^{-1}. \tag{4.129}$$

That is, the transpose of $\mathbf{T}$ is its inverse. This means

$$\mathbf{T}\mathbf{T}^{-1} = \mathbf{T}\mathbf{T}^T = \mathbf{I}_n. \tag{4.130}$$

An $n \times n$ matrix $\mathbf{T}$ satisfying (4.129) [and thus (4.128) and (4.130)] is said to be an *orthogonal matrix*. We should note, computing the inverse of any $n \times n$ matrix is trivial. In fact, one does not need to do any arithmetical operation for a matrix transposition, since it involves just some index swapping in the transpose operation. Thus, we obtain our explicit orthogonal transformation matrix $\mathbf{T}$ by using the orthonormal eigenvectors obtained from the eigenvalue problem equation of (4.120) or (4.126). Many numerical programs (e.g., Matlab has the function, eig($\mathbf{\Lambda}$); Mathematica has the function Eigensystem[$\mathbf{\Lambda}$]; etc.) can be used to solve the eigenvalue problem.

We can multiply (4.126) on the left of both sides of the equation by $\mathbf{T}^T = \mathbf{T}^{-1}$ to obtain

$$\mathbf{T}^T\mathbf{\Lambda}\mathbf{T} = \mathbf{T}^T\mathbf{T}\mathbf{D} = \mathbf{T}^{-1}\mathbf{T}\mathbf{D} = \mathbf{D}. \tag{4.131}$$

Equation (4.131) shows by multiplying on the left of $\mathbf{\Lambda}$ by $\mathbf{T}^T$ and on its right also by $\mathbf{T}$, we have diagonalized it into a diagonal matrix $\mathbf{D}$. The operation in (4.131) is called a *diagonalization* of $\mathbf{\Lambda}$ by an orthogonal transformation using the orthogonal matrix $\mathbf{T}$. Similarly, if we multiply both sides of (4.126) on the right by $\mathbf{T}^T$, then we obtain

$$\mathbf{\Lambda} = \mathbf{T}\mathbf{D}\mathbf{T}^T. \tag{4.132}$$

Given a diagonal matrix $\mathbf{D}$ of non-negative real-valued numbers, the square root $\mathbf{D}^{1/2}$ of $\mathbf{D}$ is obtained by taking the square roots of the diagonal elements of $\mathbf{D}$. Next, we show how to find the square root $\mathbf{\Lambda}^{1/2}$ immediately from $\mathbf{D}^{-1/2}$ and $\mathbf{T}$.

Lemma 4.2 *The whitening filter (or whitening matrix) $\mathbf{\Lambda}^{-1/2}$ is given by*

$$\mathbf{\Lambda}^{-1/2} = \mathbf{T}\mathbf{D}^{-1/2}\mathbf{T}^T, \tag{4.133}$$

where $\mathbf{T}$ is an orthogonal matrix formed from the eigenvectors of the eigenvalue value problem of the matrix $\mathbf{\Lambda}$.

Proof: *Show $\mathbf{\Lambda}^{-1/2}\mathbf{\Lambda}^{-1/2}\mathbf{\Lambda} = \mathbf{I}_n$. But*

$$\mathbf{\Lambda}^{-1/2}\mathbf{\Lambda}^{-1/2} = \mathbf{T}\mathbf{D}^{-1}\mathbf{T}^T\mathbf{T}\mathbf{D}^{-1/2}\mathbf{T}^T = \mathbf{T}\mathbf{D}^{-1/2}\mathbf{D}^{-1/2}\mathbf{T}^T = \mathbf{T}\mathbf{D}^{-1}\mathbf{T}^T,$$

then using (4.132), we have

$$\mathbf{\Lambda}^{-1/2}\mathbf{\Lambda}^{-1/2}\mathbf{\Lambda} = \mathbf{T}\mathbf{D}^{-1/2}\mathbf{T}^T\mathbf{T}\mathbf{D}\mathbf{T}^T = \mathbf{T}\mathbf{D}^{-1}\mathbf{D}\mathbf{T}^T = \mathbf{T}\mathbf{T}^T = \mathbf{I}_n. \tag{4.134}$$

□

Recall from (4.118) and (4.119) that $\mathbf{z}^T = \mathbf{x}^T\mathbf{\Lambda}^{-1/2}$ and $\mathbf{u} = \mathbf{\Lambda}^{-1/2}(\mathbf{s}_1 - \mathbf{s}_0)$, where $\gamma_{\mathrm{CGN}} = (\mathbf{z}, \mathbf{u}) = \mathbf{z}^T\mathbf{u}$. This means we use the inner product of $\mathbf{z}$ and $\mathbf{u}$ to find the sufficient statistic γ_{CGN}. But

$$\mathbf{z} = \mathbf{\Lambda}^{-1/2}\mathbf{x} = \begin{cases} \mathbf{\Lambda}^{-1/2}\mathbf{s}_0 + \mathbf{\Lambda}^{-1/2}\mathbf{N}, \ H_0, \\ \mathbf{\Lambda}^{-1/2}\mathbf{s}_1 + \mathbf{\Lambda}^{-1/2}\mathbf{N}, \ H_1, \end{cases} \tag{4.135}$$

where the transformed noise vector under both hypotheses in (4.135) is given by

$$\mathbf{N}_0 = \mathbf{\Lambda}^{-1/2}\mathbf{N} \tag{4.136}$$

and has the desirable property of being whitened.

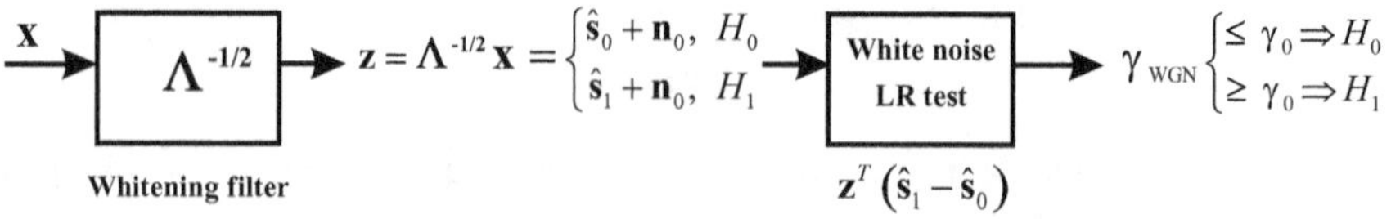

$\mathbf{x}$ → $\Lambda^{-1/2}$ (Whitening filter) → $\mathbf{z} = \Lambda^{-1/2}\mathbf{x} = \begin{cases} \hat{\mathbf{s}}_0 + \mathbf{n}_0, & H_0 \\ \hat{\mathbf{s}}_1 + \mathbf{n}_0, & H_1 \end{cases}$ → White noise LR test $\mathbf{z}^T(\hat{\mathbf{s}}_1 - \hat{\mathbf{s}}_0)$ → $\gamma_{\mathrm{WGN}} \begin{cases} \leq \gamma_0 \Rightarrow H_0 \\ \geq \gamma_0 \Rightarrow H_1 \end{cases}$

Figure 4.1 CGN receiver expressed as a whitening filter followed by a white noise LR test and a threshold comparator.

Lemma 4.3 $\mathbf{N}_0 = \mathbf{\Lambda}^{-1/2}\mathbf{N}$ *is a zero mean WGN vector.*

Proof: *First,* $\mathrm{E}\{\mathbf{N}_0\} = \mathbf{\Lambda}^{-1/2}\mathrm{E}\{\mathbf{N}\} = \mathbf{0} = \boldsymbol{\mu}_{\mathbf{N}_0}$. *Then*

$$\mathbf{\Lambda}_{\mathbf{N}_0} = \mathrm{E}\{\mathbf{N}_0\mathbf{N}_0^T\} = \mathrm{E}\{\mathbf{\Lambda}^{-1/2}\mathbf{N}\mathbf{N}^T\mathbf{\Lambda}^{-1/2}\} = \mathbf{\Lambda}^{-1/2}\mathbf{\Lambda}\mathbf{\Lambda}^{-1/2} = \mathbf{I}_n. \tag{4.137}$$

Since $\mathbf{N}$ *is a Gaussian random vector and* $\mathbf{N}_0$ *in (4.136) is a linear operation on* $\mathbf{N}$, *this shows that* $\mathbf{N}_0$ *is a Gaussian vector. From (4.137), we note that not only is* $\mathbf{N}_0$ *a WGN vector, all of the diagonal components of* $\mathbf{N}_0$ *have unit variance.* □

Lemma 4.2 shows $\mathbf{\Lambda}^{-1/2}$ is well defined and Lemma 4.3 shows $\mathbf{\Lambda}^{-1/2}$ operating on a CGN vector $\mathbf{N}$ yields a WGN vector. Thus, this justifies why we call $\mathbf{\Lambda}^{-1/2}$ a whitening filter or whitening matrix. In Fig. 4.1, we show a CGN binary detector formulated in the form of first consisting of a whitening filter $\mathbf{\Lambda}^{-1/2}$ operating on the received vector $\mathbf{x}$ as shown in (4.135). However, as $\mathbf{\Lambda}^{-1/2}$ whitens the noise to yield $\mathbf{N}_0 = \mathbf{\Lambda}^{-1/2}\mathbf{N}$, $\mathbf{\Lambda}^{-1/2}$ also transformed the two signal vectors $\mathbf{s}_0$ and $\mathbf{s}_1$ to yield

$$\hat{\mathbf{s}}_0 = \mathbf{\Lambda}^{-1/2}\mathbf{s}_0, \tag{4.138}$$

$$\hat{\mathbf{s}}_1 = \mathbf{\Lambda}^{-1/2}\mathbf{s}_1. \tag{4.139}$$

Thus, the LR test operating on the whitened vector $\mathbf{z}$ has to use the transformed vectors $\hat{\mathbf{s}}_0$ and $\hat{\mathbf{s}}_1$ instead of the original $\mathbf{s}_0$ and $\mathbf{s}_1$. Finally, the sufficient statistic $\gamma_{\mathrm{CGN}} = \mathbf{z}^T(\hat{\mathbf{s}}_1 - \hat{\mathbf{s}}_0)$ is compared to the threshold constant γ_0 in deciding the binary hypothesis.

In order to discuss the detection of two deterministic signal waveforms in CGN, we first consider the concept of a complete orthonormal series expansion of the deterministic signals over a finite interval in Section 4.5.

4.5 Complete orthonormal series expansion

Let $s(t)$, $a \leq t \leq b$, be any real-valued function with a finite energy $E = \int_a^b s^2(t)dt$. Then $s(t) \in L_2(a, b)$ is in the space of square integrable functions. A set of functions $\{\theta_i(t),\ a \leq t \leq b,\ i = 0, 1, \ldots\}$ is said to be a complete orthornormal (CON) set of functions in $L_2(a, b)$ if it meets the following two conditions:

(1) Orthonormality condition:

$$(\theta_i(t), \theta_j(t)) = \int_a^b \theta_i(t)\theta_j(t)dt = \begin{cases} 1, & i = j, \\ 0, & i \neq j, \end{cases} \quad 0 \leq i, j < \infty\,; \tag{4.140}$$

(2) Completeness condition:

For every $s(t) \in L_2(a, b)$, we have the expansion

$$s(t) = \sum_{i=0}^{\infty} s_i \theta_i(t), a \leq t \leq b. \tag{4.141}$$

Furthermore, each coefficient s_j in (4.141) is obtained from

$$s_j = (s(t), \theta_j(t)) = \int_a^b s(t)\theta_j(t)dt, \; j = 0, 1, \ldots. \tag{4.142}$$

Specifically, (4.142) is obtained from (4.141) by taking the inner product of the left- and right-hand sides of (4.141) with respect to $\theta_j(t)$ yielding

$$(s(t), \theta_j(t)) = \sum_{i=0}^{\infty} s_i(\theta_i(t), \theta_j(t)) = s_j. \tag{4.143}$$

Example 4.5 The classical Fourier series expansion function

$$\{(1/\sqrt{2\pi}) \exp(int), \; -\pi \leq t \leq \pi, \; -\infty \leq t \leq \infty\} \tag{4.144}$$

is a CON set on $L_2(-\pi, \pi)$. □

Example 4.6 The set

$$\left\{(1/\sqrt{\pi}) \sin(nt), \; -\pi \leq t \leq \pi, \; 1 \leq n \leq \infty\right\} \cup$$

$$\left\{(1/\sqrt{2\pi}), \; -\pi \leq t \leq \pi\right\} \cup \left\{(1/\sqrt{\pi}) \cos(nt), \; -\pi \leq t \leq \pi, \; 1 \leq n \leq \infty\right\} \tag{4.145}$$

is also a CON set on $L_2(-\pi, \pi)$. □

While a CON set of functions can be used to perform a series expansion of the deterministic binary signal waveforms, the Karhunen–Loève expansion needed to perform a series expansion of the random noise waveforms is considered in Section 4.6.

4.6 Karhunen–Loève expansion for random processes

Karhunen–Loève (KL) expansion is a random version of a CON expansion for a second order random process (SORP). Let $N(t)$ be a real-valued zero mean second order random process defined on a finite interval $-\infty < a \leq t \leq b < \infty$. Specifically,

$$\mu(t) = \mathrm{E}\{N(t)\} = 0, \; a \leq t \leq b, \tag{4.146}$$

$$R(t, \tau) = \mathrm{E}\{N(t)N(\tau)\}, \tag{4.147}$$

$$R(t, t) = \mathrm{E}\{N^2(t)\} < \infty, \; a \leq t \leq b. \tag{4.148}$$

We note, the second order noise process $\{N(t), -\infty < t < \infty\}$ includes the wide-sense-stationary random process (WSSRP), where the autocorrelation function $R(t, \tau) = R(t - \tau)$. Thus, the WSRP is a special case of the SORP. The KL expansion of $N(t)$ on $[a, b]$ is given by

$$N(t) = \sum_{i=0}^{\infty} N_i \theta_i(t), \ a \le t \le b, \tag{4.149}$$

where $\{\theta_i(t), \ a \le t \le b, \ i = 0, 1, \ldots\}$ is a CON set in $L_2(a, b)$, obtained from the solution of the eigenvectors in the integral eigenvalue problem of

$$\int_a^b R(t, \tau)\theta_i(\tau)d\tau = \lambda_i \theta_i(t), \ a \le t \le b, \ i = 0, 1, \ldots \tag{4.150}$$

In (4.150), $R(t, \tau)$ is called the kernel of the integral equation, $\theta_i(t)$ is called the eigenvector, and λ_i is called the eigenvalue of the integral eigenvalue problem. Symbolically, (4.150) can be written as

$$\mathfrak{R}\{\theta_i\} = \lambda_i \theta_i, \ i = 0, 1, \ldots, \tag{4.151}$$

where $\mathfrak{R}\{\cdot\}$ is an integral operator given by the l.h.s. of (4.150). The coefficient N_i is given by

$$N_i = \int_a^b N(t)\theta_i(t)dt = (N(t), \theta_i(t)), \ i = 0, 1, \ldots. \tag{4.152}$$

In particular, $\{N_i, \ i = 0, 1, \ldots\}$ is a set of uncorrelated r.v.'s.

Lemma 4.4 *Let $\{N_i, \ i = 0, 1, \ldots\}$ be defined by (4.152) and $\{\theta_i(t), \ a \le t \le b, \ i = 0, 1, \ldots\}$ is the solution of the integral eigenvalue problem, then $\{N_i, \ i = 0, 1, \ldots\}$ is a set of uncorrelated random variables satisfying*

$$\mathrm{E}\{N_i N_j\} = \begin{cases} 0, \ i \ne j, \\ \lambda_i, \ i = j, \end{cases} \ i = 0, 1, \ldots. \tag{4.153}$$

Proof:

$$\begin{aligned}
\mathrm{E}\{N_i N_j\} &= \mathrm{E}\left\{\int_a^b N(t)\theta_i(t)dt \int_a^b N(\tau)\theta_j(\tau)d\tau\right\} \\
&= \int_a^b \int_a^b \mathrm{E}\{N(t)N(\tau)\}\theta_j(\tau)\theta_i(t)d\tau dt = \int_a^b \int_a^b R(t, \tau)\theta_j(\tau)\theta_i(t)d\tau dt \\
&= \int_a^b \left\{\int_a^b R(t, \tau)\theta_j(\tau)d\tau\right\}\theta_i(t)dt = \int_a^b \lambda_j \theta_j(t)\theta_i(t)dt \\
&= \begin{cases} 0, \ i \ne j, \\ \lambda_i, \ i = j, \end{cases} \ i = 0, 1, \ldots.
\end{aligned} \tag{4.154}$$

In particular, the top line of the last equation of (4.154) states $\{N_i, \ i = 0, 1, \ldots\}$ is a set of uncorrelated random variables and the bottom line of the last equation *of (4.154) states*

$$\mathrm{E}\{N_i^2\} = \sigma_{N_i}^2 = \lambda_i,\ i = 0,\ 1,\ \ldots. \quad \square \tag{4.155}$$

Furthermore, if $\{N(t),\ a \le t \le b\}$ is a Gaussian random process, then since $\{N_i,\ i = 0,\ 1,\ \ldots\}$ is a set of uncorrelated Gaussian random variables, thus $\{N_i,\ i = 0,\ 1,\ \ldots\}$ is a set of independent Gaussian random variables.

In addition, using (4.149), we have

$$\begin{aligned} R(t,\tau) = \mathrm{E}\{N(t)N(\tau)\} &= \sum_{i=0}^{\infty}\sum_{j=0}^{\infty} \mathrm{E}\{N_i N_j\}\theta_i(t)\theta_j(\tau) \\ &= \sum_{j=0}^{\infty} \lambda_j \theta_j(t)\theta_j(\tau). \end{aligned} \tag{4.156}$$

Equation (4.156) is called the Mercer theorem on the expansion of the autocorrelation function in terms of its eigenvectors. If we take (4.156) and put it back to the l.h.s. of the integral eigenvalue problem equation (4.150), we obtain

$$\begin{aligned} \int_a^b R(t,\tau)\theta_i(\tau)d\tau &= \sum_{j=0}^{\infty} \lambda_j\theta_j(t) \int_a^b \theta_j(\tau)\theta_i(\tau)d\tau \\ &= \lambda_i\theta_i(t), \end{aligned} \tag{4.157}$$

which is consistent with the r.h.s. of (4.150).

Example 4.7 Consider a WGN process $N(t)$ on $[a, b]$. That is, let $\mathrm{E}\{N(t)\} = 0$ and $R(t,\tau) = \mathrm{E}\{N(t)N(\tau)\} = (N_0/2)\delta(t-\tau)$. Then any CON set of functions on $L_2(a, b)$ is a valid set of functions for the KL Expansion.

Proof:

We want to prove any CON set $\{\theta_i(t),\ a \le t \le b,\ i = 0,\ 1,\ \ldots,\}$ satisfies (4.150). Put $R(t,\tau) = (N_0/2)\delta(t-\tau)$ in the l.h.s. of (4.150) with any CON set $\{\theta_i(t),\ a \le t \le b,\ i = 0,\ 1\ldots,\}$, then we have

$$\int_a^b (N_0/2)\delta(t-\tau)\theta_i(\tau)d\tau = (N_0/2)\theta_i(t),\ a \le t \le b,\ i = 0,\ 1,\ \ldots. \tag{4.158}$$

Thus, the r.h.s. of (4.158) and the r.h.s. of (4.150) show $\lambda_i = N_0/2$. Thus, every CON set of functions on $L_2(a, b)$ is an eigenvector of the integral eigenvalue problem of the KL expansion when $R(t,\tau) = (N_0/2)\delta(t-\tau)$ and the eigenvalues are identical to $\lambda_i = N_0/2$. $\square$

Example 4.8 Let $N(t)$ be a real-valued zero mean process with a correlation function

$$R(t,\tau) = \sum_{i=0}^{N} \lambda_i\theta_i(t)\theta_i(\tau),\ a \le t,\ \tau \le b,\ i = 0,\ 1,\ldots,\ N, \tag{4.159}$$

where λ_i is any positive number and $\{\theta_i(t),\ a \leq t \leq b,\ i = 0,\ 1, \ldots, N\}$ is some orthonormal set of functions (but not necessarily complete) on $L_2(a, b)$. The $N(t)$ has a KL expansion

$$N(t) = \sum_{i=0}^{N} N_i \theta_i(t),\ a \leq t \leq b, \tag{4.160}$$

with

$$N_i = \int_a^b N(t)\theta_i(t)dt,\ i = 0,\ 1, \ldots, N. \tag{4.161}$$

Proof:
Consider applying the autocorrelation of $N(t)$ of (4.160) for $R(t, \tau)$ on the l.h.s. of (4.159), then

$$\begin{aligned} R(t, \tau) = \mathrm{E}\{N(t)N(\tau)\} &= \sum_{i=0}^{N}\sum_{j=0}^{N} \mathrm{E}\{N_i N_j\}\theta_i(t)\theta_j(\tau) \\ &= \sum_{j=0}^{N} \lambda_j \theta_j(t)\theta_j(\tau) \end{aligned} \tag{4.162}$$

shows the last expression of (4.162) is identical to the r.h.s. of (4.159). □

Example 4.9 Consider the Wiener random process (also called the Brownian motion random process) $N(t),\ 0 \leq t < \infty$, to have zero mean, with all of its finite dimensional distributions to be Gaussian distributed and its autocorrelation function defined by

$$\begin{aligned} R(t, \tau) &= (N_0/2)\min(t, \tau) \\ &= \frac{N_0}{2}\begin{cases} \tau,\ 0 \leq \tau \leq t, \\ t,\ 0 \leq t \leq \tau, \end{cases} \quad 0 \leq t, \tau \leq 1. \end{aligned} \tag{4.163}$$

Consider the solution of the integral eigenvalue problem of (4.150) for the Wiener process. Upon substituting (4.163) into (4.150), we have

$$\begin{aligned} &\int_0^1 R(t, \tau)\theta_i(\tau)d\tau = \frac{N_0}{2}\int_0^t \tau\theta_i(\tau)d\tau + \frac{N_0 t}{2}\int_t^1 \theta_i(\tau)d\tau \\ &= \lambda_i\theta_i(t),\ 0 \leq t \leq 1,\ i = 0,\ 1,\ \ldots. \end{aligned} \tag{4.164}$$

Differentiating (4.164) with respect to the variable t, to obtain

$$\frac{N_0 t\theta_i(t)}{2} + \frac{N_0 t(-\theta_i(t))}{2} + \frac{N_0}{2}\int_t^1 \theta_i(\tau)d\tau = \lambda_i\theta_i'(t),\ 0 \leq t \leq 1,\ i = 0,\ 1,\ \ldots. \tag{4.165}$$

Differentiating (4.165) with respect to the variable t again, we obtain

$$0 = \lambda_i \theta_i''(t) + \frac{N_0 \theta_i(t)}{2}, \ 0 \le t \le 1, \ i = 0, \ 1, \ \ldots . \quad (4.166)$$

Consider the different possible values of λ_i. If $\lambda_i = 0$, then $\theta_i(t) = 0, \ 0 \le t \le 1$, which results in degenerate eigenvectors. Thus, $\lambda_i = 0$ is not possible. Since $\lambda_i = \mathrm{E}\{n_i^2\} > 0$, we do not need to consider $\lambda_i < 0$.

The most general solution of (4.166) has the form of either

$$\theta_i(t) = Ae^{-jt\sqrt{N_0/2\lambda_i}} + Be^{jt\sqrt{N_0/2\lambda_i}}, \ 0 \le t \le 1, \ i = 0, \ 1, \ \ldots, \quad (4.167)$$

or

$$\theta_i(t) = C \sin\left(\sqrt{N_0/2\lambda_i}t\right) + D\cos\left(\sqrt{N_0/2\lambda_i}t\right), \ 0 \le t \le 1, \ i = 0, \ 1, \ \ldots . \quad (4.168)$$

Let us use $\theta_i(t)$ of (4.168). At $t = 0$, (4.164) shows

$$0 = \lambda_i \theta_i(0), \ i = 0, \ 1, \ldots, \quad (4.169)$$

and (4.168) shows

$$\theta_i(0) = D, \ i = 0, \ 1, \ \ldots . \quad (4.170)$$

From (4.169) and (4.170), since $\lambda_i \ne 0$, then $D = 0$. Thus, the most general form of $\theta_i(t)$ is given by

$$\theta_i(t) = C \sin\left(\sqrt{N_0/2\lambda_i}t\right), \ 0 \le t \le 1, \ i = 0, \ 1, \ \ldots . \quad (4.171)$$

In (4.165), set $t = 1$, then

$$0 = \lambda_i \theta_i'(1), \ i = 0, \ 1, \ \ldots . \quad (4.172)$$

Differentiate (4.171) with respect to the variable t and also set $t = 1$. Then

$$\theta_i'(t)\Big|_{t=1} = C\frac{\sqrt{N_0}}{2\lambda_i} \cos\left(\sqrt{N_0/2\lambda_i}t\right)\Big|_{t=1} = C\frac{\sqrt{N_0}}{2\lambda_i} \cos\left(\sqrt{N_0/2\lambda_i}\right), \ i = 0, 1, \ldots . \quad (4.173)$$

Substitute (4.173) into (4.172) to obtain

$$0 = \lambda_i C\frac{\sqrt{N_0}}{2\lambda_i} \cos\left(\sqrt{N_0/2\lambda_i}\right), \ i = 0, \ 1, \ \ldots . \quad (4.174)$$

Since $\lambda_i \ne 0$, then (4.174) implies

$$0 = \cos\left(\sqrt{N_0/2\lambda_i}\right), \ i = 0, \ 1, \ \ldots, \quad (4.175)$$

or

$$\sqrt{\frac{N_0}{2\lambda_i}} = \frac{(2i+1)\pi}{2}, \ i = 0, \ 1, \ \ldots . \quad (4.176)$$

Thus, the eigenvalues are given by

$$\lambda_i = \frac{2N_0}{(2i+1)^2\pi^2}, \ i = 0, \ 1, \ \ldots \quad (4.177)$$

Table 4.1 Tabulations of eigenvalues $\lambda_{\max} = \lambda'_{\max}/N$ and $\lambda_{\min} = \lambda'_{\min}/N$ vs. N for a given $\mathbf{R}$ matrix

N	$\lambda_{\max} = \lambda'_{\max}/N$	$\lambda_{\min} = \lambda'_{\min}/N$
4	4.5845	1.2330
8	4.1334	0.3146
16	3.9799	0.0784
32	3.9218	0.0195
64	3.8974	0.0049
128	3.8864	0.0012
256	3.8812	3.0518×10^{-4}
512	3.8787	7.6294×10^{-5}
1,024	3.8775	1.9074×10^{-5}

with the corresponding eigenvectors given by

$$\theta_i(t) = C\sin\left(\sqrt{N_0/2\lambda_i t}\right) = C\sin((i+1/2)\pi t),\ 0 \le t \le 1,\ i = 0, \ldots, \tag{4.178}$$

and the corresponding orthonormal eigenvectors are given by

$$\theta_i(t) = \sqrt{2}\sin((i+1/2)\pi t),\ 0 \le t \le 1,\ i = 0, 1, \ldots, \tag{4.179}$$

which are a collections of sinusoids. □

Example 4.10 Consider a wide-sense stationary random process with an autocorrelation function $R(t) = 10e^{-4|t|}$, $-\infty < t < \infty$. Example 4.9 converts the integral eigenvalue problem to a set of differential equations problem. Another approach for solving the integral eigenvalue problem of

$$\int_0^1 R(t-\tau)\theta_i(\tau)d\tau = \lambda_i\theta_i(t),\ 0 \le t \le 1, \tag{4.180}$$

is to approximate numerically the solution of (4.180) by using $(N+1)$ sums to replace the integral resulting in the form of

$$\sum_{j=0}^{N} R\left((k-j)/N\right)\theta_i(j/N)(1/N) \approx \lambda_i\theta_i(k/N),\ 0 \le k \le N,\ i = 0,\ 1,\ \ldots,\ N. \tag{4.181}$$

Then (4.181) can be solved as a standard matrix eigenvalue problem of the form

$$\mathbf{R}\boldsymbol{\theta}_i = \lambda'_i\boldsymbol{\theta}_i,\ i = 0,\ 1,\ \ldots,\ N \tag{4.182}$$

where $\lambda'_i = N\lambda_i$ and

$$\mathbf{R} = \left[R\left((k-j)/N\right)\right]_{0\le k\le N,\, 0\le j\le N}, \tag{4.183}$$

is a $(N+1) \times (N+1)$ symmetric autocorrelation matrix. Table 4.1 shows the numerical solution of the matrix eigenvalue problem of (4.182), for $4 \le N \le 1024$.

Table 4.2 Values of the 5×5 matrix of eigenvectors $\mathbf{V}_4 = [\boldsymbol{\theta}_4, \boldsymbol{\theta}_3, \ldots, \boldsymbol{\theta}_0]$.

0.2475	−0.4525	−0.5686	−0.5433	0.3398
−0.5067	0.5433	0.0711	−0.4525	0.4881
0.6034	−0.0000	0.5859	0.0000	0.5410
−0.5067	−0.5433	0.0711	0.4525	0.4881
0.2475	−0.4525	−0.5686	−0.5433	0.3398

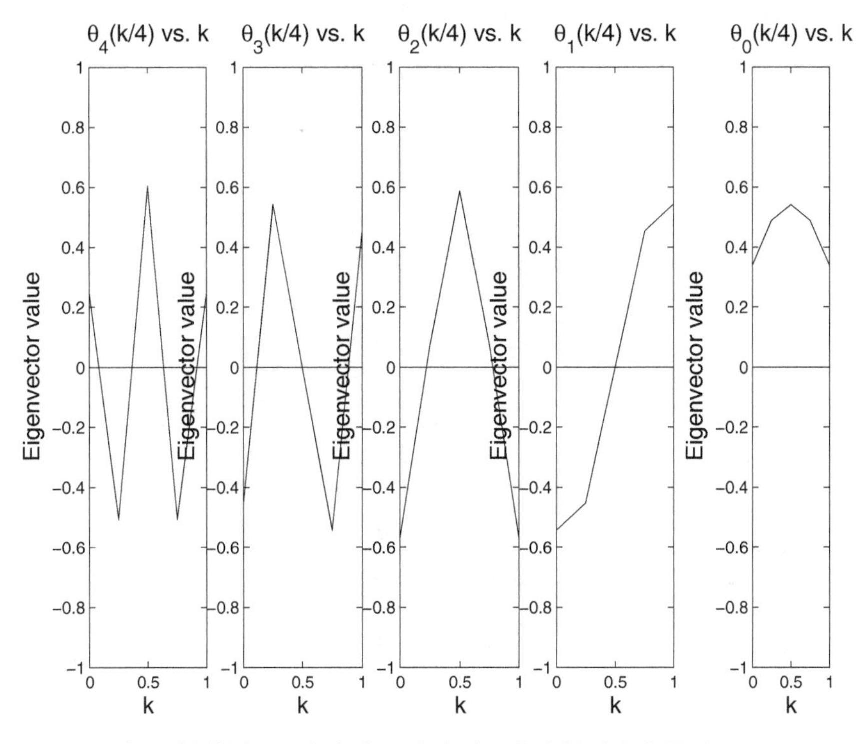

Figure 4.2 Plots of $\theta_i(k/4)$, $i = 4, 3, 2, 1, 0$, for $k = 0, 0.25, 0.5, 0.75, 1$.

From Table 4.1, we may conclude the largest eigenvalue $\lambda_{\max}$ seems to converge to a value of $3.87\ldots$, while the smallest $\lambda_{\min}$ seems to converge to zero. Other analytical methods show the above eigenvalue conclusions are valid. The $(N+1) \times (N+1)$ eigenvector matrix $\mathbf{V}_N = [\boldsymbol{\theta}_N, \boldsymbol{\theta}_{N-1}, \ldots, \boldsymbol{\theta}_0]$ for $N = 4$ is given in Table 4.2 and plotted in Fig. 4.2.

Similarly, for $N = 8$, $\mathbf{V}_8 = [\boldsymbol{\theta}_8, \boldsymbol{\theta}_7, \ldots, \boldsymbol{\theta}_0]$ is tabulated in Table 4.3 and plotted in Fig. 4.3.

From both Fig. 4.2 (for $N = 4$) and Fig. 4.3 (for $N = 8$), the eigenvector $\theta_i(k/N)$ as a function of $k = 0, \ldots, 1$, has i zeros in $[0, 1]$. The zero crossing property of the eigenvectors exhibits a similar property to the sinusoids. □

By using the CON series expansion of the binary deterministic signal waveforms and the KL expansion of the random noise process, Section 4.7 considers the detection of known binary deterministic signal waveforms in CGN.

Table 4.3 Values of the 9×9 matrix of eigenvectors $\mathbf{V}_8 = [\boldsymbol{\theta}_8, \boldsymbol{\theta}_7, \ldots, \boldsymbol{\theta}_0]$.

0.0978	−0.1918	0.2783	0.3530	0.4109	0.4444	−0.4405	−0.3756	0.2240
−0.2437	0.4149	−0.4615	−0.3666	−0.1550	0.1103	−0.3383	−0.4258	0.2987
0.3617	−0.4539	0.2071	−0.1947	−0.4464	−0.3501	−0.0245	−0.3654	0.3555
−0.4384	0.2916	0.2437	0.4507	0.0526	−0.4096	0.3046	−0.2100	0.3911
0.4649	−0.0000	−0.4634	−0.0000	0.4585	0.0000	0.4432	0.0000	0.4031
−0.4384	−0.2916	0.2437	−0.4507	0.0526	0.4096	0.3046	0.2100	0.3911
0.3617	0.4539	0.2071	0.1947	−0.4464	0.3501	−0.0245	0.3654	0.3555
−0.2437	−0.4149	−0.4615	0.3666	−0.1550	−0.1103	−0.3383	0.4258	0.2987
0.0978	0.1918	0.2783	−0.3530	0.4109	−0.4444	−0.4405	0.3756	0.2240

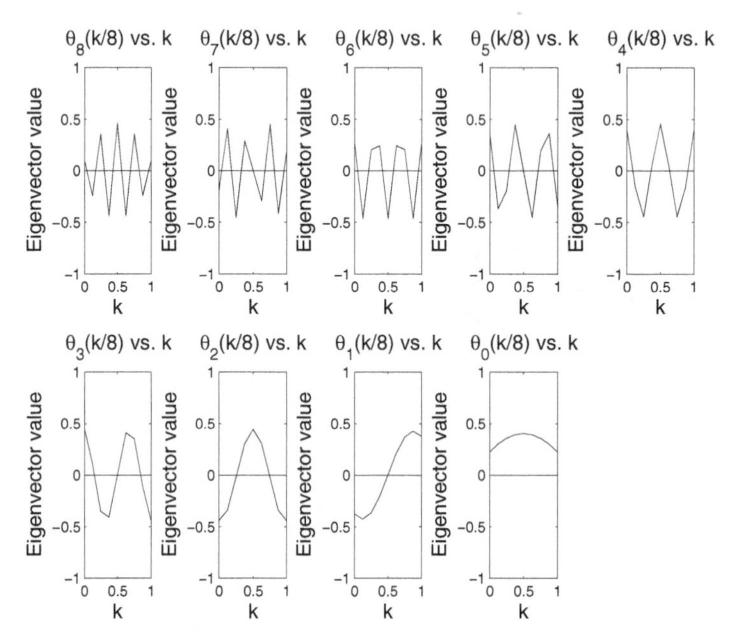

Figure 4.3 Plots of $\theta_i(k/8)$, $i = 8, 7, \ldots, 1, 0$, for $k = 0, 0.125, 0.25, \ldots, 0.75, 0.875, 1$.

4.7 Detection of binary known signal waveforms in CGN via the KL expansion method

Consider the received waveform modeled by

$$X(t) = \begin{cases} s_0(t) + N(t), & H_0, \\ s_1(t) + N(t), & H_1, \end{cases} \quad a \leq t \leq b, \tag{4.184}$$

where $s_0(t)$ and $s_1(t)$ are real-valued finite energy known waveforms on $[a, b]$, $N(t)$ is a real-valued zero mean CGN process with an arbitrary known autocorrelation function $R(t, \tau) = \mathrm{E}\{N(t)N(\tau)\}$, which need not be wide-sense stationary, but is required to satisfy $\mathrm{E}\{N^2(t)\} < \infty$. However, often the CGN is assumed to be wide-sense stationary, with its autocorrelation function denoted by

$$R(t) = \mathrm{E}\{N(t+\tau)N(\tau)\} \tag{4.185}$$

and its power spectral density function denoted by

$$S(f) = \int_{-\infty}^{\infty} \exp(-i2\pi ft)R(t)dt. \tag{4.186}$$

Then the WGN process is a special case of the CGN process, where

$$S(f) = C = \text{positive constant}, \ -\infty < f < \infty, \tag{4.187}$$

and the autocorrelation function satisfies

$$R(t) = C\delta(t), \ -\infty < f < \infty. \tag{4.188}$$

Now, back in (4.184), assume the $N(t)$ noise is an arbitrary CGN process. Let $\{\theta_i(t),\ a \leq t \leq b,\ i = 0,\ 1,\ \ldots\}$ and $\{\lambda_i,\ i = 0,\ 1,\ \ldots\}$ be the set of eigenvectors and eigenvalues obtained from the solutions of the integral eigenvalue value problem with the $R(t, \tau)$ kernel. Then the KL expansion of $N(t)$ and its coefficients are given by

$$N(t) = \sum_{i=0}^{\infty} N_i \theta_i(t), \ a \leq t \leq b, \tag{4.189}$$

$$N_i = \int_a^b N(t)\theta_i(t)dt, \ i = 0,\ 1,\ \ldots, \tag{4.190}$$

where

$$\mathrm{E}\{N_i\} = 0, \ \mathrm{E}\{N_i N_j\} = \lambda_i \delta_{ij}, \ i, j = 0, 1, \ldots. \tag{4.191}$$

Then $\{N_i,\ i = 0,\ 1,\ \ldots\}$ is a set of independent Gaussian random variables of mean zero and variance

$$\sigma_i^2 = \lambda_i, \ i = 0,\ 1, \ldots. \tag{4.192}$$

Since $\{\theta_i(t),\ a \leq t \leq b,\ i = 0,\ 1,\ \ldots\}$ is a CON set, then expanding $s_0(t)$ and $s_1(t)$ in terms of the CON functions yields

$$s_0(t) = \sum_{i=0}^{\infty} s_{0i}\theta_i(t), \ a \leq t \leq b, \tag{4.193}$$

$$s_{0i} = \int_a^b s_0(t)\theta_i(t)dt, \ i = 0,\ 1,\ \ldots, \tag{4.194}$$

$$s_1(t) = \sum_{i=0}^{\infty} s_{1i}\theta_i(t), \ a \leq t \leq b, \tag{4.195}$$

$$s_{1i} = \int_a^b s_1(t)\theta_i(t)dt, \ i = 0,\ 1,\ \ldots. \tag{4.196}$$

Combine (4.189) with (4.193) and (4.195) to obtain

$$X(t) = \sum_{i=0}^{\infty} X_i\theta_i(t) = \begin{cases} \sum_{i=0}^{\infty}(s_{0i} + N_i)\theta_i(t), & H_0, \\ \sum_{i=0}^{\infty}(s_{1i} + N_i)\theta_i(t), & H_1, \end{cases} \quad a \le t \le b, \tag{4.197}$$

where

$$X_i = \begin{cases} s_{0i} + N_i, & H_0, \\ s_{1i} + N_i, & H_1. \end{cases} \tag{4.198}$$

Then $\{X_i,\ i = 0,\ 1,\ \ldots\}$ is a set of independent Gaussian random variables with

$$\mathrm{E}\{X_i|H_0\} = \mu_{0i} = s_{0i},\ \mathrm{E}\{X_i|H_1\} = \mu_{1i} = s_{1i},\ i = 0,\ 1,\ \ldots, \tag{4.199}$$

$$\sigma_{X_i}^2 = \mathrm{E}\{(X_i - \mu_{0i})^2|H_0\} = \lambda_i = \mathrm{E}\{(X_i - \mu_{1i})^2|H_0\},\ i = 0,\ 1,\ \ldots. \tag{4.200}$$

The conditional probabilities of $\{X_i,\ i = 0,\ 1,\ \ldots\ N\}$ under H_0 and H_1 are given by

$$p(x_0, \ldots, x_N|H_0) = \prod_{i=0}^{N} p(x_i|H_0) = \prod_{i=0}^{N} \frac{1}{\sqrt{2\pi\lambda_i}} \exp(-(x_i - s_{0i})^2/(2\lambda_i)), \tag{4.201}$$

$$p(x_0, \ldots, x_N|H_1) = \prod_{i=0}^{N} p(x_i \mid H_1) = \prod_{i=0}^{N} \frac{1}{\sqrt{2\pi\lambda_i}} \exp(-(x_i - s_{1i})^2/(2\lambda_i)). \tag{4.202}$$

From (4.197) and (4.198), the LR function is given by

$$\begin{aligned} \Lambda(x(t),\ a \le t \le b) = \Lambda(x_i,\ i = 0,\ 1, \ldots) &= \lim_{N\to\infty} \left\{ \frac{p(x_0, \ldots, x_N|H_1)}{p(x_0,\ \ldots, x_N|H_0)} \right\} \\ &= \lim_{N\to\infty} \left\{ \prod_{i=0}^{N} \exp\left\{ \frac{-(x_i - s_{1i})^2}{2\lambda_i} + \frac{(x_i - s_{0i})^2}{2\lambda_i} \right\} \right\} \\ &= \lim_{N\to\infty} \left\{ \exp\left\{ \sum_{i=0}^{N} \frac{1}{2\lambda_i} [-(x_i - s_{1i})^2 + (x_i - s_{0i})^2)] \right\} \right\} \\ &= \exp\left\{ \sum_{i=0}^{\infty} \frac{1}{2\lambda_i} [-x_i^2 + 2x_i s_{1i} - s_{1i}^2 + x_i^2 - 2x_i s_{0i} + s_{0i}^2] \right\}. \end{aligned} \tag{4.203}$$

Thus,

$$\begin{aligned} \mathbb{R}_1 &= \{x_i :\ \Lambda(x_i,\ i = 0,\ 1,\ \ldots) \ge \Lambda_0\} \\ &= \{x_i :\ \ln(\Lambda(x_i, i = 0, 1, \ldots)) \ge \ln(\Lambda_0)\} \\ &= \left\{ x_i :\ \sum_{i=0}^{\infty} \frac{x_i\,(s_{1i} - s_{0i})}{\lambda_i} \ge \frac{1}{2} \sum_{i=0}^{\infty} \frac{(s_{1i}^2 - s_{0i}^2)}{\lambda_i} + \ln(\Lambda_0) \right\}, \end{aligned} \tag{4.204}$$

where we can define the CGN sufficient statistic and its threshold by

$$\Gamma^c_{\text{CGN}} = \sum_{i=0}^{\infty} \frac{X_i (s_{1i} - s_{0i})}{\lambda_i}, \tag{4.205}$$

$$\gamma^c_0 = \frac{1}{2} \sum_{i=0}^{\infty} \frac{\left(s^2_{1i} - s^2_{0i}\right)}{\lambda_i} + \ln(\Lambda_0). \tag{4.206}$$

Then the two decision regions under CGN can be expressed as

$$\mathbb{R}_1 = \left\{x_i : \Gamma^c_{\text{CGN}} \geq \gamma^c_0\right\}, \ \mathbb{R}_0 = \left\{x_i : \Gamma^c_{\text{CGN}} \leq \gamma^c_0\right\}. \tag{4.207}$$

Since Γ^c_{CGN} is derived from the realization of a Gaussian random variable, we need to find its mean and variance under H_0 and H_1 in order to obtain the performance of the receiver. The means of Γ^c_{CGN} under H_0 and H_1 are given by

$$\mu_0 = \sum_{i=0}^{\infty} \frac{s_{0i} (s_{1i} - s_{0i})}{\lambda_i}, \ \mu_1 = \sum_{i=0}^{\infty} \frac{s_{1i} (s_{1i} - s_{0i})}{\lambda_i}, \tag{4.208}$$

and the variances of Γ^c_{CGN} under H_0 and H_1 are given by

$$\begin{aligned}
\sigma_0^2 &= \mathrm{E}\left\{ \sum_{i=0}^{\infty} \frac{N_i (s_{1i} - s_{0i})}{\lambda_i} \sum_{j=0}^{\infty} \frac{N_j \left(s_{1j} - s_{0j}\right)}{\lambda_j} \right\} \\
&= \sum_{i=0}^{\infty} \sum_{j=0}^{\infty} \frac{1}{\lambda_i \lambda_j} \mathrm{E}\left\{N_i N_j\right\} (s_{1i} - s_{0i}) \left(s_{1j} - s_{0j}\right) \\
&= \sum_{i=0}^{\infty} \frac{\lambda_i}{\lambda_i^2} (s_{1i} - s_{0i})^2 = \sum_{i=0}^{\infty} \frac{(s_{1i} - s_{0i})^2}{\lambda_i} = \sigma_1^2.
\end{aligned} \tag{4.209}$$

Then the P_{FA} and the P_{D} are given by

$$P_{\text{FA}} = \text{P}(\Gamma^c_{\text{CGN}} \geq \gamma^c_0 | H_0) = \text{Q}\left(\frac{\gamma^c_0 - \mu_0}{\sigma_0}\right), \tag{4.210}$$

$$P_{\text{D}} = \text{P}(\Gamma^c_{\text{CGN}} \geq \gamma^c_0 | H_1) = \text{Q}\left(\frac{\gamma^c_0 - \mu_1}{\sigma_1}\right). \tag{4.211}$$

The detection of an appropriately chosen deterministic signal waveform for a specific CGN can lead to the mathematically feasible "singular detection" result in which with finite signal energy we can obtain the ideal $P_{\text{FA}} = 0$ and $P_{\text{D}} = 1$.

Example 4.11 Consider the following example, in which under $H_0, s_0(t) = 0, \ a \leq t \leq b$. Then from (4.208), we obtain $\mu_0 = 0$. From (4.208) and (4.209), we obtain

$$\mu_1 = \sum_{i=0}^{\infty} \frac{s^2_{1i}}{\lambda_i} = \sigma_0^2 = \sigma_1^2 = \tilde{S}, \tag{4.212}$$

and from (4.206) we obtain

$$\gamma_0^c = \frac{1}{2}\sum_{i=0}^{\infty}\frac{s_{1i}^2}{\lambda_i} + \ln(\lambda_0) = \frac{\tilde{S}}{2} + \ln(\lambda_0). \tag{4.213}$$

Now, take coefficients of the $s_1(t)$ waveform to have the following form

$$s_{10} = A < \infty,\ s_{1i} = \frac{a}{i},\ i = 1,\ 2,\ \ldots,\ 0 < a < \infty, \tag{4.214}$$

for some positive valued constants a and A. The energy E of this waveform is given by

$$E = \sum_{i=0}^{\infty} s_{1i}^2 = A^2 + \sum_{i=1}^{\infty}\frac{a^2}{i^2} = A^2 + a^2\xi(2), \tag{4.215}$$

where the Riemann zeta function [1] $\xi(x)$ evaluated at $x = 2$ yields $\xi(2) = \sum_{i=1}^{\infty} 1/i^2 = \pi^2/6$. Thus, the energy of $s_1(t)$ given by $E = A^2 + a^2\pi^2/6$ is finite. Now, consider the noise process $N(t)$ to be a zero mean WSSRP with its autocorrelation function, using the Mercer theorem of (4.156), denoted by

$$R(t-\tau) = \sum_{i=0}^{\infty}\lambda_i\theta_i(t)\theta_i(\tau),\ a \le t,\ \tau \le b. \tag{4.216}$$

For a WSSRP, $R(t,t) = \mathrm{E}\left\{N^2(t)\right\} = C < \infty$. Then

$$\int_a^b R(t-t)dt = \sum_{i=0}^{\infty}\lambda_i\int_a^b \theta_i(t)\theta_i(t)dt = \sum_{i=0}^{\infty}\lambda_i = C < \infty. \tag{4.217}$$

For this noise process, take its eigenvalues to satisfy

$$\lambda_0 = B,\ \lambda_i = \frac{1}{i^3},\ i = 1,\ 2,\ \ldots. \tag{4.218}$$

Then the eigenvalues of (4.218) satisfy the WSSRP requirement of (4.217), since

$$\sum_{i=0}^{\infty}\lambda_i = B + A_{\mathrm{Ap}} \approx B + 1.202 < \infty, \tag{4.219}$$

where $A_{\mathrm{Ap}} = \xi(3)$ is the Apéry's constant [1, 2] with an approximate value of 1.202. By using the signal coefficients of (4.214) and the noise coefficients of (4.218) in (4.212), we obtain

$$\begin{aligned}\mu_1 &= \sum_{i=0}^{\infty}\frac{s_{1i}^2}{\lambda_i} = \frac{A^2}{B} + \sum_{i=0}^{\infty}\frac{a^2 i^3}{i^2} = \frac{A^2}{B} + a^2\sum_{i=0}^{\infty} i \\ &= \sigma_0^2 = \sigma_1^2 = \tilde{S} = \infty\,. \end{aligned}\tag{4.220}$$

Then using μ_0, μ_1 of (4.220), and γ_0^c of (4.213) in the P_{FA} and P_{D} expressions of (4.210) and (4.211), we obtain

$$P_{\text{FA}} = \text{Q}\left(\frac{\gamma_0^c - 0}{\sigma_0}\right) = \text{Q}\left(\left.\frac{\tilde{S}/2 + \ln(\lambda_0)}{\sqrt{\tilde{S}}}\right|_{\tilde{S}=\infty}\right) = \text{Q}(\infty) = 0, \tag{4.221}$$

$$\begin{aligned} P_{\text{D}} &= \text{Q}\left(\frac{\gamma_0^c - \mu_1}{\sigma_1}\right) = \text{Q}\left(\left.\frac{\tilde{S}/2 + \ln(\lambda_0) - \tilde{S}}{\sqrt{\tilde{S}}}\right|_{\tilde{S}=\infty}\right) \\ &= \text{Q}\left(\left.\frac{-\tilde{S}/2 + \ln(\lambda_0)}{\sqrt{\tilde{S}}}\right|_{\tilde{S}=\infty}\right) = \text{Q}\left(-\sqrt{\tilde{S}}/2\Big|_{\tilde{S}=\infty}\right) \\ &= \text{Q}(-\infty) = 1. \end{aligned} \tag{4.222}$$

The remarkable "singular detection" results of (4.221) and (4.222) can be achieved even if both the parameters A and a in the signal energy E of the signal waveform $s_1(t)$ can be made arbitrarily small (resulting in arbitrarily small E) without affecting the $\tilde{S} = \infty$ property of (4.220), which is needed in the ideal results of (4.221) and (4.222). □

Example 4.12 Singular detection is not physically feasible, since a hypothetical assumed noise eigenvalue of the form in (4.218) is not possible due to the presence of a non-zero thermal WGN (other than at zero degrees kelvin) at the detector of the receiver (see Section 1.4). Thus, the noise autocorrelation of (4.216) has an additional WGN term resulting in a noise autocorrelation of the form

$$R(t-\tau) = \frac{N_0}{2} + \sum_{i=0}^{\infty} \lambda_i \theta_i(t)\theta_i(\tau) = \sum_{i=0}^{\infty} \lambda_i' \theta_i(t)\theta_i(\tau), \; a \le t, \; \tau \le b, \tag{4.223}$$

where the noise eigenvalue λ''s are still given by

$$\lambda_0' = B + \frac{N_0}{2}, \; \lambda_i' = \frac{1}{i^3} + \frac{N_0}{2}, \; i = 1, 2, \ldots. \tag{4.224}$$

Now, use the same signal coefficients of (4.214) as before, but use the noise eigenvalue λ''s of (4.224), then we still have, μ_0, but now $\mu_1 = \sigma_0^2 = \sigma_1^2 = \tilde{S}'$ have the same finite value given by

$$\begin{aligned} \mu_1 &= \sum_{i=0}^{\infty} \frac{s_{1i}^2}{\lambda_i'} = \frac{A^2}{B + N_0/2} + \sum_{i=1}^{\infty} \frac{a^2}{i^2\left((1/i^3) + N_0/2\right)} \\ &= \frac{A^2}{B + N_0/2} + a^2 \sum_{i=1}^{\infty} \frac{i}{1 + i^3 N_0/2} \\ &= \sigma_0^2 = \sigma_1^2 = \tilde{S}' < \frac{A^2}{B + N_0/2} + \frac{2a^2}{N_0} \sum_{i=1}^{\infty} \frac{1}{i^2} = \frac{A^2}{B + N_0/2} + \frac{2a^2 \xi(2)}{N_0} \\ &= \frac{A^2}{B + N_0/2} + \frac{a^2 \pi^2}{3 N_0} < \infty. \end{aligned} \tag{4.225}$$

Using $\mu_0 = 0$ and $\mu_1 = \sigma_0^2 = \sigma_1^2 = \tilde{S}'$ values of (4.225) in (4.210) and (4.211) results in $P_{\rm FA} > 0$ and $P_{\rm D} < 1$, and thus singular detection can not occur. □

4.8 Applying the WGN detection method on CGN channel received data (*)

Section 4.1 considered the LR receiver for the detection of binary deterministic signal vectors in the presence of WGN, while Section 4.3 dealt with CGN. Now, suppose the noise vector $\mathbf{N}$ is indeed CGN with a zero mean and a $n \times n$ covariance matrix $\mathbf{\Lambda}$, but we do not use this information, out of either ignorance or an inability to measure and estimate $\mathbf{\Lambda}$. Instead, we perform a WGN detection, where the noise is assumed to be "white" with a covariance matrix $\mathbf{\Lambda} = \sigma^2 \mathbf{I}_n$. What is the consequent detection performance degradation with this mismatched solution? For simplicity of discussions in this section, take $\mathbf{s}_0$ to be an $n \times 1$ vector of all zeros, then denote the signal vector $\mathbf{s}_1 = \mathbf{s}$, and use the N-P criterion.

First, the optimum decision algorithm is to compare the observed statistic $\gamma_{\rm CGN} = \mathbf{x}^T \mathbf{\Lambda}^{-1} \mathbf{s}$ to a threshold determined by the probability of false alarm. From (4.111) we know that the probability of false alarm is given by

$$P_{\rm FA} = \mathrm{Q}\left(\frac{\gamma_0 + (1/2)\mathbf{s}^T \mathbf{\Lambda}^{-1} \mathbf{s}}{\sqrt{\mathbf{s}^T \mathbf{\Lambda}^{-1} \mathbf{s}}} \right), \tag{4.226}$$

where γ_0 is taken from (4.105) to yield the desired probability of false alarm and then the optimum probability of detection from (4.112) is given by

$$P_{\rm D}^{\rm opt} = \mathrm{Q}\left(\frac{\gamma_0 - (1/2)\mathbf{s}^T \mathbf{\Lambda}^{-1} \mathbf{s}}{\sqrt{\mathbf{s}^T \mathbf{\Lambda}^{-1} \mathbf{s}}} \right), \tag{4.227}$$

where $\mathrm{Q}(\cdot)$ is the complementary Gaussian distribution function. If the probability of false alarm is fixed at some acceptable value (e.g., set $P_{\rm FA} = 10^{-3}$, which yields $\mathrm{Q}^{-1}(P_{\rm FA}) = \mathrm{Q}^{-1}\left(10^{-3}\right) = 3.09$), then the threshold constant γ_0 can be solved from (4.226) as

$$\gamma_0 = \sqrt{\mathbf{s}^T \mathbf{\Lambda}^{-1} \mathbf{s}}\,\mathrm{Q}^{-1}(P_{\rm FA}) - (1/2)\mathbf{s}^T \mathbf{\Lambda}^{-1} \mathbf{s}\,, \tag{4.228}$$

where $\mathrm{Q}^{-1}(\cdot)$ is the inverse of the complementary Gaussian distribution function. This γ_0 is then used to obtain the optimum probability of detection in (4.227) to yield

$$P_{\rm D}^{\rm opt} = \mathrm{Q}\left(\frac{\sqrt{\mathbf{s}^T \mathbf{\Lambda}^{-1} \mathbf{s}}\,\mathrm{Q}^{-1}(P_{\rm FA}) - \mathbf{s}^T \mathbf{\Lambda}^{-1} \mathbf{s}}{\sqrt{\mathbf{s}^T \mathbf{\Lambda}^{-1} \mathbf{s}}} \right) = \mathrm{Q}\left(\mathrm{Q}^{-1}(P_{\rm FA}) - \sqrt{\mathbf{s}^T \mathbf{\Lambda}^{-1} \mathbf{s}} \right). \tag{4.229}$$

On the other hand, if the observation noise is indeed colored with a covariance matrix $\mathbf{\Lambda}$, but we do not use this information (out of either ignorance or an inability to measure and estimate $\mathbf{R}$), and treat the noise as "white" with a covariance matrix $\mathbf{\Lambda} = \sigma^2 \mathbf{I}_M$, then the suboptimum "white matched filter" operating in colored noise has a statistic given by $\gamma_{\text{WGN}} = \mathbf{x}^T\mathbf{s}$. If we use the same decision procedure as above, then the probability of false alarm is now given by

$$P_{\text{FA}} = \text{Q}\left(\frac{\gamma_0}{\sqrt{\mathbf{s}^T\boldsymbol{\Lambda}\mathbf{s}}}\right), \tag{4.230}$$

where γ_0 is taken to yield the desired probability of false alarm, and the suboptimum probability of detection is given by

$$P_{\text{D}}^{\text{sub}} = \text{Q}\left(\frac{\gamma_0 - \mathbf{s}^T\mathbf{s}}{\sqrt{\mathbf{s}^T\boldsymbol{\Lambda}\mathbf{s}}}\right). \tag{4.231}$$

If the probability of false alarm is fixed at some acceptable value, then the threshold constant γ_0 can be solved from (4.230) as

$$\gamma_0 = \sqrt{\mathbf{s}^T\boldsymbol{\Lambda}\mathbf{s}}\text{Q}^{-1}\left(P_{\text{FA}}\right). \tag{4.232}$$

Upon substituting γ_0 from (4.232) into (4.231), we obtain

$$P_{\text{D}}^{\text{sub}} = \text{Q}\left(\frac{\sqrt{\mathbf{s}^T\boldsymbol{\Lambda}\mathbf{s}}\text{Q}^{-1}\left(P_{\text{FA}}\right) - \mathbf{s}^T\mathbf{s}}{\sqrt{\mathbf{s}^T\boldsymbol{\Lambda}\mathbf{s}}}\right) = \text{Q}\left(\text{Q}^{-1}\left(P_{\text{FA}}\right) - \frac{\mathbf{s}^T\mathbf{s}}{\sqrt{\mathbf{s}^T\boldsymbol{\Lambda}\mathbf{s}}}\right). \tag{4.233}$$

Since the expression in (4.227) represents the optimum (i.e., largest) probability of detection $P_{\text{D}}^{\text{opt}}$, we need to show that expression must be greater than or equal to the suboptimum probability of detection $P_{\text{D}}^{\text{sub}}$ expression of (4.233). That is, we need to show

$$P_{\text{D}}^{\text{sub}} = \text{Q}\left(\text{Q}^{-1}\left(P_{\text{FA}}\right) - \frac{\mathbf{s}^T\mathbf{s}}{\sqrt{\mathbf{s}^T\boldsymbol{\Lambda}\mathbf{s}}}\right) \leq \text{Q}\left(\text{Q}^{-1}\left(P_{\text{FA}}\right) - \sqrt{\mathbf{s}^T\boldsymbol{\Lambda}^{-1}\mathbf{s}}\right) = P_{\text{D}}^{\text{opt}}. \tag{4.234}$$

Since $\text{Q}(\cdot)$ is a monotonically decreasing function, for the inequality in (4.234) to hold, it is equivalent to show that for any $\mathbf{s}$ and $\boldsymbol{\Lambda}$, the parameters χ^{opt} associated with $P_{\text{D}}^{\text{opt}}$ and χ^{sub} associated with $P_{\text{D}}^{\text{sub}}$ defined in (4.235) must satisfy

$$\chi^{\text{sub}} = \frac{\mathbf{s}^T\mathbf{s}}{\sqrt{\mathbf{s}^T\boldsymbol{\Lambda}\mathbf{s}}} \leq \sqrt{\mathbf{s}^T\boldsymbol{\Lambda}^{-1}\mathbf{s}} = \chi^{\text{opt}}, \tag{4.235}$$

or equivalently,

$$\|\mathbf{s}\|^4 \leq \left(\mathbf{s}^T\boldsymbol{\Lambda}\mathbf{s}\right)\left(\mathbf{s}^T\boldsymbol{\Lambda}^{-1}\mathbf{s}\right). \tag{4.236}$$

In Appendix 4.A, by using the Schwarz Inequality, we show that the inequalities in (4.235)–(4.236) are always mathematically valid. In particular, when $\boldsymbol{\Lambda}$ is a white noise covariance matrix satisfying $\boldsymbol{\Lambda} = \sigma^2\mathbf{I}_M$, then $\boldsymbol{\Lambda}^{-1} = (1/\sigma^2)\mathbf{I}_M$, equality is attained in (4.236). We also note that any signal vector s in (4.236) is invariant to a scalar multiplication factor. In other words, if $\mathbf{s}$ satisfies (4.236), then using $\tilde{\mathbf{s}} = c\mathbf{s}$ also satisfies (4.236), since the scalar multiplication factor c that appears in the form of $|c|^4$ on both sides of the inequality of (4.236) can be cancelled.

Now, consider an explicit example to illustrate the advantage of using the colored matched filter over a white matched filter in the presence of colored noise.

Table 4.4 Values of SNR(dB) and $\chi^{\text{opt}}/\chi^{\text{sub}}$ for a Markov covariance matrix with $r = 0.1$ and 0.9 of dimensions $n = 2$, 3, and 50 for five uniform pseudo-random generated signal vectors $\mathbf{s}$ with the $P_{\text{FA}} = 10^{-3}$ constraint.

$n = 2$			SNR(dB)		
$\chi^{\text{opt}}/\chi^{\text{sub}}$	−1.96	−9.93	−1.76	−9.60	−1.51
r	#1	#2	#3	#4	#5
0.1000	1.0006	1.0021	1.0002	1.0006	1.0000
0.9000	1.2139	1.6591	1.0826	1.2158	1.0108
$n = 3$			SNR(dB)		
$\chi^{\text{opt}}/\chi^{\text{sub}}$	−2.66	−10.58	−1.48	−7.66	−3.16
r	#1	#2	#3	#4	#5
0.1000	1.0021	1.0047	1.0022	1.0047	1.0015
0.9000	1.2834	1.8340	1.1951	1.7868	1.8814
$n = 50$			SNR(dB)		
$\chi^{\text{opt}}/\chi^{\text{sub}}$	−2.66	−10.58	−1.48	−7.66	−3.16
r	#1	#2	#3	#4	#5
0.1000	1.0047	1.0094	1.0037	1.0050	1.0067
0.9000	4.5241	6.2867	3.9924	4.8558	5.2242

Example 4.13 Consider a colored Markov covariance matrix $\mathbf{\Lambda}$ of dimension $n \times n$ of the form

$$\mathbf{\Lambda} = \begin{bmatrix} 1 & r & \cdots & r^{n-1} \\ r & 1 & r & \vdots \\ \vdots & \cdots & \ddots & r \\ r^{n-1} & \cdots & r & 1 \end{bmatrix}. \tag{4.237}$$

Suppose we select five pseudo-randomly generated $n \times 1$ signal vectors $\mathbf{s}$ for $n = 2$, 3, and 50, with $r = 0.1$ and 0.9, with the constraint of $\text{Q}^{-1}(P_{\text{FA}}) = 3.09$. Now, we tabulate 30 simulation results (i.e., 5 signal vectors × 3 values of n × 2 values of r) of $\chi^{\text{opt}}/\chi^{\text{sub}}$. Table 4.4 shows one set of such 30 simulation results using the uniform pseudo-random generator rand generating values on (0, 1) with the seeds of [2008, 2007, 2006, 2005, 2004] for the five realizations.

From (4.236), we note if $r = 0$, then the CGN problem reduces to the WGN problem. Thus, for the small value of $r = 0.1$, all the $\chi^{\text{opt}}/\chi^{\text{sub}}$ values are only slightly greater than unit value for all $n = 2$, 3, and 50 cases. However, as r increases to $r = 0.9$, we note the ratios of $\chi^{\text{opt}}/\chi^{\text{sub}}$ increase greatly as the dimension increases to $n = 50$ for all five realizations. □

4.8.1 Optimization for evaluating the worst loss of performance

The results shown in Table 4.4 of Example 4.13 were obtained from specific realizations in the simulations. An interesting question is for any given $n \times n$ colored non-singular covariance matrix $\mathbf{\Lambda}$, what $n \times 1$ signal vector $\mathbf{s}$ of unit norm (i.e., $\|s\|^2 = 1$) will theoretically yield the smallest (worst performing) $P_{\mathrm{D}}^{\mathrm{sub}}$ in (4.234)? This is asking what signal vector $\mathbf{s}$ provides the largest advantage in using the colored matched detector over a white detector. Equivalently, what $\mathbf{s}$ will yield the largest ratio $\chi^{\mathrm{opt}}/\chi^{\mathrm{sub}}$ in (4.235)? Or equivalently, what $\mathbf{s}$ will attain the maximum of the product of the two factors $\left(\mathbf{s}^T\mathbf{\Lambda}\mathbf{s}\right)\left(\mathbf{s}^T\mathbf{\Lambda}^{-1}\mathbf{s}\right)$ in (4.236)?

As stated earlier, the Schwarz Inequality (Appendix 4.A) provides a lower bound on the l.h.s. of $\left(\mathbf{s}^T\mathbf{\Lambda}\mathbf{s}\right)\left(\mathbf{s}^T\mathbf{\Lambda}^{-1}\mathbf{s}\right)$ in (4.236). It turns out that the Kantorovich Inequality [3–5] provides an upper bound on $\left(\mathbf{s}^T\mathbf{\Lambda}\mathbf{s}\right)\left(\mathbf{s}^T\mathbf{\Lambda}^{-1}\mathbf{s}\right)$ in (4.236). Specifically, in order to use the Kantorovich Inequality stated in Appendix 4.B, we set the colored non-singular covariance matrix $\mathbf{\Lambda}$ to the positive-definite symmetric matrix $\mathbf{A}$ of (4.266) and (4.268). Then the maximum of $\left(\mathbf{s}^T\mathbf{\Lambda}\mathbf{s}\right)\left(\mathbf{s}^T\mathbf{\Lambda}^{-1}\mathbf{s}\right)$ is achieved when equality of (4.266) is attained by using the unit norm signal vector

$$\hat{\mathbf{s}} = (\boldsymbol{\theta}_1 + \boldsymbol{\theta}_n)/\sqrt{2} \tag{4.238}$$

specified by (4.269). An alternative approach to find the maximum of $\left(\mathbf{s}^T\mathbf{\Lambda}\mathbf{s}\right)\left(\mathbf{s}^T\mathbf{\Lambda}^{-1}\mathbf{s}\right)$ is to use the spectral representation of $\mathbf{\Lambda}$

$$\mathbf{\Lambda} = \mathbf{U}\tilde{\mathbf{\Lambda}}\mathbf{U}^T, \tag{4.239}$$

where the eigenvalues $\mathbf{\Lambda}$ are packed onto the $n \times n$ diagonal matrix $\tilde{\mathbf{\Lambda}}$, with $\mathrm{diag}(\tilde{\mathbf{\Lambda}}) = [\lambda_1, \ldots, \lambda_n]$ with their corresponding normalized eigenvectors packed onto the orthogonal matrix

$$\mathbf{U} = [\boldsymbol{\theta}_1, \ldots, \boldsymbol{\theta}_n]. \tag{4.240}$$

Then its inverse $\mathbf{\Lambda}^{-1}$ has the form of

$$\mathbf{\Lambda}^{-1} = (1/\lambda_1)\boldsymbol{\theta}_1\boldsymbol{\theta}_1^T + \ldots + (1/\lambda_n)\boldsymbol{\theta}_n\boldsymbol{\theta}_n^T = \mathbf{U}\tilde{\mathbf{\Lambda}}^{-1}\mathbf{U}^T. \tag{4.241}$$

Then

$$\begin{aligned}\left(\mathbf{s}^T\mathbf{\Lambda}\mathbf{s}\right)\left(\mathbf{s}^T\mathbf{\Lambda}^{-1}\mathbf{s}\right) &= \left(\mathbf{s}^T\mathbf{U}\tilde{\mathbf{\Lambda}}\mathbf{U}^T\mathbf{s}\right)\left(\mathbf{s}^T\mathbf{U}\tilde{\mathbf{\Lambda}}^{-1}\mathbf{U}^T\mathbf{s}\right) = \left(\mathbf{z}^T\tilde{\mathbf{\Lambda}}\mathbf{z}\right)\left(\mathbf{z}^T\tilde{\mathbf{\Lambda}}^{-1}\mathbf{z}\right) \\ &= \left(\sum_{k=1}^{n} z_k^2\lambda_k\right)\left(\sum_{k=1}^{n}\frac{z_k^2}{\lambda_k}\right) \leq \frac{((a+A)/2)^2}{aA},\end{aligned} \tag{4.242}$$

where

$$\mathbf{z} = \mathbf{U}^T\mathbf{s}. \tag{4.243}$$

To attain the maximum of $\left(\mathbf{s}^T\mathbf{\Lambda}\mathbf{s}\right)\left(\mathbf{s}^T\mathbf{\Lambda}^{-1}\mathbf{s}\right)$ in (4.242), take $\hat{\mathbf{s}}$ of (4.238) to yield

$$\hat{\mathbf{z}} = \mathbf{U}^T\hat{\mathbf{s}} = \begin{bmatrix}\boldsymbol{\theta}_1^T \\ \vdots \\ \boldsymbol{\theta}_n^T\end{bmatrix}\left(\frac{\boldsymbol{\theta}_1 + \boldsymbol{\theta}_n}{\sqrt{2}}\right) = \begin{bmatrix}1/\sqrt{2} \\ 0 \\ \vdots \\ 0 \\ 1/\sqrt{2}\end{bmatrix}. \tag{4.244}$$

Using $\hat{\mathbf{z}}$ of (4.244) on the l.h.s. of (4.242) yields

$$\left(\mathbf{s}^T\boldsymbol{\Lambda}\mathbf{s}\right)\left(\mathbf{s}^T\boldsymbol{\Lambda}^{-1}\mathbf{s}\right) = \left(\mathbf{z}^T\tilde{\boldsymbol{\Lambda}}\mathbf{z}\right)\left(\mathbf{z}^T\tilde{\boldsymbol{\Lambda}}^{-1}\mathbf{z}\right) = \left(\frac{\lambda_1+\lambda_n}{2}\right)\left(\frac{1}{2\lambda_1}+\frac{1}{2\lambda_n}\right)$$
$$= \left(\frac{A+a}{2}\right)\left(\frac{1}{2A}+\frac{1}{2a}\right) = \frac{(A+a)}{4}\left(\frac{A+a}{aA}\right) = \frac{(a+A)^2}{4aA}, \tag{4.245}$$

which equals the r.h.s. of (4.242) or equivalently attains the equality of (4.242). □

Example 4.14 Consider the $n = 2$ case where the CGN covariance matrix spectral decomposition yields an orthogonal matrix $\mathbf{U}$ and diagonal $\tilde{\boldsymbol{\Lambda}}$ given by

$$\mathbf{U} = \frac{1}{\sqrt{2}}\begin{bmatrix}1 & 1\\ 1 & -1\end{bmatrix}, \quad \tilde{\boldsymbol{\Lambda}} = \frac{\sigma^2}{100.01}\begin{bmatrix}100 & 0\\ 0 & 0.01\end{bmatrix}. \tag{4.246}$$

For $P_{\mathrm{FA}} = 10^{-3}$ at SNR(dB) $= 10$ dB, using the optimum of (4.238) or $\hat{\mathbf{z}}$ of (4.244), we obtain $P_{\mathrm{D}}^{\mathrm{opt}} = 1$ and $P_{\mathrm{D}}^{\mathrm{sub}} = 0.9165$. □

Example 4.15 Consider the $n = 3$ case where the CGN covariance matrix spectral decomposition yields an orthogonal matrix $\mathbf{U}$ and diagonal $\tilde{\boldsymbol{\Lambda}}$ given by

$$\mathbf{U} = \frac{1}{\sqrt{2}}\begin{bmatrix}1 & 0 & 1\\ 0 & \sqrt{2} & 0\\ 1 & 0 & -1\end{bmatrix}, \quad \tilde{\boldsymbol{\Lambda}} = \frac{\sigma^2}{101.01}\begin{bmatrix}100 & 0 & 0\\ 0 & 1 & 0\\ 0 & 0 & 0.01\end{bmatrix}. \tag{4.247}$$

For $P_{\mathrm{FA}} = 10^{-3}$ at SNR(dB) $= 10$ dB, using the optimum $\hat{\mathbf{s}}$ of (4.238) or $\hat{\mathbf{z}}$ of (4.244), we obtain $P_{\mathrm{D}}^{\mathrm{opt}} = 1$ and $P_{\mathrm{D}}^{\mathrm{sub}} = 0.9199$. □

Example 4.16 Consider the problem using the Markov covariance matrix $\boldsymbol{\Lambda}$ of (4.237) in Example 4.13. By using the optimum of (4.238) to evaluate the ratio of $\left(\chi^{\mathrm{opt}}/\chi^{\mathrm{sub}}\right)\big|_{\mathbf{s}=\hat{\mathbf{s}}}$, we obtained the following results: We note while the values of χ^{opt} and χ^{sub} depend on the value of the SNR, $\left(\chi^{\mathrm{opt}}/\chi^{\mathrm{sub}}\right)\big|_{\mathbf{s}=\hat{\mathbf{s}}}$ is independent of the SNR value. We also note that for any given n and r, all the $\left(\chi^{\mathrm{opt}}/\chi^{\mathrm{sub}}\right)\big|_{\mathbf{s}=\hat{\mathbf{s}}}$ in Table 4.5 are greater than all five of the pseudo-randomly generated corresponding values of $\chi^{\mathrm{opt}}/\chi^{\mathrm{sub}}$ in Table 4.4 of Example 4.13. □

Example 4.17 Consider again the problem using the 50×50 Markov covariance matrix $\boldsymbol{\Lambda}$ of (4.237) in Example 4.13.

Figure 4.4(a) shows the plots of χ^{opt} (solid curve) and χ^{sub} (dotted curve) versus a, where the signal vector $\mathbf{s} = \sqrt{1-a}\boldsymbol{\theta}_{50} + \sqrt{a}\boldsymbol{\theta}_1$, $0 \le a \le 1$. As expected χ^{opt} is greater than χ^{sub}, except when they have equal values for $a = 0$ and $a = 1$. Figure 4.4(b) shows the plot of $\chi^{\mathrm{opt}}/\chi^{\mathrm{sub}}$ versus a. Also for $a = 1/2$, $\mathbf{s} = (\boldsymbol{\theta}_{50} + \boldsymbol{\theta}_1)/\sqrt{2}$ equal $\hat{\mathbf{s}}$ of (4.238). This shows that $\chi^{\mathrm{opt}}/\chi^{\mathrm{sub}}$ achieves its maximum $\left(\chi^{\mathrm{opt}}/\chi^{\mathrm{sub}}\right)\big|_{\mathbf{s}=\hat{\mathbf{s}}}$ using the optimum $\hat{\mathbf{s}}$ as expected from theory. □

Table 4.5 Values of $\left(\chi^{\text{opt}}/\chi^{\text{sub}}\right)\big|_{\mathbf{s}=\hat{\mathbf{s}}}$ for $n = 2$, 3, and 50 for $r = 0.10$ and $r = 0.90$.

$n = 2$	$\left(\chi^{\text{opt}}/\chi^{\text{sub}}\right)\big\|_{\mathbf{s}=\hat{\mathbf{s}}}$
$r = 0.1$	1.0050
$r = 0.9$	2.2942
$n = 3$	$\left(\chi^{\text{opt}}/\chi^{\text{sub}}\right)\big\|_{\mathbf{s}=\hat{\mathbf{s}}}$
$r = 0.1$	1.0101
$r = 0.9$	3.2233
$n = 50$	$\left(\chi^{\text{opt}}/\chi^{\text{sub}}\right)\big\|_{\mathbf{s}=\hat{\mathbf{s}}}$
$r = 0.1$	1.0196
$r = 0.9$	8.7236

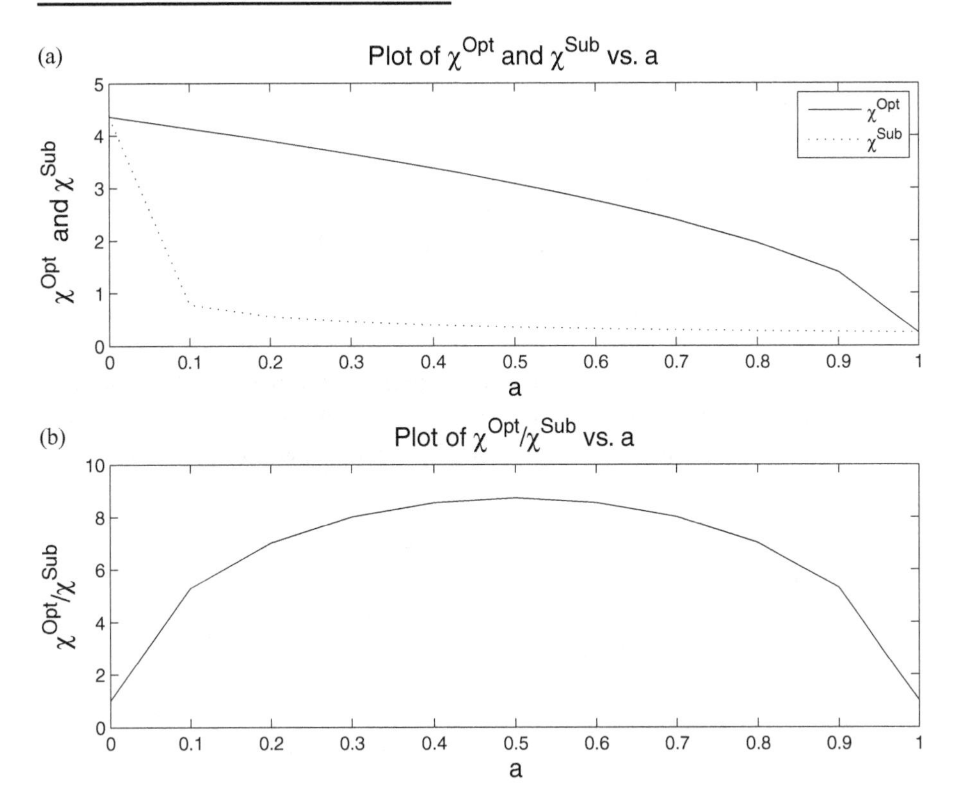

Figure 4.4 (a) Plots of χ^{opt} (solid curve) and χ^{sub} (dotted curve) versus a. (b) Plot of $\chi^{\text{opt}}/\chi^{\text{sub}}$ versus a.

4.9 Interpretation of a correlation receiver as a matched filter receiver

The sufficient statistic from (4.56) of a continuous-time binary LR receiver under the WGN observed over the $[0,\ T]$ interval is given by

$$\gamma^c_{\text{WGN}} = \int_0^T x(t)(s_1(t) - s_0(t))dt \gtrless^{H_1}_{H_0} \gamma^c_0. \tag{4.248}$$

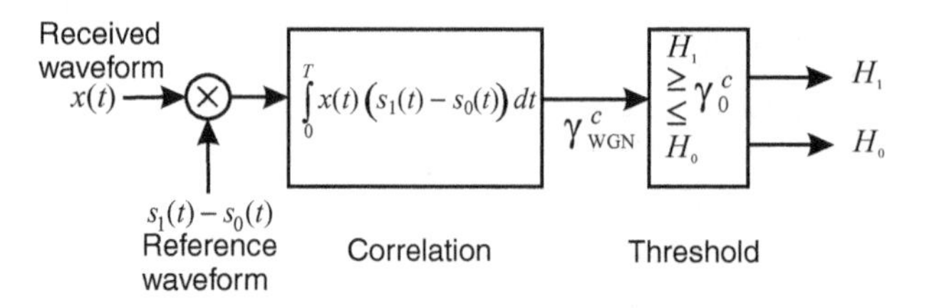

Figure 4.5 Binary waveform correlation receiver in WGN.

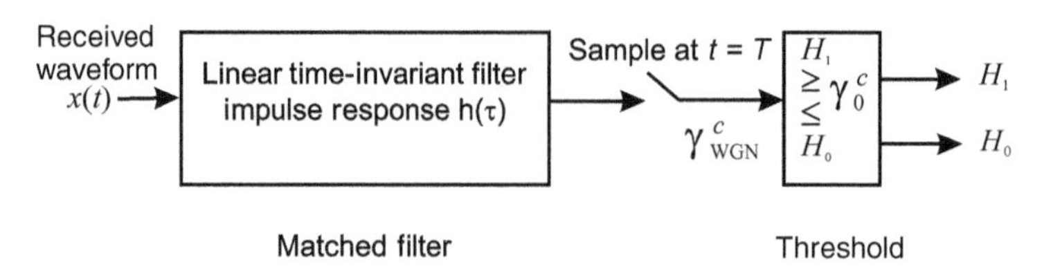

Figure 4.6 Binary waveform matched filter receiver in WGN.

A "correlation receiver" implementation of (4.248) is given in Fig. 4.5. Here the received waveform $x(t)$ is multiplied by the known reference waveform $(s_1(t) - s_0(t))$ and then integrated over the interval $[0, T]$. This "inner product" operation (in the mathematician notation) is called a "correlation" operation in communication/signal processing. Then the output of the correlation operation yields γ^c_{WGN} and is compared to the threshold γ^c_0 to make the binary decision. This implementation in Fig. 4.5 is called a *correlation receiver*. On the other hand, Fig. 4.6 shows an alternative implementation in the form of a matched filter receiver. Consider a linear time-invariant filter with an impulse response function $h(\tau)$ with an input $x(t)$ and an output $y(t)$ given by

$$y(t) = \int_{-\infty}^{\infty} h(t - \tau)x(\tau)d\tau. \tag{4.249}$$

Now, let $h(\tau)$ satisfy

$$h(\tau) = \begin{cases} s_1(T - \tau) - s_0(T - \tau), \; 0 \leq \tau \leq T, \\ \qquad\qquad 0, \; \tau < 0, \; T < \tau, \end{cases} \tag{4.250}$$

or

$$h(t - \tau) = \begin{cases} s_1(T - t + \tau) - s_0(T - t + \tau), \; t - T \leq \tau \leq t, \\ \qquad\qquad 0, \; \tau < t - T, \; t < \tau. \end{cases} \tag{4.251}$$

Take $t = T$. Then $y(T)$ in (4.249) using (4.251) becomes

$$y(T) = \int_0^T (s_1(\tau) - s_0(\tau))x(\tau)d\tau = \gamma^c_{\text{WGN}}. \tag{4.252}$$

Using a linear time-invariant filter with an impulse response function $h(\tau)$ given by (4.250) yields an output that when sampled at $t = T$ is identical to the correlation receiver output of Fig. 4.5. A filter satisfying (4.250) is called a *matched filter*. Clearly, from the theoretical point of view a binary receiver under the WGN condition, either implemented as a correlation receiver or a matched filter receiver, has the same performance. A matched filter receiver (or a correlation receiver) has to "dump" its sufficient statistic γ^c_{WGN} after the threshold test for each binary data. Since the correlation operation or the matched filter operation is an integration, both receivers are sometimes also called an *"integrate-and-dump" receiver*.

For a binary CGN receiver, the sufficient statistic γ^c_{CGN} was given in (4.205) in the form of an infinite series expansion, but can also be given in the form of an integral given by

$$\gamma^c_{\mathrm{CGN}} = \int_0^T x(t)g(t)dt = \sum_{i=0}^{\infty} \frac{x_i(s_{1i} - s_{0i})}{\lambda_i}, \tag{4.253}$$

where the deterministic $g(t)$ function can be obtained from the solution of the integral equation given by

$$\int_0^T R(t,\tau)g(\tau)d\tau = s_1(t) - s_0(t),\ 0 \le t \le T. \tag{4.254}$$

In order to show the integral form is equivalent to the infinite series form in (4.254), in (4.254) express all the functions as infinite series expansions, where $R(t,\tau)$ from the Mercer theorem is given by (4.156), $s_1(t) - s_0(t)$ is given by (4.155) and (4.157), and denote $g(t)$ by

$$g(t) = \sum_{i=0}^{\infty} g_i\theta_i(t),\ 0 \le t \le T. \tag{4.255}$$

Then (4.254) becomes

$$\begin{aligned} \int_0^T R(t,\tau)g(\tau)d\tau &= \int_0^T \sum_{j=0}^{N} \lambda_j\theta_j(t)\theta_j(\tau) \sum_{i=0}^{\infty} g_i\theta_i(\tau)d\tau \\ &= \sum_{i=0}^{N} \lambda_i g_i\theta_i(t) = \sum_{i=0}^{\infty} (s_{1i} - s_{0i})\theta_i(\tau),\ 0 \le t \le T, \end{aligned} \tag{4.256}$$

where

$$g_i = \frac{(s_{1i} - s_{0i})}{\lambda_i},\ i = 0, 1, \ldots, \tag{4.257}$$

and

$$g(t) = \sum_{i=0}^{\infty} \frac{(s_{1i} - s_{0i})\theta_i(t)}{\lambda_i},\ 0 \le t \le T. \tag{4.258}$$

Using the expression of $x(t)$ in (4.259)

$$x(t) = \sum_{j=0}^{\infty} x_j\theta_j(t),\ 0 \le t \le T, \tag{4.259}$$

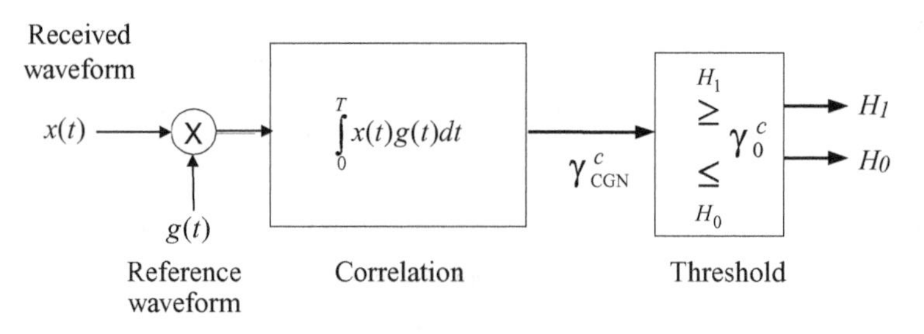

Figure 4.7 Binary waveform correlation receiver in CGN.

and the expression of $g(t)$ in (4.258) in (4.253), yields

$$\gamma^c_{\text{CGN}} = \int_0^T x(\tau)g(\tau)d\tau = \int_0^T \sum_{j=0}^{\infty} x_j\theta_j(t) \sum_{i=0}^{\infty} \frac{(s_{1i} - s_{0i})\theta_i(t)}{\lambda_i} d\tau$$
$$= \sum_{i=0}^{\infty} \frac{x_i(s_{1i} - s_{0i})}{\lambda_i}, \tag{4.260}$$

which shows the integral form and the infinite series form in (4.253) are identical. Figure 4.7 shows a correlation receiver for the binary detection of known deterministic waveforms in CGN.

4.10 Conclusions

In Chapter 3, we have shown that for a binary hypothesis test, under the Bayes, the Minimax, and the Neyman–Pearson criteria, all of these optimum decision rules result in LR tests only with different LR constants. In this chapter, we have elaborated on these issues. In Section 4.1, we considered the detection of known binary signal vectors in WGN. The optimum decision regions for the two hypotheses, the sufficient statistic, and the probabilities of false alarm and detection were obtained. Section 4.2 treated the detection of known binary signal waveforms in WGN, based on the sampling approach, and obtained similar results as in Section 4.1. Performances of BPSK and BFSK systems were given. Section 4.3 considered the detection of known deterministic binary signal vectors in CGN. Section 4.4 presents the whitening filter interpretation of the CGN detector utilizing the inverse of the square root of the CGN covariance matrix after having solved the eigenvalue problem. Section 4.5 introduces the concept of a complete orthonormal series expansion for representing deterministic waveforms. Section 4.6 introduces the Karhunen–Loève (KL) expansion for representing CGN processes. Two examples (one based on the differential equation approach and another one based on matrix theory approach) to solve the eigenvalue problem in the KL expansion were given. Section 4.7 showed the combined use of the deterministic signal series expansion and the KL noise expansion to perform the detection of binary known waveforms in CGN. An example using certain idealized CGN processes and matched finite energy deterministic signal waveforms showed the theoretical attainment of a "singular detection" system performance having zero probability of false alarm with unit value probability of detection. Another example showed in the presence of AWGN, due to non-zero degree kelvin front-end RF thermal noise, singular detection can not occur. Section 4.8 considers the robust detection issue of using a WGN detector on CGN channel received data. This scenario models the problem when the colored covariance matrix is not used in the detector (due to either ignorance or an inability to estimate it). For a specified colored noise covariance matrix, the "optimized" signal vector leading to the worst possible detection performance was presented. An analysis of this problem utilized the Schwarz Inequality and the Kantorovich Inequality to obtain the lower and upper bounds of an expression to be optimized for this problem. Section 4.9 presents some interpretations of the correlation receivers as matched filter receivers for binary deterministic waveforms in WGN and CGN. In practice, an approximate matched filter receiver may be much simpler to implement than a correlation receiver.

4.11 Comments

As discussed in Chapter 1 and Chapter 3, the concept of optimum hypothesis testing and the resulting use of likelihood ratio test were introduced in the 1930s by Fisher and Neyman–Pearson [5, 7]. Early applications of these concepts to radar, sonar, and digital communication systems appeared in the books that were published in the 1950s [8–10], although much of the research was performed earlier. Since then many books (listed in chronological order) dealing with the KL expansion, detection of binary vectors, and waveforms in WGN and CGN, and radar detection have appeared [11–19].

4.A

Use the Schwarz Inequality to obtain the lower bound of $\left(\mathbf{s}^T\boldsymbol{\Lambda}\mathbf{s}\right)\left(\mathbf{s}^T\boldsymbol{\Lambda}^{-1}\mathbf{s}\right)$ in (4.236). For any column vectors **a** and **b**, the Schwarz Inequality states that

$$\left(\mathbf{a}^T\mathbf{b}\right)^2 \le \left(\mathbf{a}^T\mathbf{a}\right)\left(\mathbf{b}^T\mathbf{b}\right). \tag{4.261}$$

Now, pick $\mathbf{a} = \boldsymbol{\Lambda}^{1/2}\mathbf{s}$ (or $\mathbf{a}^T = \mathbf{s}^T\boldsymbol{\Lambda}^{1/2}$) and $\mathbf{b} = \boldsymbol{\Lambda}^{-1/2}\mathbf{s}$ (or $\mathbf{b}^T = \mathbf{s}^T\boldsymbol{\Lambda}^{-1/2}$). Then

$$\mathbf{a}^T\mathbf{b} = \mathbf{s}^T\boldsymbol{\Lambda}^{1/2}\boldsymbol{\Lambda}^{-1/2}\mathbf{s} = \mathbf{s}^T\mathbf{s}, \tag{4.262a}$$

$$\mathbf{a}^T\mathbf{a} = \mathbf{s}^T\boldsymbol{\Lambda}^{1/2}\boldsymbol{\Lambda}^{1/2}\mathbf{s} = \mathbf{s}^T\boldsymbol{\Lambda}\mathbf{s}, \tag{4.262b}$$

$$\mathbf{b}^T\mathbf{b} = \mathbf{s}^T\boldsymbol{\Lambda}^{-1/2}\boldsymbol{\Lambda}^{-1/2}\mathbf{s} = \mathbf{s}^T\boldsymbol{\Lambda}^{-1}\mathbf{s}. \tag{4.262c}$$

Using (4.262) in (4.261) yields

$$\|\mathbf{s}\|^4 = \left(\mathbf{s}^T\mathbf{s}\right)^2 \le \left(\mathbf{s}^T\boldsymbol{\Lambda}\mathbf{s}\right)\left(\mathbf{s}^T\boldsymbol{\Lambda}^{-1}\mathbf{s}\right), \tag{4.263}$$

which is identical to (4.236) or equivalently

$$\|\mathbf{s}\|^2 = \mathbf{s}^T\mathbf{s} \leq \left(\mathbf{s}^T\mathbf{\Lambda}\mathbf{s}\right)^{1/2}\left(\mathbf{s}^T\mathbf{\Lambda}^{-1}\mathbf{s}\right)^{1/2}, \tag{4.264}$$

or equivalently

$$\frac{\mathbf{s}^T\mathbf{s}}{\sqrt{\mathbf{s}^T\mathbf{\Lambda}\mathbf{s}}} \leq \sqrt{\mathbf{s}^T\mathbf{\Lambda}^{-1}\mathbf{s}}, \tag{4.265}$$

which is identical to (4.235). □

4.B

Use the Kantorovich Inequality to obtain the upper bound of $\left(\mathbf{s}^T\mathbf{\Lambda}\mathbf{s}\right)\left(\mathbf{s}^T\mathbf{\Lambda}^{-1}\mathbf{s}\right)$ of (4.236).

The Kantorovich Inequality [3, 4] states

$$\left(\mathbf{u}^T\mathbf{A}\mathbf{u}\right)\left(\mathbf{u}^T\mathbf{A}^{-1}\mathbf{u}\right) \leq \frac{((a+A)/2)^2}{aA}, \tag{4.266}$$

where $\mathbf{u}$ is any $n \times 1$ real-valued unit norm vector, $\mathbf{A}$ is an $n \times n$ positive-definite symmetric matrix with eigenvalues denoted by

$$A = \lambda_1 > \cdots > \lambda_n = a > 0. \tag{4.267}$$

Denote the corresponding $n \times 1$ orthonormal eigenvectors of the eigenvalues in (4.267) by $\{\boldsymbol{\theta}_1, \ldots, \boldsymbol{\theta}_n\}$. The outer product form of the spectral decomposition of $\mathbf{A}$ yields

$$\mathbf{A} = \lambda_1\boldsymbol{\theta}_1\boldsymbol{\theta}_1^T + \cdots + \lambda_n\boldsymbol{\theta}_n\boldsymbol{\theta}_n^T = A\boldsymbol{\theta}_1\boldsymbol{\theta}_1^T + \cdots + a\boldsymbol{\theta}_n\boldsymbol{\theta}_n^T. \tag{4.268}$$

In particular, the unit norm vector $\mathbf{u}$ of the form

$$\mathbf{u} = (\boldsymbol{\theta}_1 + \boldsymbol{\theta}_n)/\sqrt{2} \tag{4.269}$$

satisfies

$$\begin{aligned}\mathbf{u}^T\mathbf{A}\mathbf{u} &= \left((\boldsymbol{\theta}_1 + \boldsymbol{\theta}_n)/\sqrt{2}\right)^T\left(A\boldsymbol{\theta}_1\boldsymbol{\theta}_1^T + \cdots + a\boldsymbol{\theta}_n\boldsymbol{\theta}_n^T\right)\left((\boldsymbol{\theta}_1 + \boldsymbol{\theta}_n)/\sqrt{2}\right) \\ &= (a + A)/2.\end{aligned} \tag{4.270}$$

Furthermore, if $\mathbf{A}$ is given by (4.268), then its inverse $\mathbf{A}^{-1}$ has the form of

$$\mathbf{A}^{-1} = (1/\lambda_1)\boldsymbol{\theta}_1\boldsymbol{\theta}_1^T + \cdots + (1/\lambda_n)\boldsymbol{\theta}_n\boldsymbol{\theta}_n^T = (1/A)\boldsymbol{\theta}_1\boldsymbol{\theta}_1^T + \cdots + (1/a)\boldsymbol{\theta}_n\boldsymbol{\theta}_n^T, \tag{4.271}$$

and

$$\begin{aligned}\mathbf{u}^T\mathbf{A}^{-1}\mathbf{u} &= \left((\boldsymbol{\theta}_1 + \boldsymbol{\theta}_n)/\sqrt{2}\right)^T\left((1/A)\boldsymbol{\theta}_1\boldsymbol{\theta}_1^T + \cdots + (1/a)\boldsymbol{\theta}_n\boldsymbol{\theta}_n^T\right)\left((\boldsymbol{\theta}_1 + \boldsymbol{\theta}_n)/\sqrt{2}\right) \\ &= (1/2)\left((1/A) + (1/a)\right) = \frac{a + A}{2aA}.\end{aligned} \tag{4.272}$$

By using (4.270) and (4.272), we obtain

$$\left(\mathbf{u}^T\mathbf{A}\mathbf{u}\right)\left(\mathbf{u}^T\mathbf{A}^{-1}\mathbf{u}\right) \leq \frac{((a+A)/2)^2}{aA}, \tag{4.273}$$

which attains the equality of (4.266) in the Kantorovich Inequality. □

References

[1] M. Abramowitz and I. A. Stegun. *Handbook of Mathematical Functions*. National Bureau of Standards, 1964.

[2] http://en.wikipedia.org/wiki/Ap%C3%A9ry's_constant.

[3] L. V. Kantorovich. On an effective method of solution of extremal problems for a quadratic functional. *Dokl. Akad. Nauk SSSR*, 1945.

[4] D. P. Bertsekas. *Nonlinear Programming*. Athena Scientific, 2d ed., 1999.

[5] R. A. Fisher. Two new properties of mathematical likelihood. *Proc. Royal Soc., London, Ser. A*, 1934.

[6] A. Wald. Sequential tests of statistical hypotheses. *Ann. Math. Statist.*, pp. 117–186, 1945.

[7] J. Neyman and E. S. Pearson. Contributions to the theory of testing statistical hypothesis. *Stat. Res. Memoir*, vol. 1, 1936; vol. 2, 1938.

[8] P. M. Woodward. *Probability and Information Theory, with Applications to Radar*. Pergamon Press, 1953.

[9] D. Middleton. *An Introduction to Statistical Communication Theory*. McGraw-Hill, 1960.

[10] V. A. Kotel'nikov. *Theory of Optimum Noise Immunity*. Government Power Engineering Press, 1956, translated by R. A. Silverman, McGraw-Hill, 1960.

[11] Jr. W. B. Davenport and W. L. Root. *An Introduction to the Theory of Random Signals and Noise*. McGraw-Hill, 1958.

[12] J. M. Wozencraft and I. M. Jacobs. *Principles of Communication Engineering*. Wiley, 1965.

[13] H. L. Van Trees. *Detection, Estimation, and Modulation, Part I*. Wiley, 1968.

[14] C. L. Weber. *Elements of Detection and Signal Design*. McGraw-Hill, 1968.

[15] C. W. Helstrom. *Statistical Theory of Signal Detection*. Pergamon, 2d ed., 1960. An expanded version appeared as *Elements of Signal Detection and Estimation*, Prentice-Hall, 1995.

[16] A. D. Whalen. *Detection of Signals in Noise*. Academic Press, 1st ed., 1971. R. N. McDonough and A. D. Whalen, 2d ed., 1995.

[17] J. V. DiFranco and W. L. Rubin. *Radar Detection*. Artech House, 1980.

[18] V. H. Poor. *An Introduction to Detection and Estimation*. Springer-Verlag, 1st ed., 1989, 2d ed., 1994.

[19] S. M. Kay. *Fundamentals of Statistical Signal Processing, Volume 2: Detection Theory*. Prentice Hall, 1998.

Problems

4.1 Consider a Bayes criterion receiver for

$$\mathbf{X} = \begin{cases} \mathbf{s}_0 + \mathbf{N}\,, & H_0, \\ \mathbf{s}_1 + \mathbf{N}\,, & H_1, \end{cases}$$

where $\mathbf{s}_0 = [1, 2, 3]^T = -\mathbf{s}_1$ and $\mathbf{N}$ is a 3×1 real-valued zero mean Gaussian vector with a covariance matrix $\mathbf{R} = \mathbf{I}_3$. Let $\pi_0 = 1/4$ and $c_{00} = c_{11} = 0$, but $c_{01} = 1$ and $c_{10} = 2$. Find the optimum decision regions $\mathbb{R}_1$ and $\mathbb{R}_0$ as well as P_{FA} and P_{D}. Repeat this problem when $\mathbf{s}_0 = [2, -\sqrt{6}, 2]^T = -\mathbf{s}_1$.

4.2 Consider a LR receiver for

$$X(t) = \begin{cases} \sqrt{E_0}\, s_0(t) + N(t)\,, & H_0, \\ \sqrt{E_1}\, s_1(t) + N(t)\,, & H_1, \end{cases} \quad 0 \le t \le T,$$

where E_0 and E_1 are two given positive constants, $\|s_0(t)\| = \|s_1(t)\| = 1$, $(s_0(t), s_1(t)) = \rho$, and $N(t)$ is a WGN process of zero mean and two-sided spectral density of $N_0/2$.

(a) Find P_{D} and P_{FA}.
(b) Repeat (a) for $\rho = -1$, $E_0 = E_1$, with LR constant $\Lambda_0 = 1$.
(c) Repeat (a) for $\rho = 0$, $E_0 = E_1$, with LR constant $\Lambda_0 = 1$.
(d) Why is the receiver performance better under (b) than under (c)?

4.3 Consider the received waveform given by

$$x(t) = \begin{cases} n(t) & , \ H_0, \\ s(t) + n(t)\,, & H_1, \end{cases} \quad 0 \le t \le T,$$

where $s(t) = 10 \sin(2\pi f_0 t)$, $0 \le t \le T$, $f_0 = 10^5$ Hz, and $T = 1$ sec. $n(t)$ is the realization of a WGN process of zero mean and power spectral density $N_0/2 = 1$.

(a) Consider a LR receiver with $\Lambda_0 = 1$. Find the optimum decision regions $\mathbb{R}_0$ and $\mathbb{R}_1$. Find the sufficient statistic of this receiver.
(b) Find the P_{FA} and P_{D} of this detector.
(c) Now, consider a sub-optimum receiver, where we do not use the optimum $s(t)$ in the sufficient statistic, but use a sub-optimum "hard-quantized" version $s'(t) = 10 \times \mathrm{signum}(s(t))$, $0 \le t \le T$, where $\mathrm{signum}(z) = 1$, for $z \ge 0$, and $\mathrm{signum}(z) = -1$, for $z < 0$. (Note: It is easier to implement $s'(t)$ than $s(t)$.) Find the new sufficient statistic γ' where $s(t)$ is replaced by $s'(t)$ in the receiver; but $s(t)$ is still used in the transmitter and in $x(t)$ under H_1. We assume the original γ_0 is still used here. Also find P'_{FA} and P'_{D} of this sub-optimum receiver.
(d) Continuing with part (c), suppose we take a new γ'_0 such that $P'_{\mathrm{FA}} = P_{\mathrm{FA}}$. What is the resulting $P'_{\mathrm{D}}(\gamma'_0)$? (That is, we use the new γ'_0 in the receiver using $s'(t)$.) How does this $P'_{\mathrm{D}}(\gamma'_0)$ compare to the original ideal P_{D}?

4.4 Consider the detection of a binary hypothesis originating from one source into two receiving sensors. The received 2×1 data vector at sensor A is $\mathbf{x}_{\mathrm{A}}$ and the received 2×1 data vector at sensor B is $\mathbf{x}_{\mathrm{B}}$, with

$$\mathbf{x}_{\mathrm{A}} = \begin{cases} \mathbf{s}_{0,\mathrm{A}} + \mathbf{n}_{\mathrm{A}}, & H_0, \\ \mathbf{s}_{1,\mathrm{A}} + \mathbf{n}_{\mathrm{A}}, & H_1, \end{cases} \qquad \mathbf{x}_{\mathrm{B}} = \begin{cases} \mathbf{s}_{0,\mathrm{B}} + \mathbf{n}_{\mathrm{B}}, & H_0, \\ \mathbf{s}_{1,\mathrm{B}} + \mathbf{n}_{\mathrm{B}}, & H_1. \end{cases}$$

The 2×1 signal vectors are given by

$$\mathbf{s}_{0,\mathrm{A}} = -\mathbf{s}_{1,\mathrm{A}} = \begin{bmatrix} 1 \\ 0 \end{bmatrix}, \qquad \mathbf{s}_{0,\mathrm{B}} = -\mathbf{s}_{1,\mathrm{B}} = \begin{bmatrix} 0 \\ 1 \end{bmatrix}.$$

The 2×1 noise vectors $\mathbf{n}_\mathrm{A}$ and $\mathbf{n}_\mathrm{B}$ are mutually independent Gaussian with zero means, and covariances

$$\mathbf{R}_\mathrm{A} = \begin{bmatrix} 1 & 0.5 \\ 0.5 & 1 \end{bmatrix}, \qquad \mathbf{R}_\mathrm{B} = \begin{bmatrix} 1 & 0 \\ 0 & 1 \end{bmatrix}.$$

Let $\Lambda_0 = 1$ for all the following cases.

(a) Find the LR receiver at source A. Find its P_{FA} and P_D.
(b) Repeat (a) for source B.
(c) Suppose $\mathbf{x}_\mathrm{A}$ and $\mathbf{x}_\mathrm{B}$ are combined to form the new 2×1 data vector $\mathbf{x}_\mathrm{C} = \mathbf{x}_\mathrm{A} + \mathbf{x}_\mathrm{B}$. Find the LR receiver using the data vector $\mathbf{x}_\mathrm{C}$. Find its P_{FA} and P_D.
(d) Suppose $\mathbf{x}_\mathrm{A}$ and $\mathbf{x}_\mathrm{B}$ are combined to form the new 4×1 data vector $\mathbf{x}_\mathrm{D} = [\mathbf{x}_\mathrm{A}^T, \mathbf{x}_\mathrm{B}^T]^T$. Since the inverse of the covariance matrix now is a 4×4 matrix (which you can not easily evaluate), we find for you that $P_{\mathrm{FA}} = \mathrm{Q}(1.528)$ and $P_\mathrm{D} = \mathrm{Q}(-1.528)$. Why are the P_{FA} and P_D of the system in (d) better than those in systems of (a), (b), and (c)?

4.5 Consider the detection of the signal $s(t)$ from the received waveform given by

$$X(t) = \begin{cases} N(t) & , \quad H_0, \\ s(t) + N(t), & \quad H_1, \end{cases} \quad 0 \leq t \leq T$$

using a matched filter receiver, where $s(t)$ is a known real-valued deterministic waveform and $N(t)$ is a real-valued zero mean WGN process with a two-sided spectral density $S(f) = N_0/2, -\infty < f < \infty$. Find the maximum output signal-to-noise ratio (SNR) of the matched filter. The output SNR is defined by $s_\mathrm{o}^2(t_\mathrm{o})/\sigma_\mathrm{o}^2$, where $s_\mathrm{o}(t_\mathrm{o})$ is the matched filter output signal waveform sampled at some desired time t_o and σ_o^2 is the output noise variance. Hint: You may need to use the Schwarz Inequality, which states that for any real-valued functions $f(t)$ and $h(t)$, defined on (a, b), $\left(\int_a^b f(t)h(t)dt\right)^2 \leq \left(\int_a^b f^2(t)dt\right) \cdot \left(\int_a^b h^2(t)dt\right)$, with equality if and only if $f(t) = c \cdot h(t)$ for any constant c.

4.6 Consider the detection of binary known signal vectors in WGN modeled by

$$\mathbf{X} = \begin{cases} \mathbf{s}_0 + \mathbf{N}, & H_0, \\ \mathbf{s}_1 + \mathbf{N}, & H_1, \end{cases}$$

where the signal vectors are given by $\mathbf{s}_1 = -\mathbf{s}_0 = [s_{1,1}, \ldots, s_{1,100}]^T$, with $s_{1,n} = \sin(2\pi n/50)$, $n = 1, \ldots, 100$, and the WGN vector has zero mean and $\sigma = 3$. Consider a matched filter receiver implementation of a LR receiver with $\Lambda_0 = 1$, except instead of using the true $\mathbf{s}_1 = -\mathbf{s}_0$, it uses $\tilde{\mathbf{s}}_1 = -\tilde{\mathbf{s}}_0 = [\tilde{s}_{1,1}, \ldots, \tilde{s}_{1,100}]^T$, with $\tilde{s}_{1,n} = \sin(\theta + (2\pi n/50))$, with $\theta = 0, 0.1, 0.2, 0.4$. (These errors could have been caused by an imperfect phase estimator for the true signal vectors.) Find the average error probability P_e for the four values of θ.

4.7 Consider the eigenvalue problem of

$$\int_{-\infty}^{\infty} K(t-s)f(s)ds = \lambda f(t), \ -\infty < t < \infty.$$

Suppose $K(t) = h(t)$, $-\infty < t < \infty$, where $h(t)$ is the impulse response function of a linear time-invariant system with a system transfer function $H(\omega) = \mathscr{F}\{h(t)\}$, $-\infty < \omega < \infty$. Show $f(t) = \exp(i\omega_0 t)$, $-\infty < t < \infty$ is an eigenfunction of the above eigenvalue problem for $K(t) = h(t)$. What is the eigenvalue λ for the eigenfunction $\exp(i\omega_0 t)$, $-\infty < t < \infty$?

4.8 Consider the covariance matrix

$$\mathbf{R} = \begin{bmatrix} 1 & \sqrt{2} \\ \sqrt{2} & 2 \end{bmatrix}.$$

(a) Find the eigenvalues $\lambda_1 \geq \lambda_0$ and associated eigenvectors $\boldsymbol{\theta}_1$ and $\boldsymbol{\theta}_0$ of $\mathbf{R}$.
(b) What is the Mercer theorem expansion of $\mathbf{R}$ in terms of these eigenvalues and eigenvectors?
(c) Does $\mathbf{R}^{-1}$ exist?

4.9 Consider the binary detection problem of

$$\mathbf{X} = \begin{cases} \mathbf{s}_0 + \mathbf{N}, & H_0, \\ \mathbf{s}_1 + \mathbf{N}, & H_1, \end{cases}$$

where $\mathbf{s}_0 = [30, -30]^T = -\mathbf{s}_1$ and $\mathbf{N}$ is a zero mean CGN vector with a covariance matrix

$$\mathbf{R} = \begin{bmatrix} 8 & 2 \\ 2 & 5 \end{bmatrix}.$$

(a) Find the eigenvalues and eigenvectors of $\mathbf{R}$.
(b) Find an orthogonal transformation $\mathbf{T}$ such that $\mathbf{T}^T\mathbf{R}\mathbf{T}$ is a diagonal matrix.
(c) Find the whitening filter for this $\mathbf{R}$.
(d) Find the optimum decision regions $\mathbb{R}_1$ and $\mathbb{R}_0$ as well as the optimum LR receiver with a LR constant $\Lambda_0 = 1$.

4.10 Consider a zero mean wide-sense stationary CGN process $\{X(t), \ -\infty < t < \infty\}$ with an autocorrelation function $R(t) = 10e^{-5.5|t|}$, $-\infty < t < \infty$. We want to investigate some of its eigenvalue and eigenvector properties. Start with the integral eigenvalue problem

$$\int_0^1 R(t-\tau)\theta_k(\tau)d\tau = \lambda_k \theta_k(t), \ k = 0, 1, \ldots, \ 0 \leq t \leq 1. \tag{4.274}$$

We can numerically approximate the solutions of (4.274) by using $N + 1$ term summations in (4.275):

$$\sum_{j=0}^{N} R((i-j)/N)\theta_k(j/N)(1/N) \approx \lambda_k \theta_k(i/N),$$

$$0 \le i \le N,\ 0 \le j \le N,\ k = 0, 1, \ldots, N. \tag{4.275}$$

Equation (4.275) can be solved as a standard matrix eigenvalue problem (using Matlab or other programs) in the form of

$$\mathbf{R}\boldsymbol{\theta}_k = \lambda_k' \boldsymbol{\theta}_k,\ k = 0, 1, \ldots, N. \tag{4.276}$$

where $\lambda_k' = N\lambda_k$,

$$\boldsymbol{\theta}_k = [\theta_k(0/N), \theta_k(1/N), \ldots, \theta_k(N/N)]^T,\ k = 0, 1, \ldots, N. \tag{4.277}$$

and

$$\mathbf{R} = [R((i-j)/N)]_{0 \le i \le N; 0 \le j \le N} \tag{4.278}$$

is a $(N + 1) \times (N + 1)$ symmetric correlation matrix.

(a) How is the $\mathbf{R}$ of (4.278) related for different values of $N = 4, 8, 16, \ldots$?
(b) Find an approximation to the largest eigenvalue λ_0 in (4.274). Note, the eigenvalues of eig$(\cdot)$ in Matlab are not guaranteed to be in monotonic increasing form (although most them are). We have to sort them out for ourselves.
(c) Can we find the minimum eigenvalue of λ_k of (4.274) by using the approximation of (4.275) and (4.276)?
(d) From integral eigenvalue theory, it can be shown that for the above eigenfunction, $\theta_k(t)$, $k = 0, 1, \ldots,$ has k zeros in $[0, 1]$. Does your $\theta_k(j/N)$ satisfy this condition for $N = 4, 8, 16, \ldots$?

4.11 (*) Solve Problem 4.10 using the differential equation method used in Example 4.9. Hint: For more details, see [15, pp. 133–143] and [16, pp. 361–370].

4.12 Let the binary detection of the signal vector $\mathbf{s}$ in CGN be modeled by

$$\mathbf{X} = \begin{cases} \mathbf{N}, & H_0, \\ \mathbf{s} + \mathbf{N}, & H_1, \end{cases}$$

where

$$\mathbf{X} = [x_1, x_2]^T,\ \mathbf{N} = [N_1, N_2]^T,\ \mathbf{s} = [1, 2]^T,$$

and $\mathbf{N}$ is a zero mean Gaussian vector with a covariance matrix

$$\mathbf{R} = \begin{bmatrix} 2 & 1 \\ 1 & 4 \end{bmatrix}.$$

(a) Find the eigenvalues λ_0 and λ_1 (take $\lambda_1 > \lambda_0$) and normalized eigenvectors $\boldsymbol{\theta}_0$ and $\boldsymbol{\theta}_1$.

(b) Consider the KL expansion of $\mathbf{N} = n_0\boldsymbol{\theta}_0 + n_1\boldsymbol{\theta}_1$. Find n_0, n_1, μ_{n_0}, μ_{n_1}, $\sigma^2_{n_0}$, and $\sigma^2_{n_1}$.
(c) Consider the expansion of $\mathbf{s} = s_0\boldsymbol{\theta}_0 + s_1\boldsymbol{\theta}_1$. Find s_0 and s_1.
(d) Consider the expansion of $\mathbf{X} = x_0\boldsymbol{\theta}_0 + x_1\boldsymbol{\theta}_1$. Find x_0 and x_1.
(e) Numerically compute $\lambda_0\boldsymbol{\theta}_0\boldsymbol{\theta}_0^T + \lambda_1\boldsymbol{\theta}_1\boldsymbol{\theta}_1^T$, and show it is equal to $\mathbf{R}$.
(f) Numerically compute

$$\frac{\boldsymbol{\theta}_0\boldsymbol{\theta}_0^T}{\lambda_0} + \frac{\boldsymbol{\theta}_1\boldsymbol{\theta}_1^T}{\lambda_1}$$

and show it is equal to $\mathbf{R}^{-1}$.
(g) Show the sufficient statistic

$$\Gamma^{(1)}_{\mathrm{CGN}} = \frac{X_0 s_0}{\lambda_0} + \frac{X_1 s_1}{\lambda_1}$$

is equal to

$$\Gamma^{(2)}_{\mathrm{CGN}} = \mathbf{X}^T\mathbf{R}^{-1}\mathbf{s}$$

by numerical computation.

4.13 (*) Consider a binary hypothesis testing of $s(t)$ in CGN modeled by

$$X(t) = \begin{cases} N(t) & , \quad H_0 \\ s(t) + N(t), & \quad H_1 \end{cases}, \quad 0 \le t \le T$$

where $N(t)$ is a zero-mean CGN process with an autocorrelation function $R(t)$ and $s(t)$ is a deterministic waveform.

(a) The optimum correlation function $\{g(t),\ 0 \le t \le T\}$ for this CGN problem used in the sufficient statistic of $\gamma = \int_0^T x(t)g(t)dt$ is obtained from the solution of $\int_o^T R(t-t')g(t')dt' = s(t),\ 0 \le t \le T$. Then the CGN matched filter impulse response function $h(t) = g(T-t)$, $0 \le t \le T$. If T is large, approximate the range of integration to be over the entire real line and show the transfer function $H(\omega)$ of the matched filter impulse response function $h(t)$ is then given by $H(\omega) = S^*(\omega)\exp(-j\omega T)/N(\omega)$, where $S(\omega)$ is the Fourier transform of the signal $s(t)$ and $N(\omega)$ is the power spectral density of the CGN process.
(b) Suppose $N(\omega) = K/(\omega^2 + \omega_o^2)$. Find the causal transfer function $H_{\mathrm{W}}(\omega)$ of the whitening filter. A transfer function is causal if all of its zeros and poles are in the left half plane.
(c) Show the output signal component of the matched filter under hypothesis H_1 is given by $s_0(t) = (1/2\pi)\int_{-\infty}^{\infty}[|S(\omega)|^2/N(\omega)]\exp(j\omega(t-T))d\omega$.
(d) Show singular detection with $P_{\mathrm{D}} = 1$ can occur for arbitrarily small signal energy $(1/2\pi)\int_{-\infty}^{\infty}|S(\omega)|^2 d\omega$ if the noise spectral density $N(\omega)$ is such that $(1/2\pi)\int_{-\infty}^{\infty}[|S(\omega)|^2/N(\omega)]d\omega = \infty$.
(e) Suppose the signal spectrum $S(\omega) \neq 0$, $\omega \in \{(-b,-a) \cup (a,b)\}$ but the noise spectral density $N(\omega) = 0$, $\omega \in \{(-b,-a) \cup (a,b)\}$. Then show singular detection with $P_{\mathrm{D}} = 1$ can occur.

4.14 Consider the received waveform of a binary detection receiver to be given by

$$X(t) = \begin{cases} -2 + N(t)\ ,\ H_0, \\ \ \ 2 + N(t)\ ,\ H_1, \end{cases} \quad 0 \le t \le 1, \tag{4.279}$$

where $N(t)$ is a WGN process of zero-mean and autocorrelation function of $R(t) = \delta(t)$. Suppose we use a receiver based on the statistic

$$\gamma = \int_0^1 x(t)h(t)dt \begin{matrix} > 0 \Rightarrow H_1, \\ < 0 \Rightarrow H_0. \end{matrix} \tag{4.280}$$

(a) What is the best $h(t)$ to use in (4.280) under the Bayes criterion with a LR constant $\Lambda_0 = 1$? What is the associated $P_{\rm FA}$ and $P_{\rm D}$ for this $h(t)$?

(b) Suppose we use $h(t)$ in (4.280) given by

$$h(t) = \begin{cases} \sin(\pi t), & 0 \le t \le 1, \\ 0, & \text{elsewhere}. \end{cases} \tag{4.281}$$

Find $P_{\rm FA}$ and $P_{\rm D}$ using the $h(t)$ of (4.281) in terms of the $\mathrm{Q}(\cdot)$ function.

(c) What is the equivalent loss (in dB) for the detection system using $h(t)$ of (4.281) in part (b) as compared to the ideal detection system in part (a)?

4.15 (*) Derive the upper bound of the expression on (4.242) and the associated optimum $\hat{\mathbf{s}}$ of (4.238) by using the Karish–Kuhn–Tucker (KKT) convex optimization method. Hint: Details on the KKT method can be found in [4, p. 330].

4.16 Consider the binary detection of two signal vectors in WGN process modeled by

$$\mathbf{X} = \begin{cases} -\mathbf{s} + \mathbf{N}\ ,\ H_0, \\ \ \ \mathbf{s} + \mathbf{N}\ ,\ H_1, \end{cases}$$

where $\mathbf{X} = [X_1,\ X_2]^T$, $\mathbf{s} = [s_1,\ s_2]^T = [1/2,\ 1/2]^T$, $\mathbf{N} = [N_1,\ N_2]^T$, with i.i.d. $N_i \sim \mathcal{N}(0, \sigma^2)$, $i = 1, 2$. Using the LR test, the sufficient statistic yields $\Gamma = \mathbf{X}^T(\mathbf{s} - (-\mathbf{s})) = 2\mathbf{X}^T\mathbf{s} = [X_1,\ X_2][1,\ 1]^T = X_1 + X_2$. With an LR constant of $\Lambda_0 = 1$, the sufficient statistic threshold is given by $\gamma_0 = 0$.

(a) Analytical evaluation of performance.
From theory, we know Γ is a Gaussian r.v. under both hypotheses. Define the SNR by $\mathrm{SNR(dB)} = 10\log_{10}(s_i^2/\sigma^2)$. Find the numerical value of $P_{\rm FA}$ and $P_{\rm D}$ for $\mathrm{SNR(dB)} = 0$ dB and $\mathrm{SNR(dB)} = 6.06$ dB.

(b) Monte Carlo evaluation of performance.
Suppose we do not know that Γ is a Gaussian r.v., then we can find by definition of $P_{\rm FA}$ as

$$P_{\rm FA} = \mathrm{P}(\Gamma \ge 0 | H_0). \tag{4.282}$$

Let us use MC simulation to perform the evaluation of $P_{\rm FA}$ of (4.282) as follows. In (4.282), use the appropriate Gaussian $\{X_1,\ X_2\}$ with the appropriate means and variances under H_0 for a given SNR(dB). Then generate two appropriate independent

PN sequences X_1 and X_2 of length M. Then denote by M_0 the number of $\{X_1, X_2\}$ satisfying $\Gamma = X_1 + X_2 \geq 0$. Thus a MC simulation of P_{FA} can be given by

$$P_{\text{FA}}^{(\text{MC})} = M_0/M. \tag{4.283}$$

(c) Compare the analytically evaluated P_{FA} to the MC evaluation $P_{\text{FA}}^{(\text{MC})}$ as a function of M for different SNR(dB).

5 M-ary detection and classification of deterministic signals

In Chapter 4, we considered the detection of known binary deterministic signals in Gaussian noises. In this chapter, we consider the detection and classification of M-ary deterministic signals. In Section 5.1, we introduce the problem of detecting M given signal waveforms in AWGN. Section 5.2 introduces the Gram–Schmidt orthonormalization method to obtain a set of N orthonormal signal vectors or waveforms from a set of N linearly independent signal vectors or waveforms. These orthonormal vectors or signal waveforms are used as a basis for representing M-ary signal vectors or waveforms in their detection. Section 5.3 treats the detection of M-ary given signals in AWGN. Optimum decisions under the Bayes criterion, the minimum probability of error criterion, the maximum a posteriori criterion, and the minimum distance decision rule are considered. Simple minimum distance signal vector geometry concepts are used to evaluate symbol error probabilities of various commonly encountered M-ary modulations including binary frequency-shifted-keying (BFSK), binary phase-shifted-keying (BPSK), quadra phase-shifted-keying (QPSK), and quadra-amplitude-modulation (QAM) communication systems. Section 5.4 considers optimum signal design for M-ary systems. Section 5.5 introduces linearly and non-linearly separable and support vector machine (SVM) concepts used in classification of M deterministic pattern vectors. A brief conclusion is given in Section 5.6. Some general comments are given in Section 5.7. References and homework problems are given at the end of this chapter.

5.1 Introduction

Consider the transmission of one of M messages denoted by m_i, $i = 1, \ldots, M$, or equivalently one of M hypotheses H_i, $i = 1, \ldots, M$, with each having an associated signal waveform $s_i(t)$, $i = 1, \ldots, M$, $0 \le t \le T$, observed over an AWGN channel as shown in Fig. 5.1. The observed waveform is denoted by

$$x(t) = s_i(t) + n(t),\ 0 \le t \le T,\ i \in \{1, \ldots, M\}, \tag{5.1}$$

where each $s_i(t)$ is a known deterministic signal waveform of finite energy over the interval $[0, T]$, and $n(t)$ is the realization of a zero-mean WGN process with a power spectral density of $S(f) = N_0/2$, $-\infty < f < \infty$. The M-ary detection problem is given the observed waveform $x(t)$, $0 \le t \le T$, how do we decide which message m_i, or hypothesis H_i, or signal waveform $s_i(t)$ was sent under some appropriate criterion?

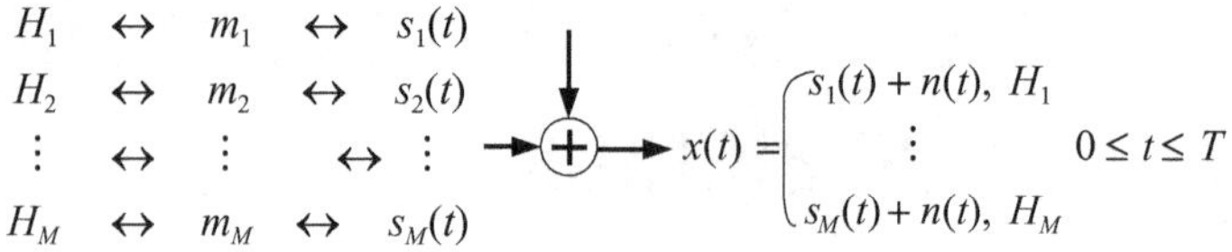

$$
\begin{array}{ccccc}
H_1 & \leftrightarrow & m_1 & \leftrightarrow & s_1(t) \\
H_2 & \leftrightarrow & m_2 & \leftrightarrow & s_2(t) \\
\vdots & \leftrightarrow & \vdots & \leftrightarrow & \vdots \\
H_M & \leftrightarrow & m_M & \leftrightarrow & s_M(t)
\end{array}
\longrightarrow \oplus \longrightarrow x(t) = \begin{cases} s_1(t) + n(t), \ H_1 \\ \vdots \\ s_M(t) + n(t), \ H_M \end{cases} \quad 0 \leq t \leq T
$$

Figure 5.1 M-ary detection over an AWGN channel.

Let us consider the Bayes criterion where the costs c_{ij} and the prior probabilities π_i are known. Then the average cost is given by

$$
C = \sum_{i=1}^{M} \sum_{i=1}^{M} c_{ij} \pi_i \mathrm{P}(\text{Decide hypothesis } j | \text{Hypothesis } i) = \sum_{i=1}^{M} \sum_{j=1}^{M} c_{ij} \pi_i P_{\mathrm{e}_j | H_i}. \quad (5.2)
$$

In particular, consider the special case of the Bayes criterion where the costs are given by $c_{ij} = 1 - \delta_{ij},\ i, j = 1, \ldots, M$. That is, $c_{ij} = 1,\ i \neq j$, and $c_{ij} = 0,\ i = j$. In other words, there is unit cost when an error occurs and no cost when there is no error. This special case of the Bayes criterion makes sense for digital communication system modeling. Furthermore, denote the conditional probability of an error given hypothesis H_i as

$$
P_{\mathrm{e}|H_i} = \mathrm{P}(\text{Error} | H_i),\ i \in \{1, \ldots, M\}. \quad (5.3)
$$

Then the average probability of error given by

$$
P_{\mathrm{e}} = \sum_{i=1}^{M} \pi_i P_{\mathrm{e}|H_i} = \sum_{j=1}^{M} \sum_{j=1}^{M} \pi_i c_{ij} P_{\mathrm{e}_j|H_i} = C|_{c_{ij} = 1 - \delta_{ij}} \quad (5.4)
$$

is equal to the average cost for the special case of $c_{ij} = 1 - \delta_{ij},\ i, j = 1, \ldots, M$. Thus, the Bayes criterion receiver which minimizes the average cost is identical to the receiver that minimizes the average probability of an error.

5.2 Gram–Schmidt orthonormalization method and orthonormal expansion

In order to perform an orthonormal expansion of the signals and noise vectors in terms of the transmitted waveforms, we need to use the *Gram–Schmidt* (GS) *Orthonormalization Method*. In the AWGN channel model, components of the received waveform along all the orthonormal vectors are independent and provide simple geometric interpretation in error probability evaluations. Consider a set $\{\mathbf{s}_1, \ldots, \mathbf{s}_N\}$ of linearly independent vectors in $\mathbb{R}^P$, $N \leq P$. These vectors are said to be linearly independent if $c_1 \mathbf{s}_1 + \ldots + c_N \mathbf{s}_N = 0$, only for $c_n = 0,\ n = 1, \ldots, N$. In other words, we can not express any of the $\mathbf{s}_n$ in terms of linear combination of the other $\mathbf{s}_m,\ m \neq n,\ m = 1, \ldots, N$.

The Gram–Schmidt Orthonormalization Method (GSOM) states that for every set $\{\mathbf{s}_1, \ldots, \mathbf{s}_N\}$ of linearly independent vectors in $\mathbb{R}^P$, $N \leq P$, there is an orthonormal set $\{\boldsymbol{\theta}_1, \ldots, \boldsymbol{\theta}_N\}$ of vectors in $\mathbb{R}^P$, $N \leq P$, such that the space spanned by the set of $\{\mathbf{s}_1, \ldots, \mathbf{s}_N\}$ is equal to the space spanned by the set of $\{\boldsymbol{\theta}_1, \ldots, \boldsymbol{\theta}_N\}$. That is, the space generated by $c_1 \mathbf{s}_1 + \ldots + c_N \mathbf{s}_N$, for all possible real-valued numbers

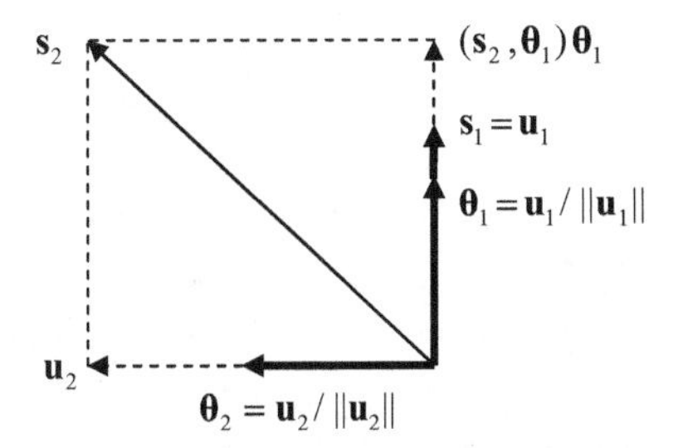

Figure 5.2 G-S orthonormalization of $\{\mathbf{s}_1, \mathbf{s}_2\}$ to obtain orthonormal $\{\boldsymbol{\theta}_1, \boldsymbol{\theta}_2\}$.

$\{c_n, n = 1, \ldots, N\}$ is equal to the space generated by $d_1\boldsymbol{\theta}_1 + \ldots + d_N\boldsymbol{\theta}_N$, for all possible real-valued numbers $\{d_n, n = 1, \ldots, N\}$. Furthermore, the GSOM provides an explicit formula for finding the set of $\{\boldsymbol{\theta}_1, \ldots, \boldsymbol{\theta}_N\}$.

Gram–Schmidt Orthonormalization Method (GSOM)

Step 1. Take $\mathbf{u}_1 = \mathbf{s}_1$. In general, $\|\mathbf{u}_1\|$ is not equal to one.

Step 2. Define $\boldsymbol{\theta}_1 = \mathbf{u}_1/\|\mathbf{u}_1\|$. Then $\|\boldsymbol{\theta}_1\| = 1$.

Step 3. Take $\mathbf{u}_2 = \mathbf{s}_2 - (\mathbf{s}_2, \boldsymbol{\theta}_1)\boldsymbol{\theta}_1$. Then $\mathbf{u}_2$ is orthogonal to $\boldsymbol{\theta}_1$. That is,

$$\begin{aligned}(\mathbf{u}_2, \boldsymbol{\theta}_1) &= (\mathbf{s}_2 - (\mathbf{s}_2, \boldsymbol{\theta}_1)\boldsymbol{\theta}_1, \boldsymbol{\theta}_1) = (\mathbf{s}_2, \boldsymbol{\theta}_1) - ((\mathbf{s}_2, \boldsymbol{\theta}_1)\boldsymbol{\theta}_1, \boldsymbol{\theta}_1), \\ &= (\mathbf{s}_2, \boldsymbol{\theta}_1) - (\mathbf{s}_2, \boldsymbol{\theta}_1)(\boldsymbol{\theta}_1, \boldsymbol{\theta}_1) = (\mathbf{s}_2, \boldsymbol{\theta}_1) - (\mathbf{s}_2, \boldsymbol{\theta}_1) = 0. \end{aligned} \qquad (5.5)$$

The orthogonal property of (5.5) can also be seen from Fig. 5.2.

Step 4. Define $\boldsymbol{\theta}_2 = \mathbf{u}_2/\|\mathbf{u}_2\|$. Clearly, $\|\boldsymbol{\theta}_2\| = 1$ and from (5.5),

$$(\boldsymbol{\theta}_1, \boldsymbol{\theta}_2) = (\mathbf{u}_2/\|\mathbf{u}_2\|, \boldsymbol{\theta}_1) = 0.$$

$\vdots$

Step $2N - 1$. Take $\mathbf{u}_N = \mathbf{s}_N - \sum_{i=1}^{N-1} (\mathbf{s}_N, \boldsymbol{\theta}_i)\boldsymbol{\theta}_i$.

Step $2N$. Finally define $\boldsymbol{\theta}_N = \mathbf{u}_N/\|\mathbf{u}_N\|$.

The set $\{\boldsymbol{\theta}_1, \ldots, \boldsymbol{\theta}_N\}$ then forms an orthonormal set of vectors in $\mathbb{R}^P$, $N \leq P$.

Example 5.1 Consider $N = P = 2$ with $\mathbf{s}_1 = [0,\ 1.8]^T$ and $\mathbf{s}_2 = [-2,\ 2.8]^T$ as shown in Fig. 5.2. From Step 1, take $\mathbf{u}_1 = \mathbf{s}_1$. Then $\|\mathbf{u}_1\| = 1.8$ and from Step 2, $\boldsymbol{\theta}_1 = \mathbf{u}_1/\|\mathbf{u}_1\| = [0,\ 1]^T$. Clearly, $\|\boldsymbol{\theta}_1\| = 1$. Next in Step 3,

$$\begin{aligned}\mathbf{u}_2 = \mathbf{s}_2 - (\mathbf{s}_2, \boldsymbol{\theta}_1)\boldsymbol{\theta}_1 &= \begin{bmatrix} -2 \\ 2.8 \end{bmatrix} - \left(\begin{bmatrix} -2 \\ 2.8 \end{bmatrix}, \begin{bmatrix} 0 \\ 1 \end{bmatrix} \right) \begin{bmatrix} 0 \\ 1 \end{bmatrix} \\ &= \begin{bmatrix} -2 \\ 2.8 \end{bmatrix} - 2.8 \begin{bmatrix} 0 \\ 1 \end{bmatrix} = \begin{bmatrix} -2 \\ 0 \end{bmatrix},\end{aligned}$$

then $\|\mathbf{u}_2\| = 2$ and from Step 4, $\boldsymbol{\theta}_2 = \mathbf{u}_2/\|\mathbf{u}_2\| = [-2,\ 0]^T/2 = [-1,\ 0]^T$. Indeed, $\|\boldsymbol{\theta}_2\| = 1$ and $(\boldsymbol{\theta}_1, \boldsymbol{\theta}_1) = ([0,\ 1]^T, [-1,\ 0]^T) = 0$. □

Example 5.2 Consider $N = 3$ and $P = 2$, where $\mathbf{s}_1 = [0,\ 1.8]^T$ and $\mathbf{s}_2 = [-2,\ 2.8]^T$ are identical to the $\mathbf{s}_1$ and $\mathbf{s}_2$ of Example 5.1. In addition, define $\mathbf{s}_3 = [-4,\ 8.3]^T = 1.5\mathbf{s}_1 + 2\mathbf{s}_2$. Then $\mathbf{s}_3$ is linearly dependent of $\mathbf{s}_1$ and $\mathbf{s}_2$. Thus, by using the two linearly independent $\mathbf{s}_1$ and $\mathbf{s}_2$ in the GSOM, $\boldsymbol{\theta}_1$ and $\boldsymbol{\theta}_2$ identical to those in Example 5.1 are obtained. However, suppose we re-label the indices of these three vectors as $\tilde{\mathbf{s}}_1 = \mathbf{s}_3 = [-4,\ 8.3]^T$, $\tilde{\mathbf{s}}_2 = \mathbf{s}_1 = [0,\ 1.8]^T$, and $\tilde{\mathbf{s}}_3 = \mathbf{s}_2 = [-2,\ 2.8]^T$, all in $\mathbb{R}^2$. Now, suppose we use the two linearly independent vectors in the GSOM to generate the corresponding $\tilde{\boldsymbol{\theta}}_1$ and $\tilde{\boldsymbol{\theta}}_2$. Direct evaluations based on GSOM show, $\tilde{\boldsymbol{\theta}}_1 = [-0.43414,\ 0.90084]^T$ and $\tilde{\boldsymbol{\theta}}_2 = [0.90084,\ 0.43414]^T$, satisfying $\|\tilde{\boldsymbol{\theta}}_1\| = \|\tilde{\boldsymbol{\theta}}_2\| = 1$ and $(\tilde{\boldsymbol{\theta}}_1, \tilde{\boldsymbol{\theta}}_2) = 0$. We note, these two orthonormal vectors $\tilde{\boldsymbol{\theta}}_1$ and $\tilde{\boldsymbol{\theta}}_2$ using $\tilde{\mathbf{s}}_1$ and $\tilde{\mathbf{s}}_2$ in the GSOM are different from the two orthonormal vectors $\boldsymbol{\theta}_1$ and $\boldsymbol{\theta}_2$ using $\mathbf{s}_1$ and $\mathbf{s}_2$ in the GSOM. Indeed, both sets of $\{\tilde{\boldsymbol{\theta}}_1, \tilde{\boldsymbol{\theta}}_2\}$ and $\{\boldsymbol{\theta}_1, \boldsymbol{\theta}_2\}$ span $\mathbb{R}^2$. This example shows the set of N orthonormal vectors obtained from a given set of N independent vectors are not unique. Simple permutations of the indices of the linearly independent vectors in using the GSOM can yield different sets of orthonormal vectors. □

One important application of the GSOM is its ability to take any set of N linearly independent vectors to generate a set of N orthonormal vectors $\{\boldsymbol{\theta}_1, \ldots, \boldsymbol{\theta}_N\}$ in $\mathbb{R}^N$, for the purpose of representing any real-valued vector $\mathbf{x} \in \mathbb{R}^N$ in an orthonormal series expansion

$$\mathbf{x} = c_1\boldsymbol{\theta}_1 + c_2\boldsymbol{\theta}_2 + \ldots + c_N\boldsymbol{\theta}_N,\ \mathbf{x} \in \mathbb{R}^N. \tag{5.6}$$

Each of the coefficients c_j, $j = 1, \ldots, N$, is given explicitly and uniquely by

$$c_j = \left(\mathbf{x}, \boldsymbol{\theta}_j\right),\ j = 1, \ldots, N. \tag{5.7}$$

Example 5.3 Consider the two vectors $\mathbf{s}_1 = [0,\ 1.8]^T$, $\mathbf{s}_2 = [-2,\ 2.8]^T$, and the corresponding orthonormal vectors $\boldsymbol{\theta}_1 = [0,\ 1]^T$ and $\boldsymbol{\theta}_2 = [-1,\ 0]^T$ of Example 5.1. By inspection, $\mathbf{s}_1$ and $\mathbf{s}_2$ have the orthonormal series expansions of

$$\mathbf{s}_1 = [0,\ 1.8]^T = 1.8\boldsymbol{\theta}_1 + 0\boldsymbol{\theta}_2 \Rightarrow c_1 = 1.8,\ c_2 = 0, \tag{5.8}$$

$$\mathbf{s}_2 = [-2,\ 2.8]^T = 2.8\boldsymbol{\theta}_1 + 2\boldsymbol{\theta}_2 \Rightarrow c_1 = 2.8,\ c_2 = 2. \quad \square \tag{5.9}$$

Example 5.4 Consider the three vectors $\mathbf{s}_1 = [0,\ 1.8]^T$, $\mathbf{s}_2 = [-2,\ 2.8]^T$, $\mathbf{s}_3 = [-4,\ 8.3]^T$, and the original corresponding orthonormal vectors of $\boldsymbol{\theta}_1 = [0,\ 1]^T$ and $\boldsymbol{\theta}_2 = [-1,\ 0]^T$ of Example 5.2. Then the orthonormal series expansions of $\mathbf{s}_1$ and $\mathbf{s}_2$ are still given by (5.8) and (5.9). The orthonormal series expansion of $\mathbf{s}_3$ is given by

$$\mathbf{s}_3 = [-4,\ 8.3]^T = 8.3\boldsymbol{\theta}_1 + 4\boldsymbol{\theta}_2, \Rightarrow c_1 = 8.3,\ c_2 = 4. \tag{5.10}$$

However, upon permuting the vector indices, consider $\tilde{\mathbf{s}}_1 = \mathbf{s}_3 = [-4,\ 8.3]^T$, $\tilde{\mathbf{s}}_2 = \mathbf{s}_1 = [0,\ 1.8]^T$, and $\tilde{\mathbf{s}}_3 = \mathbf{s}_2 = [-2,\ 2.8]^T$, with the corresponding orthonormal vectors of $\tilde{\boldsymbol{\theta}}_1 = [-0.43414,\ 0.90084]^T$ and $\tilde{\boldsymbol{\theta}}_2 = [0.90084,\ 0.43414]^T$ of Example 5.2. The orthonormal series expansions of $\tilde{\mathbf{s}}_1$, $\tilde{\mathbf{s}}_2$, and $\tilde{\mathbf{s}}_3$ are now given by

$$\tilde{\mathbf{s}}_1 = \mathbf{s}_3 = [-4,\ 8.3]^T = 9.2136\tilde{\boldsymbol{\theta}}_1 - 1.7231\tilde{\boldsymbol{\theta}}_2 \Rightarrow \tilde{c}_1 = 9.2136,\ \tilde{c}_2 = -1.7231, \tag{5.11}$$

$$\tilde{\mathbf{s}}_2 = \mathbf{s}_1 = [0,\ 1.8]^T = 1.6215\tilde{\boldsymbol{\theta}}_1 + 0.78146\tilde{\boldsymbol{\theta}}_2 \Rightarrow \tilde{c}_1 = 1.6215,\ \tilde{c}_2 = 0.78146, \tag{5.12}$$

$$\tilde{\mathbf{s}}_3 = \mathbf{s}_2 = [-2,\ 2.8]^T = 3.3906\tilde{\boldsymbol{\theta}}_1 - 0.58609\tilde{\boldsymbol{\theta}}_2 \Rightarrow \tilde{c}_1 = 3.3906,\ \tilde{c}_2 = -0.58609. \tag{5.13}$$

By comparing the coefficients c_j in (5.8)–(5.10) with the coefficients $\tilde{c}_j$ in (5.11)–(5.13), we note they are different since the orthonormal vectors are different. However, in both cases, the coefficients are given simply by the inner product of $\mathbf{x}$ with $\boldsymbol{\theta}_j$ given by (5.7). We also note, if instead of using an orthonormal series expansion, suppose we want to perform a series expansion of $\mathbf{x}$ with respect to the linearly independent set of vectors $\{\mathbf{s}_1, \ldots, \mathbf{s}_N\}$ as given by

$$\mathbf{x} = d_1\mathbf{s}_1 + d_2\mathbf{s}_2 + \ldots + d_N\mathbf{s}_N,\ \mathbf{x} \in \mathbb{R}^N. \tag{5.14}$$

The problem is that the coefficients d_j, $j = 1, 2, \ldots, N$, in general do not have any simple and explicit method for their evaluations. □

Example 5.5 Consider $\tilde{\mathbf{s}}_1 = \mathbf{s}_3 = [-4,\ 8.3]^T$, $\tilde{\mathbf{s}}_2 = \mathbf{s}_1 = [0,\ 1.8]^T$, and $\tilde{\mathbf{s}}_3 = \mathbf{s}_2 = [-2,\ 2.8]^T$ of Example 5.2. Since $\tilde{\mathbf{s}}_2 = \mathbf{s}_1 = [0,\ 1.8]^T$ and $\tilde{\mathbf{s}}_3 = \mathbf{s}_2 = [-2,\ 2.8]^T$ are linearly independent vectors in $\mathbb{R}^2$, we know from (5.14), it is possible to express $\tilde{\mathbf{s}}_1 = \mathbf{s}_3 = [-4,\ 8.3]^T$ as a linear combination of $\tilde{\mathbf{s}}_2 = \mathbf{s}_1 = [0,\ 1.8]^T$ and $\tilde{\mathbf{s}}_3 = \mathbf{s}_2 = [-2,\ 2.8]^T$ in the form of

$$\mathbf{s}_3 = [-4,\ 8.3]^T = d_1\mathbf{s}_1 + d_2\mathbf{s}_2. \tag{5.15}$$

From Example 5.2, since we defined $\mathbf{s}_3 = [-4,\ 8.3]^T = 1.5\mathbf{s}_1 + 2\mathbf{s}_2$, then of course we know $d_1 = 1.5$ and $d_2 = 2$. However, if we were given only $\mathbf{s}_1 = [0,\ 1.8]^T$, $\mathbf{s}_2 = [-2,\ 2.8]^T$, and $\mathbf{s}_3 = [-4,\ 8.3]^T$, it is not obvious how to find the coefficients d_1 and d_2 in a simple and explicit manner. □

Most importantly, all the above $2N$ steps in generating the N orthonormal vectors $\{\boldsymbol{\theta}_1, \ldots, \boldsymbol{\theta}_N\}$ from the N linearly independent vectors $\{\mathbf{s}_1, \ldots, \mathbf{s}_N\}$ in the GSOM as well as the orthonormal series expansion of (5.6), are also valid when considered to be functions. In other words, we can generate finite energy real-valued orthonormal functions $\{\theta_1(t), \ldots, \theta_N(t)\}$ over the interval $[0,\ T]$ from the set of linearly independent finite energy real-valued functions $\{s_1(t), \ldots, s_N(t)\}$ over the interval $[0,\ T]$. Thus, the space generated by $c_1 s_1(t) + \ldots + c_N s_N(t)$, for all possible real-valued numbers

$\{c_n,\ n = 1, \ldots, N\}$ is equal to the space generated by $d_1\theta_1(t) + \ldots + d_N\theta_N(t)$, for all possible real-valued numbers $\{d_n,\ n = 1, \ldots, N\}$.

5.3 M-ary detection

Let $\{s_1(t), \ldots, s_M(t)\}$ be the set of M transmission waveforms corresponding to the M hypothesis or M messages in (5.1). This set of M waveforms need not be linearly independent. In other words, some waveforms may be a linear combination of some other waveforms. Let $\{s_1(t), \ldots, s_N(t)\}$, $N \leq M$ be a subset of $\{s_1(t), \ldots, s_M(t)\}$ signal waveforms that are linearly independent, but still has the same linear span as that of $\{s_1(t), \ldots, s_M(t)\}$. Then from these $\{s_1(t), \ldots, s_N(t)\}$, $N \leq M$, waveforms, we want to use the GSOM to generate the orthonormal functions $\{\theta_1(t), \ldots, \theta_N(t)\}$ over the interval $[0,\ T]$. Thus, we can expand each of the M transmission waveforms by

$$s_i(t) = \sum_{j=1}^{N} s_{ij}\theta_j(t),\ 0 \leq t \leq T,\ 0 \leq i \leq M, \tag{5.16}$$

with the expansion coefficients given by

$$s_{ij} = \left(s_i(t), \theta_j(t)\right),\ 0 \leq i \leq M,\ 0 \leq j \leq N. \tag{5.17}$$

The noise can be expanded by

$$n(t) = \sum_{j=1}^{N} n_j\theta_j(t) + n_r(t),\ 0 \leq t \leq T, \tag{5.18}$$

where the noise remainder term $n_r(t)$ is defined by

$$n_r(t) = \sum_{k=N+1}^{\infty} \tilde{n}_k\theta_j(t),\ 0 \leq t \leq T, \tag{5.19}$$

with their coefficients given by

$$n_j = (n(t),\ \theta_j(t)),\ 0 \leq j \leq N, \tag{5.20}$$

$$\tilde{n}_k = (n(t),\ \theta_k(t)),\ N + 1 \leq k < \infty. \tag{5.21}$$

In (5.18), we note all the $\{\theta_1(t), \ldots, \theta_N(t)\}$ are obtained from the GSOM of the $\{s_1(t), \ldots, s_N(t)\}$, $N \leq M$, waveforms. In (5.19), we also know from theory there exists a set of orthonormal function $\{\theta_{N+1}(t),\ \theta_{N+2}(t), \ldots\}$ such that $\{\theta_1(t), \ldots, \theta_N(t),\ \theta_{N+1}(t),\ \theta_{N+2}(t), \ldots\}$ constitutes a complete orthonormal system of finite energy functions on $[0,\ T]$. Since all the functions in $\{\theta_1(t), \ldots, \theta_N(t),\ \theta_{N+1}(t),\ \theta_{N+2}(t), \ldots\}$ are orthonormal, then all the coefficients n_j, $0 \leq j \leq N$, are uncorrelated with all the coefficients $\tilde{n}_k$, $N + 1 \leq k < \infty$. That is,

$$\mathrm{E}\{n_j\tilde{n}_k\} = 0,\ 0 \leq j \leq N,\ N + 1 \leq k < \infty. \tag{5.22}$$

Under hypothesis H_i, the received waveform has the form of

$$x(t) = s_i(t) + n(t) = \sum_{j=1}^{N} s_{ij}\theta_j(t) + \sum_{j=1}^{N} n_j\theta_j(t) + n_r(t)$$
$$= \sum_{j=1}^{N} (s_{ij} + n_j)\theta_j(t) + n_r(t)$$
$$= \sum_{j=1}^{N} x_{ij}\theta_j(t) + n_r(t),\ 0 \leq t \leq T. \tag{5.23}$$

But the received waveform also can be written in the form of

$$x(t) = \sum_{j=1}^{N} x_j\theta_j(t) + n_r(t), \tag{5.24}$$

where the received coefficient is given by

$$x_j = (x(t),\ \theta_j(t)) = \int_0^T x(t)\theta_j(t)dt. \tag{5.25}$$

Denote the vector of the coefficients of the received waveform under hypothesis H_i by

$$\mathbf{x} = \begin{bmatrix} x_1 \\ \vdots \\ x_N \end{bmatrix} = \begin{bmatrix} s_{i1} + n_1 \\ \vdots \\ s_{iN} + n_N \end{bmatrix} = \mathbf{s}_i + \mathbf{n}. \tag{5.26}$$

Since $\{s_1(t), \ldots, s_M(t)\}$ and $\{\theta_1(t), \ldots, \theta_N(t)\}$ are known to the receiver, there is a one-to-one relationship between $\{x(t),\ 0 \leq t \leq T\}$ and $\mathbf{x}$. It is clear that $n_r(t)$ is irrelevant in the decision process, since there are no signal components in $n_r(t)$.

Now, we want to find the optimum decomposition of $\mathbb{R}^N$ into M disjoint decision regions of $\mathbb{R}_1, \ldots, \mathbb{R}_M$. For any received $\mathbf{x} \in \mathbb{R}^N$, let the probability of making a correct decision be denoted by P(Correct|$\mathbf{x}$). Then we want to maximize

$$P_c = 1 - P_e = \iiint_{\mathbb{R}^N} \mathrm{P}(\mathrm{Correct}|\mathbf{x})\mathrm{P}(\mathbf{x})d\mathbf{x}. \tag{5.27}$$

If we decide for hypothesis $\hat{H} = H_i$ or message $\hat{m} = m_i$ or the i-th signal vector $\mathbf{s}_i$ was sent when $\mathbf{x}$ was received, then

$$\mathrm{P}(\mathrm{Correct}|\mathbf{x}) = \mathrm{P}(\hat{H} = H_i|\mathbf{x}) = \mathrm{P}(\hat{m} = m_i|\mathbf{x}) = \mathrm{P}(\mathbf{s}_i \text{ was sent}|\mathbf{x}). \tag{5.28}$$

Hence, given $\mathbf{x}$, the decision rule that maximizes P_c is $\hat{H}$ such that

$$\mathrm{P}(\hat{H} = H_i|\mathbf{x}) = \mathrm{P}(H_i|\mathbf{x}) = \max_j \mathrm{P}(H_j|\mathbf{x}). \tag{5.29}$$

Thus, the decision rule that maximizes P_c is the decision rule that minimizes the average probability of error. We note, (5.29) is called *the Maximum A Posteriori (MAP) Decision Rule*. We also want to denote that the hypothesis H_i, which maximizes the conditional probability $\mathrm{P}(\cdot|\mathbf{x})$, over all M hypotheses in (5.29) with the following notation of

$$\begin{aligned}\{\hat{H} = H_i\} \Leftrightarrow & \arg\max_j \mathrm{P}(H_j|\mathbf{x}) \\ &= \arg\max_j \mathrm{P}(\mathbf{s}_j|\mathbf{x}) = \arg\max_j \mathrm{P}(\mathbf{s}_j|\mathbf{x})\mathrm{P}(\mathbf{x}) \\ &= \arg\max_j \mathrm{P}(\mathbf{s}_j \cap \mathbf{x}) = \arg\max_j \mathrm{P}(\mathbf{x}|\mathbf{s}_j)\pi_j \\ &= \arg \mathrm{P}(\mathbf{x}|\mathbf{s}_i)\pi_i, \end{aligned} \tag{5.30}$$

where π_j is the known prior probability of hypothesis $H_j,\ j = 1, \ldots, M$. We note that $\arg\max_j \mathrm{P}(H_j|\mathbf{x})$ is just a notation defining the H_i that attains the maximum of $\mathrm{P}(H_j|\mathbf{x})$ or, equivalently, the $\mathbf{s}_i$ vector that attains the maximum of $\mathrm{P}(\mathbf{x}|\mathbf{s}_j)\pi_j$. Since we have an AWGN channel, the conditional probability $\mathrm{P}(\mathbf{x}|\mathbf{s}_j)$ for the $\mathbf{s}_j$, in (5.30) is given by

$$\mathrm{P}(\mathbf{x}|\mathbf{s}_j) = \mathrm{P}_{\mathbf{N}}(\mathbf{x} - \mathbf{s}_j) = \frac{1}{(2\pi N_0/2)^{N/2}} \exp\left(\frac{-\|\mathbf{x} - \mathbf{s}_j\|^2}{2N_0/2}\right). \tag{5.31}$$

Then using (5.31) in (5.30), we have

$$\begin{aligned}\{\hat{H} = H_i\} \Leftrightarrow & \arg\max_j \frac{1}{(\pi N_0)^{N/2}} \exp\left(\frac{-\|\mathbf{x} - \mathbf{s}_j\|^2 + N_0 \ln(\pi_j)}{N_0}\right) \\ &= \arg\max_j \left\{N_0 \ln(\pi_j) - \|\mathbf{x} - \mathbf{s}_j\|^2\right\} = \arg\left\{N_0 \ln(\pi_i) - \|\mathbf{x} - \mathbf{s}_i\|^2\right\}, \end{aligned} \tag{5.32}$$

which is equivalent to

$$\{\hat{H} = H_i\} \Leftrightarrow \arg\min_j \left\{\|\mathbf{x} - \mathbf{s}_j\|^2 - N_0 \ln(\pi_j)\right\} = \arg\left\{\|\mathbf{x} - \mathbf{s}_i\|^2 - N_0 \ln(\pi_i)\right\}. \tag{5.33}$$

Equation (5.33) follows from (5.32) since $\arg\max_j\{f_j(\cdot)\} = \arg\min_j\{-f_j(\cdot)\}$. But

$$\|\mathbf{x} - \mathbf{s}_i\|^2 = (\mathbf{x} - \mathbf{s}_i, \mathbf{x} - \mathbf{s}_i) = \|\mathbf{x}\|^2 - 2(\mathbf{x}, \mathbf{s}_i) + \|\mathbf{s}_i\|^2. \tag{5.34}$$

Using (5.34) in (5.32), we have

$$\begin{aligned}\{\hat{H} = H_i\} \Leftrightarrow & \arg\max_j \{N_0 \ln(\pi_j) + 2(\mathbf{x}, \mathbf{s}_j) - \|\mathbf{s}_j\|^2\} \\ &= \arg\max_j \left\{(\mathbf{x}, \mathbf{s}_j) - \frac{E_j}{2} + \frac{N_0}{2} \ln(\pi_j)\right\} \\ &= \arg\max_j \{(\mathbf{x}, \mathbf{s}_j) + C_j\} = \arg\{(\mathbf{x}, \mathbf{s}_i) + C_i\}, \end{aligned} \tag{5.35}$$

where $E_j = \|\mathbf{s}_j\|^2$ is the energy of the j-th signal vector and $C_j = (N_0/2)\ln(\pi_j) - (E_j/2)$. But

$$(\mathbf{x}, \mathbf{s}_i) = \sum_{j=1}^{N} x_j s_{ij} = \int_0^T x(t)s_i(t)dt,\ i = 1, \ldots, M, \tag{5.36}$$

and (5.35) yields the correlation receiver structure for M-ary detection shown in Fig. 5.3. We note, there are M multipliers, M integrators, M adders, and one max comparator in Fig. 5.3. But just as in the binary detection in WGN, we can express the M-ary correlation receiver of Fig. 5.3 in the form of M matched filters, M samplers, M adders, and one

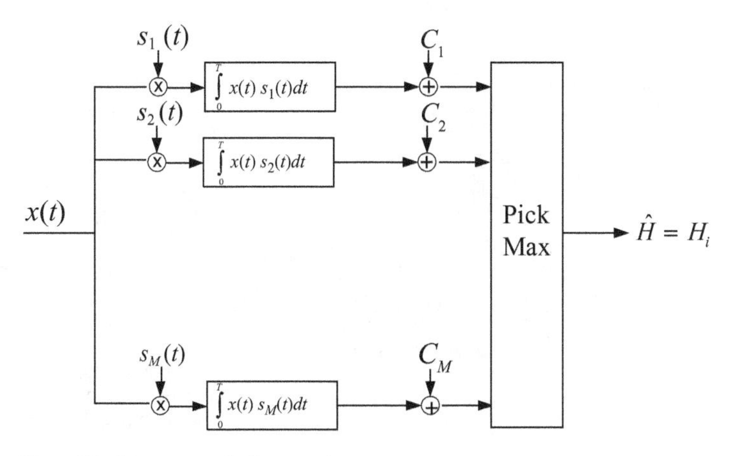

Figure 5.3 M-ary correlation receiver.

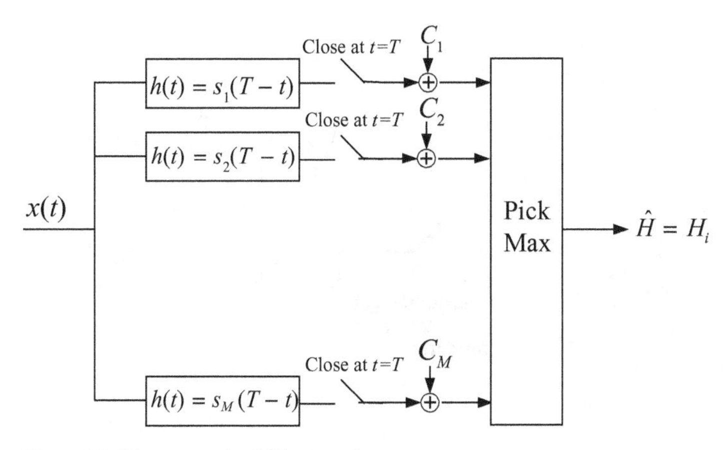

Figure 5.4 M-ary matched filter receiver.

max comparator in Fig. 5.4. Now, consider (5.33) in which all the prior probabilities $\pi_i = 1/M$, $i = 1, \ldots, M$, are equal. This is a case valid in most realistic M-ary digital communication systems. Then (5.33) becomes

$$\{\hat{H} = H_i\} \Leftrightarrow \arg\min_j \|\mathbf{x} - \mathbf{s}_j\|^2 = \arg\left\{\|\mathbf{x} - \mathbf{s}_i\|^2\right\}, \tag{5.37}$$

which is often called the M-ary minimum-distance receiver structure. Figure 5.5(a) shows the case of $M = 2$ and Fig. 5.5(b) shows the case of $M = 3$. In each case, the decision region boundaries are the perpendicular bisectors of the chords between two signal vectors.

Furthermore, if all the signal vector energies are identical, $E = \|\mathbf{s}_i\|^2$, $i = 1, \ldots, M$, then from (5.37) and (5.34) we have

$$\{\hat{H} = H_i\} \Leftrightarrow \arg\min_j \|\mathbf{x} - \mathbf{s}_j\|^2 = \arg\left\{\|\mathbf{x} - \mathbf{s}_i\|^2\right\} = \arg\max_j \left(\mathbf{x}, \mathbf{s}_j\right) = \arg\{(\mathbf{x}, \mathbf{s}_i)\}. \tag{5.38}$$

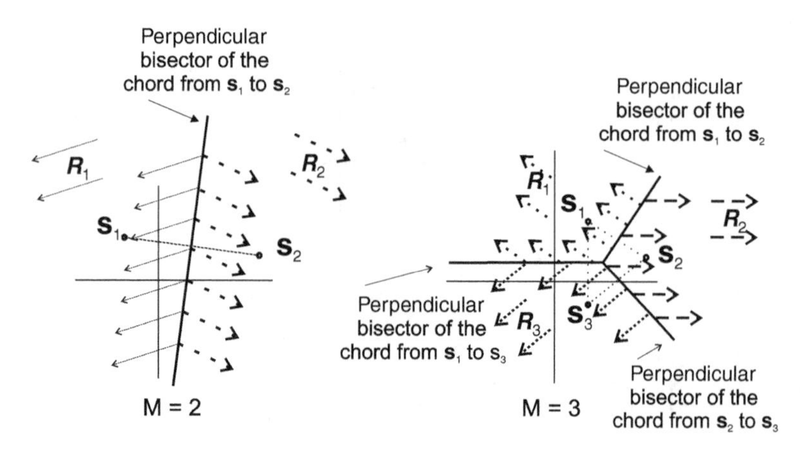

Figure 5.5 Minimum-distance receiver (a) $M = 2$. (b) $M = 3$.

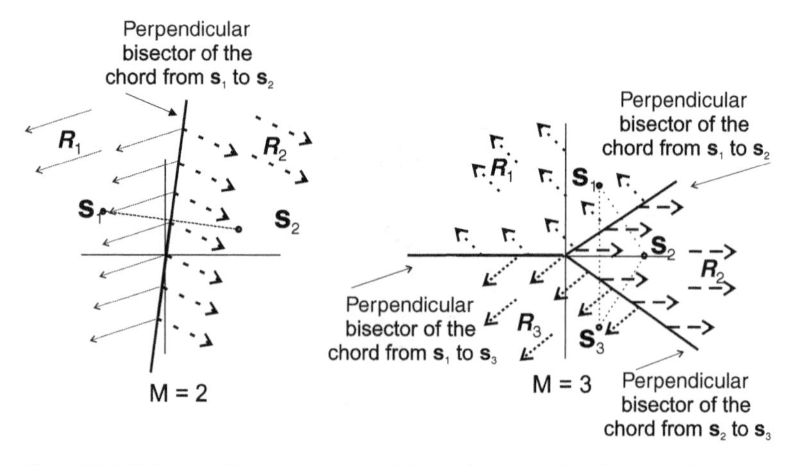

Figure 5.6 Minimum-distance receiver with equal energy signal vectors (a) $M = 2$. (b) $M = 3$.

Thus, all the perpendicular bisectors of the chords between the signal vectors pass through the origin, as can be seen in Fig. 5.6.

Example 5.6 Binary Frequency-Shifted-Keying (BFSK) Modulation
Consider a binary FSK modulation system. Assume $\pi_1 = \pi_2 = 1/2$, $c_{ij} = 1 - \delta_{ij}$, and thus $\Lambda_0 = 1$. Take the two signal waveforms $s_1(t)$ and $s_2(t)$ to be

$$s_1(t) = \sqrt{\frac{2E}{T}} \cos(2\pi f_1 t),\ 0 \le t \le T,\ s_2(t) = \sqrt{\frac{2E}{T}} \cos(2\pi f_2 t),\ 0 \le t \le T. \quad (5.39)$$

Define $\theta_1(t)$ and $\theta_2(t)$ by

$$\theta_1(t) = \sqrt{\frac{2}{T}} \cos(2\pi f_1 t),\ 0 \le t \le T,\ \theta_2(t) = \sqrt{\frac{2}{T}} \cos(2\pi f_2 t),\ 0 \le t \le T. \quad (5.40)$$

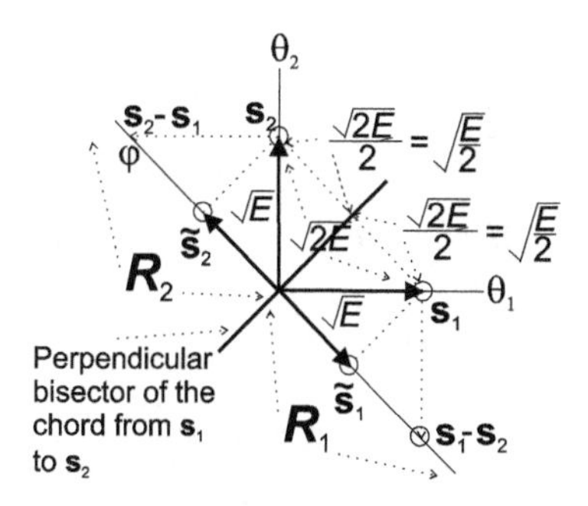

Figure 5.7 Detection of binary FSK modulated signals.

If $1/|2\pi f_1 - 2\pi f_2| \ll T$, then the two $\theta_1(t)$ and $\theta_2(t)$ functions satisfy

$$\|\theta_1(t)\|^2 \approx \|\theta_2(t)\|^2 \approx 1, \ (\theta_1(t), \theta_2(t)) = \int_0^T \theta_1(t)\theta_2(t)dt \approx 0. \tag{5.41}$$

Thus, $\theta_1(t)$ and $\theta_2(t)$ are two approximately orthonormal functions. Then

$$s_1(t) = \sqrt{E} - \theta_1(t), \ s_2(t) = \sqrt{E} - \theta_2(t), \tag{5.42}$$

and

$$\|s_1(t)\|^2 \approx \|s_2(t)\|^2 \approx E, \ (s_1(t), s_2(t)) = \int_0^T \theta_1(t)\theta_2(t)dt \approx 0. \tag{5.43}$$

Assume the standard binary known signal detection in the AWGN model of

$$x(t) = \begin{cases} s_1(t) + n(t), & H_1 \\ s_2(t) + n(t), & H_2 \end{cases}, \ 0 \le t \le T, \tag{5.44}$$

where $n(t)$ is the realization of a WGN process with zero-mean and a power spectral density of $S(f) = N_0/2, \ -\infty < f < \infty$.

The sufficient statistic of this detection problem is given by

$$\gamma_0 = \int_0^T x(t)(s_2(t) - s_1(t))dt \gtrless_{H_1}^{H_2} \gamma_0 = 0. \tag{5.45}$$

Then the optimum decision regions are given by

$$\mathbb{R}_2 = \{x(t) : \gamma_0 \ge 0\}, \ R_1 = \{x(t) : \gamma_0 \le 0\}. \tag{5.46}$$

The system performance P_{FA} and P_{D} can be evaluated.

However, by using the geometric interpretation of this BFSK modulation system, the optimum decision regions and the system performance can be evaluated readily. In Fig. 5.7, we note $\theta_1(t)$ and $\theta_2(t)$ form two orthogonal coordinates in the system and $s_1(t)$ and $s_2(t)$ are also orthogonal. From the minimum-distance receiver criterion of (5.38), all the points to the left of the perpendicular bisector of the chord connecting $\mathbf{s}_1$ to $\mathbf{s}_2$ define $\mathbb{R}_2$ and all the points to the right define $\mathbb{R}_1$. Furthermore, define a new waveform by

$$\phi(t) = \frac{s_2(t) - s_1(t)}{\|s_2(t) - s_1(t)\|}. \tag{5.47}$$

Clearly, $\|\phi(t)\|^2 = 1$ and $\|s_2(t) - s_1(t)\|^2 = 2E$. Define a new sufficient statistic

$$\tilde{\gamma} = \int_0^T n(t)\phi(t)dt. \tag{5.48}$$

Then $\tilde{\gamma}$ is the realization of a Gaussian r.v. with zero-mean and variance

$$\mathrm{E}\{\tilde{\gamma}^2\} = \mathrm{E}\left\{\int_0^T N(t)\phi(t)dt \int_0^T N(\tau)\phi(\tau)d\tau\right\} = N_0/2. \tag{5.49}$$

From Fig. 5.7, given H_2, the original signal vector $\mathbf{s}_2$ upon projection transformation onto $\boldsymbol{\varphi}$ is now denoted by $\tilde{\mathbf{s}}_2$. Then the probability of detection can be evaluated geometrically to yield

$$\begin{aligned} P_{\mathrm{D}}^{\mathrm{BFSK}} &= \mathrm{P}(\text{projected noise on } \boldsymbol{\varphi} \geq -\sqrt{E/2}|\tilde{\mathbf{s}}_2) = \mathrm{P}\left(\tilde{\gamma} \geq -\sqrt{E/2}\right) \\ &= \mathrm{P}\left(\frac{\tilde{\gamma}}{\sqrt{N_0/2}} \geq \frac{-\sqrt{E/2}}{\sqrt{N_0/2}}\right) = \mathrm{Q}(-\sqrt{E/N_0}). \end{aligned} \tag{5.50}$$

Similarly, given H_1, the probability of a false alarm can be evaluated geometrically to yield

$$\begin{aligned} P_{\mathrm{FA}}^{\mathrm{BFSK}} &= \mathrm{P}(\text{projected noise on } \boldsymbol{\varphi} \geq \sqrt{E/2}|\tilde{\mathbf{s}}_1) = \mathrm{P}(\tilde{\gamma} \geq \sqrt{E/2}) \\ &= \mathrm{P}\left(\frac{\tilde{\gamma}}{\sqrt{N_0/2}} \geq \frac{\sqrt{E/2}}{\sqrt{N_0/2}}\right) = \mathrm{Q}(\sqrt{E/N_0}). \end{aligned} \tag{5.51}$$

Example 5.7 AT&T Bell System 103A Modem using binary FSK modulation

As the number of mainframe computers increased rapidly in the early 1960s, the number of potential users wanting to use these computers at remote sites also increased significantly. In 1962, AT&T introduced the low-cost Bell 103A full-duplex (i.e., two one-way digital communication links) modem (MOdulate-DEModulate) transceiver over the analog two-wire POTS (Plain-Old-Telephone-System) system for 300 bits per second (bps) binary data communication. The binary data of a "one" (also called a "mark") and a "zero" (also called a "space"), from the old telegraph notation, can be modeled by two sinusoids (also called a "tone") of amplitude A and frequencies f_1 and f_0 of duration T. Thus, the Bell 103A modem operates as two BFSK systems as shown in Fig. 5.8. The remote modem transmitter is denoted as the originate transmitter uses a 1070 Hz tone to send a "space" and a 1270 Hz tone to send a "mark." The answer transmitter uses a 2025 Hz tone to send a "space" and a 2225 Hz tone to send a "mark." Receivers of early 103A type modems used energy detection criterion at the outputs of four simple bandpass filters centered at each of these four frequencies to decide which tone was transmitted. We note, for this system, $8 \times 10^{-4} = 1/|2\pi(1270 - 1070)| = 1/|2\pi(2225 - 2025)| <$

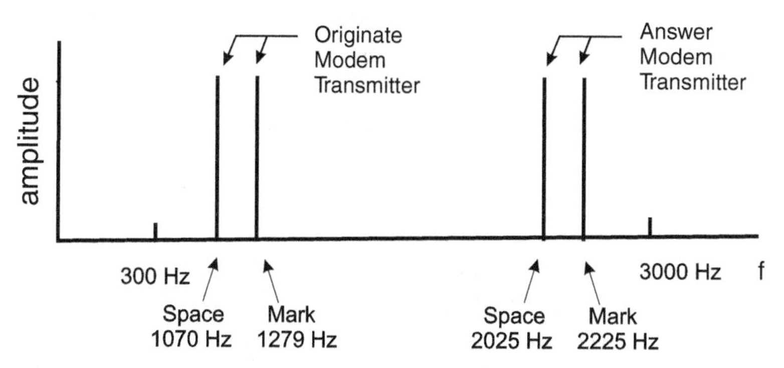

Figure 5.8 Frequencies of four tones used in the full-duplex 300 bps Bell 103A modem.

$1/300 = 3.3 \times 10^{-3}$, the two originate tones (as well as the two answer tones) are not completely orthonormal, since the condition of $1/|2\pi f_1 - 2\pi f_2| \ll T$ is not quite satisfied. □

Example 5.8 *M*-ary FSK modulation

The waveforms of a M-ary FSK system are modeled by

$$x(t) = \begin{cases} A\cos(2\pi f_1 t) + n(t), \\ A\cos(2\pi f_2 t) + n(t), \\ \vdots \\ A\cos(2\pi f_M t) + n(t), \end{cases} \quad 0 \le t \le T. \tag{5.52}$$

The M-ary FSK system is a practical form of modulation since for each transmission baud of duration T, it conveys $\log_2 M$ bits of data information, in contrast to the binary FSK system of $M = 1$, which conveys only one bit of data information. □

Example 5.9 Wireless pager using FSK modulation

Since the 1960s (almost 40 years before cellular telephone became available, wireless pager provided a simple method for a paging company to send few bits of information to a user with a low-cost and quite fade-resistant pager. In the simplest mode of a paging operation, the user only knows that he/she is "wanted." In many situations (e.g., a doctor who is on call), the user knows exactly how to respond to such a message (e.g., the doctor "on call" needs to go to the appropriate office). In a slightly more sophisticated paging system, the user receives a telephone number. In the early days (before the cellular telephone days), the user probably needed to find a POTS line to call this telephone number to find out detailed information from the caller. In the United States, the FLEX paging protocol, originated by Motorola, uses the FSK modulation system over different frequencies. For the 1660 bps system, the binary FSK format is used. For the 3200 bps and 6400 bps systems, $M = 4$ FSK format is used. In practice, various error-correcting codes have been used to improve the performance of the paging system. In the UK, the POCSAG paging system also uses the binary FSK format with a +4.5 KHz shift representing a "zero" and a −4.5 KHz shift representing a "one" relative to various carrier frequencies. In practice, various error-correcting codes have also been used to improve the performance of the paging system. □

Example 5.10 Binary Phase-Shifted-Keying (BPSK) modulation

In the BFSK modulation system, even though the original signal vectors $\mathbf{s}_1$ and $\mathbf{s}_2$ have energies $\|s_1(t)\|^2 \approx \|s_2(t)\|^2 \approx E$, because the two vectors are orthogonal, the transformed signal vectors $\tilde{\mathbf{s}}_1$ and $\tilde{\mathbf{s}}_2$ on $\boldsymbol{\varphi}$ coordinate had effective energy $\|\tilde{\mathbf{s}}_1\|^2 = \|\tilde{\mathbf{s}}_2\|^2 \approx E/2$. In the BPSK modulation system, as shown in Fig. 5.8, let the two signal vectors $\mathbf{s}_1$ and $\mathbf{s}_2$ also have equal energies $\|s_1(t)\|^2 \approx \|s_2(t)\|^2 \approx E$, but they are chosen to be antipodal (i.e., opposite) in the sense of

$$-s_2(t) = s_1(t) = \sqrt{\frac{2E}{T}}\cos(2\pi f t) = \sqrt{E}\theta_1(t),\ 0 \le t \le T, \tag{5.53}$$

where $\theta_1(t)$ is a unit norm waveform. Then we can define a unit norm vector $\phi(t)$ in the same manner as in (5.47), except now it becomes

$$\phi(t) = \frac{s_2(t) - s_1(t)}{\|s_2(t) - s_1(t)\|} = \frac{-2s_1(t)}{\| - 2s_1(t)\|} = \theta_1(t),\ 0 \le t \le T. \tag{5.54}$$

Similarly, the new sufficient statistic $\tilde{\gamma}$ is defined the same as in (5.48):

$$\tilde{\gamma} = \int_0^T n(t)\phi(t)dt. \tag{5.55}$$

$\tilde{\gamma}$ is the realization of a Gaussian r.v. with zero-mean and variance $N_0/2$. Then the probability of detection can be evaluated geometrically to yield

$$\begin{aligned} P_{\mathrm{D}}^{\mathrm{BPSK}} &= \mathrm{P}\left(\text{projected noise on } \boldsymbol{\varphi} \ge -\sqrt{E}|\mathbf{s}_2\right) = \mathrm{P}\left(\tilde{\gamma} \ge -\sqrt{E}\right) \\ &= \mathrm{P}\left(\frac{\tilde{\gamma}}{\sqrt{N_0/2}} \ge \frac{-\sqrt{E}}{\sqrt{N_0/2}}\right) = \mathrm{Q}\left(-\sqrt{2E/N_0}\right) > \mathrm{Q}\left(-\sqrt{E/N_0}\right) = P_{\mathrm{D}}^{\mathrm{BFSK}}. \end{aligned} \tag{5.56}$$

Similarly, the probability of a false alarm can be evaluated geometrically to yield

$$\begin{aligned} P_{\mathrm{FA}}^{\mathrm{BPSK}} &= \mathrm{P}\left(\text{projected noise on } \boldsymbol{\varphi} \ge \sqrt{E}|\mathbf{s}_1\right) = \mathrm{P}\left(\tilde{\gamma} \ge \sqrt{E}\right) \\ &= \mathrm{P}\left(\frac{\tilde{\gamma}}{\sqrt{N_0/2}} \ge \frac{\sqrt{E}}{\sqrt{N_0/2}}\right) = \mathrm{Q}\left(\sqrt{2E/N_0}\right) < \mathrm{Q}\left(\sqrt{E/N_0}\right) = P_{\mathrm{FA}}^{\mathrm{BFSK}}. \end{aligned} \tag{5.57}$$

Thus, the BPSK modulation system uses the same energy as the BFSK modulation system, but its performance is better since $P_{\mathrm{D}}^{\mathrm{BPSK}} > P_{\mathrm{D}}^{\mathrm{BFSK}}$ and $P_{\mathrm{FA}}^{\mathrm{BPSK}} < P_{\mathrm{FA}}^{\mathrm{BFSK}}$.

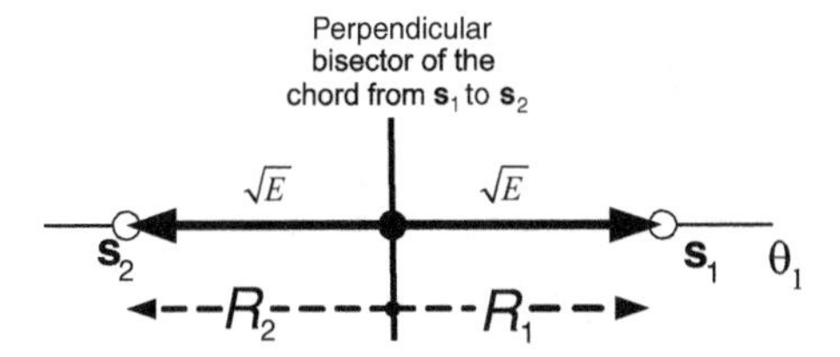

Figure 5.9 Decision regions of binary PSK modulated signals.

Finally, the average probability of an error for a BPSK system in which each of the two signal vectors has energy E is then given by

$$\begin{aligned} P_{\mathrm{e}}^{\mathrm{BPSK}} &= \frac{1}{2}\mathrm{P}(\text{declare } H_0 | H_1) + \frac{1}{2}\mathrm{P}(\text{declare } H_1 | H_0) \\ &= \frac{1}{2}\mathrm{P}\left(\tilde{\gamma} \geq \sqrt{E}\right) + \frac{1}{2}\mathrm{P}\left(\tilde{\gamma} \leq -\sqrt{E}\right) = \frac{1}{2}P_{\mathrm{FA}}^{\mathrm{BPSK}} + \frac{1}{2}P_{\mathrm{FA}}^{\mathrm{BPSK}} \\ &= \mathrm{Q}\left(\sqrt{2E/N_0}\right). \end{aligned} \tag{5.58}$$

□

Example 5.11 Quadrature-PSK (QPSK) modulation

The Quadrature (also called Quadra)-PSK (QPSK) system (also called the 4-PSK system) is one of the most commonly encountered modulation formats used in many microwave and satellite communication systems. The four equally probable transmitted waveforms given by

$$s_i(t) = \sqrt{2E/T}\cos\left(2\pi f t + (\pi/4) + (i-1)(\pi/2)\right),\ 0 \leq t \leq T,\ i = 1,\ 2,\ 3,\ 4, \tag{5.59}$$

and the received waveform given by

$$x(t) = s_i(t) + n(t),\ 0 \leq t \leq T,\ i = 1,\ 2,\ 3,\ 4, \tag{5.60}$$

where $n(t)$ is the realization of a zero-mean WGN process $N(t)$ of two-sided spectral density $N_0/2$. In the signal diagram, the transmitted signal vectors $\mathbf{s}_1$, $\mathbf{s}_2$, $\mathbf{s}_3$, $\mathbf{s}_4$, each having energy E, are located at radian values of $\{\pi/4,\ 3\pi/4,\ 5\pi/4,\ 7\pi/4\}$, as shown in Fig. 5.10. A QPSK system can be considered to be two BPSK systems, with one BPSK system (denoted as the I-component using the $-\mathbf{s}_3 = \mathbf{s}_1$ vectors) and another orthogonal BPSK system (denoted as the Q-component using the $-\mathbf{s}_4 = \mathbf{s}_2$ vectors). Using the minimum distance rule of (5.38), for the four transmitted signal vectors, their decision regions $\{\mathbb{R}_1,\ \mathbb{R}_2,\ \mathbb{R}_3,\ \mathbb{R}_4\}$ are the four quadrants of the plane. Specifically, suppose $\mathbf{s}_1$ was transmitted and consider the two uncorrelated (and thus independent) Gaussian r.v.'s

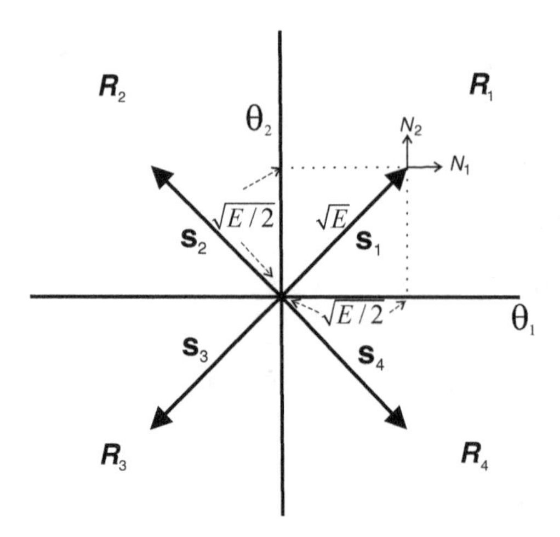

Figure 5.10 Decision regions of QPSK signal constellation.

N_1 and N_2 defined by

$$N_1 = \int_0^T N(t)\theta_1(t)dt,\ \theta_1(t) = \sqrt{2/T}\cos(2\pi ft),\ 0 \leq t \leq T, \tag{5.61}$$

$$N_2 = \int_0^T N(t)\theta_2(t)dt,\ \theta_2(t) = \sqrt{2/T}\cos\left(2\pi ft + (\pi/2)\right),\ 0 \leq t \leq T, \tag{5.62}$$

$$\begin{aligned} \mathrm{E}\{N_1 N_2\} &= (2/T)\int_0^T \int_0^T \mathrm{E}\left\{N(t)N(\tau)\right\}\cos(2\pi ft)\cos\left(2\pi f\tau + (\pi/2)\right)dtd\tau \\ &= -(N_0/T)\int_0^T \cos(2\pi ft)\sin(2\pi ft)dt \approx 0. \end{aligned} \tag{5.63}$$

Thus, N_1 and N_2 are two zero-mean uncorrelated (thus independent) Gaussian r.v.'s of variance $N_0/2$. From Fig. 5.10, consider the perturbing effect of these two r.v.'s on the transmitted signal vector $\mathbf{s}_1$. Suppose these two r.v.'s jointly satisfy $\{N_1 \geq -\sqrt{E/2}\} \cap \{N_2 \geq -\sqrt{E/2}\}$, then the perturbed received vector stays in the correct first quadrant of the decision $\mathbb{R}_1$. Thus, the probability that the correct decision conditioned on $\mathbf{s}_1$ was sent is given by

$$\begin{aligned} P_{\mathrm{c}|\mathbf{s}_1} &= \mathrm{P}\left(N_1 \geq -\sqrt{E/2},\ N_2 \geq -\sqrt{E/2}\right) = \left(1 - \mathrm{Q}(\sqrt{E/N_0})\right)^2 \\ &= P_{\mathrm{c}|\mathbf{s}_2} = P_{\mathrm{c}|\mathbf{s}_3} = P_{\mathrm{c}|\mathbf{s}_4}. \end{aligned} \tag{5.64}$$

By symmetry, the probability of correct decisions conditioned on the other three signal vectors is also given by (5.64). Since all four transmit signal vectors are equi-probable, then the average probability of an error for the "symbol" represented by the QPSK signal constellation, denoted by $P_{\mathrm{s}}^{\mathrm{QPSK}}$ is given by

$$\begin{aligned} P_{\mathrm{s}}^{\mathrm{QPSK}} &= 1 - P_{\mathrm{c}|\mathbf{s}_1} = 1 - \left(1 - \mathrm{Q}\left(\sqrt{E/N_0}\right)\right)^2 \\ &= 1 - \left(1 - 2\mathrm{Q}\left(\sqrt{E/N_0}\right) + \mathrm{Q}^2\left(\sqrt{E/N_0}\right)\right) \\ &= 2\mathrm{Q}\left(\sqrt{E/N_0}\right) - \mathrm{Q}^2\left(\sqrt{E/N_0}\right) \approx 2\mathrm{Q}\left(\sqrt{E/N_0}\right), \end{aligned} \tag{5.65}$$

where in the last approximation we have assumed for sufficient large E/N_0 that $\mathrm{Q}\left(\sqrt{E/N_0}\right) \ll 1$ and $\mathrm{Q}^2\left(\sqrt{E/N_0}\right) \ll \mathrm{Q}\left(\sqrt{E/N_0}\right)$.

Now, we want to compare the probability of an error for the BPSK case to that of the QPSK case. Since in both systems, when the signal vector $\mathbf{s}_1$ of length $\sqrt{E}$ was transmitted, the correct decision region $\mathbb{R}_1$ for the BPSK system can be considered to be the entire positive half-plane, while the correct decision region $\mathbb{R}_1$ for the QPSK system is the entire first quadrant. Thus, the additive Gaussian noise in the QPSK system has a greater probability to "escape" from the smaller correct $\mathbb{R}_1$ region as compared for the Gaussian noise in the BPSK system to "escape" from its larger correct R_1 region to make an error. Then intuitively we expect $P_{\mathrm{s}}^{\mathrm{QPSK}}$ to have a larger value than that of $P_{\mathrm{e}}^{\mathrm{BPSK}}$. Indeed by comparing (5.65) and (5.58), we obtain

$$P_{\mathrm{s}}^{\mathrm{QPSK}} \approx 2\mathrm{Q}\left(\sqrt{E/N_0}\right) > \mathrm{Q}\left(\sqrt{2E/N_0}\right) = P_{\mathrm{e}}^{\mathrm{BPSK}}. \tag{5.66}$$

We also note that in the BPSK case, the BPSK signal constellation with $M = 2$ carries $\log_2(M) = 1$ bit of information over a transmission duration of T seconds. Thus, a BPSK symbol carries 1 bit of information needing an energy of E joules in T seconds. (A symbol is sometimes also called a "baud.") Thus, the energy per bit, denoted by E_b, is the same as the energy per symbol, E in the BPSK case. On the other hand, the QPSK signal constellation with $M = 4$ carries $\log_2(M) = 2$ bits of information over a transmission duration of T seconds, but still uses an energy of E joules. Since the energy for 2 bits of information is E, then, for the QPSK case, the energy per bit $E_b = E/2$ or $2E_b = E$. Since $P_{\mathrm{s}}^{\mathrm{QPSK}}$ is the probability of an average symbol error in a QPSK system carrying two bits of information, then the probability of an average single bit error in a QPSK system can be denoted by

$$P_{\mathrm{e}}^{\mathrm{QPSK}} \approx P_{\mathrm{s}}^{\mathrm{QPSK}}/2 \approx \mathrm{Q}\left(\sqrt{E/N_0}\right) = \mathrm{Q}\left(\sqrt{2E_b/N_0}\right). \tag{5.67}$$

On the other hand, if we regard each of the I-component and Q-component parts of a QPSK system as two independent systems with each system having an energy of $E/2$, then the probability of an average bit error for these two separate components is given by

$$P_{\mathrm{e}}^{\mathrm{I,QPSK}} = P_{\mathrm{e}}^{\mathrm{Q,QPSK}} = \mathrm{Q}\left(\sqrt{2(E/2)/N_0}\right) = \mathrm{Q}\left(\sqrt{E/N_0}\right) = \mathrm{Q}\left(\sqrt{2E_b/N_0}\right). \tag{5.68}$$

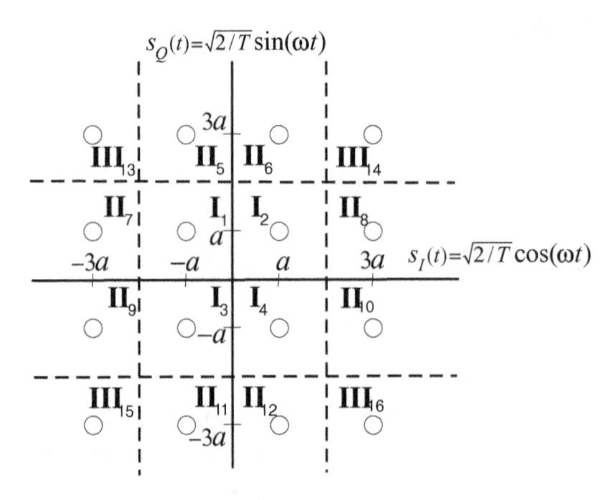

Figure 5.11 Signal constellation of a 16-QAM system.

From (5.67) and (5.68), we see the probability of an average bit error in a QPSK system is approximately the same as the probability of an average bit error in a comparable BPSK system. □

Example 5.12 AT&T Bell 201A/B modems

In 1960, AT&T introduced the Bell 201A modem using the QPSK modulation format to transmit 2000 bps data in the half-duplex mode (i.e., one way transmission) over the POTS line. The Bell 202B modem using QPSK provided 2400 bps data transmission in the full-duplex over four-wire leased lines. □

Example 5.13 Quadrature-Amplitude Modulation (QAM) system

Consider a 16-point Quadrature-Amplitude-Modulation (16-QAM) system in which the signal waveforms are given by

$$s_i(t) = \sqrt{2/T}(A_i \cos(\omega t) + B_i \sin(\omega t)),\ 0 \le t \le T,\ i = 1,\ 2,\ 3,\ 4, \tag{5.69}$$

where A_i and B_i take on the amplitude values of $\{\pm a, \pm 3a\}$. We assume all 16 of these waveforms have equal probability of occurrence and are disturbed by a WGN of zero-mean with a power spectral density of $S(f) = N_0/2,\ -\infty < f < \infty$. Of the 16 points in the QAM signal constellation, there are three distinct subclasses. Four of them are in the inside region (labeled class **I**), eight are on the outside, but not at the corners (labeled class **II**), and four are at the corners (labeled class **III**) as shown in Fig. 5.11.

Now, use the minimum distance decision rule of (5.37). For the transmitted signal vector located in region $\mathbf{I}_2$ denoted by $(s_I = a,\ s_Q = a)$, the correct received region is given by the shaded square area denoted by $[s_I - a,\ s_I + a] \otimes [s_Q - a,\ s_Q + a]$ as shown in Fig. 5.12(a). Then the correct probability of decision for the transmitted vector

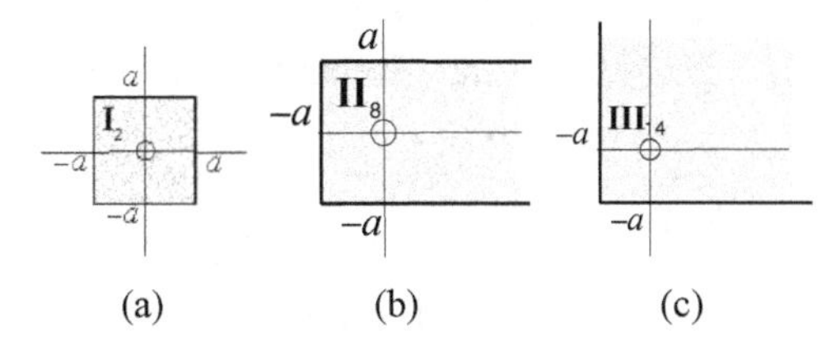

Figure 5.12 (a) Correct decision region for the class $\mathbf{I}_2$ signal vector. (b) Correct decision region for the class $\mathbf{II}_8$ signal vector. (c) Correction decision region for the class $\mathbf{III}_{14}$ signal vector.

in $\mathbf{I}_2$ is given by

$$\begin{aligned} \mathrm{P}(\mathrm{C}|\mathbf{I}_2) &= \left[\int_{-a}^{a} (1/\sqrt{\pi N_0}) \exp(-x^2/N_0) dx\right]^2 = \left[1 - 2\mathrm{Q}(\sqrt{2a^2/N_0})\right]^2 \\ &= \mathrm{P}(\mathrm{C}|\mathbf{I}_1) = \mathrm{P}(\mathrm{C}|\mathbf{I}_3) = \mathrm{P}(\mathrm{C}|\mathbf{I}_4). \end{aligned} \tag{5.70}$$

Due to the symmetrical property of white noise with respect to all directions in the $(s_\mathrm{I}, s_\mathrm{Q})$ coordinates, the probability of correct decisions for $\mathrm{P}(\mathrm{C}|\mathbf{I}_1)$, $\mathrm{P}(\mathrm{C}|\mathbf{I}_2)$, $\mathrm{P}(\mathrm{C}|\mathbf{I}_3)$, and $\mathrm{P}(\mathrm{C}|\mathbf{I}_4)$ are all given by (5.70). Similarly, for the transmitted signal vector located in region $\mathbf{II}_8$ denoted by $(s_\mathrm{I} = 3a,\ s_\mathrm{Q} = a)$, the correct received region is given by the shaded semi-infinite strip denoted by $[s_\mathrm{I} - a,\ s_\mathrm{I} + a] \otimes [s_\mathrm{Q} - a,\ s_\mathrm{Q} + a]$ as shown in Fig. 5.12(b). Then the correct probability of decisions for the transmitted vector in $\mathbf{II}_5$, $\mathbf{II}_6$, $\mathbf{II}_7$, $\mathbf{II}_8$, $\mathbf{II}_9$, $\mathbf{II}_{10}$, $\mathbf{II}_{11}$, and $\mathbf{II}_{12}$ are all given by

$$\begin{aligned} \mathrm{P}(\mathrm{C}|\mathbf{II}_8) &= \left[\int_{-a}^{\infty} (1/\sqrt{\pi N_0}) \exp(-x^2/N_0) dx\right] \left[\int_{-a}^{a} (1/\sqrt{\pi N_0}) \exp(-x^2/N_0) dx\right] \\ &= \left[1 - 2\mathrm{Q}(\sqrt{2a^2/N_0})\right] \left[1 - \mathrm{Q}(\sqrt{2a^2/N_0})\right] \\ &= \mathrm{P}(\mathrm{C}|\mathbf{II}_5) = \mathrm{P}(\mathrm{C}|\mathbf{II}_6) = \mathrm{P}(\mathrm{C}|\mathbf{II}_7) = \mathrm{P}(\mathrm{C}|\mathbf{II}_9) \\ &= \mathrm{P}(\mathrm{C}|\mathbf{II}_{10}) = \mathrm{P}(\mathrm{C}|\mathbf{II}_{11}) = \mathrm{P}(\mathrm{C}|\mathbf{II}_{12}). \end{aligned} \tag{5.71}$$

Similarly, for the transmitted signal vector located in region $\mathbf{III}_{14}$ denoted by $(s_\mathrm{I} = 3a,\ s_\mathrm{Q} = 3a)$, the correct received region is given by the shaded infinite quadrant denoted by $[s_\mathrm{I} - a,\ \infty) \otimes [s_\mathrm{Q} - a,\ \infty)$ as shown in Fig. 5.12(c). Then the correct probability of decisions for the transmitted vector in $\mathbf{III}_{13}$, $\mathbf{III}_{14}$, $\mathbf{III}_{15}$, and $\mathbf{III}_{16}$ are all given by

$$\begin{aligned} \mathrm{P}(\mathrm{C}|\mathbf{III}_{14}) &= \left[\int_{-a}^{\infty} (1/\sqrt{\pi N_0}) \exp(-x^2/N_0) dx\right]^2 = \left[1 - \mathrm{Q}(\sqrt{2a^2/N_0})\right]^2 \\ &= \mathrm{P}(\mathrm{C}|\mathbf{III}_{13}) = \mathrm{P}(\mathrm{C}|\mathbf{III}_{15}) = \mathrm{P}(\mathrm{C}|\mathbf{III}_{16}). \end{aligned} \tag{5.72}$$

Since there are four signal vectors in class **I**, eight signal vectors in class **II**, and four signal vectors in class **III**, the average error probability P_e for the 16-QAM system is given by

$$P_\mathrm{e} = 1 - [(4/16)\mathrm{P}(\mathrm{C}|\mathbf{I}_2) + (8/16)\mathrm{P}(\mathrm{C}|\mathbf{II}_8) + (4/16)\mathrm{P}(\mathrm{C}|\mathbf{III}_{14})]. \tag{5.73}$$

Finally, we need to relate the constant "a" in Fig. 5.11 and Fig. 5.12 and (5.70)–(5.72) to the average energy E of all the 16 signal waveforms. The energy of all 16 signal vectors is given by

$$[4 \times (a^2 + a^2)] + [8 \times (a^2 + 9a^2)] + [4 \times (9a^2 + 9a^2)] = 160a^2. \tag{5.74}$$

From the expression in (5.74), the average energy E is given by

$$E = 160a^2/16 = 10a^2, \tag{5.75}$$

or

$$a = \sqrt{E/10}. \tag{5.76}$$

Thus, by using a of (5.76) in (5.70)–(5.72), P_e of (5.73) can be evaluated in terms of E/N_0. □

5.4 Optimal signal design for M-ary systems

Under the minimum average probability of error criterion, how do we pick the M-ary signals $\{s_i(t),\ i = 1, \ldots, M\}$ in a WGN M-ary communication system? From (5.35), the optimum decision rule depends on the signals only through its energy $\|s_i(t)\|^2 = E_i$ and their geometric relationship through the inner product $(s_i(t), s_j(t))$. Once the signal waveforms are fixed, the probability of error of the system becomes invariant with respect to rigid body rotation and translation of all the signal waveforms. However, while rotation does not affect the energies of the signal waveforms, translation does. Thus, it seems to be reasonable to translate and rotate the signal waveforms for minimum average energy at a fixed average probability of error. Denote the average energy $\overline{E}$ of the signal waveforms by

$$\overline{E} = \sum_{i=1}^{M} \pi_i E_i = \sum_{i=1}^{M} \pi_i \|s_i(t)\|^2. \tag{5.77}$$

Translate the original signal waveforms by $a(t),\ 0 \le t \le T$, to the new signal waveforms denoted by

$$\hat{s}_i(t) = s_i(t) - a(t),\ 0 \le t \le T,\ i = 1,\ 2,\ \ldots,\ M. \tag{5.78}$$

Then the vector notation of (5.78) becomes

$$\hat{\mathbf{s}}_i = \mathbf{s}_i - \mathbf{a},\ i = 1,\ 2,\ \ldots,\ M. \tag{5.79}$$

The average energy of the new translated signal vectors is given by

$$E_{\mathbf{a}} = \sum_{i=1}^{M} \pi_i \|\hat{\mathbf{s}}_i\|^2 = \sum_{i=1}^{M} \pi_i \|\mathbf{s}_i - \mathbf{a}\|^2. \tag{5.80}$$

Now, we want to find the vector **a** to minimize $E_{\mathbf{a}}$, by setting its first partial derivatives with respect to a_j to zero for all $j = 1, \ldots, N$. Thus,

$$\frac{\partial E_{\mathbf{a}}}{\partial a_j} = \sum_{i=1}^{M} \pi_i d \frac{\partial \|s_i - \mathbf{a}\|^2}{\partial a_j} = 0, \; j = 1, \ldots, N. \tag{5.81}$$

But

$$\frac{\partial \|s_i - \mathbf{a}\|^2}{\partial a_j} = \frac{\partial \sum_{k=1}^{N} (s_{ik} - a_k)^2}{\partial a_j} = -2(s_{ij} - a_j), \; j = 1, \ldots, N. \tag{5.82}$$

Thus, (5.81) becomes

$$\frac{\partial E_{\mathbf{a}}}{\partial a_j} = -2 \sum_{i=1}^{M} \pi_i (s_{ij} - \hat{a}_j) = 0, \; j = 1, \ldots, N, \tag{5.83}$$

and

$$\hat{a}_j = \sum_{i=1}^{M} \pi_i s_{ij}, \; j = 1, \ldots, N, \tag{5.84}$$

and the optimum translating vector is given by

$$\hat{\mathbf{a}} = \sum_{i=1}^{M} \pi_i \mathbf{s}_i, \tag{5.85}$$

or the optimum translating waveform is given by

$$\hat{a}(t) = \sum_{i=1}^{M} \pi_i s_i(t), \; 0 \leq t \leq T, \tag{5.86}$$

which is the center of gravity of the signal waveforms.

For a given set of signal waveforms $\{s_i(t), \, i = 1, 2, \ldots, M\}$, if its center of gravity $\hat{a}(t) \neq 0$, then the set of the translated waveforms $\{\hat{s}_i(t) = s_i(t) - a(t), \, i = 1, 2, \ldots, M\}$ has the same average error probability as that of $\{s_i(t), \, i = 1, 2 \ldots, M\}$, but has a lower average energy. In the following three examples, we assume all the signal waveforms are equally probable, i.e., $(\pi_i = 1/M, \, i = 1, 2, \ldots, M)$ and all have the same energy (i.e., $\|s_i(t)\|^2 = E$).

Example 5.14 $M = 2$ **orthogonal signals**

From its orthogonality of $(s_1(t), s_2(t)) = 0$, then

$$\|s_1(t) - s_2(t)\|^2 = \|s_1(t)\|^2 - 2(s_1(t), s_2(t)) + \|s_1(t)\|^2 = 2E. \tag{5.87}$$

This case is identical to the FSK modulation system considered as Example 5.6. Thus, under AWGN channel for two orthogonal waveforms, $s_1(t)$ and $s_2(t)$, each of energy E in the presence of two-sided noise power spectral density of $N_0/2$, the average error probability of this $M = 2$ orthogonal signals with equal prior probability of the two

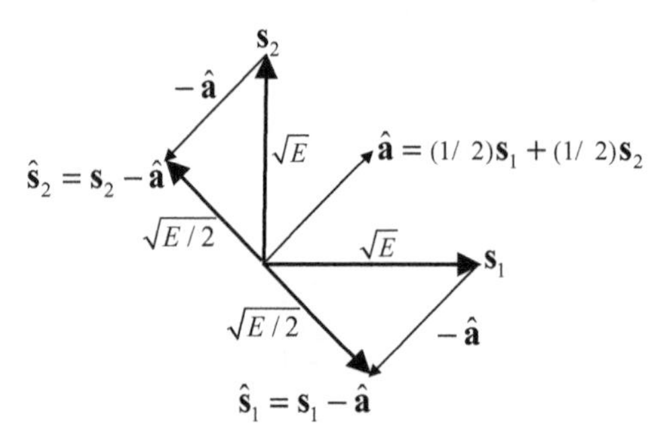

Figure 5.13 $M = 2$ orthogonal signals.

hypotheses is the same as that of (5.51) given by

$$P_\mathrm{e} = \mathrm{Q}(\sqrt{E/N_0}). \tag{5.88}$$

Geometrically, the two orthogonal signal waveforms $s_1(t)$ and $s_2(t)$ can be seen as two orthogonal signal vectors $\mathbf{s}_1$ and $\mathbf{s}_2$ in Fig. 5.11. Since the center of gravity of these two signal vectors, $\hat{\mathbf{a}} = (1/2)\mathbf{s}_1 + (1/2)\mathbf{s}_2 \neq \mathbf{0}$, we can define the two optimum translated signal vectors as

$$\hat{\mathbf{s}}_1 = \mathbf{s}_1 - \mathbf{a} = (1/2)\mathbf{s}_1 - (1/2)\mathbf{s}_2,\ \hat{\mathbf{s}}_2 = \mathbf{s}_2 - \mathbf{a} = (1/2)\mathbf{s}_2 - (1/2)\mathbf{s}_1. \tag{5.89}$$

These two optimum translated vectors $\hat{\mathbf{s}}_1$ and $\hat{\mathbf{s}}_2$ are shown in Fig. 5.13. From the geometry in Fig. 5.13, we note that the average error probability of the system using $\mathbf{s}_1$ and $\mathbf{s}_2$ is identical to the average error probability of the system using the optimum translated vectors $\hat{\mathbf{s}}_1$ and $\hat{\mathbf{s}}_2$ (the same argument has already been given in Example 5.6 and Fig. 5.7 for the FSK modulation). But the two optimum translation signal vectors are now anti-podal (i.e., $\hat{\mathbf{s}}_1 = -\hat{\mathbf{s}}_2$). However, the energy of each of the $\mathbf{s}_1$ and $\mathbf{s}_2$ is equal to E (since each vector has norm equal to $\sqrt{E}$), but the energy of each of the $\hat{\mathbf{s}}_1$ and $\hat{\mathbf{s}}_2$ is equal to $E/2$ (since each vector has the norm of $\sqrt{E/2}$), as can be seen in Fig. 5.13. Thus, the detection of $\hat{\mathbf{s}}_1$ and $\hat{\mathbf{s}}_2$ in the presence of two-sided noise power spectral density of $N_0/2$ can also be considered as that of a PSK problem in Example 5.10. The average error probability of a PSK system with two anti-podal signals of energy $\tilde{E}$ in the presence of two-sided noise power spectral density of $N_0/2$ from (5.56) is given by (5.90a). However, the energy of each the optimum translated $\hat{\mathbf{s}}_1$ and $\hat{\mathbf{s}}_2$ anti-podal signals is now given by $\tilde{E} = E/2$, resulting in the average error probability expression given by (5.90b). We note, the average error probability expression of (5.90b) obtained by considering $\hat{\mathbf{s}}_1$ and $\hat{\mathbf{s}}_2$ as a PSK problem is identical to the average error probability expression of (5.88) obtained by considering the original $\mathbf{s}_1$ and $\mathbf{s}_2$ as a FSK problem.

$$P_\mathrm{e} = \mathrm{Q}\left(\sqrt{2\tilde{E}/N_0}\right) \tag{5.90a}$$

$$= \mathrm{Q}\left(\sqrt{E/N_0}\right). \quad \square \tag{5.90b}$$

Example 5.15 $M = 2$ **Anti-podal signals**

From the anti-podal property $(s_1(t), s_2(t)) = -E$ of the two equal energy E signals, $s_1(t)$ and $s_2(t)$, we obtain

$$\|s_1(t) - s_2(t)\|^2 = \|s_1(t)\|^2 - 2(s_1(t), s_2(t)) + \|s_2(t)\|^2 = 4E. \tag{5.91}$$

But (5.91) with a value of $4E$ can be interpreted as the square of the distance between the two anti-podal signals of energy E. By comparison, from (5.87), we note it has a value of $2E$ as the square of the distance between the two orthogonal signals of energy E. Intuitively, the average error probability should decrease as the distance separating the two signals becomes larger. This intuition is fully justified since the $M = 2$ anti-podal signal case is equivalent to the PSK problem of Example 5.10, resulting in an average error probability of

$$P_e = Q(\sqrt{2E/N_0}), \tag{5.92}$$

where we assumed equal probability for the two signals in the presence of Gaussian noise with a power spectral density of $N_0/2$. □

5.5 Classification of *M* patterns

5.5.1 Introduction to pattern recognition and classification

In Chapter 4, we considered the detection of two signal vectors/waveforms, while in Section 5.4, we considered the detection of M signal vectors/waveforms observed in the presence of WGN. In this section, we consider the problem of the classification of N patterns from M possible distinct classes. Many practical problems in engineering, science, medicine, and human affairs all involve pattern recognition and the need to perform classification. At a simplistic level, a pattern recognition system can be divided into two parts – a feature extractor and a pattern classifier. Pattern recognition applications are diverse and may include: printed alphabet letter recognition (e.g., optical character recognition (OCR)); cursive script recognition; speech recognition; speaker recognition; classification of high-energy particles in various chambers; fingerprint classification; classification of vehicles; classification of diseases from X-rays and other 2D images; classification of aircraft by air frame features; classification of aircraft by engine types (e.g., jet emission signatures (JEM lines); etc.). It is clear that most pattern recognition problems are unique and the selection of the key features that characterize a given problem is crucial. The problem of feature extraction is highly problem dependent. For example, the American Bankers Association E-13B font character set used in bank checks is designed on a 9×7 zone grid. The number "2" in that character set is shown in Fig. 5.14(a). A sequence of 0's and 1's (indicating a blank or a darkened cell) can be used to characterize the number of interest. Sampled and quantized voltages at different time instants from a transducer are shown in Fig. 5.14(b). In either case, these data are used in a feature vector to characterize the particular pattern of interest. In general, the

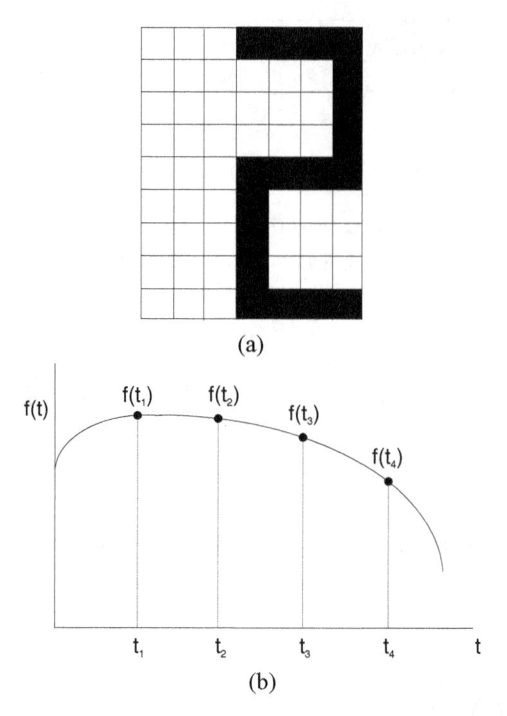

Figure 5.14 (a) Number "2" in the ABA-13B. (b) Sampled voltages at different time instants.font character set.

selection of some intrinsic properties to be used in the feature vectors depends greatly on the specific recognition problem. However, after the patterns have been collected, then we can consider some general theory of pattern classification to be presented in Section 5.2. The feature vector represented by an observable n-dimensional real-valued column vector is denoted by

$$\mathbf{x} = [x_1, x_2, \ldots, x_n]^T \in \mathbb{R}^n. \tag{5.93}$$

The feature vectors in pattern recognition problems can be considered either in statistical terms or in deterministic terms. In the statistical approach, as a simple example, consider each observed $\mathbf{x}$ as the realization of a Gaussian random vector from one of M possible classes with probability of occurrence given by π_i, with M mean vectors and M covariance matrices denoted by

$$\{\boldsymbol{\mu}_i,\ i = 1, \ldots, M\},\ \{\boldsymbol{\Lambda}_i,\ i = 1, \ldots, M\} \tag{5.94}$$

so the conditional pdf of $\mathbf{x}$ given pattern i becomes

$$p(\mathbf{x}|i) = \frac{1}{(2\pi)^{n/2}\,|\boldsymbol{\Lambda}_i|^{1/2}} e^{-\frac{1}{2}(\mathbf{x}-\boldsymbol{\mu}_i)^T \boldsymbol{\Lambda}_i^{-1}(\mathbf{x}-\boldsymbol{\mu}_i)},\ i = 1, \ldots, M. \tag{5.95}$$

Define the discriminant function by

$$D_i(\mathbf{x}) = \ln(\pi_i) - (1/2)\ln|\boldsymbol{\Lambda}_i| - (1/2)(\mathbf{x} - \boldsymbol{\mu}_i)^T \boldsymbol{\Lambda}_i^{-1}(\mathbf{x} - \boldsymbol{\mu}_i). \tag{5.96}$$

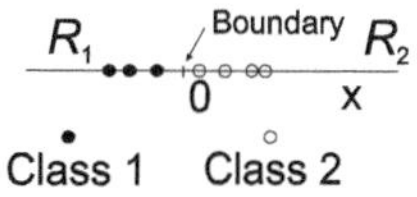

Figure 5.15 Three pattern vectors from class 1 and four pattern vectors from class 2.

Then under the minimum probability of a mis-classification criterion, the decision region for declaring pattern i is given by

$$\mathbb{R}_i = \{\mathbf{x} : \ D_i(\mathbf{x}) - D_j(\mathbf{x}) > 0, \ j \neq i\}. \tag{5.97}$$

Furthermore, suppose all the patterns are equally likely with $\pi_i = 1/M$, all the covariances are equal with $\boldsymbol{\Lambda}_i = \boldsymbol{\Lambda}$, and the observed patterns $\mathbf{x}$ are the results of observing the true patterns $\mathbf{s}_i$ in the presence of Gaussian noise $\mathbf{n}$ of zero-mean and covariance $\boldsymbol{\Lambda}$ modeled by

$$\mathbf{x} = \mathbf{s}_i + \mathbf{n}, \ i = 1, \ldots, M. \tag{5.98}$$

Then the decision region of (5.97) reduces to

$$\mathbb{R}_i = \{\mathbf{x} : \ \mathbf{x}^T \boldsymbol{\Lambda}^{-1} \mathbf{s}_i > \mathbf{x}^T \boldsymbol{\Lambda}^{-1} \mathbf{s}_j, \ j \neq i\}. \tag{5.99}$$

However, the decision rule given by (5.99) is equivalent to taking the largest value among the M matched filter outputs to declare the i-th pattern. Clearly, the pattern classification procedure in (5.97) and (5.99) is equivalent to that of M-ary detection of deterministic signal vectors in colored and white Gaussian noises, respectively.

5.5.2 Deterministic pattern recognition

In many classification problems (in the training stage), we may be given N deterministic pattern vectors, with n_1 patterns taken explicitly from class 1, n_2 taken from class 2, ..., and n_M taken from class M, such that $n_1 + n_2 + \ldots + n_M = N$.

$$\begin{aligned} &\{\mathbf{x}_1^{(1)}, \mathbf{x}_2^{(1)}, \ldots, \mathbf{x}_{n_1}^{(1)}\}, \{\mathbf{x}_1^{(2)}, \mathbf{x}_2^{(2)}, \ldots, \mathbf{x}_{n_2}^{(2)}\}, \ldots, \{\mathbf{x}_1^{(M)}, \mathbf{x}_2^{(M)}, \ldots, \mathbf{x}_{n_M}^{(M)}\}, \\ &\mathbf{x}_i^{(m)} \in \mathbb{R}^n, \ i = 1, \ldots, n_j, \ j = 1, \ldots, M, \ m = 1, \ldots, M, \end{aligned} \tag{5.100}$$

The number n_j taken from class j need not be the same as taken from another class. Of course, we hope there is a sufficient number of patterns taken from each class properly to represent the intrinsic properties of the patterns in that class for its proper classification.

Example 5.16 Consider the simple case of $N = 7$ (with $n_1 = 3$ and $n_2 = 4$) scalar pattern vectors of dimension $n = 1$ and $M = 2$ classes given by (5.101) and shown in Fig. 5.15.

$$\begin{aligned} &\{\mathbf{x}_1^{(1)} = [-2.75], \ \mathbf{x}_2^{(1)} = [-2], \ \mathbf{x}_3^{(1)} = [-1]\}, \\ &\{\mathbf{x}_1^{(2)} = [0.5], \ \mathbf{x}_2^{(2)} = [1.5], \ \mathbf{x}_3^{(2)} = [2.5], \ \mathbf{x}_4^{(2)} = [3]\}. \quad \square \end{aligned} \tag{5.101}$$

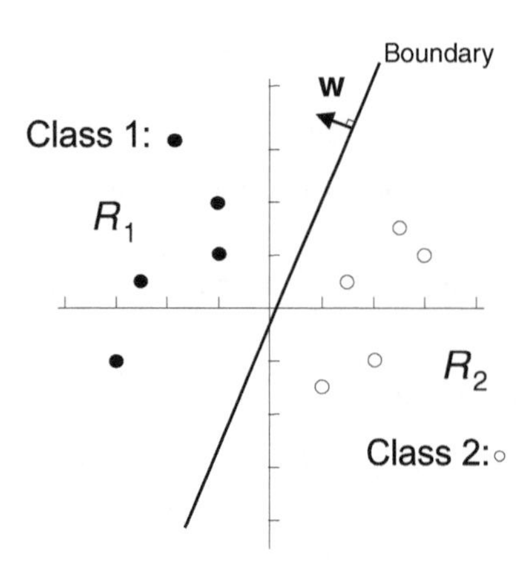

Figure 5.16 Four pattern vectors from class 1 and five pattern vectors from class 2.

Example 5.17 Consider the case of $N = 9$ (with $n_1 = 4$ and $n_2 = 5$) pattern vectors of dimension $n = 2$ and $M = 2$ classes given by (5.102) and Fig. 5.16.

$$\{\mathbf{x}_1^{(1)} = [-1,\ 1]^T,\ \mathbf{x}_2^{(1)} = [-1,\ 2]^T,\ \mathbf{x}_3^{(1)} = [-2.5,\ 0.5]^T,\ \mathbf{x}_4^{(1)} = [-3,\ -1]^T\}, \tag{5.102a}$$

$$\{\mathbf{x}_1^{(2)} = [1,\ -1.5]^T,\ \mathbf{x}_2^{(2)} = [1.5,\ 0.5]^T,\ \mathbf{x}_3^{(2)} = [2,\ -1]^T,\ \mathbf{x}_4^{(2)} = [2.5,\ 1.5]^T,$$
$$\mathbf{x}_5^{(2)} = [3,\ 1]^T\}. \quad \square \tag{5.102b}$$

In Examples 5.16 and 5.17, we want to find appropriate boundaries that separate the cluster of pattern vectors in the two classes. Intuitively, the boundaries in Fig. 5.15 and Fig. 5.16 seem to be adequate to separate the two clusters.

More formally, we want to define a discriminant function, $D(\cdot)$ for the $M = 2$ classes such that for all pattern vectors in class 1, the discriminant function yields a positive value as shown in (5.103a), while for all pattern vectors in class 2, the discriminant function yields a negative value as shown in (5.103b),

$$D(\mathbf{x}_i^{(1)}) > 0,\ \mathbf{x}_i^{(1)} \in \mathbb{R}^n,\ i = 1,\ \ldots,\ n_1, \tag{5.103a}$$

$$D(\mathbf{x}_j^{(2)}) < 0,\ \mathbf{x}_i^{(2)} \in \mathbb{R}^n,\ j = 1,\ \ldots,\ n_2. \tag{5.103b}$$

Furthermore, all the $\mathbf{x}$ in $\mathbb{R}^n$ satisfying (5.104) form the boundary in the classification between the vectors in the two classes.

$$D(\mathbf{x}) = 0,\ \mathbf{x} \in \mathbb{R}^n. \tag{5.104}$$

While there are many possible discriminant functions, let us only consider the linear discriminant function defined by

$$D(\mathbf{x}) = \mathbf{w}^T\mathbf{x} + w_{n+1}, \tag{5.105}$$

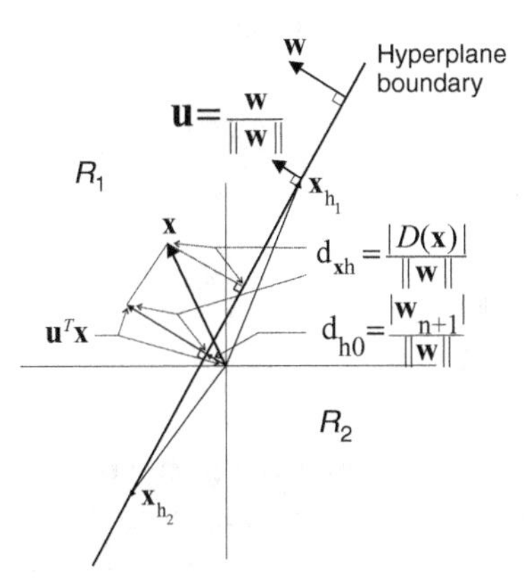

Figure 5.17 The geometries of the distance function d_{h_0} from the hyperplane to the origin and the distance function $d_{\mathbf{x}_h}$ from any vector $\mathbf{x}$ to the hyperplane.

where

$$\mathbf{w} = [w_1, \ldots, w_n]^T \tag{5.106}$$

is a $n \times 1$ column vector called the weight vector and w_{n+1} is a scalar called the threshold. While (5.105) is called a "linear discriminant" in classification, the function $D(\mathbf{x})$ in (5.105) is actually called an affine function of $\mathbf{x}$ in mathematics. However, only the function $f(\mathbf{x}) = \mathbf{w}^T\mathbf{x}$ is called a linear function of $\mathbf{x}$ in mathematics. The boundary defined by (5.104) for the linear discriminant function of (5.105) forms a $n - 1$ dimensional hyperplane. In Example 5.16, with $n = 1$, the hyperplane of dimension 0 separating the two classes is a point as shown in Fig. 5.15. In Example 5.17, with $n = 2$, the hyperplane of dimension 1 separating the two classes is a line as shown in Fig. 5.16.

From (5.104) and (5.105) and Fig. 5.17, if $\mathbf{x}_{h_1}$ and $\mathbf{x}_{h_2}$ are any two vectors on the hyperplane, then

$$\mathbf{w}^T(\mathbf{x}_{h_1} - \mathbf{x}_{h_2}) = 0. \tag{5.107}$$

Denote the normalized weight vector $\mathbf{u}$ by

$$\mathbf{u} = \frac{\mathbf{w}}{\|\mathbf{w}\|^2}. \tag{5.108}$$

Clearly $\mathbf{u}$ also satisfies

$$\mathbf{u}^T(\mathbf{x}_{h_1} - \mathbf{x}_{h_2}) = 0. \tag{5.109}$$

Since $\mathbf{x}_{h_1}$ is on the hyperplane, from (5.105) we have $\mathbf{w}^T\mathbf{x}_{h_1} = -w_{n+1}$, and thus

$$\mathbf{u}^T\mathbf{x}_{h_1} = \frac{\mathbf{w}^T\mathbf{x}_{h_1}}{\|\mathbf{w}\|} = \frac{-w_{n+1}}{\|\mathbf{w}\|}. \tag{5.110}$$

From Fig. 5.17, we note the distance d_{h_0} from the hyperplane to the origin is then given by

$$d_{h_0} = \left|\mathbf{u}^T\mathbf{x}_{h_1}\right| = \frac{|w_{n+1}|}{\|\mathbf{w}\|}. \tag{5.111}$$

From Fig. 5.17, we also note the distance $d_{\mathbf{x}_h}$ from any vector $\mathbf{x}$ to the hyperplane is given by

$$d_{\mathbf{x}_h} = \left|\mathbf{u}^T\mathbf{x} - \mathbf{u}^T\mathbf{x}_{h_1}\right| = \left|\frac{\mathbf{w}^T\mathbf{x}}{\|\mathbf{w}\|} - \frac{\mathbf{w}^T\mathbf{x}_{h_1}}{\|\mathbf{w}\|}\right| = \left|\frac{\mathbf{w}^T\mathbf{x}}{\|\mathbf{w}\|} + \frac{w_{n+1}}{\|\mathbf{w}\|}\right|$$
$$= \left|\frac{\mathbf{w}^T\mathbf{x} + w_{n+1}}{\|\mathbf{w}\|}\right| = \frac{|D(\mathbf{x})|}{\|\mathbf{w}\|}. \quad (5.112)$$

We will make use of this distance $d_{\mathbf{x}_h}$ property later.

5.5.2.1 Weight vectors for linearly separable binary class of patterns

Often the linear discriminant function of (5.105) and its weight vector of (5.106) are *augmented* to the form of

$$D(\tilde{\mathbf{x}}) = \tilde{\mathbf{w}}^T\tilde{\mathbf{x}}, \quad (5.113)$$

where now the augmented $\tilde{\mathbf{w}}$ is a $(n+1)\times 1$ column vector given by

$$\tilde{\mathbf{w}} = [w_1, \ldots, w_n, w_{n+1}]^T = [\mathbf{w}^T, w_{n+1}]^T, \quad (5.114)$$

and the new $\tilde{\mathbf{x}}$ is a $(n+1)\times 1$ column vector given by

$$\tilde{\mathbf{x}} = [x_1, \ldots, x_n, 1]^T = [\mathbf{x}^T, 1]^T. \quad (5.115)$$

Clearly, the modified definition of the linear discriminant in (5.113)–(5.115) is identical to the original definition of the linear discriminant in (5.105)–(5.106).

Now, the original collection $\{\mathbf{x}_1^{(1)}, \mathbf{x}_2^{(1)}, \ldots, \mathbf{x}_{n_1}^{(1)}\}$ of n_1 training vectors from class 1, each of $n \times 1$, will be modified to the form of $\{\tilde{\mathbf{x}}_1^{(1)}, \tilde{\mathbf{x}}_2^{(1)}, \ldots, \tilde{\mathbf{x}}_{n_1}^{(1)}\}$, where each augmented vector is of dimension $(n+1)\times 1$ given by

$$\tilde{\mathbf{x}}_i^{(1)} = [\mathbf{x}_i^{(1)T}, 1]^T, \ i = 1, \ldots, n_1. \quad (5.116)$$

Similarly, the original collection $\{\mathbf{x}_1^{(2)}, \mathbf{x}_2^{(2)}, \ldots, \mathbf{x}_{n_2}^{(2)}\}$ of n_2 training vectors from class 2, each of dimension $n \times 1$, will be modified to the form of $\{\tilde{\mathbf{x}}_1^{(2)}, \tilde{\mathbf{x}}_2^{(2)}, \ldots, \tilde{\mathbf{x}}_{n_2}^{(2)}\}$, where each augmented vector is of dimension $(n+1)\times 1$ given by

$$\tilde{\mathbf{x}}_j^{(2)} = [\mathbf{x}_j^{(2)T}, 1]^T, \ j = 1, \ldots, n_2. \quad (5.117)$$

Then each vector in $\{\tilde{\mathbf{x}}_1^{(2)}, \tilde{\mathbf{x}}_2^{(2)}, \ldots, \tilde{\mathbf{x}}_{n_2}^{(2)}\}$ is multiplied by (-1) and renamed as

$$\left\{\tilde{\tilde{\mathbf{x}}}_1^{(2)}, \tilde{\tilde{\mathbf{x}}}_2^{(2)}, \ldots, \tilde{\tilde{\mathbf{x}}}_{n_2}^{(2)}\right\} = \left\{-\tilde{\mathbf{x}}_1^{(2)}, -\tilde{\mathbf{x}}_2^{(2)}, \ldots, -\tilde{\mathbf{x}}_{n_2}^{(2)}\right\}. \quad (5.118)$$

There are two reasons for introducing the augmented weight vector and the training vectors. The first reason is that the concept of linearly separable training is best defined using the new notations. The collection of training vectors $\{\mathbf{x}_1^{(1)}, \mathbf{x}_2^{(1)}, \ldots, \mathbf{x}_{n_1}^{(1)}\}$ from class 1 and training vectors $\{\mathbf{x}_1^{(2)}, \mathbf{x}_2^{(2)}, \ldots, \mathbf{x}_{n_2}^{(2)}\}$ from class 2 is said to be *linearly separable*, if there is an augmented $\tilde{\mathbf{w}}$ such that the linear discriminant function of (5.113) satisfies

$$D(\tilde{\mathbf{x}}) = \tilde{\mathbf{w}}^T\tilde{\mathbf{x}} > 0, \ \forall\, \tilde{\mathbf{x}} = \tilde{\mathbf{x}}_i^{(1)}, \ i = 1, \ldots, n_1; \ \forall\, \tilde{\mathbf{x}} = \tilde{\tilde{\mathbf{x}}}_j^{(2)}, \ j = 1, \ldots, n_2. \quad (5.119)$$

Using the augmented weight vector and the augmented training vectors and the multiplication of (-1) on the augmented training vectors in class 2, all the new augmented and modified training vectors (from class 1 and class 2) are on the same positive side of the hyperplane as seen in (5.119). This property (having "the new augmented and all the modified training vectors on the same positive side of the hyperplane") turns out to be crucial in finding an iterative solution of the weight vector $\tilde{\mathbf{w}}$ in the Perceptron Algorithm to be shown below.

However, in order to reduce the creation of new notations from (5.113)–(5.119), we will omit the use of the single superscript tilde (i.e., $\tilde{\mathbf{z}}$, whether $\mathbf{z}$ denotes $\mathbf{w}$, $\mathbf{x}_i$, or $\mathbf{x}_j$) and double superscript tilde (i.e., $\tilde{\tilde{\mathbf{z}}}$ for $\mathbf{x}_j$) over all the $\mathbf{x}$'s and $\mathbf{w}$'s. In other words, in the rest of Section 5.5.2.1, the weight vectors and training vectors will be in the augmented form and all the tildes will be omitted.

Example 5.18 reconsiders the training vectors in Example 5.17 in the augmented modified notations.

Example 5.18 Consider the case of $N = 9$ (with $n_1 = 4$ and $n_2 = 5$) pattern vectors of dimension $n = 2$ and $M = 2$ classes from Example 5.17 expressed in the augmented modified notations.

$$\{\mathbf{x}_1^{(1)} = [-1,\ 1,\ 1]^T,\ \mathbf{x}_2^{(1)} = [-1,\ 2,\ 1]^T,\ \mathbf{x}_3^{(1)} = [-2.5,\ 0.5,\ 1]^T,\ \mathbf{x}_4^{(1)} = [-3,\ -1,\ 1]^T\}, \tag{5.120a}$$

$$\{\mathbf{x}_1^{(2)} = [-1,\ 1.5,\ -1]^T,\ \mathbf{x}_2^{(2)} = [-1.5,\ -0.5,\ -1]^T,\ \mathbf{x}_3^{(2)} = [-2,\ 1,\ -1]^T,$$

$$\mathbf{x}_4^{(2)} = [-2.5,\ -1.5,\ -1]^T,\ \mathbf{x}_5^{(2)} = [-3,\ -1,\ -1]^T\} \quad \square \tag{5.120b}$$

The Perceptron Algorithm can be used to find the weight vector $\mathbf{w}$ for a linearly separable collection of training vectors from two classes. We assume the weight vector $\mathbf{w}$ and the training vectors $\mathbf{x}$ are in the augmented and modified forms. Then the linearly separable weight $\mathbf{w}$ of (5.119) is given by the iterative solution of the Perceptron Algorithm [1].

Theorem (Perceptron Algorithm) Consider a collection of N linearly separable augmented and modified vectors $\mathbf{X} = \{\mathbf{x}_1,\ \mathbf{x}_2,\ \ldots,\ \mathbf{x}_N\}$, each of dimension $(n+1) \times 1$, taken from two known training classes.

Step 1. Let $\mathbf{w}(0)$ be an arbitrary initial real-valued vector of dimension $(n+1) \times 1$ and c be an arbitrary positive constant.
Step 2. Set $\mathbf{w}(1) = \mathbf{w}(0)$.
Step 3. For k from 1 to N, do

$$\mathbf{w}(k+1) = \begin{cases} \mathbf{w}(k), & \text{if } \mathbf{w}(k)^T \mathbf{x}_k > 0,\ \mathbf{x}_k \in \mathbf{X}, \\ \mathbf{w}(k) + c\mathbf{x}(k), & \text{if } \mathbf{w}(k)^T \mathbf{x}(k) \leq 0,\ \mathbf{x}_k \in \mathbf{X}. \end{cases} \tag{5.121}$$

Step 4. If $\mathbf{w}(k+1) = \mathbf{w}(k)$, for all $k = 1,\ \ldots,\ N$, then go to Step 5. Otherwise, set $\mathbf{w}(1) = \mathbf{w}(N)$ and go back to Step 3.

Step 5. After a finite number of cycles of Step 3, set

$$\mathbf{w} = \mathbf{w}(N), \tag{5.122}$$

as the solution of the linearly separable weight vector of (5.119). Furthermore, the $(n-1)$ dimensional hyperplane $\mathbf{x}$ is defined by the constraint equation of (5.123).

$$\mathbf{w}^T \mathbf{x} = w_1 x_1 + w_2 x_2 + \ldots + w_n x_n + w_{n+1} = 0. \quad \Box \tag{5.123}$$

The proof of the convergence of the Perceptron Algorithm is omitted here, but can be found in [2–5].

Example 5.19 Consider the original $\{\mathbf{x}_1^{(1)}, \mathbf{x}_2^{(1)}, \mathbf{x}_3^{(1)}, \mathbf{x}_4^{(1)}\}$ vectors from class 1 given by (5.102a) and the original $\{\mathbf{x}_1^{(2)}, \mathbf{x}_2^{(2)}, \mathbf{x}_3^{(2)}, \mathbf{x}_4^{(2)}, \mathbf{x}_5^{(2)}\}$ vectors from class 2 given by (5.102b) in Example 5.18. Using the augmented modified version of these vectors, they form a 3×9 matrix $\mathbf{X}$ denoted by

$$\mathbf{X} = [\mathbf{x}_1, \mathbf{x}_2, \ldots, \mathbf{x}_N] = [\mathbf{x}_1^{(1)}, \mathbf{x}_2^{(1)}, \ldots, \mathbf{x}_4^{(1)}, \mathbf{x}_1^{(2)}, \mathbf{x}_2^{(2)}, \ldots, \mathbf{x}_5^{(2)}]. \tag{5.124}$$

Start the Perceptron Algorithm of (5.121) with $\mathbf{w}_0 = [0,\ 0,\ 0]^T$. After 1 full cycle with $N = 9$ iterations, the linearly separable weight vector $\mathbf{w}$ is given by

$$\mathbf{w} = [-2.5,\ 0.5,\ 0]^T. \tag{5.125}$$

By direct evaluation,

$$\mathbf{w}^T \mathbf{X} = [3.00,\ 3.50,\ 6.50,\ 7.00,\ 3.25,\ 3.50,\ 5.50,\ 5.50,\ 7.00] \tag{5.126}$$

verifies all the training vectors satisfy the conditions of (5.121) and indeed are on the positive side of the hyperplane for this weight vector. Furthermore, the hyperplane for the weight vector in (5.123) is given by the line defined by

$$x_2 = 5x_1, \tag{5.127}$$

as shown in Fig. 5.18(a). However, if we start with $\mathbf{w}_0 = [0,\ 0,\ 1]^T$, after using the algorithm of (5.121) with 1 full cycle of $N = 9$ iterations, the linearly separable weight vector $\mathbf{w}$ is given by

$$\mathbf{w} = [-3,\ 1,\ 1]^T. \tag{5.128}$$

Direct evaluation shows

$$\mathbf{w}^T \mathbf{X} = [5.0,\ 6.0,\ 9.0,\ 9.0,\ 3.5,\ 3.0,\ 6.0,\ 5.0,\ 7.0]^T, \tag{5.129}$$

which again verifies all the training vectors satisfy the conditions of (5.121) and indeed are in the positive side of the hyperplane for this weight vector. Furthermore, the hyperplane for the weight vector in (5.123) is given by the line defined by

$$x_2 = 3x_1 - 1, \tag{5.130}$$

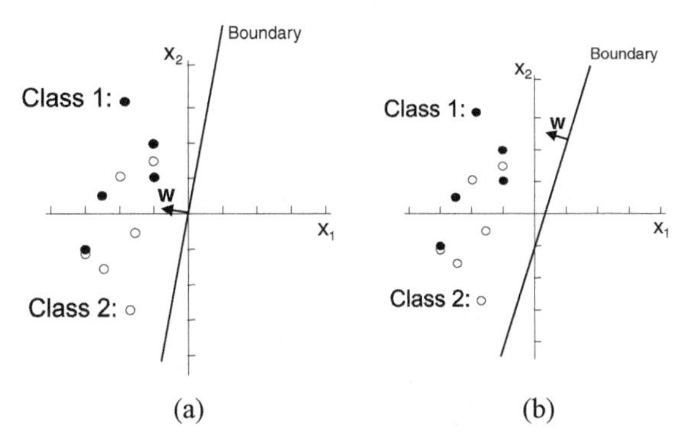

Figure 5.18 (a) Pattern vectors with a hyperplane of $x_2 = 5x_1$. (b) Pattern vectors with a hyperplane of $x_2 = 3x_1 - 1$.

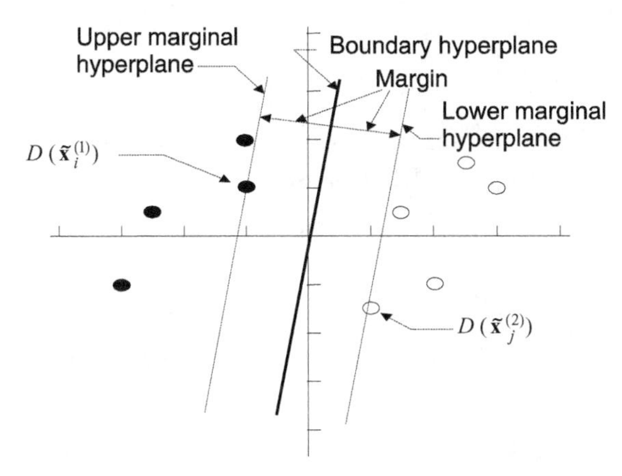

Figure 5.19 Four vectors in class 1 and five vectors in class 2 with margin, boundary, upper marginal, and lower marginal hyperplanes.

as shown in Fig. 5.18(b). From (5.124)–(5.130) and Figures 5.18(a) and (b), it is clear for a given collection of training vectors that are linearly separable the weight vector $\mathbf{w}$ and associated hyperplane are not unique. □

5.5.2.2 Margins for linearly separable binary class of patterns

In practice, the training vectors are just some specific realizations of the vectors in the two classes. There are certainly other possible vectors in the two classes that are outside of the existing cluster of vectors in the given training sets. Some other vectors from class 1 may be closer to class 2 vectors and some other vectors from class 2 may also be closer to class 1 vectors. This means we do not just want to find any arbitrary hyperplane (and associated weight vector) that can linearly separate these two given sets of vectors, but want to find the hyperplane having the largest possible margin in separating the two given sets of vectors as shown in Fig. 5.19. Consider vectors $\{\mathbf{x}_1^{(1)}, \mathbf{x}_2^{(1)}, \ldots, \mathbf{x}_{n_1}^{(1)}\}$ from class 1, then from (5.103a), we know

$$0 < D(\mathbf{x}_i^{(1)}), i = 1, \ldots, n_1. \tag{5.131}$$

Furthermore, let $\tilde{\mathbf{x}}_i^{(1)}$ be the vector in class 1 such that its discriminant function value has the lowest value satisfying

$$0 < d_1 = D\left(\tilde{\mathbf{x}}_i^{(1)}\right) = \min\{D(\mathbf{x}_k^{(1)})\} \leq D(\mathbf{x}_k^{(1)}),\ k = 1, \ldots, n_1, \tag{5.132}$$

as shown in Fig. 5.19. Similarly, for $\{\mathbf{x}_1^{(2)}, \mathbf{x}_2^{(2)}, \ldots, \mathbf{x}_{n_2}^{(2)}\}$ from class 2, (5.103b) shows

$$D(\mathbf{x}_j^{(2)}) < 0,\ j = 1, \ldots, n_2. \tag{5.133}$$

Furthermore, let $\tilde{\mathbf{x}}_j^{(2)}$ be the vector in class 2 such that its discriminant function value has the highest value satisfying

$$D(\mathbf{x}_j^{(2)}) \leq \max\{D(\mathbf{x}_k^{(2)})\} = D(\tilde{\mathbf{x}}_k^{(2)}) = d_2 < 0,\ k = 1, \ldots, n_2, \tag{5.134}$$

as shown in Fig. 5.19. We can now scale the weight vector $\mathbf{w}$ and the weight threshold w_{n+1} such that d_1 in (5.132) and d_2 in (5.134) satisfy

$$d_1 = D(\tilde{\mathbf{x}}_i^{(1)}) = 1, \tag{5.135a}$$

$$d_2 = D(\tilde{\mathbf{x}}_j^{(2)}) = -1. \tag{5.135b}$$

Then after scaling and using (5.131) and (5.135a), all the vectors in class 1 satisfy the inequality of

$$\mathbf{w}^T\mathbf{x}_i^{(1)} + b \geq 1,\ i = 1, \ldots, n_1. \tag{5.136a}$$

In particular, equality in (5.135a) is satisfied by the vector $\tilde{\mathbf{x}}_i^{(1)}$. Similarly, after scaling and using (5.134) and (5.135b), all the vectors in class 2 satisfy the inequality of

$$\mathbf{w}^T\mathbf{x}_j^{(2)} + b \leq -1,\ j = 1, \ldots, n_2. \tag{5.136b}$$

Equality in (5.136b) is satisfied by the vector $\tilde{\mathbf{x}}_j^{(2)}$. From $D(\tilde{\mathbf{x}}_i^{(1)}) = 1$ of (5.135a), then (5.112) yields the distance m from the vector $\tilde{\mathbf{x}}_i^{(1)}$ to the hyperplane, called the margin of the hyperplane, given by

$$m = \frac{1}{\|\mathbf{w}\|}. \tag{5.137}$$

Similarly, from $D(\tilde{\mathbf{x}}_j^{(2)}) = -1$ of (5.135b), then (5.112) yields the distance m from the vector $\tilde{\mathbf{x}}_j^{(2)}$ to the hyperplane also given by (5.137).

Clearly, for a linearly separable set of vectors, we want to find the hyperplane that yields the maximized margin $\hat{m}$ given by (5.137) or equivalently find the minimized $\|\hat{\mathbf{w}}\|$ subject to the constraints of (5.136a) and (5.136b). The following non-linear (quadratic)

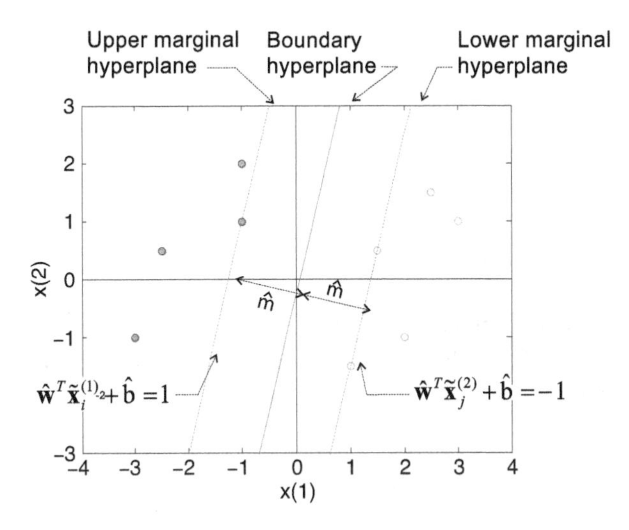

Figure 5.20 Optimized boundary hyperplane and upper marginal and lower marginal hyperplanes for linearly separable set of vectors in Example 5.20.

optimization program [6] can be used to solve this problem.

$$\text{Minimize} \quad \|\mathbf{w}\|, \tag{5.138a}$$

$$\text{subject to} \quad \mathbf{w}^T\mathbf{x}_i^{(1)} + b \geq 1, \; i = 1, \ldots, n_1, \tag{5.138b}$$

$$\mathbf{w}^T\mathbf{x}_j^{(2)} + b \leq -1, \; j = 1, \ldots, n_2. \tag{5.138c}$$

Example 5.20 Consider the four vectors from class 1 and five vectors from class 2 in Example 5.17. Using a standard cvx [6] solution of the quadratic optimization equations of (5.138), after 11 iterations, we obtained the results shown in Fig. 5.20, where the optimum weight $\hat{\mathbf{w}} = [-0.7619, \, 0.1905]^T$ yields a maximized margin $\hat{m} = 1/\|\hat{\mathbf{w}}\| = 1.273$, $\hat{b} = 0.0476$, and the associated hyperplane is given by $x_2 = 3.999x_1 - 0.2499$. From Fig. 5.20, we note, all four of the vectors from class 1 are on the positive side of the upper marginal hyperplane and all five vectors from class 2 are on the negative side of the lower marginal hyperplane. The total separation distance from the upper marginal hyperplane to the lower marginal hyperplane of $2m$ is maximized at $2 \times 1.273 = 2.546$. In other words, there is no higher separation distance between these two marginal hyperplanes for this given set of nine vectors. □

5.5.2.3 Margins for non-linearly separable binary class of patterns

Suppose a set of vectors $\{\mathbf{x}_1^{(1)}, \mathbf{x}_2^{(1)}, \ldots, \mathbf{x}_{n_1}^{(1)}\}$ from class 1 and a set of vectors $\{\mathbf{x}_1^{(2)}, \mathbf{x}_2^{(2)}, \ldots, \mathbf{x}_{n_2}^{(2)}\}$ from class 2 are not linearly separable. Then there is no hyperplane such that all the vectors from class 1 lie on one side and all the vectors from class 2 lie on the other side of the hyperplane. There are several approaches to tackle this situation. One possible approach is to note that some of the $\{\mathbf{x}_1^{(1)}, \mathbf{x}_2^{(1)}, \ldots, \mathbf{x}_{n_1}^{(1)}\}$

vectors in class 1 will not satisfy (5.138b) and/or some of the $\{\mathbf{x}_1^{(2)}, \mathbf{x}_2^{(2)}, \ldots, \mathbf{x}_{n_2}^{(2)}\}$ vectors in class 2 will not satisfy (5.138c). Thus, we can append some non-negative valued u_i, $i = 1, \ldots, n_1$, and non-negative valued v_j, $j = 1, \ldots, n_2$, such that now we have

$$\mathbf{w}^T \mathbf{x}_i^{(1)} + b \geq 1 - u_i,\ i = 1, \ldots, n_1, \tag{5.139a}$$

$$\mathbf{w}^T \mathbf{x}_j^{(2)} + b \leq -(1 - v_j),\ j = 1, \ldots, n_2. \tag{5.139b}$$

Of course, we want to find the minimum possible values of these u_i's and v_j's to satisfy the constraints in (5.139). The following linear programming (LP) program [7, p. 426] can be used to solve this problem.

$$\text{Minimize } \sum_{i=1}^{n_1} u_i + \sum_{j=1}^{n_2} v_j, \tag{5.140a}$$

$$\text{subject to } \mathbf{w}^T \mathbf{x}_i^{(1)} + b \geq 1 - u_i,\ i = 1, \ldots, n_1, \tag{5.140b}$$

$$\mathbf{w}^T \mathbf{x}_j^{(2)} + b \leq -(1 - v_j),\ j = 1, \ldots, n_2, \tag{5.140c}$$

$$0 \leq u_i,\ i = 1, \ldots, n_1;\ 0 \leq v_j,\ j = 1, \ldots, n_2. \tag{5.140d}$$

Example 5.21 Consider the four $\{\mathbf{x}_1^{(1)}, \mathbf{x}_2^{(1)}, \mathbf{x}_3^{(1)}, \mathbf{x}_4^{(1)}\}$ vectors from class 1 and the five $\{\mathbf{x}_1^{(2)}, \mathbf{x}_2^{(2)}, \mathbf{x}_3^{(2)}, \mathbf{x}_4^{(2)}, \mathbf{x}_5^{(2)}\}$ vectors from class 2 in Example 5.17. Now, only $\mathbf{x}_1^{(1)}$ is changed to $\mathbf{x}_1^{(1)} = [-0.225,\ 0.93]^T$, while $\mathbf{x}_1^{(2)}$ is changed to $\mathbf{x}_1^{(2)} = [-0.75,\ 0.93]^T$, and $\mathbf{x}_2^{(2)}$ is changed to $\mathbf{x}_2^{(2)} = [1,\ 0.8]^T$. A plot of these nine vectors is shown in Fig. 5.21. Upon inspection, it seems that there is no hyperplane (i.e., a line in this example) that can separate all the vectors in class 1 to the positive side and all the vectors in class 2 to the negative side of the hyperplane. By using the LP code of [7, p. 426] for our data, in 9 iterations, we obtained the optimized weight vector $\hat{\mathbf{w}} = [-0.7143,\ 0.4762]^T$ and an optimized margin $\hat{m} = 1/\|\hat{\mathbf{w}}\| = 1.165$, $\hat{b} = -0.6667$, and an associated hyperplane given by $\mathbf{x}_2 = 1.500\mathbf{x}_1 + 1.400$. As seen in Fig. 5.21, even for this "optimized solution," $\mathbf{x}_1^{(1)} = [-0.225,\ 0.93]^T$ is still slightly to the right of the hyperplane, while $\mathbf{x}_1^{(2)} = [-0.75,\ 0.80]^T$ is to the left of the hyperplane. Thus, there is a total mis-classification of two of the training vectors, but there is no vector inside the slab. While this hyperplane solution is not perfect (i.e., not surprisingly it was not able to achieve "linearly separable" results), it is the best we can do based on the criterion of minimizing the objective function of (5.140a) for these two non-linearly separable sets of vectors. □

The support vector machine (SVM) algorithm [8] is another well-known approach for handling non-linearly separable classification problems. SVM provides a parameter γ that allows a tradeoff between the number of possible mis-classifications versus the length of the margin m. The actual computation of the SVM algorithm uses convex optimization methods [7, p. 427]. One form of the SVM classification program for two

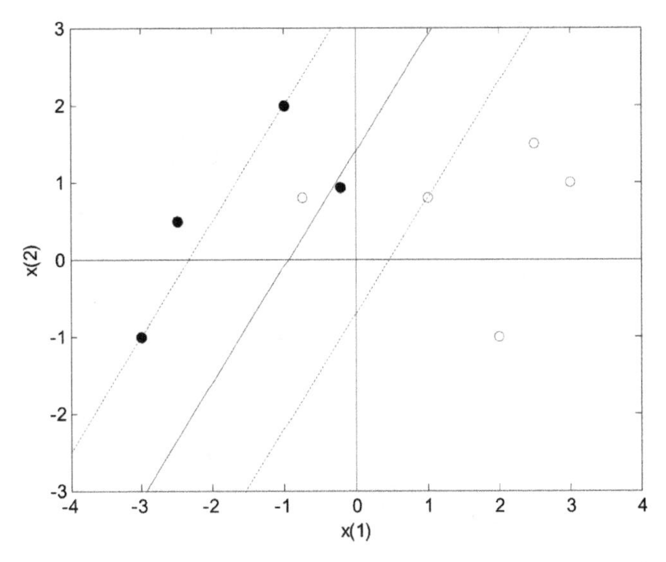

Figure 5.21 Optimized hyperplanes for non-linearly separable set of vectors in Example 5.21 with two mis-classification errors but no vector in the slab using the LP program of (5.140).

classes is given below.

$$\text{Minimize} \quad \|\mathbf{w}\| + \gamma \left(\sum_{i=1}^{n_1} u_i + \sum_{j=1}^{n_2} v_j \right), \tag{5.141a}$$

$$\text{subject to} \quad \mathbf{w}^T \mathbf{x}_i^{(1)} + b \geq 1 - u_i, \; i = 1, \ldots, n_1, \tag{5.141b}$$

$$\mathbf{w}^T \mathbf{x}_j^{(2)} + b \leq -(1 - v_j), \; j = 1, \ldots, n_2, \tag{5.141c}$$

$$0 \leq u_i, \; i = 1, \ldots, n_1; \; 0 \leq v_j, \; j = 1, \ldots, n_2. \tag{5.141d}$$

Example 5.22 We use the same four $\{\mathbf{x}_1^{(1)}, \mathbf{x}_2^{(1)}, \mathbf{x}_3^{(1)}, \mathbf{x}_4^{(1)}\}$ vectors from class 1 and the five $\{\mathbf{x}_1^{(2)}, \mathbf{x}_2^{(2)}, \mathbf{x}_3^{(2)}, \mathbf{x}_4^{(2)}, \mathbf{x}_5^{(2)}\}$ vectors from class 2 as in Example 5.21. By using the SVM algorithm of (5.141) with the parameter $\gamma = 0.10$, we obtained the following results. A plot of these nine vectors is shown in Fig. 5.22. Upon inspection, there is indeed no hyperplane (i.e., a line in this case) that can separate all the vectors in class 1 to the positive side and all the vectors in class 2 to the negative side of the hyperplane. However, the SVM algorithm shows that after 10 iterations, we obtained the optimized weight vector $\hat{\mathbf{w}} = [-0.3815, \; 0.0393]^T$ and an optimized margin $\hat{m} = 1/\|\mathbf{w}\| = 2.608$, $\hat{b} = -0.1052$, and an associated hyperplane given by $x_2 = 9.707 x_1 + 2.6777$. As seen in Fig. 5.22, even for this "optimized solution," $\mathbf{x}_1^{(2)} = [-0.75, \; 0.80]^T$ is to the left of the hyperplane. Thus, there is now only a total mis-classification of one of the training vectors, but the margin in Example 5.22 is considerably larger than the margin in Example 5.21, and now there are five vectors classified correctly but inside the slab. We also note, in the SVM program of (5.141), if the parameter γ becomes very large, then the influence of the term $\|\mathbf{w}\|$ in (5.141a) becomes negligible, and thus the solution of

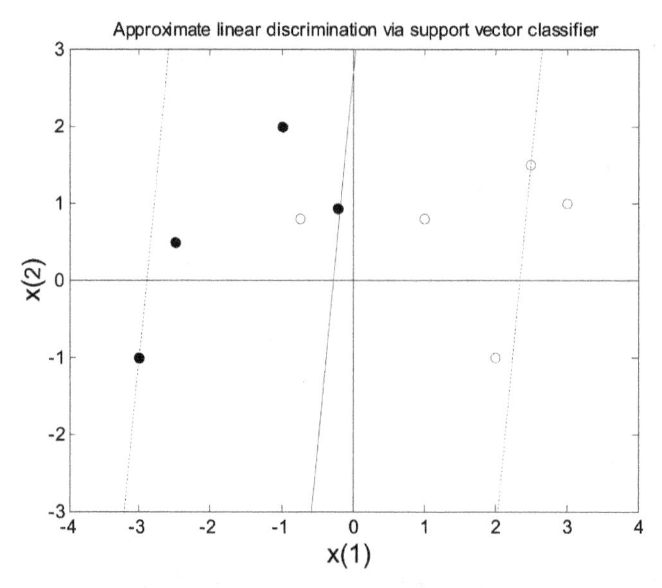

Figure 5.22 Optimized hyperplanes for non-linearly separable set of vectors in Example 5.21 with one mis-classification error but five vectors inside the slab using the SVM program of (5.141).

the SVM program becomes the same as that of the solution of the LP program with a smaller margin. Thus, by varying the parameter γ, the SVM solution yields a tradeoff in the width of the margin and the number of vectors in the slab. □

5.5.2.4 Classification of pattern vectors from M classes

Thus far, we have only considered classification of vectors from two classes. Not surprisingly, the problem of classification of M classes is more difficult and does not have an obvious generalization from the binary class. One approach is to reduce the problem to M two-class problems, where the i-th linear discriminant function separates a given pattern vector to be either in class i or not in class i, for i from 1 to M. Another more involved approach is to define $M(M-1)2$ linear discriminant functions, one for each pair of classes, for separating class i pattern vectors from class j pattern vectors. Unfortunately, in either of the above approaches, there can be regions in the pattern vector space in which classification is not defined. A resulting example, where the shaded region under the first approach is not defined, is shown in Fig. 5.23(a).

One way to avoid this problem is to define M linear discriminant functions $D_i(\mathbf{x})$, $i = 1, \ldots, M$, such that if a pattern vector $\mathbf{x}$ satisfies

$$D_i(\mathbf{x}) = \mathbf{w}_i^T \mathbf{x} + w_{i,n+1} > \mathbf{w}_j^T \mathbf{x} + w_{j,n+1} = D_j(\mathbf{x}), \Rightarrow \mathbf{x} \in \mathbb{R}_i, \; i = 1, \ldots, M, \tag{5.142}$$

where the weight vectors $\mathbf{w}_i$ for class i and $\mathbf{w}_j$ for class j are denoted by

$$\mathbf{w}_i = [w_{i,1}, \ldots, w_{i,n}], \; \mathbf{w}_j = [w_{j,1}, \ldots, w_{j,n}], \tag{5.143}$$

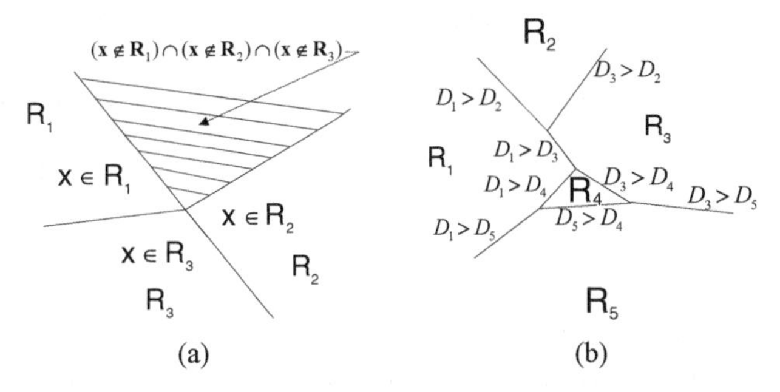

Figure 5.23 (a) Linear decision boundaries for $M = 3$ classes in $n = 2$ dimensions with an undecided shaded region. (b) Decision boundaries of a linear machine for classification of $M = 5$ classes in $n = 2$ dimensions.

then that vector $\mathbf{x}$ will be assigned to class i. If there is a tie, then the classification is not defined. The resulting classifier is called a *linear machine*. Thus, a linear machine separates the feature vectors into M decision regions, with $D_i(\mathbf{x})$ being the largest discriminant if $\mathbf{x}$ is in the decision region $\mathbb{R}_i$. If $\mathbb{R}_i$ and $\mathbb{R}_j$ are contiguous decision regions, then the boundary is a portion of the hyperplane H_{ij} defined by $D_i(\mathbf{x}) = D_j(\mathbf{x})$ or equivalently

$$(\mathbf{w}_i - \mathbf{w}_j)^T \mathbf{x} + (w_{i,n+1} - w_{j,n+1}) = 0. \tag{5.144}$$

Thus, it follows that $(\mathbf{w}_i - \mathbf{w}_j)$ is normal to H_{ij}. Therefore, in a linear machine, it is not the weight vectors but their differences that are important. In practice, for the classification of M classes, there need not be as many as $M(M-1)/2$ pairs of decision regions and associated decision hyperplanes. An example of $M = 5$ classes in $n = 2$ dimensions has only 8 decision hyperplanes (not the full $5 \times 4/2 = 10$ possible decision hyperplanes) is shown in Fig. 5.23(b).

5.6 Conclusions

In Section 5.1, we first considered the use of M-ary signaling vectors and waveforms in the idealized AWGN channel. Since each symbol associated with a M-ary signal vector/waveform carries $\log_2 M$ bits of information, modulation systems with a large M value (greater than the binary case of $M = 2$) were of great interest. Section 5.2 introduced the Gram–Schmidt orthonormalization method for finding the minimum number of vectors/waveforms as a basis for representing the M-ary signaling vectors/waveform. Section 5.3 treated the M-ary detection problem in AWGN based on various statistical decision criteria, including the Bayes criterion, the MAP criterion, and the minimum average probability of error criterion. For equal probability and equal energy M-ary signal vectors/waveforms in AWGN, the minimum distance decision criterion provided simple intuitive interpretation of system performance from the geometry of the signal vectors/waveforms. Some well-known M-ary modulation examples were treated. Section 5.4 considered optimum signal design for M-ary signal vectors/waveforms based on minimizing the signal energy. Section 5.5 introduced the pattern classification problem based on the use of a set of known training pattern vectors from M classes. Classification algorithms for linearly and non-linearly separable pattern vectors and SVM concepts were used. Some simple examples illustrated the use of these different classification algorithms.

5.7 Comments

Theoretical foundations of binary detection problems, as considered in Chapter 4, were based on the binary hypothesis testing theory developed in statistics in the 1920s–1930s, and expounded in the practical detection of targets in radar and sonar in World War II. However, for M-ary detection in AWGN, one of the early treatments based on a geometric interpretation point of view appeared in the Arthurs–Dym paper [9] and expounded in the textbook of Wozencraft–Jacobs [1]. Since then M-ary detection has been treated in [10–18] and other books. In this chapter, all the M-ary detection problems assumed an AWGN channel having an infinite bandwidth. Practical communication systems always impose realistic bandwidth constraints, which complicate the analysis of system performance and signal waveform designs [10–18]. When the channel noise can not be assumed to be WGN, then the Karhunen–Loeve expansion method (Sections 5.4–5.6) needs to be used to represent the noise. In a WGN problem, the noise components along all the orthonormal coordinates are equal, and distances among transmitted signal vectors provided simple geometric intuitions on their ability to combat the noise. However, in a colored noise problem, the simple geometric intuitions are no longer available. In this chapter (in fact in this book), we do not consider the use of coding to improve the performance and channel bandwidth of a communication system utilizing M-ary signal vectors/waveforms. Relationships of M-ary signals/waveforms to some classical aspects of coding theory were considered in [11, 13, 15–17]. The optimum M-ary signal design problems of Section 5.4 are most relevant in the context of a coded communication system [11, 13].

The fundamental theory of using known pattern vectors from M classes to perform training for classification was based on the concept of the convergence of the Perceptron Algorithm developed by Rosenblatt [2]. Early books dealing with M-ary classification included [3–5] and others. A recent detailed treatment of pattern classification and recognition was given in [19]. Solutions of various pattern classification algorithms required the use of non-linear optimization theory. Modern convex optimization theory [6] based on the Matlab cvx program [7] has been used to solve many optimum pattern classification problems. Various methods in using support vector machine (SVM) concepts for pattern classification were given in [8].

References

[1] J. M. Wozencraft and I. M. Jacobs. *Principles of Communication Engineering*. Wiley, 1965.

[2] F. Rosenblatt. *Principles of Neurodynamics: Perceptrons and the Theory of Brain Mechanisms*. Spartan Books, 1962.

[3] R. O. Duda and P. E. Hart. *Pattern Classification and Scene Analysis*. Wiley, 1973.
[4] J. T. Tou and R. C. Gonzalez. *Pattern Recognition Principles*. Addison-Wesley, 1974.
[5] T. Young and T. W. Calvert. *Classification, Estimation and Pattern Recognition*. American Elsevier, 1974.
[6] M. Grant and S. Boyd. cvx. http://stanford.edu/~boyd/cvx/citing.html.
[7] S. Boyd and L. Vandenberghe. *Convex Optimization*. Cambridge University Press, 2004.
[8] B. Schölkopf, C. J. C. Burges, and A. J. Smola, editors. *Advances in Kernel Methods: Support Vector Learning*. MIT Press, 1999.
[9] E. Arthurs and H. Dym. On the optimum detection of digital signals in the presence of white Gaussian noise – a geometric interpretation and a study of three basic data transmission systems. *IRE Trans. on Comm. System*, Dec. 1962.
[10] C. W. Helstrom. *Statistical Theory of Signal Detection*. Pergamon, 2d ed., 1960. An expanded version appeared as *Elements of Signal Detection and Estimation*, Prentice-Hall, 1995.
[11] A. J. Viterbi. *Principles of Coherent Communication*. McGraw-Hill, 1966.
[12] H. L. Van Trees. *Detection, Estimation, and Modulation, Part I*. Wiley, 1968.
[13] C. L. Weber. *Elements of Detection and Signal Design*. McGraw-Hill, 1968.
[14] A. D. Whalen. *Detection of Signals in Noise*. Academic Press, 1st ed., 1971. R. N. McDonough and A. D. Whalen, 2d ed., 1995.
[15] J. G. Proakis. *Digital Communications*. McGraw-Hill, 1983, 5th ed., 2008.
[16] E. Biglieri, S. Benedetto, and V. Castellani. *Digital Transmission Theory*. Prentice-Hall, 1987. An expanded version appeared as S. Benedetto and E. Biglieri, *Principles of Digital Transmission*, Kluwer, 1999.
[17] B. Sklar. *Digital Communications – Fundamentals and Applications*. Prentice-Hall, 1988, 2d ed., 2001.
[18] T. A. Schonhoff and A. A. Giordano. *Detection and Estimation Theory*. Pearson, 2006.
[19] S. Theodoridis and K. Koutroumbas. *Pattern Recognition*. Academic Press, 4th ed., 2009.

Problems

5.1 Consider the M-ary hypothesis decision problem based on the MAP criterion. Each hypothesis H_i, $i = 1, \ldots, M$ has a known prior probability $\mathrm{P}(H_i) = \pi_i$, $i = 1, \ldots, M$, and also a known conditional pdf $\mathrm{P}(x|H_i) = p_i(x)$, $-\infty < x < \infty$, $i = 1, \ldots, M$. Find the M-ary MAP decision rule expressed in terms of the prior probabilities and the conditional pdfs.

5.2 Consider a M-ary hypothesis decision problem based on the MAP criterion with $M = 3$, $\pi_1 = 1/2$, $\pi_2 = \pi_3 = 1/4$. Let $p_1(x) = (1/2)\exp(-|x|)$, $-\infty < x < \infty$, $p_2(x) = (1/\sqrt{2\pi})\exp(-x^2/2)$, $-\infty < x < \infty$, and $p_3(x) = (1/\sqrt{4\pi})\exp(-(x-1)^2/4)$, $-\infty < x < \infty$. Find the decision regions for the three hypotheses based on the MAP criterion.

5.3 List some known M-ary non-orthogonal communication systems. List some known M-ary orthogonal communication systems.

5.4 Consider a general M-ary orthogonal system in which all the transmitted waveforms having energy E are modeled by $s_i(t) = \sqrt{E}\theta_i(t)$, $0 \le t \le T$, $i = 1 \ldots, M$, where the set of $\{\theta_i(t),\ 0 \le t \le T,\ i = 1 \ldots, M\}$ constitutes an orthonormal system of function. Then in signal vector space, each transmitted signal waveform $s_i(t)$ can be represented by $\mathbf{s}_i = \sqrt{E}[0, \ldots, 0, 1, 0, \ldots, 0]^T$, where all the components of the vector are zeros except the ith. Assume all the signals have equi-probability of transmission and the channel is AWGN with two-sided spectral density of $N_0/2$. The coherent receiver decides the jth hypothesis from the largest value of the outputs of the M correlators (or matched filters) matched to each signal waveform, $\theta_i(t)$. If the transmitted signal is $\mathbf{s}_j$, then the outputs of all the correlators are zero-mean Gaussian r.v.'s of variance $N_0/2$, except the jth output is also a Gaussian r.v. of mean $\sqrt{E}$ and also of variance $N_0/2$. Then the probability of a correct decision when the jth signal was transmitted and conditioned on the noise by $\mathrm{P}(\mathrm{C}|\mathbf{s}_j, n_j)$ is given by

$$\begin{aligned}
\mathrm{P}(\mathrm{C}|\mathbf{s}_j, n_j) &= \mathrm{P}(\text{all } n_i \leq (\sqrt{E} + n_j),\ i \neq j) \\
&= \prod_{i=1, i \neq j}^{M} \mathrm{P}(n_i \leq (\sqrt{E} + n_j)) \\
&= \left[\left(1/\sqrt{N_0 \pi}\right) \int_{-\infty}^{\sqrt{E}+n_j} \exp(-t^2/N_0)\, dt \right]^{M-1}.
\end{aligned}$$

Then obtain $\mathrm{P}(\mathrm{C}|\mathbf{s}_j) = \mathrm{E}_{N_j}\{\mathrm{P}(\mathrm{C}|\mathbf{s}_j, n_j)\}$. Find the average probability of a symbol error $\mathrm{P}(\text{error}) = 1 - \mathrm{P}(\mathrm{C}|\mathbf{s}_j)$.

5.5 Consider a baseband M-ary pulse amplitude modulation (PAM) system (also known as the M-ary amplitude shift keyed M-ASK system) where the transmission signal set is given by $\{\pm A, \pm 3A, \ldots, \pm(M-1)A\}$. We assume M is an even integer. Specifically, denote $s_m = (2m-1)A,\ m = 1, \ldots, M/2$ and $s_{-m} = -(2m-1)A,\ m = 1, \ldots, M/2$. Let each transmitted signal be disturbed by an additive Gaussian r.v. N of zero-mean and variance σ^2. Assume all M signals are transmitted with equal probability. Based on the minimum-distance decision rule, find the decision region $\mathbb{R}_m,\ m = \pm\{1, \ldots, M/2\}$ for a transmitted s_m. Find the average symbol error probability of this system expressed in terms of the average energy of the signal and σ^2.

5.6 Formulate the equivalent M-ASK system of Problem 5.5 using a carrier at frequency f_0 denoted by $\cos(2\pi f_0 t),\ 0 \leq t \leq T$. Let the transmitted signal waveform be disturbed by an AWGN process of zero-mean and two-sided power spectral density of $N_0/2$. Find the coherent receiver based on the minimum-distance decision criterion. Find the decision regions and the average symbol error probability, P_e^s, of this system in terms of the average energy of the signal waveforms and $N_0/2$.

In Problem 5.6, take $M = 4$, and then find the decision regions and the P_e^s of this system in terms of the average energy of the signal waveforms and $N_0/2$. Compare the P_e^s of this 4-ASK system to the P_e^s of a QPSK system. Intuitively, why is the performance of the QPSK system better than that of this 4-ASK system? (Hint: What are the dimensions of signal sets in 4-ASK and QPSK?)

5.7 Consider a BPSK system. In an ideal coherent receiver (with perfect carrier synchronization), the average probability of a bit error $P_{\mathrm{e}}^{\mathrm{BPSK}}$ as shown in (5.58) is given by $P_{\mathrm{e}}^{\mathrm{BPSK}} = \mathrm{Q}(\sqrt{2E/N_0})$. Suppose the carrier synchronization at the receiver introduced a phase error in the sense that it uses a noisy reference waveform $r(t) = \cos(2\pi ft + \psi)$ in the correlation process, where the phase error ψ has a pdf denoted by $p_\Psi(\psi)$. For a small phase error ψ, its pdf can be approximated as a zero-mean Gaussian r.v. of variance σ_Ψ^2. Now, find the average probability of a bit error of the BPSK system with this synchronization error. Show this new average probability of a bit error can be approximated as the ideal $P_{\mathrm{e}}^{\mathrm{BPSK}} = \mathrm{Q}(\sqrt{2E/N_0})$ with another term which is a function of σ_Ψ^2. Hint: Use a Maclaurin power series approximation of the probability of error as a function of the phase error ψ and assume $\sigma_\Psi \ll 1$.

5.8 (*) Repeat Problem 5.7 for a QPSK system with an imperfect carrier synchronization process using $r_{\mathrm{I}}(t) = \cos(2\pi ft + \psi)$ for the I-channel correlation and $r_{\mathrm{Q}}(t) = \sin(2\pi ft + \psi)$ for the Q-channel correlation. Find the average symbol error probability of the QPSK system with this synchronization error.

5.9 Consider a baseband digital signal denoted by $s(t) = \sum_{n=-N}^{N} B_n g(t - nT)$, where $\{B_n, -N < n < N\}$ is a set of random data and $g(t)$ is a deterministic waveform of duration T with a Fourier transform of $G(f) = \mathscr{F}\{g(t)\}$.

(a) Derive the power spectral density function (PSD) $S_s(f)$ of $s(t)$ by taking the statistical average over $\{B_n, -N < n < N\}$ and time average over $|C(f)|^2$, where $C(f) = \mathscr{F}\{s(t)\}$. Assume B_n is a wide-stationary sequence with an autocorrelation function $R(m) = \mathrm{E}\{B_n B_{n+m}\}$. Show $S_s(f) = (|G(f)|^2/T)\sum_{m=-\infty}^{\infty} R(m)\exp(-im2\pi fT)$.
(b) Suppose $\mathrm{E}\{B_n\} = \mu$ and $\mathrm{E}\{B_n B_{n+m}\} = \mu^2$ (i.e., r.v.'s in $\{B_n, -N < n < N\}$ are uncorrelated). Find the PSD $S_s(f)$. Hint: Use the Poisson formula $\sum_{m=-\infty}^{\infty} \exp(-im2\pi fT) = (1/T)\sum_{m=\infty}^{\infty} \delta(f - m/T)$.

5.10 Consider the two classes of pattern vectors given by (5.102a)–(5.102b).

(a) Show the convergence of the Perceptron Algorithm with five arbitrarily chosen different initial vectors for these two classes of pattern vectors.
(b) Suppose only $\mathbf{x}_1^{(1)}$ can be modified. Find a new $\mathbf{x}_1^{(1)}$ so that these two classes of pattern vectors are not linearly separable. Furthermore, find another new $\mathbf{x}_1^{(1)}$ such that it has minimum distance from the original $\mathbf{x}_1^{(1)}$ so that these two classes of pattern vectors are not linearly separable.

5.11 Duplicate Example 5.21 and verify the results in that example.

5.12 (*) Duplicate Example 5.22 and verify the results in that example. Hint: There are various ways to perform the necessary convex optimization needed to solve for this SVM problem. The cvx routine is a Matlab plug-in public domain program available for downloading from the Internet.

6 Non-coherent detection in communication and radar systems

In the previous chapters, we always assumed that from the received waveform

$$x(t) = s_i(t) + n(t),\ 0 \le t \le T,\ i = 1, \ldots, M, \tag{6.1}$$

we do not know the realization of the random noise $n(t)$ (otherwise we do not need any detection theory at all), but we have available the deterministic transmitted waveforms $\{s_i(t),\ 0 \le t \le T,\ i = 1, \ldots, M\}$ completely except for the index i. In reality, there may be various parameters of the waveforms that are not available at the receiver. In Section 6.1, we assume under H_0 the received waveform contains only a WGN and under H_1 the received waveform consists of a sinusoidal function with known amplitude and frequency but having random phase in the presence of a WGN. We find the optimum receiver structure of this non-coherent communication system. In Section 6.2, we study the performance of this system by finding its probability of a false alarm and its probability of detection. In Section 6.3, we deal with non-coherent detection problems encountered in various radar systems. Section 6.3.1 considers coherent integration in a radar system. In Section 6.3.2, we treat post-detection integration of a received train of n pulses with identical amplitudes with independent uniformly distributed initial random phases of the pulses. For a small SNR, the LR receiver reduces to a quadratic receiver, while for a large SNR, the LR receiver reduces to a linear receiver. In Section 6.3.3, the double-threshold receiver in a radar system is presented. Section 6.3.4 deals with the constant false alarm issue under unknown background noises.

6.1 Binary detection of a sinusoid with a random phase

In various communication systems, the received waveform may be modeled by

$$x(t) = \begin{cases} n(t), & H_0, \\ A\cos(2\pi f_0 t + \theta) + n(t), & H_1, \end{cases} \quad 0 \le t \le T, \tag{6.2}$$

where $n(t)$ is the realization of a WGN process with zero-mean and two-sided power spectral density value of $N_0/2$, A is a known amplitude, f_0 is a known frequency with $(1/f_0) \ll T$, and θ is the realization of a uniformly distributed random variable Θ on

$[0,\ 2\pi)$. That is, Θ has a pdf given by

$$p_\Theta(\theta) = \begin{cases} 1/(2\pi), & 0 \leq \theta \leq 2\pi, \\ 0, & \text{elsewhere}. \end{cases} \tag{6.3}$$

Some possible reasons for using the uniformly distributed phase pdf model include:

1. In a radar system, the range from the transmitter to the target is unknown. A uniformly distributed random variable model for Θ implies complete lack of information about the phase;
2. An unstable transmitter oscillator with no control over the phase;
3. A transmission medium that introduced a random phase;
4. A receiver that does not track the phase θ of the transmitted sinusoid.

Consider the LR test using the sampling approach for the received waveform modeled by (6.1). Denote the conditional probabilities of the realization of the random vector $\mathbf{x}$ obtained from sampling of the waveform $\{x(t),\ 0 \leq t \leq T\}$, under H_0 by $p_0(\mathbf{x})$, and under H_1 by $p_1(\mathbf{x} \mid \theta)$ for a given θ realization. Then the LR function for $\mathbf{x}$ is given by

$$\begin{aligned} \Lambda(\mathbf{x}) &= \frac{p_1(\mathbf{x})}{p_0(\mathbf{x})} = \frac{\int_0^{2\pi} p_1(\mathbf{x}|\theta) p_\Theta(\theta) d\theta}{p_0(\mathbf{x})} = \frac{\frac{1}{2\pi}\int_0^{2\pi} p_1(\mathbf{x}|\theta) d\theta}{p_0(\mathbf{x})} \\ &= \frac{\frac{1}{2\pi}\int_0^{2\pi} \exp\left(-\frac{\|\mathbf{x}-\mathbf{s}(\theta)\|^2}{2\sigma^2}\right) d\theta}{\exp\left(-\frac{\|\mathbf{x}\|^2}{2\sigma^2}\right)}. \end{aligned} \tag{6.4}$$

Take the limit as the number of samples goes to infinity in the last expression of (6.4). Then the LR function for $\{x(t),\ 0 \leq t \leq T\}$ is given by

$$\begin{aligned} &\Lambda(x(t),\ 0 \leq t \leq T) \\ &= \frac{\frac{1}{2\pi}\int_0^{2\pi} \exp\left(-\frac{1}{N_0}\int_0^T (x(t) - A\cos(2\pi f_0 t + \theta))^2\, dt\right) d\theta}{\exp\left(-\frac{1}{N_0}\int_0^T x^2(t) dt\right)} \\ &= \frac{1}{2\pi}\int_0^{2\pi} \exp\left(-\frac{A^2}{N_0}\int_0^T \cos^2(2\pi f_0 t + \theta) dt + \frac{2A}{N_0}\int_0^T x(t)\cos(2\pi f_0 t + \theta) dt\right) d\theta. \end{aligned} \tag{6.5}$$

We note that

$$\cos^2(2\pi f_0 t + \theta) = \frac{1}{2}\left[1 + \cos\left(2(2\pi f_0 t + \theta)\right)\right] \tag{6.6}$$

and

$$\begin{aligned} &\int_0^T \cos(2(2\pi f_0 t + \theta)) dt \\ &= \left.\frac{\sin\left(2(2\pi f_0 t + \theta)\right)}{4\pi f_0}\right|_0^T = \frac{\sin\left(2(2\pi f_0 T + \theta)\right) - \sin(2\theta)}{4\pi f_0} \ll T. \end{aligned} \tag{6.7}$$

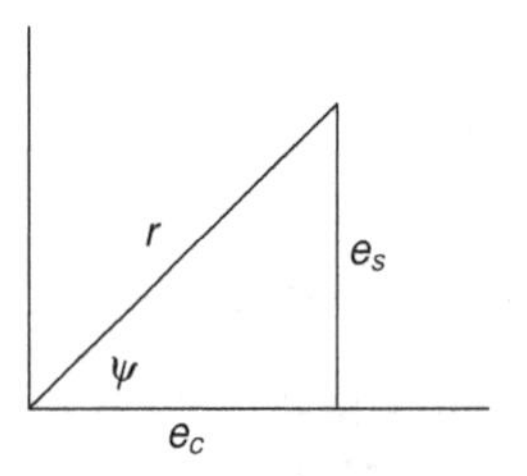

Figure 6.1 Trigonometric relationships on e_c, e_s, r, and ψ.

Thus, using (6.6) and (6.7) in the first integral in the exp(·) expression of (6.5), we obtain

$$A^2 \int_0^T \cos^2(2\pi f_0 t + \theta) dt = \frac{A^2}{2}\left[T + \int_0^T \cos(2(2\pi f_0 t + \theta)) dt\right] \simeq \frac{A^2 T}{2} = E. \tag{6.8}$$

Then (6.5) can be rewritten as

$$\begin{aligned}&\Lambda(x(t),\ 0 \le t \le T)\\&= \exp(-E/N_0) \cdot \frac{1}{2\pi}\int_0^{2\pi} \exp\left(\frac{2A}{N_0}\int_0^T x(t)\cos(2\pi f_0 t + \theta) dt\right) d\theta.\end{aligned} \tag{6.9}$$

Thus, the decision regions $\mathbb{R}_1$ and $\mathbb{R}_0$ are given by

$$\begin{aligned}\mathbb{R}_1 &= \{x(t),\ 0 \le t \le T : \Lambda(x(t),\ 0 \le t \le T)\} \ge \Lambda_0\},\\ \mathbb{R}_0 &= \{x(t),\ 0 \le t \le T : \Lambda(x(t),\ 0 \le t \le T)\} \le \Lambda_0\}.\end{aligned} \tag{6.10}$$

Clearly, the observed $\{x(t),\ 0 \le t \le T\}$ is used in the LR function through the integral with $\cos(2\pi f_0 t + \theta)$ in the exp(·) expression part of (6.9). But we note,

$$\cos(2\pi f_0 t + \theta) = \cos(2\pi f_0 t)\cos(\theta) - \sin(2\pi f_0 t)\sin(\theta). \tag{6.11}$$

Denote the I-component of the received waveform by

$$e_c = \int_0^T x(t)\cos(2\pi f_0 t) dt = r\cos(\psi) \tag{6.12}$$

and the Q-component of the received waveform by

$$e_s = \int_0^T x(t)\sin(2\pi f_0 t) dt = r\sin(\psi), \tag{6.13}$$

where the envelope of the I-Q components is denoted by r and the associated angle ψ defined as the arc-tangent of e_s/e_c is shown in Fig. 6.1 and defined by

$$r = \sqrt{e_c^2 + e_s^2},\ \psi = \tan^{-1}(e_s/e_c),\ \cos(\psi) = e_c/r,\ \sin(\psi) = e_s/r. \tag{6.14}$$

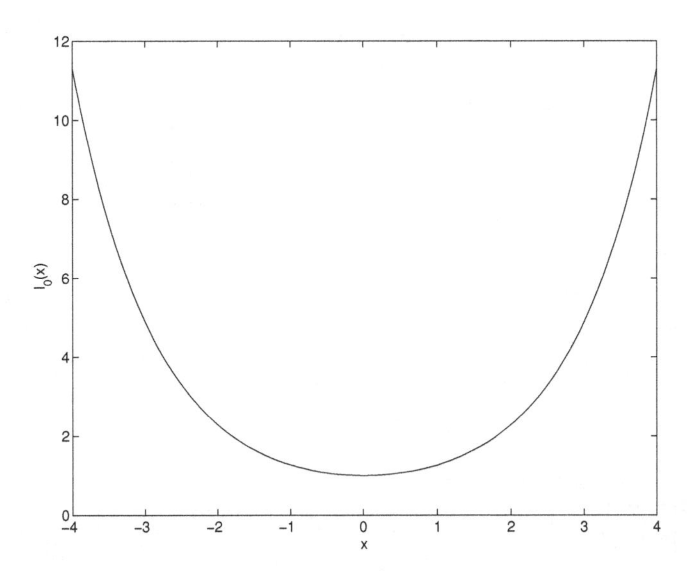

Figure 6.2 Plot of $\mathrm{I}_0(x)$ vs x.

Thus, the integral in (6.9) can be expressed in terms of the I-Q components or r and ψ as

$$\int_0^T x(t)\cos(2\pi f_0 t + \theta)dt = e_{\mathrm{c}}\cos(\theta) - e_{\mathrm{s}}\sin(\theta)$$

$$= \sqrt{e_{\mathrm{c}}^2 + e_{\mathrm{s}}^2}\cos(\theta + \psi) = r\cos(\theta + \psi). \tag{6.15}$$

From (6.9), (6.10), and (6.15), we have

$$\mathrm{I}_0(2Ar/N_0)$$

$$= \frac{1}{2\pi}\int_0^{2\pi} \exp\left(\frac{2Ar\cos(\theta+\psi)}{N_0}\right) d\theta \underset{\leq_{H_0}}{\overset{\geq^{H_1}}{}} \Lambda_0 \exp(E/N_0) = \gamma, \tag{6.16}$$

where the modified Bessel function of the first kind of order zero, $\mathrm{I}_0(x)$, is defined by

$$\mathrm{I}_0(x) = \frac{1}{2\pi}\int_{\theta_0}^{\theta_0+2\pi} \exp x\cos\theta d\theta, \tag{6.17}$$

and is shown in Fig. 6.2 [1, p. 374]. We note, the integral in (6.17) has the same value for any initial θ_0. We note that $\mathrm{I}_0(x)$ is a symmetric function on the entire real line and has a minimum with a value of 1 for $x = 0$. In some sense, $\mathrm{I}_0(x)$ behaves like the cosh(x) function. (Specifically, $\mathrm{I}_0(0) = 1$, $\cosh(0) = 1$, $\mathrm{I}_0(1) = 1.27$, $\cosh(1) = 1.54$, $\mathrm{I}_0(x) \simeq (1/\sqrt{2\pi x})\exp(x)$, $x \to \infty$ and $\cosh(x) = (1/2)\exp(x)$, $x \to \infty$.) We note, $\mathrm{I}_0(x)$ is a monotonically increasing function for $0 \leq x < \infty$. Thus, for $1 \leq y < \infty$, the inverse $\mathrm{I}_0^{-1}(y)$ exists. Applying $\mathrm{I}_0^{-1}(y)$ to both sides of (6.16), we obtain

$$2Ar/N_0 \underset{\leq_{H_0}}{\overset{\geq^{H_1}}{}} \mathrm{I}_0^{-1}(\gamma), \tag{6.18}$$

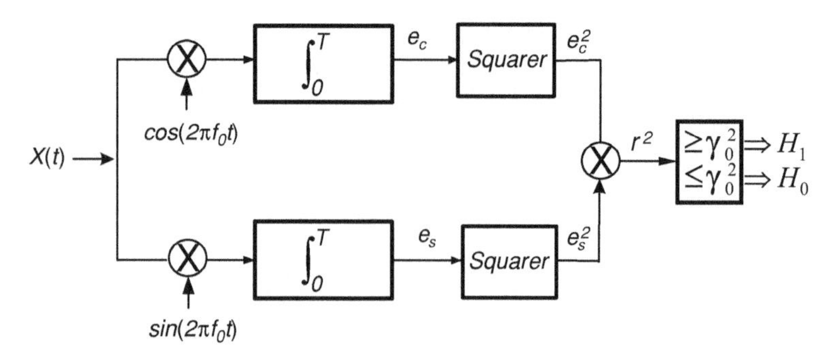

Figure 6.3 A quadrature (integration) form of a non-coherent binary receiver.

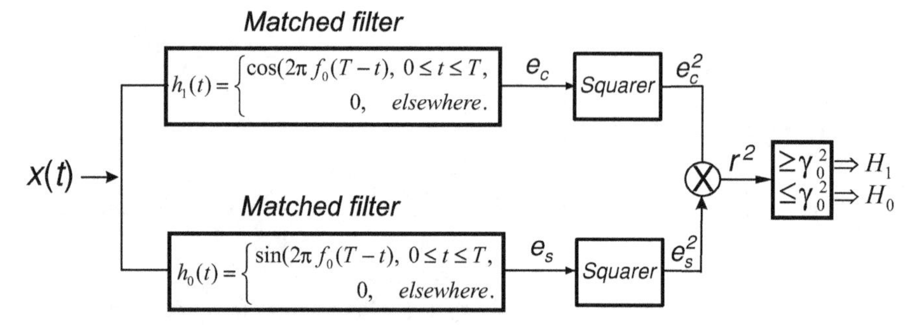

Figure 6.4 A quadrature (matched filter) form of a non-coherent binary receiver.

or equivalently the envelope r is given by

$$r \underset{\leq_{H_0}}{\overset{\geq^{H_1}}{}} \frac{N_0 I_0^{-1}(\gamma)}{2A} = \gamma_0, \tag{6.19}$$

or the square of the envelope is given by

$$r^2 \underset{\leq_{H_0}}{\overset{\geq^{H_1}}{}} \gamma_0^2. \tag{6.20}$$

Then the decision regions $\mathbb{R}_1$ and $\mathbb{R}_0$ of (6.10) can be expressed in terms of r^2 as

$$\mathbb{R}_1 = \{x(t),\ 0 \leq t \leq T : r^2 \geq \gamma_0^2\},\ \mathbb{R}_0 = \{x(t),\ 0 \leq t \leq T : r^2 \leq \gamma_0^2\}. \tag{6.21}$$

Figure 6.3 shows the quadrature receiver (in the integration form) of the decision rule given in (6.21). Figure 6.4 shows an alternative quadrature (matched filter) form of the decision rule of (6.20). Furthermore, the decision rule based on (6.19) can be approximated by a non-coherent bandpass-filter-envelope-detector receiver given by Fig. 6.5.

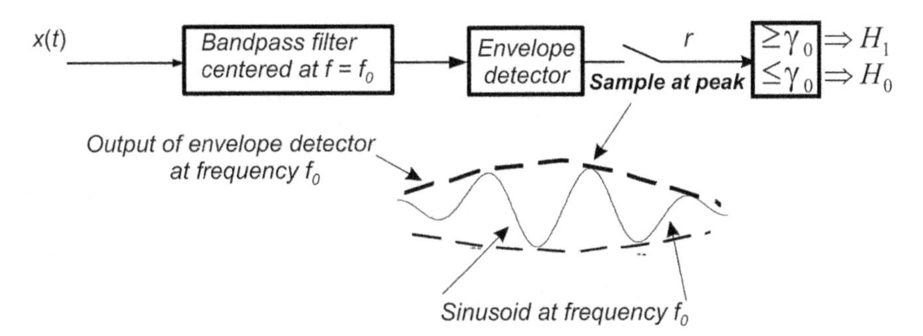

Figure 6.5 Binary non-coherent bandpass-filter-envelope-detector receiver.

6.2 Performance analysis of the binary non-coherent detection system

Now, we want to find the P_{FA} and P_{D} of the binary non-coherent detection system. Since $N(t)$ is a WGN process, then e_{c} and e_{s} are realizations of two Gaussian random variables under H_0. Furthermore, upon explicit evaluations

$$\text{E}\{e_{\text{c}}|H_0\} = \text{E}\{e_{\text{s}}|H_0\} = 0, \tag{6.22}$$

$$\text{E}\{e_{\text{c}}^2|H_0\} = \frac{N_0}{2}\int_0^T \cos^2(2\pi f_0 t)dt \simeq \frac{N_0 T}{4}, \tag{6.23}$$

$$\text{E}\{e_{\text{s}}^2|H_0\} = \frac{N_0}{2}\int_0^T \sin^2(2\pi f_0 t)dt \simeq \frac{N_0 T}{4}, \tag{6.24}$$

$$\text{E}\{e_{\text{c}}e_{\text{s}}|H_0\} = \frac{N_0}{2}\int_0^T \cos(2\pi f_0 t)\sin(2\pi f_0 t)dt \simeq 0. \tag{6.25}$$

Thus, under H_0, $\{e_{\text{c}},\ e_{\text{s}}\}$ are two independent zero-mean Gaussian random variables with the same variance of $N_0 T/4$. Then from (6.20) we have

$$P_{\text{FA}} = P\{R \geq \gamma_0\} = \iint_{R\geq\gamma_0} \left(\frac{1}{\sqrt{2\pi}\sqrt{N_0 T/4}}\right)^2 \exp\left(\frac{-4e_{\text{c}}^2 - 4e_{\text{s}}^2}{2N_0 T}\right) de_{\text{c}}de_{\text{s}}. \tag{6.26}$$

But from (6.12)–(6.14), we have

$$e_{\text{c}} = r\cos(\psi),\ e_{\text{s}} = r\sin(\psi),\ r^2 = e_{\text{c}}^2 + e_{\text{s}}^2. \tag{6.27}$$

Thus, replace the rectangular coordinate system using $\{e_{\text{c}},\ e_{\text{s}}\}$ by the polar coordinate system using $\{r,\ \psi\}$ in the double integral of (6.26), then the differentials of the two coordinate systems are related by

$$de_{\text{c}}de_{\text{s}} = r\,dr\,d\psi. \tag{6.28}$$

Then (6.26) is equivalent to

$$\begin{aligned} P_{\text{FA}} &= \frac{2}{N_0 \pi T} \int_{\psi=0}^{2\pi} \int_{r=\gamma_0}^{\infty} \exp\left(-2r^2/(N_0 T)\right) r dr d\psi \\ &= \frac{4}{N_0 T} \int_{R=\gamma_0}^{\infty} \exp\left(-2r^2/(N_0 T)\right) r dr = \frac{4}{N_0 T} \frac{\exp\left(-2r^2/(N_0 T)\right)}{-4/(N_0 T)} \bigg|_{\gamma_0}^{\infty} \\ &= \exp\left(-2\gamma_0^2/(N_0 T)\right). \end{aligned} \tag{6.29}$$

Recall the definition of the Rayleigh random variable R. Let X $\sim \mathcal{N}(0, \sigma^2)$ and Y $\sim \mathcal{N}(0, \sigma^2)$ be two independent Gaussian random variables. Then $R = \sqrt{X^2 + Y^2}$ is a Rayleigh random variable with a pdf given by

$$f_R(r) = \begin{cases} \frac{r}{\sigma^2} \exp\left(-\frac{r^2}{2\sigma^2}\right), & 0 \leq r < \infty, \\ 0, & r < 0. \end{cases} \tag{6.30}$$

Comparing the integrand in the last integral of (6.29) with (6.30), we see that

$$P_{\text{FA}} = \int_{r=\gamma_0}^{\infty} f_R(r) dr = \exp(-2\gamma_0^2/(N_0 T)), \tag{6.31}$$

where $\sigma^2 = (N_0 T)/4$. We also note the threshold constant γ_0 can be obtained from (6.31) in terms of P_{FA} as

$$\gamma_0 = \sqrt{(N_0 T/2) \ln(1/P_{\text{FA}})} = \sqrt{-2\sigma^2 \ln P_{\text{FA}}}. \tag{6.32}$$

Now, we want to evaluate P_{D}. Thus, under H_1, we have

$$e_{\text{c}} = \int_0^T x(t) \cos(2\pi f_0 t) dt = \int_0^T [A \cos(2\pi f_0 t + \theta) + n(t)] \cos(2\pi f_0 t) dt, \tag{6.33}$$

$$e_{\text{s}} = \int_0^T x(t) \sin(2\pi f_0 t) dt = \int_0^T [A \cos(2\pi f_0 t + \theta) + n(t)] \sin(2\pi f_0 t) dt. \tag{6.34}$$

From evaluation, we note $\cos(2\pi f_0 t)$ and $\sin(2\pi f_0 t)$ are essentially orthogonal

$$\int_0^T \cos(2\pi f_0 t) \sin(2\pi f_0 t) dt \simeq 0, \tag{6.35}$$

and

$$\int_0^T \cos^2(2\pi f_0 t) dt \simeq \int_0^T \sin^2(2\pi f_0 t) dt \simeq \frac{T}{2}. \tag{6.36}$$

Then the means of e_c and e_s conditioned on θ and under H_1 are given by

$$\begin{aligned} \mathrm{E}\{e_c|\theta,\ H_1\} &= A\int_0^T \cos(2\pi f_0 t)[\cos(2\pi f_0 t)\cos(\theta) - \sin(2\pi f_0 t)\sin(\theta)]\,dt \\ &\simeq \frac{AT\cos(\theta)}{2}, \end{aligned} \tag{6.37}$$

$$\begin{aligned} \mathrm{E}\{e_s|\theta,\ H_1\} &= A\int_0^T \sin(2\pi f_0 t)[\cos(2\pi f_0 t)\cos(\theta) - \sin(2\pi f_0 t)\sin(\theta)]\,dt \\ &\simeq \frac{-AT\sin(\theta)}{2}. \end{aligned} \tag{6.38}$$

Next, the variance of e_c conditioned on θ and under H_1 is given by

$$\sigma^2_{e_c|\theta,\,H_1} = \mathrm{E}\left\{e_c^2|\theta,\ H_1\right\} - [\mathrm{E}\{e_c|\theta,\ H_1\}]^2. \tag{6.39}$$

The second moment of e_c conditioned on θ and under H_1 is given by

$$\begin{aligned} \mathrm{E}\left\{e_c^2|\theta,\ H_1\right\} &= \mathrm{E}\left\{\left(\int_0^T [A\cos(2\pi f_0 t + \theta) + n(t)]\cos(2\pi f_0 t)dt\right)\cdot\right. \\ &\quad \left.\left(\int_0^T [A\cos(2\pi f_0 \tau + \theta) + n(\tau)]\cos(2\pi f_0 \tau)d\tau\right)\right\} \\ &= \int_0^T\int_0^T \cos(2\pi f_0 t)\cos(2\pi f_0 \tau)\cdot \\ &\quad \left[A^2\cos(2\pi f_0 t + \theta)\cos(2\pi f_0 \tau + \theta) + \frac{N_0}{2}\delta(t-\tau)\right] dt d\tau \\ &\simeq \left(\frac{AT\cos(\theta)}{2}\right)^2 + \frac{N_0}{2}\int_0^T \cos^2(2\pi f_0 t)dt \\ &\simeq \left(\frac{AT\cos(\theta)}{2}\right)^2 + \frac{N_0 T}{4}. \end{aligned} \tag{6.40}$$

Then using (6.40) and (6.37) in (6.39), we have

$$\sigma^2_{e_c|\theta,\,H_1} = \left(\frac{AT\cos(\theta)}{2}\right)^2 + \frac{N_0 T}{4} - \left(\frac{AT\cos(\theta)}{2}\right)^2 = \frac{N_0 T}{4}. \tag{6.41}$$

Similarly, the variance of e_s conditioned on θ and under H_1 is given by

$$\sigma^2_{e_s|\theta,H_1} = \mathrm{E}\{e_s^2|\theta,\ H_1\} - [\mathrm{E}\{e_s|\theta,\ H_1\}]^2, \tag{6.42}$$

and the second moment of e_s conditioned on θ and under H_1 is given by

$$\mathrm{E}\{e_s^2|\theta,\ H_1\} \simeq \left(\frac{AT\sin(\theta)}{2}\right)^2 + \frac{N_0 T}{4}, \tag{6.43}$$

and the variance of e_s conditioned on θ and under H_1 is given by

$$\sigma^2_{e_s|\theta,H_1} = \frac{N_0 T}{4}. \tag{6.44}$$

Furthermore, the cross covariance of

$$\mathrm{E}\{(e_{\mathrm{c}} - \mathrm{E}\{e_{\mathrm{c}}|\theta, H_1\})(e_{\mathrm{s}} - \mathrm{E}\{e_{\mathrm{s}}|\theta, H_1\})|\,\theta, H_1\} \simeq 0. \tag{6.45}$$

Thus, e_{c} and e_{s} conditioned on θ and under H_1 are two independent Gaussian r.v.'s with joint pdf given by

$$f_{e_{\mathrm{c}},e_{\mathrm{s}}|\theta,H_1}(e_{\mathrm{c}}, e_{\mathrm{s}}|\theta, H_1)$$
$$= \frac{1}{2\pi(N_0T/4)} \exp\left(-\frac{1}{2(N_0T/4)}\left[\left(e_{\mathrm{c}} - \frac{AT\cos(\theta)}{2}\right)^2 + \left(e_{\mathrm{s}} + \frac{AT\sin(\theta)}{2}\right)^2\right]\right). \tag{6.46}$$

Then the pdf of e_{c} and e_{s} under H_1 is given by

$$\begin{aligned} &f_{e_{\mathrm{c}},e_{\mathrm{s}}|H_1}(e_{\mathrm{c}}, e_{\mathrm{s}}|H_1) \\ &= \frac{1}{2\pi}\int_0^{2\pi} f_{e_{\mathrm{c}},e_{\mathrm{s}}|\theta,H_1}(e_{\mathrm{c}}, e_{\mathrm{s}}|\theta, H_1)d\theta \\ &= \frac{2}{\pi N_0T} \exp\left(\frac{-2}{N_0T}\left(e_{\mathrm{c}}^2 + e_{\mathrm{s}}^2 + \frac{A^2T^2}{4}\left(\cos^2(\theta) + \sin^2(\theta)\right)\right)\right) \cdot \\ &\quad \frac{1}{2\pi}\int_0^{2\pi} \exp\left(\frac{-2AT}{N_0T}(-e_{\mathrm{c}}\cos(\theta) + e_{\mathrm{s}}\sin(\theta))\right)d\theta. \end{aligned} \tag{6.47}$$

But we note

$$f_{e_{\mathrm{c}},e_{\mathrm{s}}|H_1}(e_{\mathrm{c}}, e_{\mathrm{s}}|H_1) = \frac{2}{\pi N_0T}\exp\left(\frac{-2r^2}{N_0T}\right)\exp\left(\frac{-E}{N_0}\right)\mathrm{I}_0\left(\frac{2Ar}{N_0}\right). \tag{6.48}$$

Using (6.48) in (6.47), we have

$$f_{e_{\mathrm{c}},e_{\mathrm{s}}|H_1}(e_{\mathrm{c}}, e_{\mathrm{s}}|H_1) = \frac{2}{\pi N_0T}\exp\left(\frac{-2r^2}{N_0T}\right)\exp\left(\frac{-E}{N_0}\right)\mathrm{I}_0\left(\frac{2Ar}{N_0}\right). \tag{6.49}$$

Then

$$\begin{aligned} \mathrm{P_D} = \mathrm{P}\{R \geq \gamma_0\} &= \iint_{r=\gamma_0} f_{e_{\mathrm{c}},e_{\mathrm{s}}|H_1}(e_{\mathrm{c}}, e_{\mathrm{s}}|H_1)de_{\mathrm{c}}de_{\mathrm{s}} \\ &= \int_{\psi=0}^{2\pi}\int_{r=\gamma_0}^{\infty} \frac{2}{\pi N_0T}\exp\left(\frac{-2r^2}{N_0T}\right)\exp\left(\frac{-E}{N_0}\right)\mathrm{I}_0\left(\frac{2Ar}{N_0}\right)r\,dr\,d\psi \\ &= \frac{4}{N_0T}\exp\left(\frac{-E}{N_0}\right)\int_{r=\gamma_0}^{\infty} r\exp\left(\frac{-2r^2}{N_0T}\right)\mathrm{I}_0\left(\frac{2Ar}{N_0}\right)dr \\ &= \int_{x=\gamma_0}^{\infty} p_{R_i}(x)dx, \end{aligned} \tag{6.50}$$

where the pdf $p_{R_i}(x)$ is the Rician pdf [2] (named after S. O. Rice) defined by

$$p_{R_i}(x) = \begin{cases} \frac{x}{\sigma^2}\exp\left(-\frac{(x^2+\mu^2)}{2\sigma^2}\right)\mathrm{I}_0\left(\frac{\mu x}{\sigma^2}\right), & 0 \leq x < \infty,\ 0 < \sigma,\ 0 < \mu, \\ 0 & , x < 0, \end{cases} \tag{6.51}$$

with

$$\sigma^2 = N_0 T/4, \ \mu = \sqrt{ET/2} = AT/2, \text{ and } E = A^2 T/2. \tag{6.52}$$

Up the normalization of $\mu/\sigma = a$ and $x = \sigma y$, then the normalized Rician pdf of (6.51) can be expressed as

$$p_{R_i}(y) = \begin{cases} y \exp\left(-\frac{(y^2+a^2)}{2}\right) \mathrm{I}_0(ay) \ , \ 0 \le y < \infty, \ 0 < a, \\ 0 \ , \ y < 0. \end{cases} \tag{6.53}$$

Equivalently, the probability of detection versus the probability of false alarm from (6.50) can be expressed as

$$\mathrm{P_D} = \mathrm{Q}\left(\sqrt{2E/N_0}, \ \sqrt{-2\ln(\mathrm{P_{FA}})}\right), \tag{6.54}$$

where the Marcum-Q function [10], is the complementary cdf of the normalized Rician pdf of (6.53), given by

$$\mathrm{Q}(a, b) = \int_{y=b}^{\infty} y \exp\left[-\frac{(y^2 + a^2)}{2}\right] \mathrm{I}_0(ay)dy. \tag{6.55}$$

We should not confuse the Marcum-Q function (which is a function of two arguments) with the complementary Gaussian cdf Q function (which is a function of a single argument). Both the Marcum-Q function and the conventional Q function do not have closed form expressions. Thus, tables of Marcum-Q function and conventional Q function have been tabulated. The marcumq.m function in Matlab can be used to evaluate numerically the Marcum-Q function.

Example 6.1 Suppose we want to constrain $P_{\mathrm{FA}} = 10^{-4}$. Then from (6.32), we obtain $\gamma_0/\sigma = 4.29193$. Figure 6.6 plots the Rayleigh pdf $p_R(x)$ of (6.30) with $\sigma = 1$ and the Rician pdf $p_{R_i}(x)$ of (6.51) with $\sigma = 1$ and $\mu = 6$. Thus, we note the P_{D} is the area under the Rician pdf $p_{R_i}(x)$ and the P_{FA} is the area under the Rayleigh pdf $p_R(x)$, both to the right of the threshold $\gamma_0 = 4.29193$ for this example. From this figure, it is clear that if the parameters of the two pdf's are fixed, then by increasing the threshold value of γ_0, both the values of P_{D} and P_{FA} will decrease, while by decreasing γ_0, both of their values will increase. □

Example 6.2 In Fig. 6.7, we plot the non-coherent detection P_{D} versus SNR(dB) from 5 dB to 17 dB as four dashed curves for values of $P_{\mathrm{FA}} = 10^{-2}$, 10^{-4}, 10^{-6}, and 10^{-8} using the Marcum-Q function expression of (6.54). Here we define $\mathrm{SNR} = 2E/N_0$ or SNR in dB given by

$$\mathrm{SNR(dB)} = 10\log_{10}(\mathrm{SNR}) = 10\log_{10}(2E/N_0) = 20\log_{10}(AT/(2\sigma)). \tag{6.56}$$

Others have defined $\mathrm{SNR}_0 = E/N_0$. Thus, our $\mathrm{SNR} = 2\mathrm{SNR}_0$ or our

$$\mathrm{SNR(dB)} = 10\log_{10}(2\mathrm{SNR}_1) = 3.01\mathrm{dB} + \mathrm{SNR}_0(\mathrm{dB}). \tag{6.57}$$

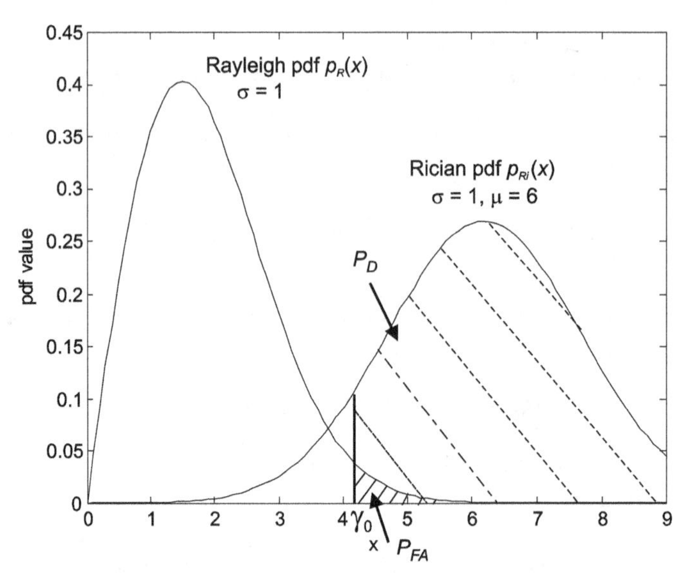

Figure 6.6 P_{D} is the area under the Rician pdf $p_{\mathrm{R_i}}(x)$ and P_{FA} is the area under the Rayleigh pdf $p_R(x)$, both to the right of the threshold γ_0.

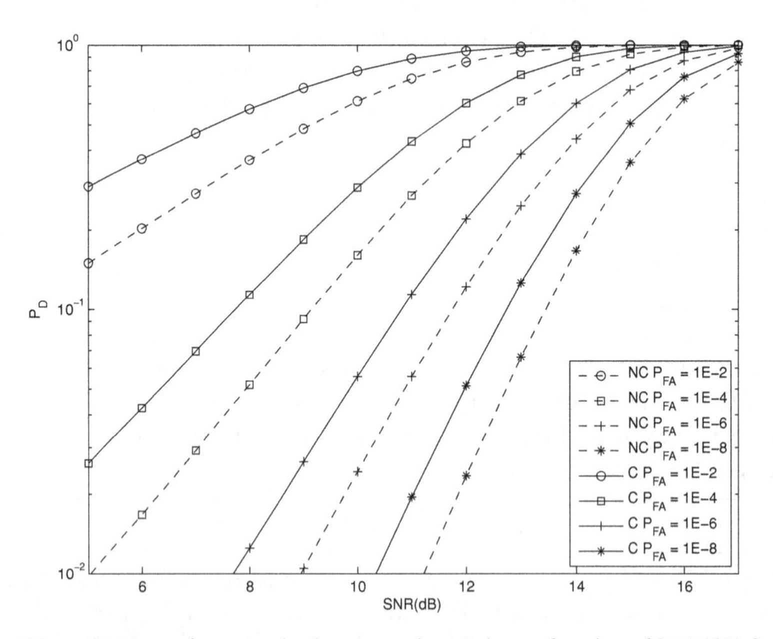

Figure 6.7 Non-coherent and coherent receiver P_{D}'s as a function of SNR(dB) for four different values of P_{FA} for the binary communication system model of (6.2).

In other words, every value of SNR(dB) in Fig. 6.7 is 3.01 dB higher than $\mathrm{SNR_0}$(dB). For comparison, we also plot the corresponding coherent detection P_{D}'s versus the same SNR(dB) as four solid curves for the four P_{FA}'s in Fig. 6.7. We note, for the non-coherent binary communication system modeled by (6.2), the phase θ is uniformly distributed and thus not known to the receiver. For the coherent receiver, the phase θ is assumed to be fully known to the receiver. Equivalently, for the coherent receiver, θ is absent in the model of (6.2). In all cases, for a given SNR(dB) and a given P_{FA}, the solid P_{D} curve of the coherent receiver is always higher (i.e., having a larger value) compared to that of the dashed P_{D} non-coherent receiver. For high values of P_{D} (e.g., near 0.9) and high values of P_{FA} (e.g., near 10^{-2}), the non-coherent detection system suffers a degradation of about 1 dB in SNR relative to the coherent detection system (e.g., in the upper middle part of Fig. 6.7). However, for high values of P_{D} (e.g., near 0.9) and low values of P_{FA} (e.g., 10^{-8}), the non-coherent detection system suffers a degradation of only about 0.2 dB in the SNR relative to the coherent detection system (e.g., in the upper right side of Fig. 6.7). Thus, for a relatively high-quality system, the loss of performance using a non-coherent detection system (which is simpler) is quite modest relative to the coherent detection system (which is more complex since it needs phase synchronization at the receiver) for the binary communication system model of (6.2). □

6.3 Non-coherent detection in radar receivers

A radar is a system that emits electromagnetic pulses and upon examining the reflected returned waveform may detect the presence of a target and possibly estimate and/or track its range, its relative velocity, and its radar cross-section, etc. There are many different kinds of radar, such as land-based, ship-based, airborne, and satellite-based, each with its own defining purpose and characteristic. It is our intention in this section only to study some basic aspects of the non-coherent detection methods used in some generic radar receivers and certainly not deal with a myriad of advanced radar concepts. A classical radar system has a rotating antenna that typically scans the azimuth angle in a periodic manner. Each rotating antenna is called a scan and one or more pulses may hit a target of interest. A more modern radar may not use a mechanically rotating antenna, but may use an electronically scanning beam to search over the angles of interest. The scanning rate and the number of pulses per scan can vary greatly depending on the radar hardware and its applications. In Section 6.3.1, we first consider the issue of coherent integration in a radar receiver.

6.3.1 Coherent integration in radar

Suppose in a given scan, the target return is given by n identical pulses each of duration T. Ideally, we assume a train of n pulses of total duration nT, with a known initial phase, is received in the presence of a WGN as modeled by (6.58) and sketched in Fig. 6.8. In (6.58)

$$x(t) = \begin{cases} n(t), & 0 \leq t \leq nT, \ H_0, \\ \sum_{k=1}^{n} x_k(t), & 0 \leq t \leq nT, \ H_1, \end{cases} \tag{6.58}$$

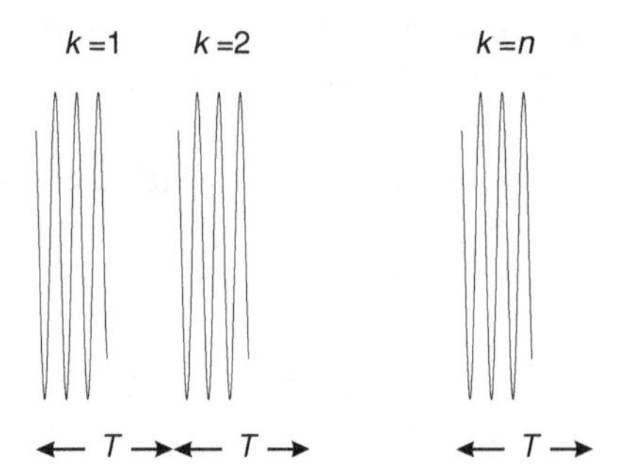

Figure 6.8 Train of a deterministic waveform of k-pulses.

$n(t)$ is the realization of a zero-mean WGN of power spectral density of $N_0/2$, and in (6.59)

$$\begin{aligned} x_k(t) &= s_k(t) + n(t) \\ &= A\cos(2\pi f_0 t + \theta) + n(t),\ (k-1)T \le t \le kT,\ k = 1, \ldots, n, \end{aligned} \tag{6.59}$$

where the phase θ is known and is the same for all the pulses. In this model, all the pulses in the received train experience the same medium propagation distortion. Then the optimum LR radar receiver can treat this problem identically to that of the binary detection of an arbitrary completely known waveform $s_1(t) = \sum_{k=1}^{n} s_k(t)$ over a finite interval $[0,\ nT]$ in WGN as in (4.38), except now we assume the waveform is of duration nT. As shown in Chapter 4.9, the optimum receiver then uses a matched filter, matched to the full train of pulses in $[0,\ nT]$. Since this operation is performing *coherent integration* of the received voltage from n pulses in the waveform, then the total energy of the entire train of pulses is proportional to n^2 of the energy of a single pulse, while the noise variance of the train is proportional to n. Thus, the SNR at the output of the matched filter increases linearly with n. Ideally, this coherent integration radar problem is in principle identical to that of a binary communication problem treated in Chapter 4. In practice, achieving a coherent integration in a radar receiver over this $[0,\ nT]$ period may be difficult, since during this entire period of the train of pulses, the target may not stay fully stationary and the propagation channel may be changing, and thus the amplitudes and phases of the return train of pulses may not be fully known.

6.3.2 Post detection integration in a radar system

Next, in the received train of n pulses, we assume each $x_k(t)$ pulse waveform is now modeled by (6.60)

$$\begin{aligned} x_k(t) &= s_k(t) + n(t) \\ &= A\cos(2\pi f_0 t + \Theta_k) + n(t),\ (k-1)T \le t \le kT,\ k = 1, \ldots, n. \end{aligned} \tag{6.60}$$

That is, the k-th received pulse over $[0,\ T]$ is a sinusoid with known amplitude A and frequency f_0, but with a uniformly distributed phase random variable Θ_k on $[0,\ 2\pi)$, $k =$

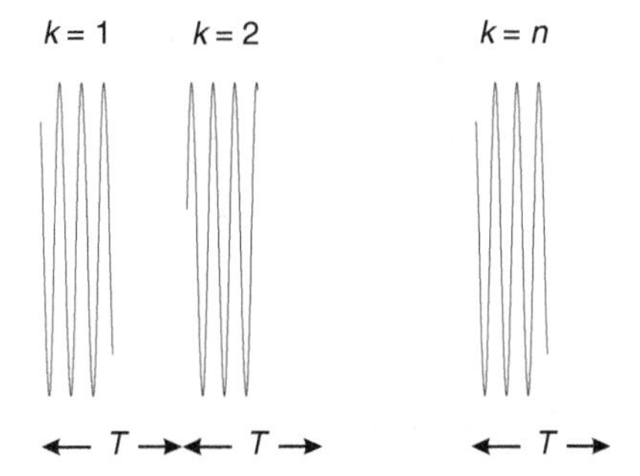

Figure 6.9 Train of a waveform of k-pulses with a uniformly distributed initial phase for each pulse.

$1, \ldots, n$, plus $n(t)$. Here we assume all the received pulses have the same amplitude and frequency, but each phase random variable Θ_k is assumed to be independent, as shown in Fig. 6.9. Then denote the sampled version of $\{x(t),\ 0 \le t \le nT\}$ by $\mathbf{x} = [\mathbf{x}_1, \ldots, \mathbf{x}_n]$, then corresponding to (6.4), the LR for the sampled waveform is given by

$$\Lambda(\mathbf{x}) = \frac{p_1(\mathbf{x})}{p_0(\mathbf{x})} = \frac{\int_0^{2\pi}\int_0^{2\pi}\cdots\int_0^{2\pi} p_1(\mathbf{x}_1|\theta_1)\ldots p_1(\mathbf{x}_n|\theta_n)p_{\Theta_1}(\theta_1)\ldots p_{\Theta_n}(\theta_n)d\theta_1\ldots d\theta_n}{p_0(\mathbf{x})}. \tag{6.61}$$

Taking the limit as the number of samples goes to infinity (similar to (6.4)–(6.5)), (6.61) becomes

$$\begin{aligned}
&\Lambda(x(t),\ 0 \le t \le nT) \\
&= \frac{\prod_{k=1}^{n}\left\{\frac{1}{2\pi}\int_0^{2\pi}\exp\left(-\frac{1}{N_0}\int_{(k-1)T}^{kT}(x_k(t) - A\cos(2\pi f_0 t + \theta_k))^2\,dt\right)d\theta_k\right\}}{\prod_{k=1}^{n}\exp\left(-\frac{1}{N_0}\int_0^T x_k^2(t)dt\right)} \\
&= \exp(-nE/N_0)\prod_{k=1}^{n}\frac{1}{2\pi}\int_0^{2\pi}\exp\left(\frac{2A}{N_0}\int_{(k-1)T}^{kT}x_k(t)\cos(2\pi f_0 t + \theta_k)dt\right)d\theta_k, \\
&= \exp(-nE/N_0)\prod_{k=1}^{n}\mathrm{I}_0(2Ar_k/N_0),
\end{aligned} \tag{6.62}$$

where $E = A^2T/2$, $\mathrm{I}_0(\cdot)$ is the modified Bessel function of the first kind of order zero, and

$$r_k = \sqrt{e_{c_k}^2 + e_{s_k}^2},\ e_{c_k} = \int_{(k-1)T}^{kT} x_k(t)\cos(2\pi f_0 t)dt,$$
$$e_{s_k} = \int_{(k-1)T}^{kT} x_k(t)\sin(2\pi f_0 t)dt,\ k = 1, \ldots, n. \tag{6.63}$$

The arguments going from the first line to the third line of the r.h.s. in (6.62) are identical to the arguments going from (6.5) to (6.16). The LR test, using (6.62), can be modified

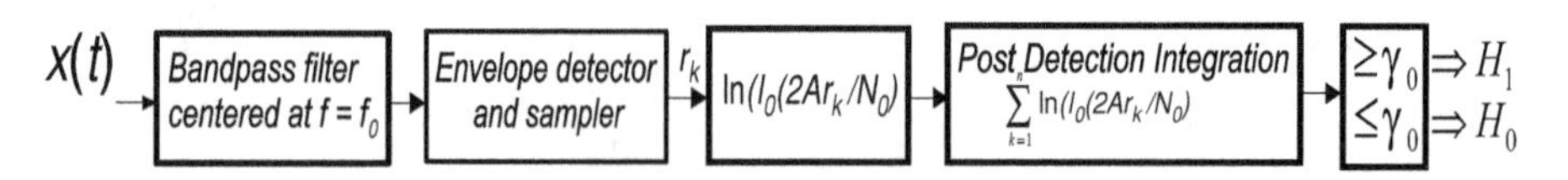

Figure 6.10 Post detection integration of n pulses.

after taking the $\ln(\cdot)$ operation to yield

$$\gamma = \sum_{k=1}^{n} \ln\left(\mathrm{I}_0(2Ar_k/N_0)\right) \underset{\leq H_0}{\overset{\geq H_1}{}} \exp(nE/N_0)\Lambda_0 = \gamma_0. \tag{6.64}$$

By comparing (6.64) to (6.16), the sufficient statistic of (6.64) can be implemented using the LR receiver structure given by the non-coherent bandpass-filter-envelope-detector receiver of Fig. 6.5, followed by a post detection integration block diagram as shown in Fig. 6.10.

Now, we want to evaluate the P_{FA} and P_{D} based on the LR receiver of Fig. 6.10 with the sufficient statistic of (6.64). Unfortunately, the performance analysis based on (6.64) is considerably more difficult than that of the non-coherent receiver of Fig. 6.5 in Section 6.3.1. We can consider the two special cases of the argument of $\ln(\mathrm{I}_0(x))$, when x is small and when x is large. For small values of x, we know from [5, p. 213]

$$\ln(\mathrm{I}_0(x)) = x^2/4 - x^4/64 + O(x^6). \tag{6.65}$$

Then, using just the first term in the approximation of (6.65), (6.64) yields

$$\gamma_{\mathrm{S}} = \sum_{k=1}^{n} r_k^2 \underset{\leq H_0}{\overset{\geq H_1}{}} (N_0/A)^2 \exp(nE/N_0)\Lambda_0 = \gamma_{S0}. \tag{6.66}$$

Equation (6.66) shows for small SNR conditions, the LR receiver reduces to a quadratic receiver consisting of the sum of the squares of the envelope r_k, given by (6.63), of the n pulses, compared to the threshold γ_{S0}.

For large values of x, also from [5, p. 213], we have

$$\ln(\mathrm{I}_0(x)) = x - \ln(2\pi x)/2 + 1/(8x) + O(x^{-2}). \tag{6.67}$$

Then using just the first term in the approximation of (6.67), (6.64) yields

$$\gamma_{\mathrm{L}} = \sum_{k=1}^{n} r_k \underset{\leq H_0}{\overset{\geq H_1}{}} (N_0/2A) \exp(nE/N_0)\Lambda_0 = \gamma_{L0}. \tag{6.68}$$

Equation (6.68) shows for large SNR conditions the LR receiver reduces to a linear receiver consisting of the sum of the envelope r_k, given by (6.63), of the n pulses, compared to the threshold γ_{L0}.

Unfortunately, the exact analysis of the P_{FA} and P_{D} of the original LR receiver with the sufficient statistic of (6.64) appears to be intractable. Performance analysis for small SNR approximation of (6.66) and for large SNR approximation of (6.68) is complicated, but various results are known [4–7]. For the small SNR approximation of (6.66) under the additional constraint of $n \gg 1$, an approximate relationship [5, p. 224], [6, p. 296]

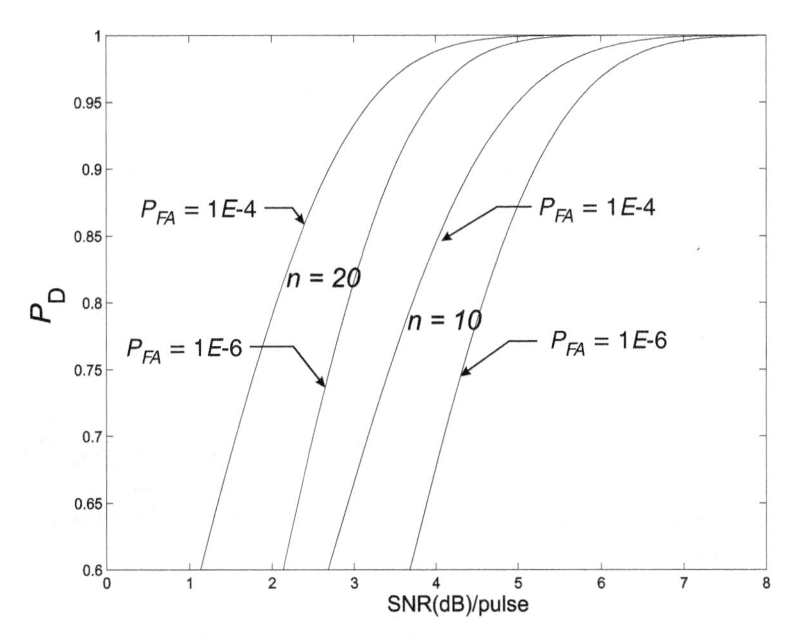

Figure 6.11 P_D vs. SNR(dB)/pulse for different parameters of P_{FA} and n for a quadratic receiver.

between P_D and P_{FA} is given by

$$P_D = Q\left(\sqrt{\frac{1}{\mathrm{SNR}+1}}\left(Q^{-1}(P_{FA}) - \left(\frac{\sqrt{n}\mathrm{SNR}}{\sqrt{2}}\right)\right)\right), \tag{6.69}$$

where $\mathrm{SNR} = A^2T/N_0$ for a single pulse, $Q(\cdot)$ is the complementary Gaussian distribution of a zero-mean and unit variance random variable, and $Q^{-1}(\cdot)$ is the inverse of the $Q(\cdot)$ function. Several plots of P_D versus SNR(dB)/pulse for different values of P_{FA} and n are given in Fig. 6.11. Detailed system performance analysis [4], [6, p. 301], [8, p. 370] comparing the quadratic receiver of (6.66) obtained under the small SNR assumption and the linear receiver of (6.68) obtained under the large SNR assumption showed under most practical applications their performances (i.e., P_D vs. P_{FA} as a function of SNR for a fixed n) are quite similar (e.g., within 0.2 dB of each other).

6.3.3 Double-threshold detection in a radar system

In Section 6.3.1, given a pulse train with a fixed amplitude and fully known relative phases between pulses, the LR receiver is given by coherent integration using a matched filtering. In Section 6.3.2, for a pulse train of n pulses with a fixed amplitude but independent uniformly distributed initial phases, the LR receiver yields a post detection integrator. Depending on the SNR condition, the post detection integration results in a quadratic receiver (for small SNR) or a linear receiver (for large SNR).

In the study of double-threshold detection, consider the received waveform to be modeled by either the returns of n independent scans or a train of n pulses as modeled in Section 6.3.2. In either case, the waveform in each scan or in each pulse is detected by the non-coherent method of Sections 6.1 and 6.2. The threshold γ_0 in (6.32) and in Fig. 6.7 is chosen to meet the desired P_{D} and P_{FA} requirement of (6.54). Since the above γ_0 threshold is the first threshold in the double-threshold detector, for this section, we shall denote this threshold as γ_1. After this first threshold stage, the detector output takes a binary value of 1 if there is a detection, and a value of 0 if there is no detection. This detector output is placed in a binary counter. After processing the n scans or a train of n pulses, if this counter value k is greater than or equal to the second integral threshold of γ_2, then a target in the n scans or the train of n pulses is declared to be detected. Thus, this double-threshold detector has a simpler implementation than the coherent integration detector of Section 6.3.1 and the post detection integration detector of Section 6.3.2. The analysis of a double-threshold detector [8, pp. 498–503] is also relatively simple. Let p be the probability that a single sample exceeds the first threshold of γ_1 and $(1 - p)$ be the probability that it does not exceed this threshold. Then exactly k out of the n decisions exceeding this threshold is given by the binomial probability distribution of

$$p(k) = \mathrm{C}_k^n p^k (1 - p)^{n-k}, \tag{6.70}$$

where $\mathrm{C}_k^n = n!/(k!(n-k)!)$ is the binomial coefficient in which one can have k successes in n independent trials, and its mean is given by $\mu_k = np$ and its variance is given by $\sigma_k^2 = np(1-p)$. Then, the probability that k, the number of times the first threshold γ_1 is exceeded, is greater than or equal to the second integral threshold γ_2 in the binary counter is given by

$$\mathrm{P}(k \geq \gamma_2) = \sum_{k=\gamma_2}^{n} p(k) = \sum_{k=\gamma_2}^{n} \mathrm{C}_k^n p^k (1-p)^{n-k}. \tag{6.71}$$

The probability $\mathrm{P}(k \geq \gamma_2)$ in (6.71) depends on n, γ_2, and p. Since each sampled value r is given by the envelope of the non-coherent detector of Section 6.1, then in the absence of a target (denote it as hypothesis H_0), it has a Rayleigh pdf given by (6.30) and in the presence of a target (denote it as hypothesis H_1), it has a Rician pdf given by (6.51). Then we can denote the P_{FA} of the double-threshold detector by

$$P_{\text{FA}} = P_{\text{FA}}(p_0, \gamma_2, n) = \sum_{k=\gamma_2}^{n} \mathrm{C}_k^n p_0^k (1-p_0)^{n-k}, \tag{6.72}$$

where from (6.31)

$$p_0 = \int_{\gamma_1}^{\infty} (r/\sigma^2) \exp(-r^2/2\sigma^2) dr = \exp\left(-2\gamma_1^2/(N_0 T)\right), \tag{6.73}$$

with $\sigma^2 = N_0 T/4$. Similarly, we can denote the P_{D} of the double-threshold detector by

$$P_{\text{D}} = P_{\text{D}}(p_1, \gamma_2, n) = \sum_{k=\gamma_2}^{n} \mathrm{C}_k^n p_1^k (1-p_1)^{n-k}, \tag{6.74}$$

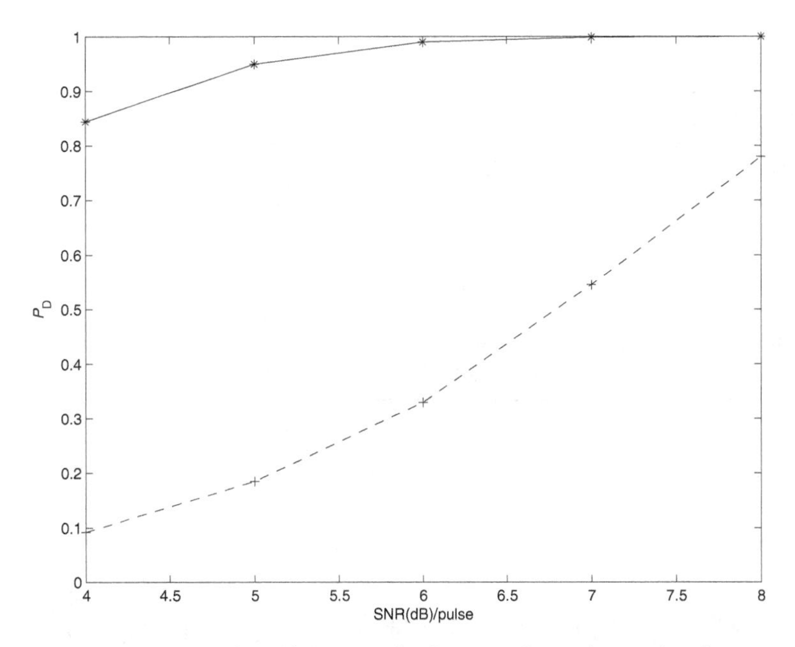

Figure 6.12 P_{D} versus SNR(dB) curves for the non-coherent integration detector and the double threshold detector for $P_{\mathrm{FA}} = 10^{-4}$ and $n = 10$.

where from (6.50)

$$p_1 = (4/N_0T)\exp(-E/N_0)\int_{r=\gamma_1}^{\infty} r\exp\left(-2r^2/(N_0T)\right)\mathrm{I}_0(2Ar/N_0)dr, \tag{6.75}$$

with $E = A^2T/2$. Furthermore, denote the single sample SNR $= 2E/N_0$, then p_1 and p_0 from (6.53) are related by

$$p_1 = \mathrm{Q}\left(\sqrt{\mathrm{SNR}}, \sqrt{-2\ln(p_0)}\right), \tag{6.76}$$

where $\mathrm{Q}(\cdot,\cdot)$ is the Marcum Q-function.

In Fig. 6.12, we compare the solid P_{D} versus SNR(dB) curve for fixed $P_{\mathrm{FA}} = 10^{-4}$ with $n = 10$ for the non-coherent integration detector with the corresponding dashed P_{D} curve of the double threshold detector. We note this solid non-coherent P_{D} versus SNR(dB) curve is the second curve from the right in Fig. 6.11. As expected, the value of the P_{D} of the double threshold detector is significantly lower than that under the non-coherent integration detector.

6.3.4 Constant False Alarm Rate (CFAR)

In Section 6.2, for a non-coherent binary detection receiver, if P_{FA} is specified, then from (6.31), the normalized threshold is given explicitly by

$$\frac{\gamma_0}{\sigma} = \sqrt{-2\ln(P_{\mathrm{FA}})}. \tag{6.77}$$

Table 6.1 Original P_D versus SNR(dB) and new noisier P_{D1} versus SNR1 (dB)

SNR(dB)	P_D	SNR1(dB)	P_{D1}
10dB	0.1604	7dB	0.2734
11dB	0.2695	8dB	0.3666
12dB	0.4252	9dB	0.4819
13dB	0.6141	10dB	0.6134
14dB	0.7963	11dB	0.7466
15dB	0.9242	12dB	0.8615
16dB	0.9827	13dB	0.9410
17dB	0.9980	14dB	0.9821

Thus, the P_D can be evaluated from the Marcum Q-function expression of (6.54) given by

$$P_D = Q\left(\sqrt{\text{SNR}}, \sqrt{-2\ln(P_{FA})}\right)$$
$$= Q\left(AT/(2\sigma), \sqrt{-2\ln(P_{FA})}\right) = Q(AT/(2\sigma), \gamma_0/\sigma). \qquad (6.78)$$

In practice, in many communication and radar scenarios, even if the noise variance σ^2 can be estimated, due to ever-changing conditions the prior estimated variance may not be accurate.

Example 6.3 Consider again Example 6.1 in Section 6.2, where we constrained the desired $P_{FA} = 10^{-4}$. From (6.77), the normalized threshold is given by $\gamma_0/\sigma = 4.29193$. Then the associated P_D as a function of the SNR(dB) $= 10\log_{10}(2E/N_0) = 10\log_{10}(A^2T^2/(4\sigma^2))$ are listed on the left-hand side of Table 6.1. Suppose the noise variance is doubled yielding $\sigma_1^2 = 2\sigma^2$ or the noise standard deviation becomes $\sigma_1 = \sqrt{2}\sigma$. If we assume the threshold γ_0 as well as the signal parameters A and T are unchanged, then the new normalized threshold yields $\gamma_0/\sigma_1 = 3.0349$ resulting in a new $P_{FA1} = 10^{-2}$. The resulting associated P_{D1} as a function of the new SNR1(dB) $= 10\log_{10}(A^2T^2/(4\sigma_1^2))$ are given on the right-hand side of Table 6.1.

Due to the lowering of the normalized threshold value of $\gamma_0/\sigma_1 = 3.0349$, the resulting $P_{FA1} = 10^{-2}$ was increased by two orders of magnitude from the original $P_{FA} = 10^{-4}$, which may not be acceptable in some situations. Even though the new SNR1 (dB) is decreased by 3 dB (due to the doubling of the new noise with the same signal energy), P_{D1} is greater than the corresponding P_D (at least for low-medium values of SNR1 (dB)). This example shows that in many practical communication and radar problems (operating under the NP criterion), in order to maintain the desired low values of P_{FA}, the background noise variance has to be estimated and updated frequently. □

Now, consider the simple cell-averaging constant false alarm rate (CA-CFAR) method for the adaptive updating of the background noise variance to maintain the desired PFA

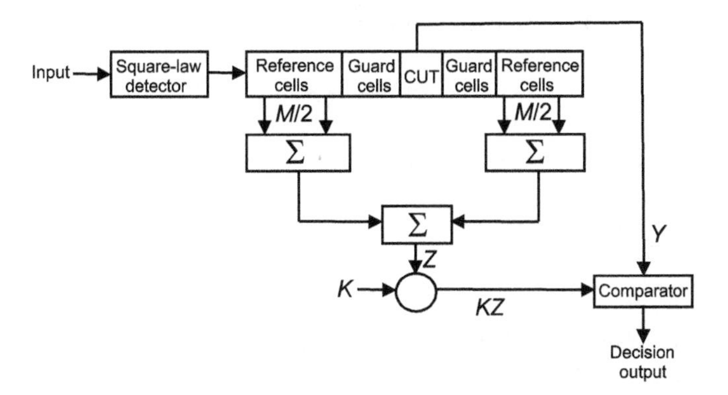

Figure 6.13 Cell-averaging constant false alarm rate (CA-CFAR) processor.

in a radar receiver. Figure 6.13 shows a generic CA-CFAR processor, where the CA is performed over a series of either range and/or Doppler cells (bins). The radar return from a given pulse is detected by a square-law detector. In a generic fashion, the samples are fed to a tapped delayed line. The cell under test (CUT) is shown in the center, and the immediate neighboring guard cells of the CUT are excluded from averaging due to possible leakages from the CUT cell. Then M reference cells (i.e., $M/2$ to the left and $M/2$ to the right of the guard cells) is used for averaging. The threshold is obtained by using the sum of envelope estimates Z of all M reference cells times a scaling constant K. In this model, the target signal is present in CUT and absent in all the reference cells. A target detection is then declared if the CUT value Y satisfies

$$Y \geq KZ, \tag{6.79}$$

where the constant K is selected to control the value of the P_{FA}. Since the target signal is absent in all the reference cells, then all such cells contain zero-mean independent Gaussian noise of variance σ^2. Then the sum of the envelopes Z has a gamma distribution of M degrees of freedom given by

$$P_Z(z) = \frac{z^{M-1}\exp\{-\frac{z}{2\sigma^2}\}}{(2\sigma^2)^M(M-1)!}, \quad 0 \leq z. \tag{6.80}$$

The conditional P_{FA} for $\gamma_0^2 = y$ corresponds to the earlier fixed threshold P_{FA} result of (6.31) rewritten as

$$\mathrm{P}(\mathrm{FA}|\gamma_0^2 = y) = e^{-\frac{y}{2\sigma^2}}, \quad 0 \leq y. \tag{6.81}$$

Then the unconditional P_{FA} is given by

$$P_{\mathrm{FA}} = \int_0^{\infty} \mathrm{P}(\mathrm{FA}|\gamma_0^2 = y)p_Y(y)dy, \tag{6.82}$$

where $p_Y(y)$ is given by

$$p_Y(y) = \frac{y^{M-1}\exp\{-\frac{y}{2K\sigma^2}\}}{(2K\sigma^2)^M(M-1)!}, \quad 0 \leq y. \tag{6.83}$$

Direct evaluation of (6.82) using (6.83) yields

$$P_{\text{FA}} = \frac{1}{(1+K)^M}, \tag{6.84}$$

which shows now the P_{FA} is independent of the noise variance σ^2.

6.4 Conclusions

In this chapter, we considered the non-coherent detection of a sinusoidal waveform with a uniformly distributed random phase in the presence of an AWGN. Section 6.1 showed the sufficient statistic for the non-coherent detector is the envelope of the I-channel and Q-channel detector given either in the correlation or matched filter form. An equivalent approximate detector consists of a bandpass filter centered at the frequency of the sinusoidal waveform followed by an envelope detector. Section 6.2 first derived the statistic of the Rayleigh pdf of the envelope in the absence of a signal and the associated P_{FA}. Then in the presence of a sinusoidal signal, the envelope had a Rician pdf, which led to the P_{D} given in terms of the Marcum Q-function. Section 6.3 considered several non-coherent detection issues in some radar receivers. In Section 6.3.1, a train of known n identical pulses with a known initial phase is received in the presence of an AWGN. Then the optimum receiver uses a matched filter to the entire waveform and this operation is called coherent integration. In Section 6.3.2, all of the n identical pulses have independent uniformly distributed phases. The optimum receiver performs post detection integration as the sum of the $\ln(\cdot)$ of the $\text{I}_0(\cdot)$ function of the envelope of each of the n envelope statistics. For low SNR, the approximate optimum receiver results in the sum of the square of the n envelope statistics. For high SNR, the approximate optimum receiver results in the sum of the n envelope statistics. Section 6.3.3 considered the double-threshold receiver, where each of the n envelope statistics is compared to the first threshold of a non-coherent detector, resulting in a zero or a unity value to a binary counter. The value of the binary counter is then compared to a second threshold to decide the presence of a signal. As expected the performance of the double-threshold receiver is inferior to the performance of the receiver in Section 6.3.2. Section 6.3.4 addressed the issue of the sensitivity of the P_{FA} to the decision threshold determined by an estimate of the background noise. For fixed threshold receiver, a sudden increase of the noise may result in significant increase of the P_{FA}. While various CFAR methods are available, a simple cell-averaging CFAR method is shown to maintain the P_{FA} under varying noise conditions.

6.5 Comments

The introduction of the Rician pdf needed in the analysis of a performance of a non-coherent receiver for the detection of a sinusoidal waveform in the presence of an AWGN was first presented by S. O. Rice in [2]. The early radar book by P. M. Woodward

[3, p. 131] and early communication theory books by W. B. Davenport, Jr. and W. L. Root [4, p. 357] and the treatise by D. Middleton [5, p. 847] have all summarized Rice's results of [2]. Since then materials in Sections 6.1 and 6.2 on non-coherent detection can be found in many communication and radar detection theory books including [6–9]. Materials in Sections 6.3.1 and 6.3.4 on non-coherent detection in radar receivers can be found in [6–12]. Various Matlab programs on radar non-coherent detection and CFAR equations appeared in [13].

References

[1] M. Abramowitz and I. A. Stegun. *Handbook of Mathematical Functions*. National Bureau of Standards, 1964.

[2] S. O. Rice. Mathematical analysis of random noise. *Bell Labs Tech. J.*, 1944.

[3] P. M. Woodward. *Probability and Information Theory, with Applications to Radar*. Pergamon Press, 1953.

[4] W. B. Davenport Jr. and W. L. Root. *An Introduction to the Theory of Random Signals and Noise*. McGraw-Hill, 1958.

[5] D. Middleton. *An Introduction to Statistical Communication Theory*. McGraw-Hill, 1960.

[6] C. W. Helstrom. *Statistical Theory of Signal Detection*. Pergamon, 2d ed., 1968.

[7] A. D. Whalen. *Detection of Signals in Noise*. Academic Press, 1st ed., 1971. R. N. McDonough and A. D. Whalen, 2d ed., 1995.

[8] C. W. Helstrom. *Elements of Signal Detection and Estimation*. Prentice-Hall, 1995.

[9] J. V. DiFranco and W. L. Rubin. *Radar Detection*. Artech House, 1980.

[10] J. I. Marcum. A statistical theory of target detection by pulse radar. Technical Report RM-753, Rand Corp., 1948.

[11] M. I. Skolnik. *Introduction to Radar Systems*. McGraw-Hill, 1962.

[12] N. Levanon. *Radar Principles*. Wiley, 1988.

[13] B. R. Mahafza. *Radar Systems Analysis and Design using Matlab*. Chapman and Hall, 2008.

Problems

6.1 Let $X \sim \mathcal{N}(0, \sigma^2)$ and $Y \sim \mathcal{N}(0, \sigma^2)$ be two independent Gaussian r.v.'s. Show analytically that $R = \sqrt{X^2 + Y^2}$ is a Rayleigh r.v. and find its pdf.

6.2 Let $X \sim \mathcal{N}(a\cos(\alpha), \sigma^2)$ and $Y \sim \mathcal{N}(a\sin(\alpha), \sigma^2)$, $0 \le \alpha < 2\pi$, be two independent Gaussian r.v.'s. Show analytically that $R = \sqrt{X^2 + Y^2}$ is a Rician r.v. and find its pdf.

6.3 Consider the Rician pdf $p_{\mathrm{Ri}}(r) = (r/\sigma^2)e^{-(\mu^2+r^2)/(2\sigma^2)}\mathrm{I}_0(\mu r/\sigma^2)$, $0 \le r < \infty$. Plot the following 5 Rician pdfs on the same figure.

(a) Case 1: $\mu = 0,\ \sigma = 1$
(b) Case 2: $\mu = \sigma = 1$.
(c) Case 3: $\mu = 2,\ \sigma = 1$.
(d) Case 4: $\mu = 1,\ \sigma = 2$.
(e) Case 5: $\mu = \sigma = 2$.

6.4 Consider the Rayleigh pdf $p_R(r) = (r/\sigma^2)e^{-r^2/(2\sigma^2)}$, $0 \leq r < \infty$.

(a) Analytically find the cdf $F_R(r)$ of this Rayleigh pdf.
(b) Analytically find the mean $\mu_R = \mathrm{E}\{R\}$ of this Rayleigh pdf as a function of σ^2.
(c) Analytically find the variance $\sigma_R^2 = \mathrm{E}\{(R - \mu_R)^2\}$ of this Rayleigh pdf as a function of σ^2.

6.5 Consider the generation of a $N \times 1$ pseudo-random Rayleigh vector as follows. Set randn('state', 49). Let x = randn(N, 1) and y = randn(N, 1), with $N = 1000$. Then let r = sqrt($x. \wedge 2 + y. \wedge 2$) . Show r is Rayleigh distributed.

(a) Find the estimated mean $\tilde{\mu}_R$ from the generated vector r.
(b) Find the estimated variance $\tilde{\sigma}_R^2$ from the generated vector r.
(c) Find the estimated standard deviation $\tilde{\sigma}_1$ for the σ of the original Gaussian parameter used in the randn(.,.) pseudo-random generation of the Rayleigh r.v. based on the estimated mean $\tilde{\mu}_R$ result of part (a) and the formula in part (b) of Problem 6.4.
(d) Find the estimated standard deviation $\tilde{\sigma}_2$ for the σ of the original Gaussian parameter used in the randn(.,.) pseudo-random generation of the Rayleigh r.v. based on the estimated variance $\tilde{\sigma}_R^2$ result of part (b) and the formula in part (c) of Problem 6.4.
(e) The two estimated values of $\tilde{\sigma}$ in part (c) and part (d) should be very close. Use either of the $\tilde{\sigma}$ to plot the cdf of a Rayleigh cdf for $x = 0 : 0.01 : 4$, using the Matlab function raylcdf($x, \tilde{\sigma}$).
(f) Use the generated vector r and the Matlab function cdfplot(r) to plot the empirical cdf obtained from r. Place this empirical cdf on top of the cdf obtained in part (e). Do these two cdfs match each other well?

6.6 Consider a simulation verification of the behavior of a non-coherent matched filter.

(a) Construct a narrow bandpass filter with unity gain centered at frequency $[0.35 - \delta, 0.35 + \delta]$ and approximately zero gain elsewhere. Use Matlab's fir1 function to construct a FIR filter of order 200. Specifically, use b = fir1(200, [0.345, 0.355]) to construct the FIR coefficients. Check the frequency response of this filter with $[h, w]$ = freqz(b, 1, 128). Then plot the frequency response with plot((w/π), (20 $*$ log 10(abs(h)))). Verify your filter has the passband around $f_0 = 0.35$.
(b) Evaluate the impulse response of your bandpass filter. Take x_1 = [1, zeros(1, 200)], y_1 = filter(b, 1, x_1). Plot your impulse response function for $\delta = 0.005$. Vary the δ. If the δ is very small, what can you say about the impulse response function?
(c) Consider a transmitter signal $s(t) = \sin(2\pi f_0 t)$, $0 \leq t \leq T$. Discretize $s(t)$ by sampling every $\tau = 1/f_s$ second. In particular, the highest allowable frequency is set to $f_s/2$ (i.e., the Nyquist condition). By taking $t = n\tau = n/f_s$, $n = 1, \ldots, N = 100$, the sampled signal is given by the vector $s = \sin(0.35 * \pi * n)$, $n = 1, \ldots, 100$. We have set $0.35 = f_0/(f_s/2)$. Plot x = [sin(0.35 $* \pi *$ [1 : 100], zeros(1, 100))]. Here we have padded 100 zeros after the signal for the following desired convolution in the bandpass filter.
(d) Obtain the filter output y = filter(b, 1, x). Plot the output y. Does this output (in particular its peak location) agree with theory? Give some justification.

6.7 It is known that if the parameter μ in the Rician pdf $p_{\mathrm{Ri}}(r) = (r/\sigma^2)e^{-(\mu^2+r^2)/(2\sigma^2)}$ $\mathrm{I}_0(ar)$, $0 \leq r < \infty$, is set to 0, the Rician pdf becomes the Rayleigh pdf $p_{\mathrm{Ray}}(r) = (r/\sigma^2)e^{-r^2/(2\sigma^2)}$, $0 \leq r < \infty$. However, for large values of SNR $= \mu/\sigma > 3$, the Rician pdf can be approximated by a Gaussian pdf. We take $a = 2$ and $\sigma = 0.5$ resulting in an $\mu/\sigma = 4$. First, plot $p_{\mathrm{Ri}}(r)$ using $\mu = 2$ and $\sigma = 0.5$. Next, plot the Gaussian pdf with a mean of $\mu = 2$ and $\sigma = 0.5$. Finally, plot the Gaussian pdf with a modified mean of $\sqrt{\mu^2 + \sigma^2}$ and $\sigma = 0.5$, all on the same figure. How do the last two Gaussian pdfs compare to the Rician pdf?

6.8 Show the three different expressions of the Rician pdf are equivalent:

(1) $p_{\mathrm{Ri}}^{\{1\}}(x) = (x/\sigma^2)\exp(-(\mu^2 + x^2)/(2\sigma^2))\mathrm{I}_0(\mu x/\sigma^2), \ 0 < x < \infty, \ 0 < \mu$

(2) $p_{\mathrm{Ri}}^{\{2\}}(y) = y\exp(-(a^2 + y^2)/2)\mathrm{I}_0(ay), \ 0 < y < \infty, \ 0 < a, \ a = \mu/\sigma, \ x = \sigma y.$

(3) $p_{\mathrm{Ri}}^{\{3\}}(x) = (2x(1+K)/s)\exp(-K)\exp(-(1+K)x^2/s)\mathrm{I}_0(2x\sqrt{K(1+K)/s}),$
$0 < x < \infty, \ K = \mu^2/2\sigma^2, \ s = (1+K)2\sigma^2.$

6.9 The Rician pdf has two parameters (e.g., K and s in the Rician pdf expression of (3) of Problem 6.8). On the other hand, the Nakagami-m pdf, given by $p_{\mathrm{Na}}(x) = 2(m/s)^m x^{2m-1}\exp(-mx^2/s)/\Gamma(m), \ 0 < x < \infty, \ 0 < m$, has only one parameter m. However, by setting the parameter $m = (1+K)^2/(1+2K)$, the first and second moments of the Rician pdf and Nakagami-m pdf are identical. Plot the Rician pdf using the expression (3) of Problem 6.8 and the Nakagami-m pdf with $m = (1+K)^2/(1+2K)$ on the same figure using $K = 10$ and $s = 2$ for $0 \leq x \leq 3$. Repeat with another plot using $K = 2$ and $s = 1$ for $0 \leq x \leq 5$. Show the Nakagami-m pdf curve approximates the Rician pdf curve well (at least for these two sets of parameter values).

6.10 Let the received waveform be given by

$$x(t) = \begin{cases} n(t), \ H_0, \\ \sqrt{2E/T}\sin(wt + \theta) + n(t), \ H_1, \end{cases} \quad 0 \leq t \leq T,$$

where $n(t)$ is the realization of a zero-mean WGN process with two-sided spectral density $N_0/2$, θ is the realization of a binary valued r.v. independent of the noise with equal probabilities of $\theta = 0$ and $\theta = \pi$. Derive the optimum decision regions under the LR test.

7 Parameter estimation

7.1 Introduction

Thus far, in the previous chapters, we have considered decision theory (or hypothesis testing or detection theory) in which the number of signals (or messages or hypotheses) is finite. The goal was to determine which of the signals was transmitted. For the binary case with $M = 2$, we dealt with BPSK, BFSK, target present or absent in the presence of noise, etc. For the M-ary case, we dealt with QPSK ($M = 4$ PSK), 8PSK ($M = 8$ PSK), 16-QAM ($M = 16$ quadrature-amplitude-modulation), etc.

In this chapter, we consider the estimation of parameters needed in the study of communication/radar systems, as well as in various aspects of signal and array processing.

Example 7.1 Consider a received waveform in a radar receiver given by

$$x(t) = A\cos(2\pi f t + \phi) + n(t),\ 0 \le t \le T. \tag{7.1}$$

We may want to estimate the following parameters which take on real-valued numbers and not quantized to one of M possible values in detection theory.

Amplitude A: A may be a function of: the strength of the transmitted radar waveform; the total distance of the transmitter-target receiver; size and scattering property of the target cross section.

Frequency f: f may be a function of: the Doppler shift of the transmitted frequency f_0, which depends on the relative velocity between the transmitter/receiver and target.

Phase ϕ: ϕ may be a function of: the total distance of the transmitter-target-receiver; random propagation effects; imperfect receiver phase control and processing elements. □

Section 7.2 considers one of the oldest and most commonly used parameter estimation methods based on the mean-square (m.s.) error criterion. Using the m.s. estimation technique, first we treat the estimation of a single r.v. by its mean, then we consider the estimation of a single r.v. by two other r.v.'s, and then using vector-matrix notations extend the above approach to the estimation of a single r.v. by n other r.v.'s. Section 7.2.1 shows that non-linear m.s. estimation is equivalent to the evaluation of a conditional expectation. Section 7.2.2 introduces the geometry of the orthogonal principle and its relevancy to linear m.s. estimation. Section 7.2.3 treats block versus recursive m.s.

estimations. Section 7.2.3.1 summarizes the disadvantages of the block m.s. estimation, while Section 7.2.3.2 introduces a simple scalar version of the Kalman recursive filter. In Section 7.3, we consider the least-square-error and least-absolute-error estimation problem, where the parameter of interest is assumed to be deterministic, but observed in the presence of random disturbances. Section 7.4 introduces various basic statistical parameter estimation concepts. Properties of an estimator such as unbiasedness, efficiency, consistency, and sufficiency and their applications to communication and radar system examples are treated. Section 7.4.1 shows the Cramér–Rao bound provides the minimum possible estimation variance of an unbiased estimator. Section 7.4.2 treats the maximum-likelihood estimator. Section 7.4.3 treats the maximum a posteriori estimator. Section 7.4.4 presents the Bayes estimator. A brief conclusion is given in Section 7.5 and some comments are given in Section 7.6. Two appendices and some references are given at the end of the chapter. In this chapter, the optimum estimator (optimum under some criterion) is denoted by the symbol of a circumflex ˆ over the estimator.

7.2 Mean-square estimation

There are many forms of parameter estimations. The simplest and computationally most tractable form of estimation is based on the *mean-square* (m.s.) error criterion. The geometry and insight of the m.s. estimation method are provided by the orthogonal principle in an inner product vector space.

First, we consider the optimum m.s. estimation of a single r.v. by its mean. Let X be a real-valued r.v. with a pdf $p_X(x)$ (or $p(x)$ if there is only one r.v. under consideration). We want to find an optimum estimate (i.e., a deterministic number) $\hat{\theta}$ that minimizes the m.s. error of the r.v. X and $\hat{\theta}$. That is, denote the m.s. error by

$$\varepsilon^2 = \mathrm{E}\{(X-\theta)^2\} = \int (x-\theta)^2 p(x)dx. \tag{7.2}$$

To find the minimum of ε^2, set

$$\frac{\partial \varepsilon^2}{\partial \theta} = -2\int (x-\theta)p(x)dx = 0. \tag{7.3}$$

We know the mean of X is given by

$$\mu = \int xp(x)dx \tag{7.4}$$

and

$$\int p(x)dx = 1. \tag{7.5}$$

Then using (7.4)–(7.5) in (7.3) to solve for $\hat{\theta}$ yields

$$\hat{\theta} = \mu = \int xp(x)dx. \tag{7.6}$$

Furthermore, we note

$$\left.\frac{\partial^2 \varepsilon^2}{\partial \theta^2}\right|_{\theta=\hat{\theta}} = 2\int p(x)dx = 2 > 0. \tag{7.7}$$

Thus, the estimate $\hat{\theta}$ given by the stationary solution of (7.3), in light of (7.7), shows it is a local minimum. Since ε^2 in (7.3) is a quadratic function of $\hat{\theta}$, then the local minimum solution is a global minimum solution. Thus, the optimum m.s. estimate $\hat{\theta}$ is just the mean of the r.v. X.

In practice, if we have n observations of the r.v. X with values $\{x_1, \ldots, x_n\}$, then the sample mean $\bar{x}$ of these observations is defined by

$$\bar{x} = (1/n)\sum_{i=1}^{n} x_i. \tag{7.8}$$

Furthermore, with the assumption of ergodicity in the mean, we use the sample mean $\bar{x}$ to estimate the true mean μ of the r.v. X.

Next, we want to find the optimum m.s. estimate of one r.v. by two other r.v.'s. Consider three r.v.'s $\{X_1, X_2, X_3\}$ taken from a real-valued zero-mean wide-sense stationary random sequence $\{X_n, -\infty < n < \infty\}$ with an autocorrelation sequence $R(n) = \mathrm{E}\{X_{n+m}X_m\}$. We want to use X_1 and X_2 linearly to estimate X_3 in the m.s. sense. That is, we want to find coefficient parameters $\{a_1, a_2\}$ to minimize

$$\begin{aligned}
\varepsilon^2 &= \mathrm{E}\{(X_3 - (a_1X_1 + a_2X_2))^2\} \\
&= \mathrm{E}\{X_3^2\} - 2a_1\mathrm{E}\{X_3X_1\} - 2a_2\mathrm{E}\{X_3X_2\} \\
&\quad + a_1^2\mathrm{E}\{X_1^2\} + 2a_1a_2\mathrm{E}\{X_1X_2\} + a_2^2\mathrm{E}\{X_2^2\} \\
&= R(0) - 2a_1R(2) - 2a_2R(1) + a_1^2R(0) + 2a_1a_2R(1) + a_2^2R(0).
\end{aligned} \tag{7.9}$$

Then taking the partial derivatives of ε^2 w.r.t. a_1 and a_2 respectively, and setting them to zeros yields

$$\frac{\partial \varepsilon^2}{\partial a_1} a_1 = \hat{a}_1, a_2 = \hat{a}_2 = -2R(2) + 2\hat{a}_1R(0) + 2\hat{a}_2R(1) = 0, \tag{7.10}$$

$$\frac{\partial \varepsilon^2}{\partial a_2} a_1 = \hat{a}_1, a_2 = \hat{a}_2 = -2R(1) + 2\hat{a}_1R(1) + 2\hat{a}_2R(0) = 0. \tag{7.11}$$

Now, (7.10) and (7.11) can be expressed as

$$\mathbf{R}\hat{\mathbf{a}} = \mathbf{J}\mathbf{r}, \tag{7.12}$$

where

$$\mathbf{R} = \begin{bmatrix} R(0) & R(1) \\ R(1) & R(0) \end{bmatrix}, \tag{7.13}$$

$$\hat{\mathbf{a}} = \begin{bmatrix} \hat{a}_1 \\ \hat{a}_2 \end{bmatrix}, \tag{7.14}$$

$$\mathbf{J} = \begin{bmatrix} 0 & 1 \\ 1 & 0 \end{bmatrix}, \tag{7.15}$$

$$\mathbf{r} = \begin{bmatrix} R(1) \\ R(2) \end{bmatrix}. \tag{7.16}$$

Equation (7.12) is called the *normal equation* in m.s. estimation. (7.13) is the autocorrelation matrix for this sequence, (7.14) is the optimum coefficient vector, (7.15) is a rotation matrix, and (7.16) is the autocorrelation vector for this m.s. estimation problem. Assume $\mathbf{R}$ is positive-definite. Then $\mathbf{R}^{-1}$ exists and the minimum m.s. error coefficient vector $\mathbf{a}$ now denoted by $\hat{\mathbf{a}}$ is given by

$$\hat{\mathbf{a}} = \mathbf{R}^{-1}\mathbf{J}\mathbf{r}. \tag{7.17}$$

The m.s. error of (7.9) can now be expressed in matrix-vector form (for any coefficient vector $\mathbf{a}$) as

$$\varepsilon^2 = R(0) + \mathbf{a}^T\mathbf{R}\mathbf{a} - 2\mathbf{a}^T\mathbf{J}\mathbf{r}. \tag{7.18}$$

The minimum m.s. error of (7.18) using the optimum coefficient vector $\hat{\mathbf{a}}$ of (7.17) becomes

$$\begin{aligned} \varepsilon^2_{\text{min}} &= R(0) + \hat{\mathbf{a}}^T\mathbf{R}\hat{\mathbf{a}} - 2\hat{\mathbf{c}}^T\mathbf{J}\mathbf{r} \\ &= R(0) + \mathbf{r}^T\mathbf{J}\mathbf{R}^{-1}\mathbf{R}\mathbf{R}^{-1}\mathbf{J}\mathbf{r} - 2\mathbf{r}^T\mathbf{J}\mathbf{R}^{-1}\mathbf{J}\mathbf{r}. \end{aligned} \tag{7.19}$$

But we note

$$\mathbf{J}\mathbf{J} = \mathbf{I}_2, \ \mathbf{J}\mathbf{R}^{-1}\mathbf{J} = \mathbf{R}^{-1}. \tag{7.20}$$

Using (7.20) in (7.12) yields

$$\varepsilon^2_{\text{min}} = R(0) - \mathbf{r}^T\mathbf{R}^{-1}\mathbf{r}. \tag{7.21}$$

Next, consider the optimum m.s. estimation of one r.v. from n other r.v.'s. The purpose of using the matrix-vector notations from (7.12)–(7.21) allows us to generalize the previously obtained results to the present problem readily. Let $\{X_1, \ldots, X_n, Y\}$ be $(n+1)$ r.v.'s. Again, we assume $\{X_1, \ldots, X_n\}$ as taken from a real-valued zero-mean wide-sense stationary random sequence $\{X_n, -\infty < n < \infty\}$ with an autocorrelation sequence $R(n) = \mathrm{E}\{X_{n+m}X_m\}$. The zero mean r.v. Y is correlated to the $\{X_1, \ldots, X_n\}$ in some manner to be specified shortly. We want to estimate Y using a linear combination of the $\{X_1, \ldots, X_n\}$ in the m.s. error sense. Denote an arbitrary linear combination of $\{X_1, \ldots, X_n\}$ using coefficients of $\{a_1, \ldots, a_n\}$ given by

$$\tilde{Y} = a_1X_1 + \cdots + a_nX_n, \tag{7.22}$$

as an estimate of Y. Then the m.s. error between Y and $\tilde{Y}$ becomes

$$\begin{aligned} \varepsilon^2 &= \mathrm{E}\{(Y - \tilde{Y})^2\} = \mathrm{E}\{Y^2\} - 2\sum_{i=1}^{n} a_i\mathrm{E}\{YX_i\} + \sum_{i=1}^{n}\sum_{j=1}^{n} a_ia_j\mathrm{E}\{X_iX_j\} \\ &= \sigma_Y^2 + \mathbf{a}_n^T\mathbf{R}_n\mathbf{a}_n - 2\mathbf{a}_n^T\mathbf{r}_{Y\mathbf{X}_n}, \end{aligned} \tag{7.23}$$

where

$$\mathbf{R}_n = \begin{bmatrix} R(0) & R(1) & \cdots & R(n-1) \\ R(1) & R(0) & \cdots & R(n-2) \\ \vdots & \vdots & \ddots & \vdots \\ R(n-1) & R(n-2) & \cdots & R(0) \end{bmatrix}, \tag{7.24}$$

$$\begin{aligned} &\sigma_Y^2 = \mathrm{E}\{Y^2\}, \ \mathbf{a}_n = [a_1, \ \ldots, \ a_n]^T, \\ &\mathbf{X}_n = [X_1, \ \ldots, \ X_n]^T, \ \mathbf{r}_{Y\mathbf{X}_n} = [\mathrm{E}\{YX_1\}, \ \ldots, \ \mathrm{E}\{YX_n\}]^T. \end{aligned} \tag{7.25}$$

Define the $n \times 1$ gradient $\nabla\{\cdot\}$ of the scalar ε^2 w.r.t. the $n \times 1$ column vector $\mathbf{a}_n$ by

$$\nabla_{\mathbf{a}_n}(\varepsilon^2) = \left[\frac{\partial(\varepsilon^2)}{\partial a_1}, \ldots, \frac{\partial(\varepsilon^2)}{\partial a_n}\right]^T, \tag{7.26}$$

then direct evaluations show

$$\nabla_{\mathbf{a}_n}(\mathbf{a}_n^T \mathbf{R}_n \mathbf{a}_n) = 2\mathbf{R}_n \mathbf{a}_n, \ \nabla_{\mathbf{a}_n}(\mathbf{a}_n^T \mathbf{r}_{Y\mathbf{X}_n}) = \mathbf{r}_{Y\mathbf{X}_n}. \tag{7.27}$$

By taking the gradient of the scalar ε^2 in (7.23) w.r.t. the column vector $\mathbf{a}_n$ and using (7.26)–(7.27), we obtain

$$\nabla_{\mathbf{a}_n}(\varepsilon^2) = 2\mathbf{R}_n \hat{\mathbf{a}}_n - 2\mathbf{r}_{Y\mathbf{X}_n} = 0. \tag{7.28}$$

Furthermore, we can define the $n \times n$ Hessian $\nabla^2\{\cdot\}$ of the scalar ε^2 w.r.t. the $n \times 1$ column vector $\mathbf{a}_n$ by

$$\nabla^2_{\mathbf{a}_n}\{\varepsilon^2\} = \begin{bmatrix} \dfrac{\partial^2(\varepsilon^2)}{\partial a_1^2} & \cdots & \dfrac{\partial^2(\varepsilon^2)}{\partial a_1 \partial a_n} \\ \vdots & \ddots & \vdots \\ \dfrac{\partial^2(\varepsilon^2)}{\partial a_n \partial a_1} & \cdots & \dfrac{\partial^2(\varepsilon^2)}{\partial a_n^2} \end{bmatrix}. \tag{7.29}$$

Then application of (7.29) to ε^2 yields

$$\nabla^2_{\mathbf{a}_n}(\varepsilon^2) = 2\mathbf{R}_n. \tag{7.30}$$

For a positive-definite $\mathbf{R}_n$, then the stationary solution $\hat{\mathbf{a}}_n$ of (7.28), and in light of (7.30), shows $\hat{\mathbf{a}}_n$ is a global minimum solution of the quadratic form of ε^2 of (7.23). From (7.28)

$$\mathbf{R}_n \hat{\mathbf{a}}_n = \mathbf{r}_{Y\mathbf{X}_n}. \tag{7.31}$$

Then we can solve for the optimum coefficient vector

$$\hat{\mathbf{a}}_n = \mathbf{R}_n^{-1} \mathbf{r}_{Y\mathbf{X}_n}, \tag{7.32}$$

with the associated optimum estimate given by

$$\hat{y} = \mathbf{R}_{Y\mathbf{X}_n}^T \mathbf{R}_n^{-1} \mathbf{x}_n, \tag{7.33}$$

and the associated minimum m.s. error given by

$$\varepsilon^2_{\min} = \sigma_Y^2 - \mathbf{R}_{Y\mathbf{X}_n}^T \mathbf{R}_n^{-1} \mathbf{R}_{Y\mathbf{X}_n}. \tag{7.34}$$

7.2.1 Non-linear mean-square estimation and conditional expectation

Consider two scalar r.v.'s Y and X. We want to find a function $g(\cdot)$ of the r.v. X such that $g(X)$ minimizes Y in the m.s. sense. Let

$$\begin{aligned} \varepsilon^2 = \mathrm{E}\{(Y - g(X))^2\} &= \int\int (y - g(x))^2 p_{X,Y}(x, y) dx dy \\ &= \int p_X(x) \left[\int (y - g(x))^2 p_{Y|X}(y|x) dy\right] dx. \end{aligned} \tag{7.35}$$

Since $p_X(x) \geq 0$, for each fixed (i.e., conditioned) x, we want to minimize the bracketed term in (7.35) in order to minimize ε^2. But for a fixed x, $g(x)$ is a constant. Thus, we can use the result of (7.6) on the optimum estimate of a r.v. by a constant in the m.s. sense to yield

$$\min_{\text{fixed}\,x}\left[\int (y - g(x))^2 p_{Y|X}(y|x)dy\right]$$
$$= \int \left(y - \int y' p_{Y|X}(y'|x)dy'\right)^2 p_{Y|X}(y|x)dy. \quad (7.36)$$

Thus, the optimum estimate $\hat{Y}$ of X is given by

$$\hat{Y} = \hat{g}(x) = \int y' p_{Y|X}(y'|x)dy' = \mathrm{E}\{Y|X = x\}. \quad (7.37)$$

Then (7.37) shows the optimum estimate $\hat{Y}$ of X is the conditional mean of Y given (i.e., conditioned on) $X = x$. Unfortunately, this conditional expectation is generally not linear in x.

If X and Y are independent, then

$$\mathrm{E}\{Y|X = x\} = \mathrm{E}\{Y\} = \mu_Y, \quad (7.38)$$

and the optimum estimate $\hat{Y}$ of X in the m.s. sense is a constant not depending on the value of x. Now, consider r.v.'s $\{X_1, X_2, Y\}$. We want to find the optimum function $g(x_1, x_2)$ to minimize

$$\varepsilon^2 = \mathrm{E}\{(Y - g(X_1, X_2))^2\}. \quad (7.39)$$

Using the approach as considered in (7.35)–(7.37), the optimum estimate $\hat{Y}$ using $g(x_1, x_2)$ in the m.s. sense is again given as a conditional expectation of Y conditioned on $\{x_1, x_2\}$ given by

$$\hat{Y} = \hat{g}(x_1, x_2) = \mathrm{E}\{Y|X_1 = x_1, X_2 = x_2\}. \quad (7.40)$$

In general, given $\{X_1, \ldots, X_n, Y\}$, the optimum m.s. estimate of Y using $g(x_1, \ldots, x_n)$ is again a conditional expectation given by

$$\hat{Y} = \hat{g}(x_1, \ldots, x_n) = \mathrm{E}\{Y|X_1 = x_1, \ldots, X_n = x_n\}. \quad (7.41)$$

The actual evaluation of (7.41) for some given large integer n is generally computationally costly.

Next, consider real-valued jointly Gaussian r.v.'s $\{X_1, \ldots, X_n, Y\}$. Denote $\mathbf{X}_n = [X_1, \ldots, X_n]^T$, its mean by $\boldsymbol{\mu}_{\mathbf{X}_n} = \mathrm{E}\{\mathbf{X}_n\}$, and its covariance matrix by $\boldsymbol{\Lambda}_{\mathbf{X}_n} = \mathrm{E}\{(\mathbf{X}_n - \boldsymbol{\mu}_{\mathbf{X}_n})(\mathbf{X}_n - \boldsymbol{\mu}_{\mathbf{X}_n})^T\}$. Denote the mean of Y by $\mu_Y = \mathrm{E}\{Y\}$ and its variance by $\sigma_Y^2 = \mathrm{E}\{(Y - \mu_Y)^2\}$. Denote the cross covariance between Y and $\mathbf{X}_n$ by $\boldsymbol{\lambda}_{Y\mathbf{X}_n} = \mathrm{E}\{(Y - \mu_Y)(\mathbf{X}_n - \boldsymbol{\mu}_{\mathbf{X}_n})\} = \boldsymbol{\lambda}_{Y\mathbf{X}_n^T}^T$. Then the minimum m.s. estimate of Y given the observation vector of $\mathbf{X}_n = \mathbf{x}_n$ from (7.41) yields

$$\hat{Y} = \hat{g}(\mathbf{x}_n) = \mathrm{E}\{Y|\mathbf{X}_n = \mathbf{x}_n\} = \int y p_{Y|\mathbf{X}_n}(y|\mathbf{x}_n)dy. \quad (7.42)$$

However, when $\{X_1, \ldots, X_n, Y\}$ are jointly Gaussian, then the integral on the right-hand side of (7.42) can be explicitly evaluated and has the form of

$$\hat{Y} = \hat{g}(\mathbf{x}_n) = \mathrm{E}\{Y|\mathbf{X}_n = \mathbf{x}_n\} = \int y p_{Y|\mathbf{X}_n}(y|\mathbf{x}_n)dy$$
$$= \mu_Y + \boldsymbol{\lambda}_{Y\mathbf{X}_n}^T \boldsymbol{\Lambda}_{\mathbf{X}_n}^{-1}(\mathbf{x}_n - \boldsymbol{\mu}_{\mathbf{X}_n}). \tag{7.43}$$

Furthermore, the minimum m.s. error of this estimation is given by

$$\varepsilon_{\min}^2 = \sigma_Y^2 - \boldsymbol{\lambda}_{Y\mathbf{X}_n}^T \boldsymbol{\Lambda}_{\mathbf{X}_n}^{-1} \boldsymbol{\lambda}_{Y\mathbf{X}_n}. \tag{7.44}$$

The crucial observation is that the optimum m.s. estimate of Y given the observation $\mathbf{X}_n = \mathbf{x}_n$ is *linear* in $\mathbf{X}_n$ as shown in the r.h.s. of (7.43) when $\{X_1, \ldots, X_n, Y\}$ are jointly Gaussian. We also note that when $\{X_1, \ldots, X_n, Y\}$ are jointly Gaussian, (7.43) includes (7.34) as a special case.

7.2.2 Geometry of the orthogonal principle and mean-square estimation

Consider a real-valued vector space $\mathbb{V}$ with a well-defined inner product $(\cdot, \cdot)$ satisfying

1. $(a_1\mathbf{x}_1 + a_2\mathbf{x}_2, \mathbf{y}) = a_1(\mathbf{x}_1, \mathbf{y}) + a_2(\mathbf{x}_2, \mathbf{y})$,
2. $(\mathbf{x}, \mathbf{y}) = (\mathbf{y}, \mathbf{x})$,
3. $(\mathbf{x}, \mathbf{x}) = \|\mathbf{x}\|^2 \geq 0;\ \|\mathbf{x}\|^2 = 0 \Leftrightarrow \mathbf{x} = 0$.

Example 7.2 Consider two vectors $\mathbf{x} = [x_1, \ldots, x_n]^T$ and $\mathbf{y} = [y_1, \ldots, y_n]^T$ in the n-dimensional Euclidean space $\mathbb{R}^n$. Then the inner product is defined by

$$(\mathbf{x}, \mathbf{y}) = \sum_{i=1}^{n} x_i y_i$$

and the norm is defined by

$$\|\mathbf{x}\| = \sqrt{\sum_{i=1}^{n} x_i^2}.$$

It is clear that this inner product and norm satisfy all the above three properties. □

Suppose we want to approximate the vector $\mathbf{y}$ by the vector $\mathbf{x}$ linearly when both of these vectors are elements of $\mathbb{V}$. Denote the norm squared of the approximation error ε^2 by

$$\varepsilon^2 = \|\mathbf{y} - a\mathbf{x}\|^2 = (\mathbf{y} - a\mathbf{x}, \mathbf{y} - a\mathbf{x})$$
$$= (\mathbf{y}, \mathbf{y}) - 2a(\mathbf{y}, \mathbf{x}) + a^2(\mathbf{x}, \mathbf{x}). \tag{7.45}$$

Taking the partial derivative of ε^2 w.r.t. a shows the optimum $\hat{a}$ must satisfy

$$-2(\mathbf{y}, \mathbf{x}) + 2\hat{a}(\mathbf{x}, \mathbf{x}) = 0, \tag{7.46}$$

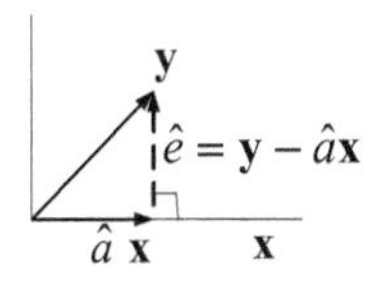

Figure 7.1 Orthogonal principle showing the orthogonal projection of $\mathbf{y}$ onto $\mathbf{x}$ to yield $\hat{a}\mathbf{x}$.

or

$$\hat{a} = (\mathbf{y}, \mathbf{x})/(\mathbf{x}, \mathbf{x}). \tag{7.47}$$

But (7.46) or (7.47) can also be expressed as

$$(\mathbf{y}, \mathbf{x}) - \hat{a}(\mathbf{x}, \mathbf{x}) = 0, \tag{7.48}$$

or

$$(\mathbf{y} - \hat{a}\mathbf{x}, \mathbf{x}) = 0. \tag{7.49}$$

Denote the approximation error of the vector $\mathbf{y}$ by the vector $\mathbf{x}$ linearly as

$$\mathbf{e} = \mathbf{y} - a\mathbf{x} \tag{7.50}$$

and the optimum approximation error $\hat{\mathbf{e}}$ as

$$\hat{\mathbf{e}} = \mathbf{y} - \hat{a}\mathbf{x}. \tag{7.51}$$

Then (7.49) and (7.51) show that

$$(\hat{\mathbf{e}}, \mathbf{x}) = 0. \tag{7.52}$$

Any two vectors $\mathbf{x}$ and $\mathbf{y}$ that satisfy

$$(\mathbf{x}, \mathbf{y}) = 0 \tag{7.53}$$

are said to be *orthogonal*. Thus, (7.52) shows that the optimum error $\hat{\mathbf{e}}$ must be orthogonal to the observed data vector $\mathbf{x}$. Geometrically, two vectors that are orthogonal are at a right angle to each other. Thus, to approximate the vector $\mathbf{y}$ by the vector $\mathbf{x}$ linearly, we have the geometric orthogonal relationship given in Fig. 7.1. $\hat{a}\mathbf{x}$ is said to be the *orthogonal projection* (commonly just called the *projection* of $\mathbf{y}$ onto $\mathbf{x}$). Thus, the above discussion on approximation was for an arbitrary vector $\mathbf{y}$ by the vector $\mathbf{x}$ linearly when both of these vectors are elements of $\mathbb{V}$. Now, define the vector space $\mathbb{V}$ to be relevant to our r.v. mean-square estimation problem. Consider a sequence $\{Y, X_1, X_2, \ldots\}$ of real-valued zero mean r.v.'s with finite second moments as a vector space with each r.v. X_i considered as a vector and the inner product defined by

$$(X_i, X_j) = \mathrm{E}\{X_i X_j\} \tag{7.54}$$

and its norm defined by

$$\|X_i\| = \sqrt{(X_i, X_i)} = \sqrt{\mathrm{E}\{X_i^2\}} < \infty. \tag{7.55}$$

Now, consider the linear estimation of a r.v. Y by a r.v. X in the m.s. error sense. Then using the inner product of r.v.'s defined by (7.54), the optimum $\hat{a}$ of (7.47) is now expressible as

$$\hat{a} = \mathrm{E}\{YX\}/\mathrm{E}\{X^2\} \tag{7.56}$$

and the optimum error $\hat{e}$ must satisfy the orthogonal property of (7.52) expressible now as

$$\mathrm{E}\{\hat{e}X\} = \mathrm{E}\{(Y - \hat{a}X)X\} = 0. \tag{7.57}$$

Equation (7.57) shows the *orthogonal principle* in the theory of linear m.s. estimation. It states the optimal error $\hat{e}$ r.v. must be orthogonal to the observation r.v. X. From the orthogonal principle of (7.57), we can derive the optimum coefficient $\hat{a}$ of (7.56). The associated minimum m.s. estimation error $\varepsilon^2_{\mathrm{min}}$ is given by

$$\begin{aligned}\varepsilon^2_{\mathrm{min}} &= \mathrm{E}\{(Y - \hat{a}X)^2\} = \mathrm{E}\{Y^2\} - 2\hat{a}\mathrm{E}\{YX\} + \hat{a}^2\mathrm{E}\{X^2\} \\ &= \sigma_Y^2 - 2\frac{\mathrm{E}\{YX\}\mathrm{E}\{YX\}}{\mathrm{E}\{X^2\}} + \left[\frac{\mathrm{E}\{YX\}}{\mathrm{E}\{X^2\}}\right]^2 \mathrm{E}\{X^2\} \\ &= \sigma_Y^2 - \frac{(\mathrm{E}\{YX\})^2}{\mathrm{E}\{X^2\}}.\end{aligned} \tag{7.58}$$

In general, given real-valued zero-mean r.v.'s $\{Y, X_1, X_2, \ldots\}$, the optimum linear estimation of Y using $\{X_1, \ldots, X_n\}$ with the optimum coefficients $\{\hat{a}_1, \ldots, \hat{a}_n\}$ is denoted by

$$\hat{Y} = \hat{a}_1 X_1 + \cdots + \hat{a}_n X_n. \tag{7.59}$$

Then the optimum error $\hat{e}$ defined by

$$\hat{e} = Y - \hat{Y} = Y - (\hat{a}_1 X_1 + \cdots + \hat{a}_n X_n) \tag{7.60}$$

must satisfy

$$\mathrm{E}\{\hat{e}X_i\} = \mathrm{E}\{(Y - \hat{Y})X_i\} = 0, \ i = 1, 2, \ldots, n. \tag{7.61}$$

Equation (7.61) states the optimum error $\hat{e}$ must be orthogonal to all the $\{X_1, \ldots, X_n\}$. (7.61) is called the *general orthogonal principle* (commonly just called the *orthogonal principle*) for the estimation of Y using $\{X_1, \ldots, X_n\}$ linearly in the m.s. sense. We note, (7.61) yields n linear equations with n unknowns of the optimum coefficients $\{\hat{a}_1, \ldots, \hat{a}_n\}$. In the matrix-vector form, (7.61) becomes

$$\mathbf{R}_n \hat{\mathbf{a}}_n = \mathbf{r}_{Y\mathbf{X}_n}, \tag{7.62}$$

where random vector $\mathbf{X}_n$ used for the linear estimation is denoted by

$$\mathbf{X}_n = [X_1, \ldots, X_n]^T, \tag{7.63}$$

the optimum coefficient $\hat{\mathbf{a}}_n$ is denoted by

$$\hat{\mathbf{a}}_n = [\hat{a}_1, \ldots, \hat{a}_n]^T, \tag{7.64}$$

the autocorrelation matrix $\mathbf{R}_n$ of $\{X_1, \ldots, X_n\}$ is denoted by

$$\mathbf{R}_n = \begin{bmatrix} R_{X_1X_1} & R_{X_2X_1} & \ldots & R_{X_nX_1} \\ R_{X_1X_2} & R_{X_2X_2} & \ldots & R_{X_nX_2} \\ \vdots & \vdots & \ddots & \vdots \\ R_{X_1X_n} & R_{X_2X_n} & \ldots & R_{X_nX_n} \end{bmatrix}, \tag{7.65}$$

and the cross-correlation vector $\mathbf{r}_{Y\mathbf{X}_n}$ is denoted by

$$\mathbf{r}_{Y\mathbf{X}_n} = [\mathrm{E}\{YX_1\}, \ldots, \mathrm{E}\{YX_n\}]^T. \tag{7.66}$$

Note that the autocorrelation matrix $\mathbf{R}_n$ of (7.65) is similar to that in (7.24), except now we do not assume the $\{X_1, \ldots, X_n\}$ have autocorrelation values satisfying the wide-sense stationarity conditions in (7.24). As before, (7.62) is called the *normal equation* for the estimation of Y using $\{X_1, \ldots, X_n\}$ linearly in the m.s. sense. If $\mathbf{R}_n$ is a positive-definite matrix, then its inverse exists and we can find the optimum coefficient $\hat{\mathbf{a}}_n$ is formally given by solving

$$\hat{\mathbf{a}}_n = \mathbf{R}_n^{-1}\mathbf{r}_{Y\mathbf{X}_n}. \tag{7.67}$$

In practice, from a numerical point of view, using the inverse $\mathbf{R}_n^{-1}$ to solve for $\hat{\mathbf{a}}_n$ in (7.67) is not advisable, particularly when n is large. There are various numerically stable methods for solving a normal equation.

The m.s. estimation concepts considered above can be generalized in two directions.

1. Suppose the r.v.'s under consideration do not have zero means. Let $\{Z, W_1, \ldots, W_n\}$ have means denoted by $\mu_Z = \mathrm{E}\{Z\}$ and $\mu_{W_i} = \mathrm{E}\{W_i\}$, $i = 1, \ldots, n$, then we can define

$$Y = Z - \mu_Z,\ X_i = W_i - \mu_{W_i},\ i = 1, \ldots, n, \tag{7.68}$$

where now $\{Y, X_1, \ldots, X_n\}$ all have zero means and all the above results on linear m.s. estimation for zero-mean r.v.'s are applicable. Thus, in all the above equations, we replace each Y by $Z - \mu_Z$ and each X_i by $W_i - \mu_{W_i}$ and use the cross-correlation values of $\{Z, W_1, \ldots, W_n\}$ and $\left\{\mu_Z = \mathrm{E}\{Z\},\ \mu_{W_i} = \mathrm{E}\{W_i\},\ i = 1, \ldots, n\right\}$.

2. Suppose the r.v.'s are complex-valued. For a complex-valued vector space, the inner product satisfies

$$\begin{aligned} &(a_1\mathbf{x}_1 + a_2\mathbf{x}_2,\ \mathbf{y}) = a_1(\mathbf{x}_1, \mathbf{y}) + a_2(\mathbf{x}_2, \mathbf{y}), \\ &(\mathbf{x}, \mathbf{y}) = (\mathbf{y}, \mathbf{x})^*, \\ &(\mathbf{x}, \mathbf{x}) = \|\mathbf{x}\|^2 \geq 0;\ \|\mathbf{x}\|^2 = 0 \Leftrightarrow \mathbf{x} = 0, \end{aligned} \tag{7.69}$$

where $(\cdot)^*$ denotes the complex conjugate of the expression in $(\cdot)$. In particular,

$$\begin{aligned} (\mathbf{y}, a_1\mathbf{x}_1 + a_2\mathbf{x}_2) &= (a_1\mathbf{x}_1 + a_2\mathbf{x}_2, \mathbf{y})^* = a_1^*(\mathbf{x}_1, \mathbf{y})^* + a_2^*(\mathbf{x}_2, \mathbf{y})^* \\ &= a_1^*(\mathbf{y}, \mathbf{x}_1) + a_2^*(\mathbf{y}, \mathbf{x}_2). \end{aligned} \tag{7.70}$$

For a sequence of complex-valued r.v.'s $\{Y, X_1, X_2, \ldots\}$, the inner product and the norm are defined by

$$(X, Y) = \mathrm{E}\{XY^*\},\ \|X\| = \sqrt{(X, X)} = \sqrt{\mathrm{E}\left\{|X|^2\right\}} < \infty. \tag{7.71}$$

Example 7.3 Let X_1, X_2, and X_3 be three zero mean real-valued r.v.'s with finite variances and covariances $R_{ij} = \mathrm{E}\{X_i X_j\},\ i,\ j = 1,\ 2,\ 3.$

a. Find $\hat{a}_1$ and $\hat{a}_2$ that minimize $\varepsilon^2 = \mathrm{E}\left\{[X_3 - (a_1 X_1 + a_2 X_2)]^2\right\}$.

From the orthogonal principle,

$$\mathrm{E}\{(X_3 - \hat{a}_1 X_1 - \hat{a}_2 X_2)X_1\} = 0 \Rightarrow R_{31} = \hat{a}_1 R_{11} + \hat{a}_2 R_{21},$$

$$\mathrm{E}\{(X_3 - \hat{a}_1 X_1 - \hat{a}_2 X_2)X_2\} = 0 \Rightarrow R_{32} = \hat{a}_1 R_{12} + \hat{a}_2 R_{22}.$$

Using the matrix-vector form of the normal equation yields

$$\begin{bmatrix} R_{31} \\ R_{32} \end{bmatrix} = \begin{bmatrix} R_{11} & R_{21} \\ R_{21} & R_{22} \end{bmatrix} \begin{bmatrix} \hat{a}_1 \\ \hat{a}_2 \end{bmatrix}$$

or

$$\begin{bmatrix} \hat{a}_1 \\ \hat{a}_2 \end{bmatrix} = \mathbf{R}^{-1} \begin{bmatrix} R_{31} \\ R_{32} \end{bmatrix} = \frac{1}{R_{11}R_{22} - R_{21}^2} \begin{bmatrix} R_{22} & -R_{21} \\ -R_{21} & R_{11} \end{bmatrix} \begin{bmatrix} R_{31} \\ R_{32} \end{bmatrix}.$$

Thus,

$$\hat{a}_1 = \frac{R_{22}R_{31} - R_{21}R_{32}}{R_{11}R_{22} - R_{21}^2}, \quad \hat{a}_2 = \frac{-R_{21}R_{31} + R_{11}R_{32}}{R_{11}R_{22} - R_{21}^2}.$$

b. Find $\varepsilon^2_{\min}$.

$$\begin{aligned}
\varepsilon^2_{\min} &= \mathrm{E}\{(X_3 - \hat{a}_1 X_1 - \hat{a}_2 X_2)(X_3 - \hat{a}_1 X_1 - \hat{a}_2 X_2)\} \\
&= \mathrm{E}\{(X_3 - \hat{a}_1 X_1 - \hat{a}_2 X_2)X_3\} - \hat{a}_1 \mathrm{E}\{(X_3 - \hat{a}_1 X_1 \\
&\quad - \hat{a}_2 X_2)X_1\} - \hat{a}_2 \mathrm{E}\{(X_3 - \hat{a}_1 X_1 - \hat{a}_2 X_2)X_2\} \\
&= \mathrm{E}\{(X_3 - \hat{a}_1 X_1 - \hat{a}_2 X_2)X_3\} - 0 - 0 = R_{33} - \hat{a}_1 R_{13} - \hat{a}_2 R_{23} \\
&= \frac{R_{33}R_{11}R_{22} - R_{33}R_{21}^2 - R_{22}R_{31}^2 + R_{21}R_{32}R_{13} + R_{21}R_{31}R_{23} - R_{11}R_{32}^2}{R_{11}R_{22} - R_{21}^2} \\
&= \frac{R_{33}R_{11}R_{22} - R_{33}R_{21}^2 - R_{22}R_{31}^2 + 2R_{21}R_{32}R_{13} - R_{11}R_{32}^2}{R_{11}R_{22} - R_{21}^2}.
\end{aligned}$$

Note, in Example 7.3 we do not assume any wide-sense stationary property of the random sequences. Thus, R_{21} and R_{32} are not necessarily the same. However, $R_{21} = \mathrm{E}\{X_2 X_1\} = \mathrm{E}\{X_1 X_2\} = R_{12}$ and $R_{32} = \mathrm{E}\{X_3 X_2\} = \mathrm{E}\{X_2 X_3\} = R_{23}$. □

Example 7.4 Let $\mathbf{X} = [X_1,\ X_2]^T$ be a zero-mean Gaussian random vector with a covariance matrix given by

$$\mathbf{R} = \mathrm{E}\left\{\mathbf{X}\mathbf{X}^T\right\} = \begin{bmatrix} R_{11} & R_{12} \\ R_{21} & R_{22} \end{bmatrix}.$$

a. Find the $\hat{a}_1$ that minimizes

$$\varepsilon^2 = \mathrm{E}\left\{[X_2 - a_1 X_1]^2\right\}.$$

From the orthogonal principle,

$$\mathrm{E}\{(X_2 - \hat{a}_1 X_1) X_1\} = 0 \Rightarrow R_{21} = \hat{a}_1 R_{11},$$

or

$$\hat{a}_1 = R_{12}/R_{11}.$$

b. Show the conditional mean of X_2 given x_1 (which is the optimum non-linear m.s. estimate of X_2 based on $g(x_1)$) has the form of $\mu_{X_2|X_1} = \mathrm{E}\{X_2|x_1\} = \hat{a}_1 x_1$.

We know

$$p_{X_1}(x_1) = (2\pi R_{11})^{-1/2} \exp(-x_1^2/2R_{11}), \ -\infty < x_1 < \infty,$$

$$p_{X_1,X_2}(x_1, x_2) = ((2\pi)^2 D)^{-1/2} \exp(-(x_1^2 R_{22} - 2x_1 x_2 R_{12} + x_2^2 R_{11})/2D),$$

$$-\infty < x_1 < \infty, \ -\infty < x_2 < \infty,$$

where $D = R_{11} R_{22} - R_{12}^2$. Then

$$\begin{aligned} & p_{X_2|X_1}(x_2|x_1) \\ &= p_{X_1,X_2}(x_1, x_2)/p_{X_1}(x_1) \\ &= (2\pi D/R_{11})^{-1/2} \exp(-(x_1^2 R_{22} - 2x_1 x_2 R_{12} + x_2^2 R_{11} - (x_1^2 D/R_{11}))/2D) \\ &= (2\pi D/R_{11})^{-1/2} \exp(-((x_2 - x_1 R_{12}/R_{11})^2 R_{11} + B)/2D), \end{aligned}$$

where

$$B = x_1^2(-R_{12}^2 - (R_{11} R_{22} - R_{12}^2) + R_{11} R_{22})/R_{11} = 0.$$

Then

$$\begin{aligned} \mathrm{E}\{X_2|X_1\} &= (2\pi D/R_{11})^{-1/2} \int_{-\infty}^{\infty} x_2 \exp(-(x_2 - x_1 R_{12}/R_{11})^2/2(D/R_{11})) dx_2 \\ &= x_1 R_{12}/R_{11} = \mu_{X_2|X_1}. \end{aligned}$$

Thus, the optimum non-linear estimator of X_2 based on $g(X_1 = x_1)$ when both $\{X_1, \ X_2\}$ are jointly Gaussian is a linear estimator given by $x_1 R_{12}/R_{11}$.

c. Find $\varepsilon_{\min}^2$

$$\begin{aligned} & \varepsilon_{\min}^2 \\ &= \mathrm{E}\left\{\left(X_2 - \mu_{X_2|X_1}\right)^2\right\} \\ &= (2\pi D/R_{11})^{-1/2} \int_{-\infty}^{\infty} \left(x_2 - \mu_{X_2|X_1}\right)^2 \exp\left(-\left(x_2 - x_1 R_{12}\right)/R_{11}\right)^2 /2\left(D/R_{11}\right)) dx_2 \\ &= (2\pi D/R_{11})^{-1/2} \int_{-\infty}^{\infty} \left(x_2 - \mu_{X_2|X_1}\right)^2 \exp\left(-\left(x_2 - \mu_{X_2|X_1}\right)^2 /2(D/R_{11})\right) dx_2 \\ &= D/R_{11} = \left(R_{11} R_{22} - R_{12}^2\right)/R_{11}. \end{aligned}$$

Since the optimum non-linear estimator of X_2 based on $g(X_1 = x_1)$, when both $\{X_1, \ X_2\}$ are jointly Gaussian, is a linear estimator given by $x_1 R_{12}/R_{11} = \hat{a}_1 x_1$, then from the m.s.

linear estimator theory point of view using the orthogonal principle, its $\varepsilon^2_{\min}$ is given by

$$\varepsilon^2_{\min} = \mathrm{E}\{(X_2 - \hat{a}_1 X_1) X_2\} = R_{22} - \hat{a}_1 R_{12}$$
$$= R_{22} - (R_{12}/R_{11})/R_{12} = (R_{11}R_{22} - R_{12}^2)/R_{11} = D/R_{11}.$$

Thus, $\varepsilon^2_{\min}$ obtained from the two methods are identical. □

7.2.3 Block and recursive mean-square estimations

7.2.3.1 Block mean-square estimation of a r.v.

In (7.59–7.67), we considered the linear mean-squares estimator problem. That is, given real-valued zero-mean r.v.'s $\{Y, X_1, X_2, \ldots\}$, the optimum linear estimator of Y using $\{X_1, \ldots, X_n\}$ in the m.s. error sense with optimum coefficients $\{\hat{a}_1, \ldots, \hat{a}_n\}$ as shown in (7.59) is denoted by

$$\hat{Y} = \hat{a}_1 X_1 + \cdots + \hat{a}_n X_n. \tag{7.72}$$

Then as shown in (7.67), the optimum coefficient vector $\hat{\mathbf{a}}_n$ is given by

$$\hat{\mathbf{a}}_n = \mathbf{R}_n^{-1} \mathbf{r}_{Y\mathbf{X}_n}, \tag{7.73}$$

with the optimum estimate of $\hat{Y}$ given by

$$\hat{Y} = \hat{\mathbf{a}}_n^T \mathbf{X}_n = \mathbf{r}_{Y\mathbf{X}_n}^T \mathbf{R}_n^{-1} \mathbf{X}_n, \tag{7.74}$$

where the autocorrelation matrix $\mathbf{R}_n$ of $\mathbf{X}_n = [X_1, \ldots, X_n]^T$ is denoted by

$$\mathbf{R}_n = \begin{bmatrix} R_{X_1X_1} & R_{X_2X_1} & \cdots & R_{X_nX_1} \\ R_{X_1X_2} & R_{X_2X_2} & \cdots & R_{X_nX_2} \\ \vdots & \vdots & \ddots & \vdots \\ R_{X_1X_n} & R_{X_2X_n} & \cdots & R_{X_nX_n} \end{bmatrix}, \tag{7.75}$$

the cross-correlation vector is denoted by

$$\mathbf{r}_{Y\mathbf{X}_n} = [\mathrm{E}\{YX_1\}, \ldots, \mathrm{E}\{YX_n\}]^T. \tag{7.76}$$

Then the minimum m.s. error is given by

$$\varepsilon^2_{\min} = \sigma_Y^2 - \mathbf{r}_{Y\mathbf{X}_n}^T \mathbf{R}_n^{-1} \mathbf{r}_{Y\mathbf{X}_n}. \tag{7.77}$$

The operation needed for the solution of the optimum estimate $\hat{Y}$ in (7.74) is denoted as a *block operation*. Possible objections to the block operation may include:

(a) Need to evaluate the $n \times n$ autocorrelation matrix $\mathbf{R}_n$.
(b) Need to evaluate the $n \times n$ inverse matrix $\mathbf{R}_n^{-1}$.
(c) Need to evaluate the $1 \times n$ vector $\mathbf{r}_{Y\mathbf{X}_n}^T \mathbf{R}_n^{-1}$.
(d) If n is increased to $(n+1)$, we need to re-evaluate $\mathbf{r}_{Y\mathbf{X}_{n+1}}^T \mathbf{R}_{n+1}^{-1}$. Computations previously used to perform $\mathbf{r}_{Y\mathbf{X}_n}^T \mathbf{R}_n^{-1}$ are not used in any manner.
(e) The processing cost for $\mathbf{r}_{Y\mathbf{X}_n}^T \mathbf{R}_n^{-1} \mathbf{X}_n$ becomes large as n increases.

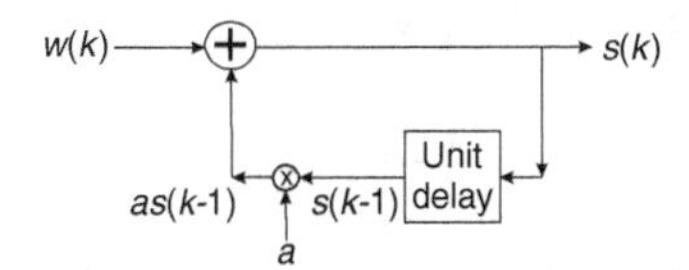

Figure 7.2 First-order autoregressive sequence $\{s(k)\}$.

7.2.3.2 Scalar recursive mean-square estimation – Kalman filtering

Consider a real-valued random signal sequence $\{S(k),\ -\infty < k < \infty\}$ generated as the output of a *linear dynamical system* driven by a real-valued white sequence $\{W(k),\ -\infty < k < \infty\}$. Then the realization of the random signal sequences is shown in Fig. 7.2 and modeled by

$$s(k) = as(k-1) + w(k),\ -\infty < k < \infty, \tag{7.78}$$

where $w(k)$ is the realization of a zero-mean white sequence of variance σ_W^2 satisfying

$$\mathrm{E}\{W(k)\} = 0,\ \mathrm{E}\{W(k)W(j)\} = \delta_{kj}\sigma_W^2,\ -\infty < j, k < \infty, \tag{7.79}$$

and a is a real number satisfying $0 < |a| < 1$. $s(k)$ is called the *state* of this linear dynamical system. Now, find the mean and covariance of $S(k)$. From (7.78), we have

$$\begin{aligned} s(k) &= as(k-1) + w(k) \\ &= a\{as(k-2) + w(k-1)\} + w(k) \\ &= a\{a[as(k-3) + w(k-2)] + w(k-1)\} + w(k) \\ &= \cdots \\ &= w(k) + aw(k-1) + a^2 w(k-2) + \cdots . \end{aligned} \tag{7.80}$$

From (7.80), since $\mathrm{E}\{W(k-i)\} = 0,\ i = 0, 1, \ldots,$ then

$$\mathrm{E}\{S(k)\} = 0. \tag{7.81}$$

From (7.80), we have

$$\begin{aligned} R_{SS}(k+l, k) &= \mathrm{E}\{S(k+l)S(k)\} \\ &= \mathrm{E}\{[W(k+l) + aW(k+l-1) + a^2 W(k+l-2) + \cdots \\ &\quad + a^l W(k) + a^{l+1} W(k-1) + \cdots] \\ &\quad \times [W(k) + aW(k-1) + a^2 W(k-2) + \cdots]\} \\ &= \mathrm{E}\{a^l W(k)W(k)\} + \mathrm{E}\{a^{l+1} W(k-1)aW(k-1)\} \\ &\quad + \mathrm{E}\{a^{l+2} W(k-2)a^2 W(k-2)\} + \cdots \\ &= a^l \sigma_W^2 + a^{l+2}\sigma_W^2 + a^{l+4}\sigma_W^2 + \cdots \\ &= \sigma_W^2 a^l \{1 + a^2 + a^4 + \cdots\}. \end{aligned} \tag{7.82}$$

Since the covariance function of $S(k)$ is given by (7.82), we have

$$R_{SS}(k+l,k) = R_{SS}(l) = \frac{\sigma_W^2 a^{|l|}}{1-a^2}, \tag{7.83}$$

which shows $S(k)$ is a WSS random sequence. Furthermore, we want to show this sequence is a *first-order autoregressive sequence* having a *memory of order* 1.

Suppose we consider the conditional probability $P(S(k)|s(k-1), s(k-2), \ldots)$. From (7.78), by conditioning on $S(k-1)$, then $as(k-1) = c$ is a constant, and the randomness of $S(k)$ in (7.78) depends only on $W(k)$. Similarly, by conditioning on $S(k-2)$, the randomness of $S(k-1)$ depends only on $W(k-1)$. Furthermore, $W(k)$ is uncorrelated to $W(k-1)$, $W(k-2), \ldots$. Thus, by conditioning on $S(k-1)$, $S(k)$ is not a function of $s(k-2)$, $s(k-3), \ldots$. That is,

$$P(S(k)|s(k-1), s(k-2), \ldots) = P(S(k)|s(k-1)). \tag{7.84}$$

Any random sequence that satisfies (7.84) is called a *Markov sequence of order one*. Such a sequence, as modeled by (7.78), has a *past memory of only one sample*.

Now, consider the filtering problem with the observed data $x(k)$ modeled by the *observation equation* of

$$x(k) = s(k) + v(k), \tag{7.85}$$

where the realization of the signal $s(k)$ is modeled by (7.78) and $v(k)$ is the realization of a zero mean white noise $V(k)$ independent of the signal and of $W(k)$. Denote the variance of the noise by $\sigma_V^2 = \mathrm{E}\{V(k)V(k)\}$.

Now, we want to form a *recursive estimator* for $s(k)$ that uses only the last estimate and the newly observed data. That is,

$$\hat{s}(k) = a_k s(k-1) + b_k x(k), \tag{7.86}$$

where a_k and b_k are to be defined. Equation (7.86) constitutes a linear and a memory one estimator. This linear estimator for $\hat{s}(k)$ modeled as having a memory of one is reasonable since we have just shown that $S(k)$ has a memory of only one past sample. If $s(k)$ and $v(k)$ are realizations of Gaussian random sequences, then the optimum linear estimator is the optimum non-linear estimator in the m.s. sense.

In order to find the optimum linear estimator of (7.86), we need to find the optimum coefficients $\hat{a}_k$ and $\hat{b}_k$ that minimize the m.s. error expression of

$$\sigma_k^2 = \mathrm{E}\{(\hat{S}(k) - S(k))^2\}. \tag{7.87}$$

From the variational or the orthogonal principle argument, we know

$$\mathrm{E}\{(\hat{S}(k) - S(k))X(j)\} = 0, \ j = k, k-1, \ldots. \tag{7.88}$$

After some algebraic manipulation using (7.88) with $j = k-1$, we obtain the optimum estimate for $\hat{a}_k$ as

$$\hat{a}_k = a(1 - \hat{b}_k). \tag{7.89}$$

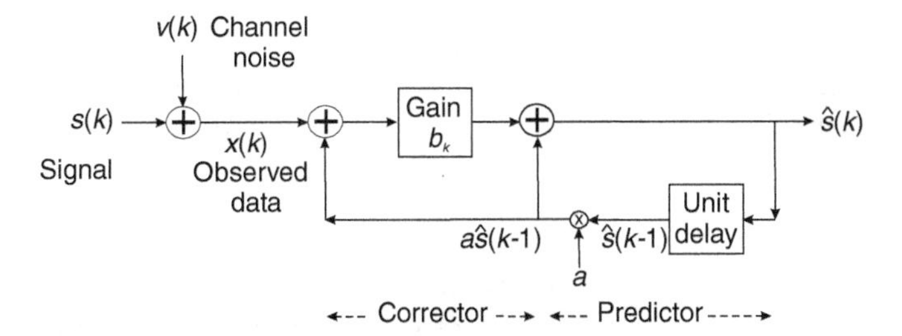

Figure 7.3 Kalman filter for a first-order scalar-valued autoregressive signal in AWN.

Thus, (7.86) shows the minimum m.s. error criterion state estimator has the recursive form of

$$\hat{s}(k) = \overbrace{a s(k-1)}^{1} + \overbrace{\hat{b}_k}^{2} \underbrace{[x(k) - a\hat{s}(k-1)]}_{3}, \; k = 1, 2, \ldots . \tag{7.90}$$

In (7.90), there are three bracketed terms, labeled 1, 2, and 3.

1. This term represents the *optimum predictor* for $s(k)$ without using the newly observed data.
2. This term represents the *Kalman gain* of the estimator.
3. This term represents the *correction term*, which is the difference of the newly observed data and the optimum predictor for $s(k)$.

Equation (7.90) is the Kalman filter for estimating a first-order scalar-valued autoregressive signal sequence in the presence of an AWN sequence [1]. The Kalman filter can be implemented as shown in Fig. 7.3. We note the predictor part of the Kalman filter is identical to the signal generator as shown in Fig. 7.2.

In order to find the optimum coefficient $\hat{b}_k$, we use (7.88) with $j = k$ and $\sigma_k^2 = \mathrm{E}\{(\hat{S}(k) - S(k))S(k)\}$. Then the Kalman gain is given by

$$\hat{b}(k) = \frac{\sigma_k^2}{\sigma_V^2} = p_k. \tag{7.91}$$

Upon some algebraic manipulations, using (7.91), (7.87), and (7.88), a non-linear recursive difference equation (i.e., commonly called the *Riccati equation*) for p_k is given by

$$p_k = \frac{\mathrm{SNR} + a^2 p_{k-1}}{\mathrm{SNR} + 1 + a^2 p_{k-1}}, \; k = 1, 2, \ldots, \tag{7.92}$$

where $\mathrm{SNR} = \sigma_W^2/\sigma_V^2$ is the SNR of the filter. The interesting point is that for known signal and channel noise statistics, the Kalman gain $\hat{b}_k = p_k$ can be computed off-line before any data $x(k)$ are received. Thus, the real-time processing part of the recursive filter as given in Fig. 7.2 is quite straightforward.

Example 7.5 Now, consider the special case when the SNR $\to \infty$. Then in (7.92), $\hat{b}_k = p_k \to 1$. But, SNR $\to \infty$ means $\sigma_V^2 \to 0$ or no channel noise is present. Thus, $x(k) \approx s(k)$. But $\hat{b}_k \to 1$ means in (7.90), $\hat{s}(k) \approx x(k) \approx s(k)$. Thus, in this idealized case, the Kalman filter is able to estimate $s(k)$ perfectly.

Example 7.6 Now, we consider the variation of p_k with k. In steady state, we can set $p_k = p_{k-1} = p$ for large k in (7.92) and solve for p. Suppose we take the SNR $= 2$ and $a^2 = 0.5$. Then from (7.92) we obtain $\hat{b}_\infty = p = 0.7016$. However, starting with $\hat{s}(0) = 0$, which is a reasonable estimate to take, all signals and noises have zero means. By using (7.87), (7.83), $a^2 = 0.5$, and SNR $= 2$, we obtain $\sigma_0^2 = \mathrm{E}\{(S(0))^2\} = R_{SS}(0) = \sigma_W^2/(1-a^2) = 2\sigma_W^2 = 2\sigma_V^2$. Then from (7.91) with $k = 0$, $p_0 = 2$. Thus, from (7.92), for $k = 1$, we obtain $\hat{b}_1 = p_1 = 0.8$. Iteratively, for $k = 2$, $\hat{b}_2 = p_2 = 0.7059$, and for $k = 3$, $\hat{b}_3 = p_3 = 0.7017$. Thus, after only three samples with $k = 3$, $\hat{b}_k$ has almost converged to $\hat{b}_\infty = p = 0.7016$. □

In practice, the scalar-valued signal dynamical equation of (7.78) and the scalar-valued observation equation of (7.85) can be generalized to form a vector-valued Kalman filtering problem. The recursive equations for the state estimator and for the Kalman gain now involve more complicated matrix-valued equations. The solutions of these equations are computationally more demanding. Kalman filtering has been used extensively in communication, control, and system identification problems. Modern aerospace, avionic, GPS, manufacturing, biological, and economic problems all use Kalman filtering to track and estimate the states of these dynamical systems.

7.3 Linear LS and LAE estimation and related robustness and sparse solutions

In all the previous mean-square estimation problems, the signal as well as the noise are modeled as random sequences. In *least-square* (LS) and *least-absolute-error* (LAE) estimations, the unknown parameters of interest to be estimated are assumed to be deterministic, but these parameters are observed in the presence of additive noises.

7.3.1 LS estimation

Example 7.7 Suppose we want to estimate a deterministic parameter θ observed ideally from N data x_i, $i = 1, \ldots, N$, modeled by

$$x_i = \theta, \; i = 1, \ldots, N. \tag{7.93}$$

From (7.93), it is clear that we need only one sample of x_i to estimate θ with no error. However, in practice due to noisy measurements, our observation data are more realistically modeled by

$$x_i = \theta + e_i,\ i = 1, \ldots, N, \tag{7.94}$$

where we may not have much information (statistical or deterministic) about the *residual* $e_i,\ i = 1, \ldots, N$. From (7.94), the residual term is given by

$$e_i = x_i - \theta,\ i = 1, \ldots, N. \tag{7.95}$$

One reasonable criterion to estimate θ is to minimize the square of the residual of

$$\varepsilon^2 = \sum_{i=1}^{N} e_i^2 = \sum_{i=1}^{N} (x_i - \theta)^2 . \tag{7.96}$$

This results in the LS estimation method as follows. The first and second partial derivatives of ε^2 w.r.t. θ are given by

$$\frac{\partial \varepsilon^2}{\partial \theta} = -2 \sum_{i=1}^{N} (x_i - \theta), \tag{7.97}$$

$$\frac{\partial^2 \varepsilon^2}{\partial \theta^2} = 2N > 0 . \tag{7.98}$$

Since ε^2 in (7.96) is a quadratic function of θ and the second partial derivative of ε^2 w.r.t. θ is positive, then by setting the first partial derivative of ε^2 w.r.t. θ to zero, the optimum estimator $\hat{\theta}_{\mathrm{LS}}$ in the least-square (LS) criterion is given by

$$\hat{\theta}_{\mathrm{LS}} = (1/N) \sum_{i=1}^{N} x_i = \bar{x}. \tag{7.99}$$

We note, $\hat{\theta}_{\mathrm{LS}}$ is just the sample mean of the observed data. □

A more general model of (7.93) is given by the simple linear regression model of

$$x_i = \alpha + \beta u_i + e_i,\ i = 1, \ldots, N, \tag{7.100}$$

where ideally the observation data should be $x_i = \alpha + \beta u_i$, but in the presence of measurement error, the observed data are given by (7.100). Under the LS criterion, the square of the residual becomes

$$\varepsilon^2 = \sum_{i=1}^{N} e_i^2 = \sum_{i=1}^{N} (x_i - \alpha - \beta u_i)^2. \tag{7.101}$$

Then using the standard variational argument, the optimum LS estimator $\hat{\alpha}_{\mathrm{LS}}$ for α and the optimum LS estimator $\hat{\beta}_{\mathrm{LS}}$ for β are given by

$$\hat{\alpha}_{\mathrm{LS}} = \bar{x} - \hat{\beta}_{\mathrm{LS}} \bar{u} \tag{7.102}$$

$$\hat{\beta}_{\mathrm{LS}} = \frac{\sum_{i=1}^{N} (x_i - \bar{x})(u_i - \bar{u})}{\sum_{i=1}^{N} (u_i - \bar{u})^2}, \tag{7.103}$$

where $\bar{x}$ is the sample mean of the x_i and $\bar{u}$ is the sample mean of the u_i given by

$$\bar{x} = (1/N)\sum_{i=1}^{n} x_i, \tag{7.104}$$

$$\bar{u} = (1/N)\sum_{i=1}^{N} u_i. \tag{7.105}$$

Example 7.8 A vehicle moves in a straight line with the observation equation given by

$$x_i = \alpha + \beta i,\ i = 1, \ldots, N. \tag{7.106}$$

Suppose $N = 10$ and the observed data are given by $x = [2, 2, 3, 4, 4, 8, 9, 10, 11, 12]^T$. Then (7.105) shows $\bar{u} = 5.5$ and (7.104) shows $\bar{x} = 6.5$. Then $\hat{\alpha}_{\mathrm{LS}} = 6.5 - 5.5\hat{\beta}_{\mathrm{LS}}$ and $\hat{\beta}_{\mathrm{LS}} = 1.254$ and thus $\hat{\alpha}_{\mathrm{LS}} = -0.39975$. □

7.3.2 Robustness to outlier (*) of LAE solution relative to LS solution

In Sections 1.6.1–1.6.4, we considered a simple linear estimation problem solving for the unknown scalar a_0 in $\mathbf{y} = \mathbf{x}\,a_0 + \mathbf{n}$, where the observed $M \times 1$ vector $\mathbf{y} = [y(1), \ldots, y(M)]^T$ and the deterministic $M \times 1$ vector $\mathbf{x} = [x(1), \ldots, x(M)]^T$ are known, and $\mathbf{n} = [n(1), \ldots, n(M)]^T$ is the realization of an $M \times 1$ noise vector. By using the LS criterion, its optimum solution $\hat{a}_0^{\mathrm{LS}}$ was readily given by the explicit formula of (1.68). A more robust solution $\hat{a}_0^{\mathrm{LAE}}$ using the LAE criterion was also considered. The solution of $\hat{a}_0^{\mathrm{LAE}}$ involves a complicated optimization procedure. The specific Example 1.7 showed the advantage of the robustness in the $\hat{a}_0^{\mathrm{LAE}}$ solution relative to the $\hat{a}_0^{\mathrm{LS}}$ solution in the presence of a large outlier perturbation in the $\mathbf{y}$ data.

Consider a generalization of the problems presented in Sections 1.6.1–1.6.4. Now, we have

$$\mathbf{y} = \mathbf{X}\mathbf{a} + \mathbf{n}, \tag{7.107}$$

where $\mathbf{y} = [y(1), \ldots, y(M)]^T$ is an observed (known) real-valued $M \times 1$ vector, $\mathbf{X} = [\mathbf{x}_1, \mathbf{x}_2, \ldots, \mathbf{x}_P]$ is a known deterministic $M \times P$ matrix, with each of its $M \times 1$ column vectors denoted by $\mathbf{x}_p = [x_{p1}, x_{p2}, \ldots, x_{pM}]^T$, $p = 1, 2, \ldots, P$, $\mathbf{n} = [n(1), \ldots, n(M)]^T$ is the realization of a $M \times 1$ zero-mean random vector, and we seek the solution of the $P \times 1$ unknown vector $\mathbf{a} = [a(1), a(2), \ldots, a(P)]^T$. Denote the residual of (7.107) by

$$\mathbf{r} = \mathbf{y} - \mathbf{X}\mathbf{a}. \tag{7.108}$$

Then under the square-error criterion, the square of the l_2 norm of the residual of (7.108) is given by

$$\varepsilon_{\mathrm{SE}}(\mathbf{a}) = \|\mathbf{r}\|_2^2 = \sum_{m=1}^{M} \left\{ \left| y(m) - \sum_{p=1}^{p} x_{pm} a(p) \right|^2 \right\} \tag{7.109}$$

$$= (\mathbf{y} - \mathbf{X}\mathbf{a})^T(\mathbf{y} - \mathbf{X}\mathbf{a}) = \mathbf{y}^T\mathbf{y} - \mathbf{y}^T\mathbf{a}\mathbf{X} - \mathbf{a}^T\mathbf{X}^T\mathbf{y} + \mathbf{a}^T\mathbf{X}^T\mathbf{x}\mathbf{a}. \tag{7.110}$$

Upon taking the gradient $\nabla_{\mathbf{a}}$ of $\varepsilon^2_{\mathrm{SE}}(\mathbf{a})$ w.r.t. $\mathbf{a}$ and setting it to zero, we obtain

$$\nabla_{\mathbf{a}}\{\varepsilon^2_{\mathrm{SE}}(\mathbf{a})\} = -2\mathbf{X}^T\mathbf{y} + 2\mathbf{X}^T\mathbf{X}\mathbf{a} = 0, \tag{7.111}$$

which yields the *normal equation* of

$$\mathbf{X}^T\mathbf{X}\mathbf{a} = \mathbf{X}^T\mathbf{y}. \tag{7.112}$$

If $\mathbf{X}$ has rank P, then the $P \times P$ matrix $(\mathbf{X}^T\mathbf{X})$ is non-singular, and the *least-square* (LS) solution $\hat{\mathbf{a}}_0^{\mathrm{LS}}$ of the normal equation of (7.111) has the explicit form of

$$\hat{\mathbf{a}}_0^{\mathrm{LS}} = \left(\mathbf{X}^T\mathbf{X}\right)^{-1}\mathbf{X}^T\mathbf{y}. \tag{7.113}$$

From (7.111) the normal equation solution $\hat{\mathbf{a}}_0^{\mathrm{LS}}$ of (7.113) shows it is a stationary point of $\varepsilon^2_{\mathrm{SE}}(\mathbf{a})$. However, since $\varepsilon^2_{\mathrm{SE}}(\mathbf{a})$ is a quadratic form in $\mathbf{a}$ and the Hessian $\nabla^2_{\mathbf{a}}\{\varepsilon_{\mathrm{SE}}(\mathbf{a})\} = 2\mathbf{X}\mathbf{X}^T$ is a positive-definite form, thus the $\hat{\mathbf{a}}_0^{\mathrm{LS}}$ solution of (7.113) is a global minimum solution of $\varepsilon^2_{\mathrm{SE}}(\mathbf{a})$.

Now, suppose we use the absolute-error criterion, then the l_1 norm of the residual of (7.108) is given by

$$\varepsilon_{\mathrm{AE}}(\mathbf{a}) = \|\mathbf{r}\|_2 = \sum_{m=1}^{M}\left\{\left|y(m) - \sum_{p=1}^{P} x_{pm}a(p)\right|\right\}. \tag{7.114}$$

Then the *least-absolute-error* (LAE) solution $\hat{\mathbf{a}}_0^{\mathrm{LAE}}$ of (7.114) is formally given by

$$\varepsilon_{\mathrm{AE}}(\hat{\mathbf{a}}_0^{\mathrm{LAE}}) = \min_{\mathbf{a}\in\mathbb{R}^P}\left\{\sum_{m=1}^{M}\left\{\left|y(m) - \sum_{p=1}^{P} x_{pm}a(p)\right|\right\}\right\}. \tag{7.115}$$

Unfortunately, the $\hat{\mathbf{a}}_0^{\mathrm{LAE}}$ solution of (7.115) requires a complicated optimization operation and does not have a simple explicit solution like that of $\hat{\mathbf{a}}_0^{\mathrm{LS}}$ in (7.113). However, the least-absolute-error criterion solution $\hat{\mathbf{a}}_0^{\mathrm{LAE}}$ can yield a more robust solution when large outliers may be present.

Example 7.9 In (7.107), consider a $M = 6$ and $P = 3$ matrix $\mathbf{X}$ given by

$$\mathbf{X} = \begin{bmatrix} 1 & 5 & 10 \\ 2 & 7 & 12 \\ 3 & 8 & 13 \\ 4 & 9 & 15 \\ 5 & 8 & 9 \\ 6 & 11 & 15 \end{bmatrix}. \tag{7.116}$$

Suppose $\mathbf{a} = [3,\ 2,\ 1]^T$, then in the absence of $\mathbf{n}$, $\mathbf{y}_0 = [23,\ 32,\ 38,\ 45,\ 40,\ 55]^T = \mathbf{X}\mathbf{a}$. Let $\mathbf{n} = [0.05402,\ 0.04725,\ -0.01577,\ -0.16591,\ -0.01742,\ -0.12226\ ]^T$ be the realization of 6 Gaussian $\mathcal{N}(0, 0.1)$ r.v.'s. Now, the observed vector in (7.107) becomes $\mathbf{y} = \mathbf{y}_0 + \mathbf{n} = \mathbf{X}\mathbf{a} + \mathbf{n}$. Using the square-error criterion of (7.110) the normal equation solution of (7.113) yields

$$\hat{\mathbf{a}}_0^{\mathrm{LAE}} = [2.8013,\ 2.2195,\ 0.9107]^T, \tag{7.117}$$

which is not very different from the ideal solution in the absence of noise given by $\mathbf{a} = [3,\ 2,\ 1]^T$. Now, using the least-absolute-error criterion solution of (7.114) yields

$$\hat{\mathbf{a}}_0^{\mathrm{LAE}} = [2.8151,\ 2.22093,\ 0.9127]^T. \tag{7.118}$$

We note that this $\hat{\mathbf{a}}^{\mathrm{LAE}}$ solution of (7.118) is also not very different from the ideal solution of $\mathbf{a} = [3,\ 2,\ 1]^T$ and in fact is also very close to the $\hat{\mathbf{a}}^{\mathrm{LS}}$ solution of (7.117). Suppose an unexpected outlier at the second component $y(2)$ of $\mathbf{y}$ due to the noise at $n(2)$, which originally had a value of 0.04725, now becomes 100 times larger with a value of 4.725. Then the least-square-error criterion solution, using (7.113), yields a new

$$\tilde{\mathbf{a}}_0^{\mathrm{LS}} = [-0.5734,\ 5.9207,\ -0.4634]^T, \tag{7.119}$$

which is very different from the non-outlier solution of $\hat{\mathbf{a}}_0^{\mathrm{LS}}$ of (7.117). This new $\tilde{\mathbf{a}}_0^{\mathrm{LS}}$ solution may be considered to be essentially useless in approximating the ideal solution of $\mathbf{a} = [3,\ 2,\ 1]^T$. By using the least-absolute-error criterion solution of (7.115) for the outlier problem, the new $\tilde{\mathbf{a}}_0^{\mathrm{LAE}}$ is given by

$$\tilde{\mathbf{a}}_0^{\mathrm{LAE}} = [2.6906,\ 2.3436,\ 0.8646]^T. \tag{7.120}$$

This new $\tilde{\mathbf{a}}_0^{\mathrm{LAE}}$ solution is fairly close to the original non-outlier solution of $\hat{\mathbf{a}}_0^{\mathrm{LAE}}$ of (7.118), and may perhaps still be useful to approximate the ideal solution of $\mathbf{a} = [3,\ 2,\ 1]^T$. This example illustrates the robustness of the optimized solution under the absolute-error criterion relative to the optimized solution under the square-error criterion in the presence of a large outlier. Of course, when the outliers become larger, the optimized absolute-error criterion solution may not be too robust either. □

It is interesting to note that the LS solution of (7.113) has the following two theoretical properties. The Gauss–Markov theorem ([2], p. 51) states that if the unknown $\mathbf{a}$ solution vector is assumed to be a random vector and if the components of the random vector $\mathbf{n}$ are assumed to be i.i.d., the LS estimator $\hat{\mathbf{a}}_0^{\mathrm{LS}}$ has the minimum variance within the class of all linear unbiased estimators. Furthermore, if the random vector $\mathbf{n}$ is assumed to be Gaussian, then the LS estimator $\hat{\mathbf{a}}_0^{\mathrm{LS}}$ is also a ML estimator. On the other hand, the LAE estimator $\hat{\mathbf{a}}_0^{\mathrm{LAE}}$ has the interesting property that if the components of the noise vector $\mathbf{n}$ in (7.107) are assumed to be i.i.d. Laplacian r.v.'s, then the LAE estimator $\hat{\mathbf{a}}_0^{\mathrm{LAE}}$ is also a ML estimator (Section 2.6 of [3]).

7.3.3 Minimization based on l_2 and l_1 norms for solving linear system of equations (*)

If we examine the model of $\mathbf{y} = \mathbf{X}\mathbf{a} + \mathbf{n}$ in (7.107), we need not use any information about the noise vector $\mathbf{n}$. Suppose the model of (7.107) assumes that the observed vector $\mathbf{y}$ is a linear combination of the P $M \times 1$ vectors $\{\mathbf{x}_1,\ \mathbf{x}_2, \ldots, \mathbf{x}_P\}$ of the $M \times P$ matrix $\mathbf{X}$, weighted by the P components of $\{a(1),\ a(2), \ldots,\ a(P)\}$ of the $P \times 1$ vector $\mathbf{a}$, plus some not-specified disturbances labeled $\mathbf{n}$. Since we claim only the $M \times 1$ $\mathbf{y}$ vector and the $M \times P$ $\mathbf{X}$ matrix are given and we want to find the $P \times 1$ unknown weighting vector solution $\mathbf{a}$, then we may just as well formulate this problem as a linear system of equations with M equations and P unknowns in the form of

$$\mathbf{y} = \mathbf{X}\mathbf{a}. \tag{7.121}$$

When $M > P$, this is an *over-determined system of equations problem*, where there are more equations than unknowns, and generally there is no solution of $\mathbf{a}$ that solves (7.121) exactly. As in (7.108), we can define the residual by $\mathbf{r} = \mathbf{y} - \mathbf{X}\mathbf{a}$, and under the square-error criterion, the square of the l_2 norm of the residual of (7.108) is then given by (7.110). If all the P column vectors of $\{\mathbf{x}_1, \mathbf{x}_2, \ldots, \mathbf{x}_p\}$ are linearly independent, then the $M \times P$ matrix $\mathbf{X}$ has rank P and then the $P \times P$ matrix $\mathbf{X}^T\mathbf{X}$ is non-singular, thus (7.121) has a least-square-error criterion solution $\hat{\mathbf{a}}_0^{\mathrm{LS}}$ given by

$$\hat{\mathbf{a}}_0^{\mathrm{LS}} = \left(\mathbf{X}^T\mathbf{X}\right)^{-1}\mathbf{X}^T\mathbf{y}, \tag{7.122}$$

which is identical to the expression of $\hat{\mathbf{a}}_0^{\mathrm{LS}}$ given by (7.113). The $\hat{\mathbf{a}}_0^{\mathrm{LS}}$ given by (7.122) yields the smallest $\varepsilon_{\mathrm{LS}}(\mathbf{a})$ of (7.110) among all possible $\mathbf{a}$ vectors.

Example 7.10 Consider $\mathbf{y} = [1.01,\ 2.03]^T$ and $\mathbf{x} = [1,\ 2]^T$. Then the $\hat{\mathbf{a}}_0^{\mathrm{LS}}$ solution based on (7.97) is given by $\hat{\mathbf{a}}_0^{\mathrm{LS}} = \left(\mathbf{x}^T\mathbf{x}\right)^{-1}\mathbf{x}^T\mathbf{y} = \left([1,\ 2][1,\ 2]^T\right)^{-1}[1,\ 2][1.01,\ 2.02]^T = 1.014$ with $\varepsilon_{\mathrm{LS}}(\hat{\mathbf{a}}_0^{\mathrm{LS}}) = 2.0000 \times 10^{-5}$. On the other hand, consider $\mathbf{a}_1 = 1.01 < \hat{\mathbf{a}}_0^{\mathrm{LS}} = 1.014 < 1.02 = \mathbf{a}_2$. Then $\varepsilon_{\mathrm{SE}}(\mathbf{a}_1) = 1.0000 \times 10^{-4} > \varepsilon_{\mathrm{SE}}(\hat{\mathbf{a}}_0^{\mathrm{LS}})$ and $\varepsilon_{\mathrm{SE}}(\mathbf{a}_2) = 2.0000 \times 10^{-4} > \varepsilon_{\mathrm{SE}}(\hat{\mathbf{a}}_0^{\mathrm{LS}})$, consistent with the property that $\mathbf{a}_0^{\mathrm{LS}}$ attains the minimum of $\varepsilon_{\mathrm{SE}}(\mathbf{a})$ among all other possible $\mathbf{a}$. □

When $M = P$, the number of equations equals the number of unknowns, then assume $\mathbf{x}$ is a square non-singular matrix with an inverse $\mathbf{x}^{-1}$, thus $\hat{\mathbf{a}}_0$ of (7.123) solves (7.121) exactly.

$$\hat{\mathbf{a}}_0 = \mathbf{X}^{-1}\mathbf{y}. \tag{7.123}$$

When $M < P$, then this is the *under-determined system of equations problem*, where there are fewer equations than unknowns, and thus there is an uncountably infinite number of possible solutions of $\mathbf{a}$ that solve (7.121) exactly.

Example 7.11 Consider the case of $M = 1$ and $P = 3$, where $\mathbf{x} = [2,\ 4,\ 6]$ and $y = [6]$. One possible solution of (7.121) is given by $\mathbf{a}_1 = [0,\ 0,\ 1]^T$, since $\mathbf{x}\mathbf{a}_1 = [2,\ 4,\ 6][0,\ 0,\ 1]^T = [6] = \mathbf{y}$. Other solutions include: $\mathbf{a}_2 = [0,\ 1.5,\ 0]^T$, since $\mathbf{x}\mathbf{a}_2 = [2,\ 4,\ 6][0,\ 1.5,\ 0]^T = [6] = y$; $\mathbf{a}_3 = [3\ 0\ 0]^T$, since $\mathbf{x}\mathbf{a}_3 = [2,\ 4,\ 6][3,\ 0,\ 0]^T = [6] = y$; $\mathbf{a}_4 = [0.5,\ 0.5,\ 0.5]^T$, since $\mathbf{x}\mathbf{a}_4 = [2,\ 4,\ 6][0.5,\ 0.5,\ 0.5]^T = [6] = y$; $\mathbf{a}_5 = (3/14)[1,\ 2,\ 3]^T$, since $\mathbf{x}\mathbf{a}_5 = (3/14)[2,\ 4,\ 6][1,\ 2,\ 3]^T = [6]$. Clearly, there is an uncountable number of possible $\mathbf{a}$ solutions satisfying (7.121). □

Table 7.1 Comparisons of the l_p, $p = 1, 2, 4,$ and 6 norms of $\{\mathbf{a}_1, \mathbf{a}_2, \mathbf{a}_3, \mathbf{a}_4\}$.

	$\mathbf{a}_1 = \begin{bmatrix} 1 \\ 0 \\ 0 \\ 0 \end{bmatrix}$	$\mathbf{a}_2 = \begin{bmatrix} 0.5 \\ \frac{\sqrt{3}}{2} \\ 0 \\ 0 \end{bmatrix}$	$\mathbf{a}_3 = \begin{bmatrix} 0.5 \\ 0.5 \\ \frac{1}{\sqrt{2}} \\ 0 \end{bmatrix}$	$\mathbf{a}_4 = \begin{bmatrix} 0.5 \\ 0.5 \\ 0.5 \\ 0.5 \end{bmatrix}$
$l_1(\cdot)$	1	1.3660	1.7071	2
$l_2(\cdot)$	1	1	1	1
$l_4(\cdot)$	1	0.8891	0.7828	0.7070
$l_6(\cdot)$	1	0.8713	0.7339	0.6300

Theorem 7.1 *Now, assume the $M \times P$ matrix $\mathbf{X}$ of (7.121) satisfies $M < P$ with rank M (i.e., all M rows of $\mathbf{X}$ are linearly independent). Then the normal equation solution $\hat{\mathbf{a}}_0$ given by (7.123)*

$$\hat{\mathbf{a}}_0 = \mathbf{X}^T \left(\mathbf{X}\mathbf{X}^T\right)^{-1} \mathbf{y} \tag{7.124}$$

yields the minimum energy (i.e., minimum square of the l_2 norm) solution among all possible solutions of the under-determined system of equations satisfying (7.121).

Proof: The Proof is given in Appendix 7.A.

Example 7.12 Consider $\mathbf{x} = [2, 4, 6]$ and $y = [6]$ of Example 7.11. Then using (7.124), we obtain $\hat{\mathbf{a}}_0 = [2, 4, 6]^T \left([2, 4, 6][2, 4, 6]^T\right)^{-1} [6] = (3/14)[1, 2, 3]^T = \mathbf{a}_5$ of Example 7.11. By direct evaluations, the energy of $\|\mathbf{a}_1\|^2 = 1$, the energy of $\|\mathbf{a}_2\|^2 = 1.5$, the energy of $\|\mathbf{a}_3\|^2 = 3$, the energy of $\|\mathbf{a}_4\|^2 = 0.75$, and the energy of $\|\hat{\mathbf{a}}_0\|^2 = \|\mathbf{a}_5\|^2 = 0.6409$. Thus, the energies of $\mathbf{a}_1$, $\mathbf{a}_2$, $\mathbf{a}_3$, and $\mathbf{a}_4$ solutions are all larger than the energy of the $\hat{\mathbf{a}}_0 = \mathbf{a}_5$ solution given by (7.124) as expected from Theorem 7.1. □

Next, consider an under-determined system of equations in which the minimum l_1 norm solution also satisfies the "*most sparseness*" condition. The *sparseness* of a vector is defined as the number of its non-zero elements. Define the l_p norm of any $N \times 1$ vector $\mathbf{a}$ as

$$\|\mathbf{a}\|_p = \left(\sum_{m=1}^{N} |a_m|^p\right)^{1/p}, \quad 0 < p \leq \infty. \tag{7.125}$$

Consider the following example with four 4×1 vectors all having the same l_2 norm but different l_p norms.

Example 7.13 Let vectors $\{\mathbf{a}_1, \mathbf{a}_2, \mathbf{a}_3, \mathbf{a}_4\}$ in Table 7.1 all have the same l_2 norm of unity value. Some of their l_p norm values are also given in Table 7.1.

Only one of the components of vector $\mathbf{a}_1$ is non-zero, only two of the components of vector $\mathbf{a}_2$ are non-zero, only three of the components of vector $\mathbf{a}_3$ are non-zero, while all four of the components of the vector $\mathbf{a}_4$ are non-zero. Thus, $\mathbf{a}_1$ (with a sparseness of 1) is more sparse than $\mathbf{a}_2$ (with a sparseness of 2); $\mathbf{a}_2$ (with a sparseness of 2) is more sparse than $\mathbf{a}_3$ (with a sparseness of 3), and $\mathbf{a}_3$ is more sparse than $\mathbf{a}_4$ (with a sparseness of 4). By using the l_1 norm, we note that $l_1(\mathbf{a}_1) < l_1(\mathbf{a}_2) < l_1(\mathbf{a}_3) < l_1(\mathbf{a}_4)$. This observation gives a hint on why we use the minimization under the l_1 norm, rather than some other l_p, $p \neq 1$, norm in the following lemma. □

Lemma 7.1 *Consider an under-determined system of equations of (7.121), with $M < P$. Among all the* **a** *solutions of (7.121), the minimum l_1 norm solution given by*

$$\min \|\mathbf{a}\|_2\,, \text{ subject to } \mathbf{y} = \mathbf{X}\mathbf{a} \tag{7.126}$$

also satisfies the "most sparseness" condition (i.e., having the least number of non-zero elements of **a***).*

Lemma 7.1 as stated above is not very precise, since some quite sophisticated technical details need to be imposed on the conditions of the matrix **X** and some restrictions need to be imposed on how small M can be relative to P, etc. We note that the minimization in (7.125) must use the l_1 norm. If the minimization uses l_p, $p \neq 1$, norm, then Lemma 7.1 is not valid.

Example 7.14 Consider $\mathbf{x} = [2,\ 4,\ 6]$ and $y = [6]$ as in Examples 7.11 and 7.12. Use the convex optimization program cvx to perform (7.126) given by:

```
%%
cvx_setup
x = [2, 4, 6]; y = [6];
cvx_begin
variable a(3);
minimize(norm(a,1));
subject to
y == x*a;
cvx_end
a
%%
```

yields

```
a =
0.0000
0.0000
1.0000
```

The minimum l_1 norm solution of $\mathbf{a} = [0,\ 0,\ 1]^T$ obtained above based on (7.126) is equal to that of $\mathbf{a}_1 = [0,\ 0,\ 1]^T$ in Example 7.11 with sparseness of 1. While $\mathbf{a}_2$ and $\mathbf{a}_3$ in Example 7.11 both also have sparseness of 1, their l_1 norms are greater than that of

$\mathbf{a}_1$. Both the solutions of $\mathbf{a}_4$ and $\mathbf{a}_5$ have l_1 norms greater than that of $\mathbf{a}_1$ and also have sparseness of 3. □

7.4 Basic properties of statistical parameter estimation

In estimation theory, the parameter of interest can be modeled as a deterministic but unknown scalar often denoted by θ, or as a r.v. Θ (with a pdf $p_\Theta(\theta)$). We also denote the realization of a r.v. parameter by θ. An *estimator* ψ is a deterministic function $\psi(g(x))$ (or $\psi(g(x_1, \ldots, x_n))$) used to process and find the value of the parameter θ by operating on some observed scalar data x (or vector data $(x_1, \ldots, x_n)$) embedded in some possibly complicated deterministic function $g(\cdot)$. Similarly, an estimator Ψ is a r.v. function $\Psi(G(X))$ (or $\Psi(G(X_1, \ldots, X_n))$) operating on some data modeled as a r.v. X (or a random vector $\mathbf{X} = (X_1, \ldots, X_n)$) embedded in some possibly complicated random function $G(\cdot)$ that incorporates the presence of random disturbances in the observations. An *estimate* denotes the deterministic value given by the estimator ψ function or the realization of the estimator Ψ function. In estimation theory, the optimum estimate of θ (optimum under some criterion) taken from the parameter space of Ω is often denoted by the symbol of a circumflex over θ (i.e., $\hat{\theta}$). Similarly, the optimum of Θ and Ψ, are denoted by $\hat{\Theta}$ and $\hat{\Psi}$, respectively. Due to historical reasons, a *statistic* is also used to denote either an estimator or an estimate. Then various classical statistical concepts such as unbiasedness, minimum variance, efficiency, consistency, sufficiency, etc. can be used to characterize the goodness of various estimators.

In statistical analysis, the collection of all elements in a sample space under consideration is often called the *population*. The actual selected r.v. is called a *sample*. Let a sample of size n from the population be taken from the r.v.'s $X_1, \ldots, X_n$. Each observed X_i has a value denoted by x_i. A *random sample of size n* is a set of n independent r.v.'s $X_1, \ldots, X_n$ drawn from a common distribution function $F(x)$. That is, $F_{X_i}(x_i) = F(x_i)$, $i = 1, \ldots, n$. An estimator of a scalar parameter θ using $X_1, X_2, \ldots, X_n$, is a statistic denoted by $\Psi(X_1, X_2, \ldots, X_n)$, while an estimate of θ using $x_1, x_2, \ldots, x_n$, is denoted by $\psi(x_1, x_2, \ldots, x_n)$.

The *sample mean* $\bar{X}$ is the estimator for the mean $\theta = \mu$ of a random sample of size n defined by

$$\bar{X} = \frac{1}{n} \sum_{i=1}^{n} X_i. \tag{7.127}$$

We will show that the sample mean possesses many useful statistical properties.

The values given by an estimator need not be close to the desired parameter θ for every realization taken from the random sample. We want to characterize the goodness of an estimator in various statistical senses. A statistic $\Psi(X_1, X_2, \ldots, X_n)$ is said to be an *unbiased estimator* for θ if

$$\mathrm{E}\{\Psi(X_1, X_2, \ldots, X_n)\} = \theta. \tag{7.128}$$

Example 7.15 From (7.127), the sample mean $\bar{X} = (1/n)\sum_{i=1}^{n} X_i$ of (7.127) is an unbiased estimator for the true mean μ since $\mathrm{E}\{\bar{X}\} = (1/n)\sum_{i=1}^{n} \mathrm{E}\{X_i\} = (n\mu/n) = \mu$. □

Consider the variance of the sample mean $\bar{X}$ where the r.v. X has a mean of μ and variance σ^2. Then

$$\begin{aligned}\mathrm{Var}\{\bar{X}\} &= \mathrm{E}\{\bar{X}^2\} - \left(\mathrm{E}\{\bar{X}\}\right)^2 = \mathrm{Var}\left\{\frac{1}{n}\sum_{i=1}^{n}\{X_i\}\right\} \\ &= \sum_{i=1}^{n} \mathrm{Var}\left\{\frac{X_i}{n}\right\} = n\mathrm{Var}\left\{\frac{X_i}{n}\right\} = \frac{n\mathrm{Var}\{X\}}{n^2} = \frac{\sigma^2}{n}.\end{aligned} \quad (7.129)$$

As $n \to \infty$, the variance of the sample mean converges to zero. Thus, the sample mean is a good estimator for the true mean using a large number of samples.

Example 7.16 Let the sample variance estimator S^2 be defined by

$$S^2 = \frac{1}{n-1}\sum_{i=1}^{n}(X_i - \bar{X})^2. \quad (7.130)$$

Then S^2 is an unbiased estimator of σ^2, since

$$\begin{aligned}\mathrm{E}\{S^2\} &= \mathrm{E}\left\{\frac{\sum_{i=1}^{n}(X_i - \bar{X})^2}{n-1}\right\} = \frac{1}{n-1}\mathrm{E}\left\{\sum_{i=1}^{n}\left[(X_i - \mu) - (\bar{X} - \mu)\right]^2\right\} \\ &= \frac{1}{n-1}\mathrm{E}\left\{\sum_{i=1}^{n}\left[(X_i - \mu)^2 - 2(X_i - \mu)(\bar{X} - \mu) + (\bar{X} - \mu)^2\right]\right\} \\ &= \frac{1}{n-1}\left\{\sum_{i=1}^{n}\mathrm{E}\{(X_i - \mu)^2\} - 2\sum_{i=1}^{n}\mathrm{E}\{(X_i - \mu)(\bar{X} - \mu)\}\right. \\ &\quad \left. + \sum_{i=1}^{n}\mathrm{E}\{(\bar{X} - \mu)^2\}\right\} \\ &= \frac{1}{n-1}\left\{\sum_{i=1}^{n}\mathrm{E}\{(X_i - \mu)^2\} - 2\sum_{i=1}^{n}\mathrm{E}\left\{(X_i - \mu)\left(\frac{\sum_{j=1}^{n}(X_j - \mu)}{n}\right)\right\}\right. \\ &\quad \left. + \sum_{i=1}^{n}\mathrm{E}\{(\bar{X} - \mu)^2\}\right\} \\ &= \frac{1}{n-1}\left\{n\sigma^2 - 2n\frac{\sigma^2}{n} + n\frac{\sigma^2}{n}\right\} = \frac{1}{n-1}\left\{n\sigma^2 - 2\sigma^2 + \sigma^2\right\} \\ &= \frac{1}{n-1}\{n\sigma^2 - \sigma^2\} = \sigma^2.\end{aligned} \quad (7.131)$$

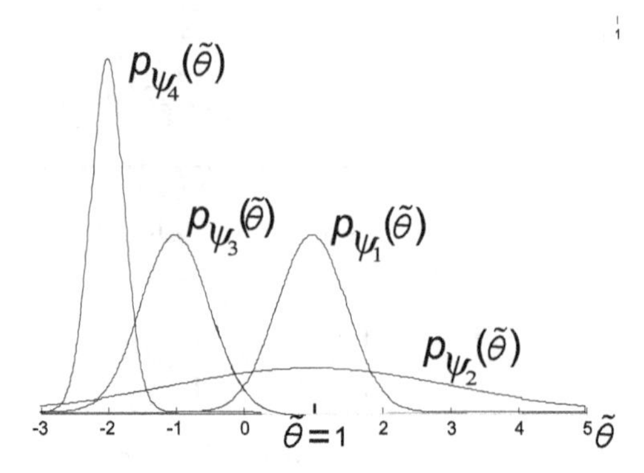

Figure 7.4 Four estimators for estimating the parameter $\theta = 1$.

We note, since S^2 defined in (7.130) is an unbiased estimator, then clearly the intuitively reasonable estimator using a normalization of n given by $(1/n)\sum_{i=1}^{n}(X_i - \bar{X})^2$ is a biased estimator. Of course, there is not much difference when n is large whether we normalize the sum in (7.130) by $(n-1)$ or by n. □

For a good estimator, we want the estimator Ψ to be close to θ in some statistical sense. If Ψ is an unbiased estimator, then $\mathrm{E}\{\Psi(X_1, X_2, \ldots, X_n)\} = \theta$, and the Chebyshev inequality shows

$$\mathrm{P}\left(|\Psi - \theta|^2 \geq \varepsilon^2\right) \leq \mathrm{Var}\{\Psi\}/\varepsilon^2. \tag{7.132}$$

Thus, a small estimator variance $\mathrm{Var}\{\Psi\}$ guarantees the estimator Ψ is close to θ in probability. Then a good estimator is one with a small variance. A possible comparison of the goodness of estimators is to compare their variances. An estimator Ψ_1 is a more *efficient* estimator for θ than an estimator Ψ, if Ψ_1 and Ψ_2 are both unbiased estimators of θ and also satisfying $\mathrm{Var}\{\Psi_1\} < \mathrm{Var}\{\Psi_2\}$. Figure 7.4 shows four estimators and their associated pdfs for the estimation of the parameter $\theta = 1$. Both $\Psi_1(\cdot)$ and $\Psi_2(\cdot)$ are unbiased (i.e., their means are at $\theta = 1$) and the variance of $\Psi_1(\cdot)$ is smaller than the variance of $\Psi_2(\cdot)$. Thus, $\Psi_1(\cdot)$ is a more efficient estimator than $\Psi_2(\cdot)$. On the other hand, $\Psi_3(\cdot)$ appears to have the same variance as $\Psi_1(\cdot)$, but appears to have a mean of $\theta = -1$. In comparing $\Psi_3(\cdot)$ to $\Psi_2(\cdot)$, even though $\Psi_3(\cdot)$ is biased, it is more concentrated in the neighborhood about the true $\theta = 1$ value than $\Psi_2(\cdot)$ with such a large variance. Thus, in some sense perhaps $\Psi_3(\cdot)$ is a more useful estimator than $\Psi_2(\cdot)$. On the other hand, $\Psi_4(\cdot)$ has a much smaller variance than all the other three estimators, but its bias is quite large to make it not a very desirable estimator for estimating the true $\theta = 1$ value.

Consider a linear estimator Ψ of the form

$$\Psi = \sum_{i=1}^{n} a_i X_i. \tag{7.133}$$

A linear estimator with real non-negative valued weights $\sum_{i=1}^{n} a_i = 1$ is an unbiased estimator since $\mathrm{E}\{\Psi\} = \sum_{i=1}^{n} a_i \mathrm{E}\{X_i\} = \sum_{i=1}^{n} a_i \mu = \mu$.

Example 7.17 The sample mean $\bar{X}$ is the *most efficient* (i.e., minimum variance) linear estimator with real non-negative-valued weights satisfying $\sum_{i=1}^{n} a_i = 1$. Consider $\alpha = \text{Var}\{\bar{X}\} = \text{Var}\{\sum_{i=1}^{n} a_i X_i\} = \sum_{i=1}^{n} a_i^2 \text{Var}\{X_i\} = \sum_{i=1}^{n} a_i^2 \sigma^2$ and denote $\beta = 1 - \sum_{i=1}^{n} a_i$. In order to minimize α subject to $\beta = 0$, we use the Lagrange multiplier method. Denote $\mathbf{a} = [a_1, \ldots, a_n]^T$ and consider

$$f(\mathbf{a}) = \sigma^2 \sum_{i=1}^{n} a_i^2 + \lambda \left(1 - \sum_{i=1}^{n} a_i \right). \quad (7.134)$$

Then set

$$\left. \frac{\partial f(\mathbf{a})}{\partial a_k} \right|_{a_k = \hat{a}_k} = 2\sigma^2 \hat{a}_k - \lambda = 0, \; k = 1, \ldots, n. \quad (7.135)$$

Solving $\hat{a}_k$ in (7.135) yields

$$\hat{a}_k = \frac{\lambda}{2\sigma^2}, \; k = 1, \ldots, n. \quad (7.136)$$

Thus, $\sum_{i=1}^{n} \hat{a}_i = 1 = \frac{n\lambda}{2\sigma^2} \Rightarrow \lambda = \frac{2\sigma^2}{n}$.

Then (7.136) yields

$$\hat{a}_k = \frac{1}{n}, \; k = 1, \ldots, n. \quad (7.137)$$

From (7.134) and (7.135), we also note

$$\frac{\partial^2 f(\mathbf{a})}{\partial a_k^2} = 2\sigma^2 > 0, \; k = 1, \ldots, n. \quad (7.138)$$

Thus, the solution in (7.135) is a local minimum. However, since $f(\mathbf{a})$ in (7.134) is a quadratic function of the a_k's, a local minimum is a global minimum. This shows a linear estimator of the form of (7.133) with $\hat{a}_k = 1/n$, $k = 1, \ldots, n$, yields a sample mean. Therefore, the sample mean has the smallest estimation variance among all linear estimators with real non-negative valued weights $\sum_{i=1}^{n} a_i = 1$. □

In general, as the number n increases, we want the estimator $\Psi(X_1, X_2, \ldots, X_n)$ of the parameter θ to be *consistent* if Ψ approaches θ in probability. That is,

$$\lim_{n \to \infty} \mathrm{P}(|\Psi(X_1, X_2, \ldots, X_n) - \theta| \geq \varepsilon) = 0 \quad (7.139)$$

for any $\varepsilon > 0$.

Example 7.18 Consider a population with a r.v. X having a finite mean μ and a finite variance σ^2. Then the sample mean $\bar{X}$ is a consistent estimator of the mean μ. Equation (7.129) shows $\text{Var}\{\bar{X}\} = \sigma^2/n$. Then by the Chebyshev Inequality, we have

$$\mathrm{P}\left(|\bar{X} - \mu| \geq \varepsilon\right) \leq \sigma^2/(n\varepsilon^2). \quad (7.140)$$

Thus, as $n \to \infty$, (7.140) yields (7.139) and $\bar{X}$ is a consistent estimator of μ. □

Indeed, for any unbiased estimator Ψ of θ, if $\lim_{n\to\infty} \mathrm{Var}\{\Psi(X_1, X_2, \ldots, X_n)\} = 0$, then with the Chebyshev Inequality of (7.140), Ψ is a consistent estimator of θ.

An estimator of a parameter θ is said to be *sufficient* (denoted by Ψ_S) if Ψ_S contains all the information about θ and no other estimator computed from the same outcomes $\{x_1, x_2, \ldots, x_n\}$ can provide additional information about θ. Let $p(x;\theta)$ be the value of the pdf of the r.v. X with a deterministic parameter θ. We use the notation that $p(\cdot;\theta)$ is defined as a *pdf* with parameter θ, while $p(x;\cdot)$ for a given x is a function defined on the parameter space of $\theta \in \Omega$. $p(x;\cdot)$ is called the *likelihood function* for an observed $X = x$. Of course, if we observed $\{x_1, x_2, \ldots, x_n\}$, then the likelihood function has the form of $p(x_1, x_2, \ldots, x_n;\cdot)$.

Theorem 7.2 ***(Factorization Theorem).*** *An estimator ψ_S is sufficient for the parameter θ if and only if the likelihood function has the form of*

$$p(x_1, x_2, \ldots, x_n;\theta) = f(\psi_S;\theta)h(x_1, x_2, \ldots, x_n), \tag{7.141}$$

where $f(\psi_S;\theta)$ is a pdf of ψ_S and depends on $\{x_1, x_2, \ldots, x_n\}$ only through $\psi_S = \psi_S(x_1, x_2, \ldots, x_n)$ and $h(x_1, x_2, \ldots, x_n)$ does not depend on θ. We note, (7.141) can be rephrased as follows. ψ_S is a sufficient statistic for θ if and only if the pdf of $\{X_1, X_2, \ldots, X_n\}$ conditioned on $\psi_S(x_1, x_2, \ldots, x_n)$ is not a function of θ. That is,

$$\begin{aligned} p(x_1, x_2, \ldots, x_n|\psi_S;\theta) &= p(x_1, x_2, \ldots, x_n, \psi_S;\theta)/f(\psi_S;\theta) \\ &= p(x_1, x_2, \ldots, x_n;\theta)/f(\psi_S;\theta) = h(x_1, x_2, \ldots, x_n). \end{aligned} \tag{7.142}$$

Example 7.19 Consider n independent Gaussian r.v.'s $\{X_1, X_2, \ldots, X_n\}$ with $\mathcal{N}(\theta, \sigma^2)$. Then the sample mean $\bar{X}$ is a sufficient statistic for the parameter $\theta = \mu$ of the r.v. X where the variance σ^2 is assumed to be known. From Example 7.15 and (7.128), recall that $\mathrm{E}\{\bar{X}\} = \mu$, $\mathrm{Var}\{\bar{X}\} = \sigma^2/n$, and $\bar{X}$ is a Gaussian r.v. Thus, the pdf of $\bar{X}$ is given by

$$f(\bar{x};\theta) = \sqrt{\frac{n}{2\pi\sigma^2}} \exp\left(\frac{-n(\bar{x}-\theta)^2}{2\sigma^2}\right), \quad -\infty < \bar{x} < \infty. \tag{7.143}$$

The pdf of $\{X_1, X_2, \ldots, X_n\}$ for the given $\theta = \mu$ has the form of

$$\begin{aligned} &p(x_1, x_2, \ldots, x_n;\theta) \\ &= \frac{1}{\left(2\pi\sigma^2\right)^{n/2}} \exp\left(\frac{-1}{2\sigma^2}\sum_{i=1}^{n}(x_i-\theta)^2\right), \quad -\infty < x_i < \infty,\ i = 1, \ldots, n. \end{aligned} \tag{7.144}$$

However, we note

$$\sum_{i=1}^{n}(x_i-\theta)^2 = \sum_{i=1}^{n}((x_i-\bar{x})+(\bar{x}-\theta))^2 = \sum_{i=1}^{n}(x_i-\bar{x})^2 + n(\bar{x}-\theta)^2. \tag{7.145}$$

By using (7.145) in (7.144), we obtain

$$p(x_1, x_2, \ldots, x_n; \theta)$$
$$= \left[\sqrt{\frac{n}{2\pi\sigma^2}} \exp\left(\frac{-n(\bar{x}-\theta)^2}{2\sigma^2} \right) \right] \left[\frac{1}{\sqrt{n}(2\pi\sigma^2)^{(n-1)/2}} \exp\left(\frac{-1}{2\sigma^2} \sum_{i=1}^{n} (x_i - \bar{x})^2 \right) \right]. \tag{7.146}$$

Thus, (7.146) satisfies (7.141) of the Factorization Theorem where $f(\bar{x};\theta)$ is given by (7.143) and

$$h(x_1, x_2, \ldots, x_n) = \left[\frac{1}{\sqrt{n}(2\pi\sigma^2)^{(n-1)/2}} \exp\left(\frac{-1}{2\sigma^2} \sum_{i=1}^{n} (x_i - \bar{x})^2 \right) \right] \tag{7.147}$$

does not depend explicitly on θ. □

7.4.1 Cramér–Rao Bound

For an unbiased estimator, earlier we stated that a small estimation variance is desirable. The question is how small can the variance be? The *Cramér–Rao Bound* (CRB) answers this question.

Theorem 7.3 ***Cramér–Rao Bound*** *Let $\Psi(\mathbf{X}, \boldsymbol{\theta}) = \boldsymbol{\Psi}$ be an $m \times 1$ random unbiased estimator of the $m \times 1$ deterministic parameter $\boldsymbol{\theta} = [\theta_1, \ldots, \theta_m]^T$ and the observation is modeled by an $n \times 1$ random vector $\mathbf{X} = [X_1, \ldots, X_n]^T$ with a pdf $p_{\mathbf{X}}(\mathbf{x};\boldsymbol{\theta}) = p(\mathbf{x})$. Then under the regularity condition, the estimation covariance of $\boldsymbol{\Psi}$ is lower bounded by*

$$E_{\mathbf{X}}\{(\boldsymbol{\Psi} - \boldsymbol{\theta})(\boldsymbol{\Psi} - \boldsymbol{\theta})^T\} \geq \mathbf{I}_{\mathrm{F}}^{-1}, \tag{7.148}$$

where $\mathbf{I}_{\mathrm{F}}$ is the $m \times m$ non-singular Fisher information matrix given by

$$\mathbf{I}_{\mathrm{F}} = \mathrm{E}_{\mathbf{X}}\{[\nabla_{\boldsymbol{\theta}}\{\ln p(\mathbf{X})\}][\nabla_{\boldsymbol{\theta}}\{\ln p(\mathbf{X})\}]^T\} \tag{7.149a}$$
$$= -\mathrm{E}_{\mathbf{X}}\{\nabla_{\boldsymbol{\theta}}^2\{\ln p(\mathbf{X})\}\}. \tag{7.149b}$$

Proof: The Proof is given in Appendix 7.B (pp. 93–96 of [42]).

Lemma 7.2 *In Theorem 7.3, consider a deterministic scalar parameter with $m = 1$ denoted by θ and let the n elements of $\mathbf{X}$ be i.i.d. with a common pdf denoted by $p(x;\theta)$. Then the unbiased scalar random estimator $\Psi(X, \theta) = \Psi$ has a minimum estimation variance given by*

$$\mathrm{E}_X\left\{(\Psi - \theta)^2\right\} \geq \frac{1}{n \int_{-\infty}^{\infty} \left(\frac{\partial \ln p(x;\theta)}{\partial \theta} \right)^2 dx}. \tag{7.150}$$

Example 7.20 Let $\bar{X}$ be the sample mean from n independent Gaussian r.v.'s $\{X_1, X_2, \ldots, X_n\}$ with $\mathcal{N}(\theta, \sigma^2)$. Assume σ^2 is known. We want to show the variance of $\bar{X}$ satisfies the CRB. The pdf of X satisfies the following properties:

$$p(x;\theta) = \frac{1}{\sqrt{2\pi\sigma^2}} \exp\left(\frac{-(x-\theta)^2}{2\sigma^2}\right), \quad (7.151)$$

$$\ln(p(x;\theta)) = \frac{-\ln(2\pi)}{2} - \ln(\sigma) - \frac{(x-\theta)^2}{2\sigma^2}, \quad (7.152)$$

$$\frac{\partial \ln(p(x;\theta))}{\partial\theta} = \frac{x-\theta}{\sigma^2}, \quad (7.153)$$

$$\begin{aligned}\int \left(\frac{\partial \ln(p(x;\theta))}{\partial\theta}\right)^2 p(x;\theta)dx &= \mathrm{E}\left\{\left(\frac{X-\theta}{\sigma^2}\right)^2\right\} \\ &= \frac{1}{\sigma^4}\mathrm{E}\left\{(X-\theta)^2\right\} = \frac{\sigma^2}{\sigma^4} = \frac{1}{\sigma^2}. \end{aligned} \quad (7.154)$$

But using (7.152) on the r.h.s. of the CRB in (7.148) shows

$$\frac{1}{n\displaystyle\int \left(\frac{\partial \ln(p(x;\theta))}{\partial\theta}\right)^2 p(x;\theta)dx} = \frac{\sigma^2}{n} = \mathrm{Var}\left\{\bar{X}\right\} \quad (7.155)$$

and the equality of the CRB in (7.148) is attained. □

Lemma 7.3 *If ψ_S is an unbiased estimator of θ, then Ψ_S is the most efficient estimator (i.e., attains the equality of the* CRB*) if and only if the following two conditions are both satisfied:*

a. ψ_S is a sufficient statistic;
b. The $f(\psi_S;\theta)$ function in the definition of a sufficient statistic in the Factorization Theorem of (7.141) satisfies

$$\partial\ln(f(\psi_S;\theta))/\partial\theta = k(\psi_S - \theta), \text{ for } f(\psi_S;\theta) > 0, \quad (7.156)$$

where k does not depend on θ.

Example 7.21 Verify the results in Example 7.20 by using the results from Lemma 7.3. The pdf of $\bar{X}$ given in (7.143) is restated below

$$f(\bar{x};\theta) = \sqrt{\frac{n}{2\pi\sigma^2}} \exp\left(\frac{-n(\bar{x}-\theta)^2}{2\sigma^2}\right), -\infty < \bar{x} < \infty.$$

Then we have

$$\ln(f(\bar{x};\theta)) = \ln\left(\sqrt{\frac{n}{2\pi\sigma^2}}\right) - \frac{n(\bar{x}-\theta)^2}{2\sigma^2},$$

$$\frac{\partial\ln(f(\bar{x};\theta))}{\partial\theta} = \frac{n(\bar{x}-\theta)}{\sigma^2}, \quad (7.157)$$

where we can take $k = n/\sigma^2$, which is not a function of θ as required by Lemma 7.3. Since $\bar{X}$ is a sufficient statistic for the parameter $\theta = \mu$ from Example 7.20 and (7.157) satisfies (7.156), thus both conditions of Lemma 7.3 are satisfied. Thus, $\bar{X}$ attains the equality of the CRB. □

Example 7.22 Consider $x(t) = A\cos(2\pi f t + \phi) + n(t)$, $0 \leq t \leq T$, where $n(t)$ is the realization of a zero-mean WGN with two-sided spectral density of $N_0/2$, and A, f, and ϕ are deterministic parameters.

Case a. Assume f and ϕ are known and we want to find the CRB for the unbiased estimator of the parameter $\theta = A$. The conditional pdf $p(x(t)|A)$ of $\{x(t), 0 \leq t \leq T\}$ conditioned on A has the form of

$$p(x(t)|A) = C\exp\left\{-(1/N_0)\int_0^T (x(t) - A\cos(2\pi f t + \phi))^2\,dt\right\}, \tag{7.158}$$

where C is a constant not a function of A and

$$\frac{\partial \ln p(x(t)|A)}{\partial A} = \frac{-2}{N_0}\int_0^T (x(t) - A\cos(2\pi f t + \phi))\cos(2\pi f t + \phi)dt \tag{7.159a}$$

$$= \frac{2}{N_0}\int_0^T n(t)\cos(2\pi f t + \phi)dt. \tag{7.159b}$$

Equating (7.159a) to zero and solving for the estimator $\hat{A}$ yields

$$\hat{A} = \frac{\int_0^T x(t)\cos(2\pi f t + \phi)dt}{\int_0^T \cos^2(2\pi f t + \phi)dt} = (2/T)\int_0^T x(t)\cos(2\pi f t + \phi)dt. \tag{7.160}$$

For the estimator $\hat{A}$ of (7.160),

$$\mathrm{E}\left\{\hat{A}\right\} = (2/T)A\int_0^T \cos(2\pi f t + \phi)\cos(2\pi f t + \phi)dt = (2/T)A(T/2) = A, \tag{7.161}$$

and thus it is an unbiased estimator. Furthermore,

$$\begin{aligned}
&\mathrm{E}\left\{\left(\frac{\partial \ln \mathrm{p}(x(t)|A)}{\partial A}\right)^2\right\} \\
&= \left(\frac{2}{N_0}\right)^2 \mathrm{E}\left\{\left(\int_0^T n(t)\cos(2\pi f t + \phi)dt\right)\left(\int_0^T n(\tau)\cos(2\pi f \tau + \phi)d\tau\right)\right\} \\
&= \left(\frac{2}{N_0}\right)^2 \frac{N_0}{2}\int_0^T\int_0^T \delta(t-\tau)\cos(2\pi f t + \phi)\cos(2\pi f \tau + \phi)d\tau dt \\
&= \left(\frac{2}{N_0}\right)\int_0^T \cos^2(2\pi f t + \phi)dt = \frac{T}{N_0}.
\end{aligned} \tag{7.162}$$

Thus, the CRB for the unbiased estimator $\hat{A}$ of A is given by

$$\mathrm{E}_X\left\{(\Psi - A)^2\right\} \geq \frac{1}{\mathrm{E}_X\left\{\left(\dfrac{\partial \ln \mathrm{p}\,(x(t)|A)}{\partial A}\right)^2\right\}} = \frac{N_0}{T}\,. \tag{7.163}$$

Case b. Assume A and f are known and we want to find the CRB for the unbiased estimator of the parameter $\theta = \phi$. Now, we have

$$\frac{\partial \ln \mathrm{p}\,(x(t)|A)}{\partial \phi} = \frac{2A}{N_0}\int_0^T (x(t) - A\cos(2\pi f t + \phi))\sin(2\pi f t + \phi)dt \tag{7.164a}$$

$$= \frac{2A}{N_0}\int_0^T n(t)\sin(2\pi f t + \phi)dt. \tag{7.164b}$$

Equating (7.164a) to zero and solving for the estimator $\hat{\phi}$ yields

$$\int_0^T x(t)\sin(2\pi f t + \hat{\phi})dt = A\int_0^T \cos(2\pi f t + \hat{\phi})\sin(2\pi f t + \hat{\phi})dt\,. \tag{7.165}$$

Since $\cos(2\pi f t + \hat{\phi})\sin(2\pi f t + \hat{\phi}) = (1/2)\sin(2(2\pi f t + \hat{\phi}))$ and $\int_0^T \sin(2(2\pi f t + \hat{\phi}))dt = -\cos(2(2\pi f t + \hat{\phi}))/(4\pi f)|_0^T = 0$, for $f > 0$ and $fT \gg 1$, then (7.165) becomes

$$\int_0^T x(t)\sin(2\pi f t + \hat{\phi})dt = 0\,. \tag{7.166}$$

From $\sin(2\pi f t + \hat{\phi}) = \sin(2\pi f t)\cos(\hat{\phi}) + \cos(2\pi f t)\sin(\hat{\phi})$, then (7.166) becomes

$$\cos(\hat{\phi})\int_0^T x(t)\sin(2\pi f t)dt = -\sin(\hat{\phi})\int_0^T x(t)\cos(2\pi f t)dt. \tag{7.167}$$

$$-\tan(\hat{\phi}) = \frac{-\sin(\hat{\phi})}{\cos(\hat{\phi})} = \frac{\int_0^T x(t)\sin(2\pi f t)dt}{\int_0^T x(t)\cos(2\pi f t)dt} = \frac{e_{\mathrm{s}}}{e_{\mathrm{c}}}, \tag{7.168}$$

or

$$\hat{\phi} = -\tan^{-1}\left\{e_{\mathrm{s}}/e_{\mathrm{c}}\right\}. \tag{7.169}$$

From (7.169) and (7.168), it is clear that $\hat{\phi}$ is a non-linear function of $x(t)$. It is also clear that $x(t) = A\cos(2\pi f t + \phi) + n(t)$ depends on ϕ in a non-linear manner. However, under high SNR conditions, $\hat{\phi}$ should be near ϕ with high probability. Thus, consider a first-order Taylor series linear approximation ([56, p. 408]) of $x(t, \phi)$ about $\phi = \hat{\phi}$ given by

$$x(t;\phi) = x(t;\hat{\phi}) + (\phi - \hat{\phi})\left.\frac{\partial x(t;\phi)}{\partial \phi}\right|_{\phi=\hat{\phi}}\,. \tag{7.170}$$

Apply (7.170) to (7.166) to yield

$$\int_0^T x(t;\hat{\phi})\sin(2\pi ft+\phi)dt = (\hat{\phi}-\phi)\int_0^T A\sin(2\pi ft+\hat{\phi})\sin(2\pi ft+\phi)dt$$
$$\simeq (\hat{\phi}-\phi)\int_0^T A\sin^2(2\pi ft+\phi)dt = (\hat{\phi}-\phi)AT/2 \tag{7.171}$$

or

$$(\hat{\phi}-\phi) = (2/AT)\int_0^T x(t;\hat{\phi})\sin(2\pi ft+\phi)dt. \tag{7.172}$$

Then $\hat{\phi}$ is an unbiased estimator since

$$\mathrm{E}\left\{(\hat{\phi}-\phi)\right\} = (2/AT)\int_0^T A\cos(2\pi ft+\hat{\phi})\sin(2\pi ft+\phi)dt \approx 0. \tag{7.173}$$

From (7.164b), we obtain

$$\mathrm{E}\left\{\left(\frac{\partial \ln \mathrm{p}\left(x(t)|A\right)}{\partial \phi}\right)^2\right\} = \left(\frac{2A^2}{N_0}\right)\int_0^T \sin^2(2\pi ft+\phi)dt = \frac{A^2T}{N_0}. \tag{7.174}$$

Thus, the CRB for the unbiased estimator $\hat{\phi}$ of ϕ, with the approximation of (7.170), is given by

$$\mathrm{E}_X\left\{(\hat{\phi}-\phi)^2\right\} \geq \frac{1}{\mathrm{E}_X\left\{\left(\frac{\partial \ln \mathrm{p}\left(x(t)|\phi\right)}{\partial \phi}\right)^2\right\}} = \frac{N_0}{A^2T}. \tag{7.175}$$

7.4.2 Maximum likelihood estimator

Recall in the discussion on sufficient statistic, we denote $p(x;\theta)$ to be the value of a pdf of a r.v. X with a deterministic parameter having the value of θ. We use the notation that $p(\cdot;\theta)$ is defined as a pdf with a parameter, while $p(x;\cdot)$ is a function on the parameter space. $p(x;\cdot)$ is called the likelihood function for an observed $X=x$. To avoid possible ambiguity, some people denote the value of the likelihood function by $L(x;\theta)=p(x;\theta)$. We note, for a fixed θ, since $p(x;\theta)$ is a pdf, then $\int p(x;\theta)dx = 1$. However, for a fixed x, $L(x;\theta)$ as a function of θ is not a pdf and its integral over θ is generally not equal to 1,

$$\int p(x;\theta)d\theta \neq 1. \tag{7.176}$$

The *maximum likelihood (ML)* estimator $\hat{\Psi}(X)$ is the estimator in the space of admissible estimators $\mathbb{P}$, such that for an observed $X=x$, $\hat{\Psi}$ satisfies

$$p(x;\hat{\Psi}(X)) = \max_{\Psi(X)\in\mathbb{P}} p(x;\Psi(X)). \tag{7.177}$$

For n independent observations with values $\{x_1, \ldots, x_n\}$, (7.177) generalizes to

$$p(x_1, \ldots, x_n; \hat{\psi}(X_1, \ldots, X_n)) = \max_{\Psi(X_1,\ldots,X_n)\in\mathbb{P}} p(x_1, \ldots, x_n; \Psi(X_1, \ldots, X_n)). \quad (7.178)$$

The intuition behind the ML estimation method is that $\hat{\Psi}(X_1, \ldots, X_n)$ is that value of θ that makes the observed values $\{x_1, \ldots, x_n\}$ most likely to occur.

Example 7.23 Consider a r.v. $X \sim \mathcal{N}(\theta, \sigma^2)$. Suppose we observe the r.v. taking the value of $x = 1.3$. What is the ML estimate of this parameter θ? The pdf of X is given by

$$p(x;\theta) = \frac{1}{\sqrt{2\pi\sigma^2}} \exp\left(\frac{-(x-\theta)^2}{2\sigma^2}\right). \quad (7.179)$$

Given the observed value of $x = 1.3$, the likelihood function is given by

$$L(1.3;\theta) = p(1.3;\theta) = \frac{1}{\sqrt{2\pi\sigma^2}} \exp\left(\frac{-(1.3-\theta)^2}{2\sigma^2}\right). \quad (7.180)$$

Then what θ maximizes $p(1.3, \theta)$? Since a Gaussian pdf in (7.179) has a peak at the value of $\theta = x = 1.3$, thus the ML estimate yields the value of $\hat{\Psi}(x = 1.3) = 1.3$. Formally, we have

$$\ln(L(1.3;\theta)) = -\frac{1}{2}\ln(2\pi\sigma^2) - \frac{(1.3-\theta)^2}{2\sigma^2}. \quad (7.181)$$

Then set

$$\frac{\partial \ln(L(1.3;\theta))}{\partial\theta} = \frac{2(1.3-\theta)}{2\sigma^2} = 0 \quad (7.182)$$

and evaluate

$$\frac{\partial^2 \ln(L(1.3;\theta))}{\partial\theta^2} = \frac{-1}{\sigma^2} < 0. \quad (7.183)$$

The solution of (7.182) yields $\hat{\Psi}(x = 1.3) = 1.3$. In light of (7.183) and that the expression in (7.181) is a quadratic expression of θ, this shows $\Psi(x = 1.3) = 1.3$ is the ML estimate of the mean of the r.v. X. □

Example 7.24 Consider a Bernoulli trial of n independent binary valued r.v.'s $\{X_1, \ldots, X_n\}$. Let $\theta = p$ be the probability of a "success" resulting in $X_i = 1$ and $q = 1 - \theta$ be the probability of a "failure" resulting in $X_i = 0$, for $i = 1, \ldots, n$. Suppose we observe k successes, then the probability of k successes is given by

$$p(k;\theta) = \mathrm{C}_k^n \theta^k (1-\theta)^{n-k},\ k = 0, \ldots, n,\ 0 \leq \theta \leq 1. \quad (7.184)$$

The ML estimate for $\theta = p$ can be obtained as follows. Consider

$$\ln(p(k;\theta)) = \ln(\mathrm{C}_k^n) + k\ln(\theta) + (n-k)\ln(1-\theta). \quad (7.185)$$

Set

$$\frac{\partial \ln(p(k;\theta))}{\partial \theta} = \frac{-k}{\theta} - \frac{n-k}{1-\theta} = 0 \tag{7.186}$$

and evaluate

$$\frac{\partial^2 \ln(p(k;\theta))}{\partial \theta^2} = \frac{-k}{\theta^2} - \frac{n-k}{(1-\theta)^2} < 0, \; 0 < \theta < 1. \tag{7.187}$$

Solving for $\hat{\theta}$ in (7.186) yields $k(1-\hat{\theta}) = \hat{\theta}(n-k)$ or equivalently $k = \hat{\theta}(n-k+k)$. Thus, the ML estimate of the probability of a success is given by

$$\hat{\theta} = k/n. \tag{7.188}$$

The ML estimate of (7.188) shows if k successes out of n trials were observed, then declaring the probability of a "success" as the ratio of k/n seems to be quite intuitively reasonable. □

Example 7.25 Consider a sample of n independent Gaussian r.v.s denoted by $\mathbf{x} = [x_1, \ldots, x_n]^T$ taken from a normal distribution $\mathcal{N}(\mu, \sigma^2)$. We want to use the ML estimation criterion to find the two unknown deterministic vector parameters $\theta = [\theta_1,\ \theta_2]^T$, where $\theta_1 = \mu$ and $\theta_2 = \sigma^2$. The likelihood function is given by

$$L(\mathbf{x};\theta_1,\theta_2) = \left((2\pi)^{-n/2}\theta_2^{-n/2}\right)\exp\left(-\frac{1}{2}\sum_{i=1}^{n}\frac{(x_i-\theta_1)^2}{\theta_2}\right). \tag{7.189}$$

Taking the $\ln(\cdot)$ of the likelihood function in (7.189) yields

$$\ln(L(\mathbf{x};\theta_1,\ \theta_2)) = -\frac{n\ln(2\pi)}{2} - \frac{n\ln(\theta_2)}{2} - \frac{1}{2}\sum_{i=1}^{n}\frac{(x_i-\theta_1)^2}{\theta_2}. \tag{7.190}$$

Taking the first and second partial derivatives of (7.190) w.r.t. θ_1 yields

$$\frac{\partial \ln(L(\mathbf{x};\theta_1,\ \theta_2))}{\partial \theta_1} = \sum_{i=1}^{n}\frac{(x_i-\theta_1)}{\theta_2}, \tag{7.191}$$

and

$$\frac{\partial^2 \ln(L(\mathbf{x};\theta_1,\ \theta_2))}{\partial \theta_1^2} = -\frac{n}{\theta_2} < 0. \tag{7.192}$$

Setting the first partial derivative in (7.191) to zero yields

$$\hat{\theta}_1 = \hat{\mu}_{\mathrm{ML}} = (1/n)\sum_{i=1}^{n} x_i = \bar{x}. \tag{7.193}$$

Since the second partial derivative in (7.192) is negative, the solution in (7.193) is a local maximum. However, since (7.190) is a quadratic function of θ_1, thus the local maximum is a global maximum for the ML estimator of the mean of Gaussian pdf. We note this ML estimator is just the sample mean of the data.

Taking the first partial derivative of (7.190) w.r.t. θ_2 yields

$$\frac{\partial \ln(L(\mathbf{x};\theta_1,\ \theta_2))}{\partial \theta_2} = -\frac{n}{2\theta_2} + \frac{1}{2}\sum_{i=1}^{n}\frac{(x_i-\theta_1)^2}{\theta_2^2}. \tag{7.194}$$

Setting the partial derivative of (7.194) to zero yields

$$\left.\frac{\partial \ln(L(\mathbf{x};\theta_1,\ \theta_2))}{\partial \theta_2}\right|_{\theta_1=\hat{\theta}_1,\ \theta_2=\hat{\theta}_2} = -n + \sum_{i=1}^{n}\frac{(x_i-\hat{\theta}_1)^2}{\hat{\theta}_2} = 0, \tag{7.195}$$

and using (7.193), we obtain

$$\hat{\theta}_2 = \hat{\sigma}^2_{\mathrm{ML}} = (1/n)\sum_{i=1}^{n}(x_i-\bar{x})^2. \tag{7.196}$$

While (7.190) is not a quadratic function of θ_2, the ML estimator for σ^2 given in (7.196) is the global maximum in (7.190). We also note, the ML estimator in (7.196) is not an unbiased estimator, since the sample variance of σ^2 is similar to (7.196) except n is replaced by $(n-1)$. Asymptotically, for large n, the ML estimator in (7.196) becomes unbiased. □

Example 7.26 A ML estimator has an important invariance property that most other estimators do not possess. Namely, if θ_{ML} is a ML estimator for some parameter θ, then $h(\theta_{\mathrm{ML}})$ is the ML estimator for the parameter $h(\theta)$. We will not give a proof of this result.

Consider a received continuous-time waveform given by

$$x(t) = s(t,\theta) + n(t),\ 0 \le t \le T, \tag{7.197}$$

where $s(t,\theta)$ is a deterministic function with a deterministic parameter θ and $n(t)$ is the realization of a zero-mean WGN process with a two-sided spectral density of $N_0/2$. We want to use the ML estimation method for estimating θ. In order to use the ML method, we need to obtain the likelihood function for a given θ (i.e., we need to find the conditional probability of $p(x(t),\ 0 \le t \le T|\theta)$).

We can use the approach based on the limit of the sampled version of the waveform as considered in Section 4.2. Then the log-likelihood function of $\{x(t),\ 0 \le t \le T\}$ given θ is given by

$$\ln(p(x(t),\ 0 \le t \le T|\theta)) = \ln(C) - (1/N_0)\int_0^T (x(t)-s(t,\theta))^2 dt, \tag{7.198}$$

where C is a constant depending on neither $\{x(t),\ 0 \le t \le T\}$ nor $\{s(t,\theta)\}$. Thus, for ML estimation, consider

$$\frac{\partial \ln(p(x(t)|\theta))}{\partial \theta} = \frac{-2}{N_0}\int_0^T (x(t)-s(t,\theta))\frac{\partial s(t,\theta)}{\partial \theta}dt. \tag{7.199}$$

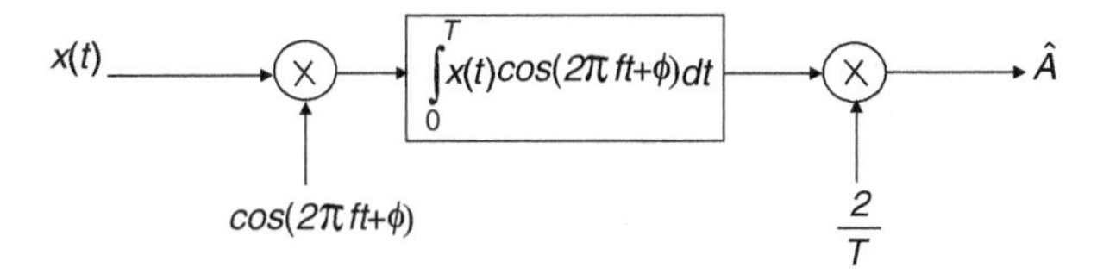

Figure 7.5 ML estimation of the amplitude A of a sinusoid in WGN.

Thus, we need to solve

$$\int_0^T (x(t) - s(t, \hat{\theta})) \frac{\partial s(t, \theta)}{\partial \theta} \bigg|_{\theta = \hat{\theta}} dt = 0. \tag{7.200}$$

□

Example 7.27 In (7.197), let $s(t, \theta) = A\cos(2\pi f t + \phi)$. Thus, $x(t) = A\cos(2\pi f t + \phi) + n(t),\ 0 \leq t \leq T$, is identical to the waveform in (7.1) and that considered in Example 7.22.

Case a. Consider the ML estimation of the amplitude parameter $\theta = A$, where f and ϕ are assumed to be known. Then (7.200) becomes

$$\int_0^T (x(t) - \hat{A}\cos(2\pi f t + \phi))\cos(2\pi f t + \phi) dt = 0, \tag{7.201}$$

and

$$\int_0^T x(t)\cos(2\pi f t + \phi) dt = \hat{A} \int_0^T \cos^2(2\pi f t + \phi) dt = \hat{A}T/2. \tag{7.202}$$

Thus, the ML estimator $\hat{A}$ of A is given by

$$\hat{A} = (2/T) \int_0^T x(t)\cos(2\pi f t + \phi) dt, \tag{7.203}$$

which is identical to (7.160). We note, $\hat{A}$ is a normalized correlation of the known $\cos(2\pi f t + \phi)$ with the received waveform $x(t)$ as shown in Fig. 7.5. Furthermore, we know from (7.161) that $\hat{A}$ is an unbiased estimator.
Now, evaluate the variance of $\hat{A}$. Since

$$\begin{aligned}(\hat{A} - A) &= (2/T) \int_0^T (x(t) - A\cos(2\pi f t + \phi))\cos(2\pi f t + \phi) dt \\ &= (2/T) \int_0^T n(t)\cos(2\pi f t + \phi) dt,\end{aligned} \tag{7.204}$$

then

$$\begin{aligned}\sigma_{\hat{A}}^2 &= \mathrm{E}\left\{(\hat{A} - A)^2\right\} = (2/T)^2 (N_0/2) \int_0^T \cos^2(2\pi f t + \phi) dt \\ &= (2/T)^2 (N_0/2)(T/2) = N_0/T\end{aligned} \tag{7.205}$$

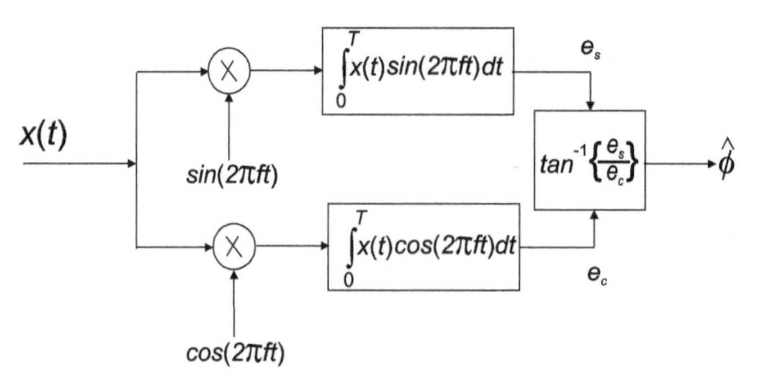

Figure 7.6 ML estimation of the phase of a sinusoid in WGN.

Comparing (7.205) with the right-hand side of the CRB expression of (7.161), we see both expressions are equal. Thus, the equality of the CRB is attained and $\hat{A}$ is an efficient estimator.

Case b. Consider the ML estimation of the phase parameter $\theta = \phi$, where A and f are assumed to be known. Then (7.200) yields

$$\int_0^T \left(x(t) - A\cos(2\pi f t + \hat{\phi})\right)(-A\sin(2\pi f t + \hat{\phi}))dt \approx 0, \tag{7.206}$$

which is identical to (7.165). Similar to the arguments of (7.166)–(7.169), we obtain

$$-\tan(\hat{\phi}) = \frac{-\sin(\hat{\phi})}{\cos(\hat{\phi})} = \frac{\int_0^T x(t)\sin(2\pi f t)dt}{\int_0^T x(t)\cos(2\pi f t)dt} = \frac{e_s}{e_c}, \tag{7.207}$$

or

$$\hat{\phi} = -\tan^{-1}\{e_s/e_c\}. \tag{7.208}$$

A block diagram for the ML estimate $\hat{\phi}$ of (7.208) is given in Fig. 7.6.

Using the same assumption of Case b of Example 7.25 of Section 7.4.1, under the high SNR condition and a first-order Taylor series linear approximation of $x(t, \phi)$ about $\phi = \hat{\phi}$, (7.170)–(7.172) show

$$(\hat{\phi} - \phi) = (2/AT)\int_0^T x(t; \hat{\phi})\sin(2\pi f t + \phi)dt, \tag{7.209}$$

$$\mathrm{E}\{(\hat{\phi} - \phi)\} = (2/AT)\int_0^T A\cos(2\pi f t + \hat{\phi})\sin(2\pi f t + \phi)dt = 0. \tag{7.210}$$

Then $\hat{\phi}$ is an unbiased estimator and the variance of $\hat{\phi}$ is then given by

$$\begin{aligned}
&\mathrm{E}\{(\hat{\phi} - \phi)^2\} \\
&= (2/AT)^2 E\left\{\left[\int_0^T x(t; \hat{\phi})\sin(2\pi f t + \phi)dt\right]\left[\int_0^T x(\tau; \hat{\phi})\sin(2\pi f \tau + \phi)d\tau\right]\right\} \\
&\approx (2/AT)^2(N_0/2)\int_0^T \sin^2(2\pi f t + \phi)dt = (2/AT)^2(N_0/2)(T/2) \\
&= N_0/(A^2 T).
\end{aligned} \tag{7.211}$$

Comparing (7.211) with the right-hand side of the CRB expression (7.175), we see both expressions are equal. Thus, the equality of the CRB is attained and $\hat{\phi}$ is an efficient estimator under the assumption of the approximation in (7.211) and under high SNR conditions, when $\hat{\phi}$ should be near ϕ with high probability. □

We note, the ML estimation of the frequency parameter $\theta = f$ in $s(t, \theta) = A\cos(2\pi f t + \phi)$ of (7.197), where A and ϕ are assumed to be known, does not seem to have a simple analytical expression.

7.4.3 Maximum a posteriori estimator

Suppose we have some information about the possible value of the parameter θ. There are various ways to incorporate this information for maximum usefulness. Suppose we assume the parameter Θ is a r.v. with some known prior pdf denoted by $f_\Theta(\theta)$. From the Bayes theorem, the a posteriori (also called the posterior) pdf of Θ given the realizations of $\{x_1, \ldots, x_n\}$ is given by

$$p(\theta|x_1, \ldots, x_n) = \frac{f_\Theta(\theta)p(x_1, \ldots, x_n|\theta)}{\int f(u)p(x_1, \ldots, x_n|u)du}. \tag{7.212}$$

The value of the *Maximum A Posteriori* (MAP) estimator Θ_{MAP} is that θ which maximizes the a posteriori pdf of (7.212). That is,

$$p(\theta_{\text{MAP}}|x_1, \ldots, x_n) = \max_\theta p(\theta|x_1, \ldots, x_n). \tag{7.213}$$

Example 7.28 Consider the Bernoulli trial of Example 7.24 in Section 7.4.2, except now θ is not a deterministic parameter, but is a r.v. denoted by Θ, with a prior pdf $f_\Theta(\theta)$ given by

$$f_\Theta(\theta) = 6\theta(1-\theta),\ 0 \le \theta \le 1. \tag{7.214}$$

Then

$$f'_\Theta(\theta) = 6 - 12\theta,\ 0 \le \theta \le 1, \tag{7.215}$$

$$f''_\Theta(\theta) = -12 < 0,\ 0 \le \theta \le 1. \tag{7.216}$$

Since $f_\Theta(\theta)$ of (7.214) is a quadratic function in θ and its second derivative $f''_\Theta(\theta)$ of (7.216) is negative for $0 \le \theta \le 1$, then its stationary solution obtained from $f'_\Theta(\theta) = 0$ in (7.215), given by $\hat{\theta}_{\max} = 6/12 = 1/2$, is the global maximum of the prior pdf $f_\Theta(\theta)$. The conditional probability, conditioned on $\Theta = \theta$, of a specific realization of $\{x_1, \ldots, x_n\}$ with k successes in a Bernoulli trial of length n is given by

$$p(x_1, \ldots, x_n|\theta) = \theta^k(1-\theta)^{n-k},\ k = \sum_{i=1}^n x_i. \tag{7.217}$$

Furthermore,

$$\int_0^1 f_\Theta(t)p(x_1, \ldots, x_n|t)dt = 6\int_0^1 t^{k+1}(1-t)^{n-k+1}dt = \frac{6\Gamma(k+2)\Gamma(n-k+2)}{\Gamma(n+4)}, \tag{7.218}$$

where $\Gamma(\cdot)$ is the gamma function defined by

$$\Gamma(x) = \int_0^\infty t^{x-1} \exp(-t)dt, \;\; \Gamma(n+1) = n! = \text{factorial}(n), \tag{7.219}$$

for non-negative integer n.
Using (7.214) and (7.217) in (7.212), we obtain

$$p(\theta|x_1, \ldots, x_n) = \frac{6\theta^{k+1}(1-\theta)^{n-k+1}}{6\Gamma(k+2)\Gamma(n-k+2)/\Gamma(n+4)} = \frac{\Gamma(n+4)\theta^{k+1}(1-\theta)^{n-k+1}}{\Gamma(k+2)\Gamma(n-k+2)}. \tag{7.220}$$

From (7.220), we can find the MAP estimate of θ given $k = \sum_{i=1}^{n} x_i$. Set

$$\frac{\partial \ln(p(\theta|x_1, \ldots, x_n))}{\partial\theta} = \frac{k+1}{\theta} - \frac{n-k+1}{1-\theta} = 0, \tag{7.221}$$

and evaluate

$$\frac{\partial^2 \ln(p(\theta|x_1, \ldots, x_n))}{\partial\theta^2} = \frac{-(k+1)}{\theta^2} - \frac{n-k+1}{(1-\theta)^2} < 0, k = 0, \, 1, \ldots, n. \tag{7.222}$$

Solving for θ in (7.221) yields $(k+1)(1-\theta) = \theta(n-k+1)$ or equivalently $k+1 = \theta(n-k+1+k+1)$ or equivalently $\theta = (k+1)/(n+2)$. Since $p(\theta|x_1, \ldots, x_n)$ of (7.220) is a concave function of θ, (7.222) shows the local maximum obtained from (7.221) yields a global maximum. Thus, the MAP estimate of the probability of a "success" is now given by

$$\hat{\theta}_{\text{MAP}} = \frac{k+1}{n+2}. \tag{7.223}$$

If $n = 0$ (i.e., no trial), then $k = 0$ (i.e., no success). Then (7.223) shows $\hat{\theta}_{\text{MAP}} = 1/2$, which is the same as $\hat{\theta}_{\text{Max}} = 1/2$, which yields the maximum of the prior pdf of (7.214). If n and k both become large, then

$$\hat{\theta}_{\text{MAP}} \to \frac{k}{n} = \hat{\theta}_{\text{ML}}. \tag{7.224}$$

Thus, for large n and large k, the information in the prior pdf $f_\Theta(\theta)$ becomes irrelevant and the MAP estimate approaches that of the ML estimate of the probability of k successes in a Bernoulli trial of length n. Similarly, if the prior pdf $f_\Theta(\theta)$ is uniformly distributed in $[a, \; a+A]$, then

$$f_\Theta(\theta) = 1/A, \theta \infty \, [a, \; a+A], \; 0 < A, \; -\infty < a < \infty, \tag{7.225}$$

and the a posteriori probability of (7.212) becomes

$$p(\theta|x_1, \ldots, x_n) = \frac{(1/A)p(x_1, \ldots, x_n|\theta)}{(1/A)\int p(x_1, \ldots, x_n|u)du} = \frac{p(x_1, \ldots, x_n|\theta)}{p(x_1, \ldots, x_n)}. \tag{7.226}$$

Thus, the θ that maximizes the left-hand side of (7.226) yielding $\hat{\theta}_{\text{MAP}}$ is identical to the θ that maximizes the right-hand side of (7.226) yielding $\hat{\theta}_{\text{ML}}$. Thus, for a uniformly distributed prior probability of θ, the MAP estimate is identical to the ML estimate. □

Example 7.29 Suppose an observation is known to be taken from a Poisson distribution with a mean λ equal to either 2 or 4. The prior distribution of Λ for $\Lambda = 2$ is four times more likely than $\Lambda = 4$. Thus, the prior probability of Λ is given by $\mathrm{P}(\Lambda = 2) = 0.8$ and $\mathrm{P}(\Lambda = 4) = 0.2$. A measurement is made yielding a value of $x = 6$. We can evaluate the two probabilities $\mathrm{P}(X = 6|\Lambda = 2) = \exp(-2) \times 2^6/6! = 0.012$ and $\mathrm{P}(X = 6|\Lambda = 4) = \exp(-4) \times 4^6/6! = 0.104$. Then the posterior probability of $\Lambda = 2$ given an observed $x = 6$ is given by

$$\begin{aligned}\mathrm{P}(\Lambda = 2|X = 6) &= \frac{\mathrm{P}(\Lambda = 2, X = 6)}{\mathrm{P}(X = 6)} \\ &= \frac{\mathrm{P}(X = 6|\Lambda = 2)\mathrm{P}(\Lambda = 2)}{\mathrm{P}(X = 6|\Lambda = 2)\mathrm{P}(\Lambda = 2) + \mathrm{P}(X = 6|\Lambda = 4)\mathrm{P}(\Lambda = 4)} \\ &= \frac{0.012 \times 0.8}{0.012 \times 0.8 + 0.104 \times 0.2} = 0.316.\end{aligned}$$

Similarly, the posterior probability of $\lambda = 4$ given an observed $x = 6$ is given by

$$\begin{aligned}\mathrm{P}(\Lambda = 4|X = 6) &= \frac{\mathrm{P}(\Lambda = 4, X = 6)}{\mathrm{P}(X = 6)} \\ &= \frac{\mathrm{P}(X = 6|\Lambda = 4)\mathrm{P}(\Lambda = 4)}{\mathrm{P}(X = 6|\Lambda = 2)\mathrm{P}(\Lambda = 2) + \mathrm{P}(X = 6|\Lambda = 4)\mathrm{P}(\Lambda = 4)} \\ &= \frac{0.104 \times 0.2}{0.012 \times 0.8 + 0.104 \times 0.2} = 0.684.\end{aligned}$$

By comparing the two posterior pdfs, we note

$$0.684 = \mathrm{P}(\Lambda = 4|X = 6) > \mathrm{P}(\Lambda = 2|X = 6) = 0.316,$$

thus by the MAP estimation criterion, the optimum MAP estimator is given by

$$\hat{\Lambda}_{\mathrm{MAP}} = 4.$$

It is interesting to note, while the prior probability $\mathrm{P}(\lambda = 2) = 0.8$, after the observation with a value of $x = 6$, the posterior probability decreased to $\mathrm{P}(\Lambda = 2|X = 6) = 0.316$. On the other hand, while the prior probability $\mathrm{P}(\Lambda = 4) = 0.2$, after the observation with a value of $x = 6$, the posterior probability increased to $\mathrm{P}(\Lambda = 4|X = 6) = 0.684$. This example shows the usefulness of measured data in correcting our possibly incorrect prior beliefs. □

7.4.4 Bayes estimator

Consider a parameter θ modeled as a r.v. Let $\Theta = \Theta(\mathbf{X})$ be an estimator of the parameter θ with observation random vector $\mathbf{X} = [X_1, \ldots, X_n]^T$. Let $C(\Theta, \theta)$ denote a cost function of using the Θ estimator to estimate the parameter θ. Some possible examples of a cost function include:

a. $C(\Theta, \theta) = \begin{cases} |\Theta - \theta| > A > 0, \\ |\Theta - \theta| \leq A. \end{cases}$ thresholding cost function. (7.227)

b. $C(\Theta, \theta) = (\Theta - \theta)$, error cost function. (7.228)

c. $C(\Theta, \theta) = |\Theta - \theta|$, absolute-error cost function. (7.229)

d. $C(\Theta, \theta) = (\Theta - \theta)^2$, square-error cost function. (7.230)

The Bayes cost conditioned on the realization of the observed data vector $\mathbf{x}$ is defined by

$$R(\Theta|\mathbf{x}) = \int_{-\infty}^{\infty} C(\Theta, \theta) p(\theta|\mathbf{x}) d\theta, \tag{7.231}$$

$p(\theta|\mathbf{x})$ is the posterior probability density, which can be evaluated from

$$p(\theta|\mathbf{x}) = \frac{f_\Theta(\theta) p(\mathbf{x}|\theta)}{\int f_\Theta(u) p(\mathbf{x}|u) du}, \tag{7.232}$$

where $f_\Theta(\theta)$ is the pdf of the r.v. Θ.

Then the Bayes cost (average cost) is given by

$$R(\Theta) = \int_{-\infty}^{\infty} \cdots \int_{-\infty}^{\infty} R(\Theta|\mathbf{x}) p(\mathbf{x}) d\mathbf{x}, \tag{7.233}$$

where $p(\mathbf{x})$ is the n-dimensional pdf of $\mathbf{X}$. The Bayes estimator $\hat{\Theta}$ is the estimator that attains the minimum of the Bayes cost of (7.233).

Example 7.30 Consider the square-error cost function $C(\Theta, \theta) = |\Theta - \theta|^2$. Then the Bayes estimator $\hat{\Theta}$ that minimizes the Bayes cost in (7.233) with the square cost function of (7.230) is also denoted as the minimum mean-square estimator $\hat{\Theta}_{\text{MS}}$.

Since for each fixed pdf $p(\mathbf{x})$, which is non-negative valued, the $R(\Theta)$ in (7.233) is minimized if $R(\Theta|\mathbf{x})$ in (7.231) is minimized. From

$$R(\Theta|\mathbf{x}) = \int_{-\infty}^{\infty} (\Theta - \theta)^2 p(\theta|\mathbf{x}) d\theta, \tag{7.234}$$

set

$$\frac{\partial R(\Theta|\mathbf{x})}{\partial \Theta} = \int_{-\infty}^{\infty} 2(\Theta - \theta) p(\theta|\mathbf{x}) d\theta = 0,$$

or

$$\hat{\Theta} = \int_{-\infty}^{\infty} p(\theta|\mathbf{x}) d\theta = \int_{-\infty}^{\infty} \theta p(\theta|\mathbf{x}) d\theta,$$

and thus we obtain

$$\hat{\Theta} = \mathrm{E}\{\boldsymbol{\theta}|\mathbf{x}\} = \mathrm{E}_{\Theta|\mathbf{x}}\{\boldsymbol{\theta}\}. \tag{7.235}$$

Furthermore, since

$$\frac{\partial^2 R(\Theta|\mathbf{x})}{\partial\Theta^2} = 2\int_{-\infty}^{\infty} p(\theta|\mathbf{x})d\theta > 0,$$

and $R(\Theta|\mathbf{x})$ in (7.234) is a quadratic function of Θ, the local minimum solution in (7.235) is a global minimum. Thus, the Bayes estimator $\hat{\Theta}$ of (7.235) is the conditional mean of Θ conditioned on $\mathbf{x}$. Earlier, we showed the minimum mean-square estimator $\hat{\Theta}_{\text{MS}}$ is $\text{E}_{\Theta|\mathbf{x}}\{\Theta\}$. □

Example 7.31 Consider absolute error cost function $C(\Theta, \theta) = |\Theta - \theta|$. Now, we want to find the Bayes estimator $\hat{\Theta}$ that minimizes the Bayes cost in (7.233) with the absolute error cost function of (7.229). The $R(\Theta|\mathbf{x})$ of (7.231) is now given by

$$\begin{aligned} R(\Theta|\mathbf{x}) &= \int_{-\infty}^{\infty} |\Theta - \theta|\, p(\theta|\mathbf{x})d\theta \\ &= -\int_{-\infty}^{\Theta} (\Theta - \theta)\, p(\theta|\mathbf{x})d\theta + \int_{\Theta}^{\infty} (\Theta - \theta)\, p(\theta|\mathbf{x})d\theta. \end{aligned}$$

Then

$$\begin{aligned} \frac{\partial\{R(\Theta|\mathbf{x})\}}{\partial\Theta} &= \frac{\partial\left\{-\int_{-\infty}^{\Theta}(\Theta-\theta)p(\theta|\mathbf{x})d\theta\right\}}{\partial\Theta} + \frac{\partial\left\{\int_{\Theta}^{\infty}(\Theta-\theta)p(\theta|\mathbf{x})d\theta\right\}}{\partial\Theta} \\ &= -\int_{-\infty}^{\Theta} p(\theta|\mathbf{x})d\theta + \int_{\Theta}^{\infty} p(\theta|\mathbf{x})d\theta = 0. \end{aligned} \tag{7.236}$$

Since $p(\theta|\mathbf{x})$ is a conditional pdf (and thus a valid pdf) satisfying

$$\int_{-\infty}^{\infty} p(\theta|\mathbf{x})d\theta = 1, \tag{7.237}$$

then (7.236) and (7.237) show

$$\int_{-\infty}^{\Theta} p(\theta|\mathbf{x})d\theta = \int_{\Theta}^{\infty} p(\theta|\mathbf{x})d\theta = 1/2. \tag{7.238}$$

Equation (7.238) shows the Bayes estimator under the absolute error criterion $\hat{\Theta}_{\text{AE}}$ is given by the *median* of the posterior pdf of the r.v. θ conditioned on the observed data vector $\mathbf{x}$. By definition, the median of a r.v. is such that half of the probability is to the left of the median and half of the probability is to the right of the median. □

Example 7.32 It can be shown that for a symmetric posterior pdf $p(\theta|\mathbf{x})$ symmetric in θ, the minimum mean-square estimator $\hat{\Theta}_{\text{MS}}$ is equal to the minimum absolute-error estimator $\hat{\Theta}_{\text{AE}}$. For a unimodal (i.e., single peak) and symmetric posterior pdf $p(\theta|\mathbf{x})$, the minimum mean-square estimator $\hat{\Theta}_{\text{MS}}$ is equal to the maximum a posteriori estimator $\hat{\Theta}_{\text{MAP}}$. In particular, a Gaussian posterior pdf $p(\theta|\mathbf{x})$ is both symmetric and unimodal, thus all the estimators, $\hat{\Theta}_{\text{MS}}$, $\hat{\Theta}_{\text{AE}}$, and $\hat{\Theta}_{\text{MAP}}$, are identical. □

7.5 Conclusions

In this chapter, we considered various types of estimators optimized using different criteria. Section 7.1 introduces the estimation of the amplitude, frequency, and phase parameters of a received radar waveform. Section 7.2 dealt with the optimum estimation of r.v.'s and random vectors based on the mean-square-error criterion. In particular, the geometric interpretation of orthogonal principle provides a simple intuition for linear mean-square estimation problems. Concepts of block and recursive estimations, including the scalar Kalman filtering problem, were presented. Section 7.3 discussed least-square and least-absolute-error criteria estimation problems. An example was used to demonstrate the robustness of the optimum solution of an over-determined linear system of equations based on the least-absolute-error criterion compared to the optimum solution based on the least-square error criterion. The sparseness property of the optimum l_1 norm-base solution of an under-determined linear system of equations was also discussed. Section 7.4 introduced various basic statistical parameter estimation methods including: unbiasedness, efficiency, consistency, sufficiency, Factorization Theorem, the Cramér–Rao Bound, maximum-likelihood estimation, maximum a posteriori estimation, and Bayes estimation. Various examples were provided to illustrate some of these concepts.

7.6 Comments

Parameter estimation methods have been used in various branches of science, statistics, and engineering since the seventeenth century by Galileo Galilei in his astronomical orbit computations as mentioned in Section 1.6.2. The mean-square estimation method considered in Section 7.2 was probably the first parameter estimation technique encountered in radio and radar engineering. H. Wold [4] studied stationary time series and their representations in 1938. A. N. Kolmogorov [5] in 1941 formulated the optimum statistical interpolation and extrapolation of stationary random sequences. In 1942, N. Wiener [6] used the mean-squared-error criterion to address the optimum solution of a statistical extrapolation problem motivated by a practical anti-aircraft fire-control problem in World War II. The papers of H. W. Bode and C. E. Shannon [7] and L. A. Zadeh and J. R. Ragazzini [8] interpreted Wiener's difficult-to-understand results and introduced the linear mean-squares estimation concept to the general communication and control engineering community. Some of the earliest books in the 1950s–1960s on this topic included [9–11]. Since then there have been dozens of engineering-oriented books dealing with mean-square estimation; some references for materials presented in Section 7.2 may be found in [12–20]. The reformulation of the linear mean-square state estimation in the recursive form is generally called a Kalman filter [21–23]. Some details on the simple scalar version of the Kalman filtering problem as considered in Section 7.2.3.2 can be found in [1, 24]. The general matrix-vector form of the Kalman filtering problem, its extensions, and applications were not discussed in this chapter; but some books dealing with this problem include [25–31].

The least-square estimation problem considered in Section 7.3 had its origin in the nineteenth century. C. F. Gauss [32] and S. Laplace [33] independently introduced the least-square-error criterion to create analytical models for errors encountered in practical astronomical orbit data analysis. By using the least-square error criterion, they were able to perform hand calculations of solutions of over-determined linear system of equations. As noted in Section 1.6.2, Galilei [34] proposed to use the least-absolute-error criterion in his astronomical orbit data tabulation. Unfortunately, optimum solutions based on this criterion required non-linear optimizations, which were not possible until digital computers. It appears that Gauss and Laplace (and others before the digital computer era) appreciated the advantage of using the least-square-error criterion over the least-absolute-error criterion for practical data analysis. In the twentieth century, least-square estimation was studied formally by Pearson [35] and R. A. Fisher [36, 37]. Extensive statistical literatures and books exist on this topic, including [2, 38–42]. Section 7.3.2 showed the optimum least-absolute-error criterion solution of an over-determined system of equations with data outliers can be more robust than the optimum least-square-error criterion solution [43–45]. Section 7.3.3 dealt with the issues of using l_2 and l_1 norms in optimum linear estimations [45, 46]. Theory of error based on the least-square criterion, using the l_2 norm, resulted in the Gaussian error model, while that based on the least-absolute-error criterion, using the l_1 norm, resulted in the Laplacian error model [3]. Sparseness solutions of the under-determined system of equations under constrained l_1 norm optimization play important roles in compressive sensing problems [47]. Optimization based on l_1 norm is also heavily used in convex programming [48, 49].

Section 7.4 first defined and then introduced various basic properties of statistical parameter estimators. R. A. Fisher [37] introduced the concepts of maximum-likelihood, sufficiency, efficiency, and consistency. Since then these concepts can be found in essentially all statistics books, including [2, 38–41]. The Factorization Theorem in the theory of sufficient statistics was obtained by P. M. Halmos and L. J. Savage [50]. However, since its derivation used the Radon–Nikodym derivative concept, the proof of this theorem is omitted in this section. Since the Cramér–Rao Bound [38, 40, 51] yields the minimum mean-square error of an unbiased estimator, it provides a simple method to compare the goodness of any proposed unbiased estimator to the optimum one. There are again extensive statistical literature and books [2, 38–41] and also a large number of communication and radar engineering books on this topic [20, 42, 52–60]. The concepts of maximum a posteriori estimator in Section 7.4.3 and Bayes estimator in Section 7.4.4 both depend on the use of the Bayes rule proposed in 1763 by T. P. Bayes [61]. Extensive use of these two estimators for engineering applications has appeared in the previously referenced books [11], [14–20], [42], and [52–60].

7.A Proof of Theorem 7.1 of Section 7.3.3

Consider using the Lagrangian multiplier method of minimizing the energy $\mathbf{a}^T\mathbf{a}$ of the solution $\mathbf{a}$, subject to (7.121). Define

$$f(\mathbf{a}) = \mathbf{a}^T\mathbf{a} + \boldsymbol{\lambda}^T\mathbf{X}\mathbf{a} = \mathbf{a}^T\mathbf{a} + \mathbf{a}^T\mathbf{X}^T\boldsymbol{\lambda}, \tag{7.239}$$

where $\boldsymbol{\lambda} = [\lambda_1, \ldots, \lambda_M]^T$ is the vector of the Lagrange multipliers. Taking the gradient $\nabla_{\mathbf{a}}$ w.r.t. $\mathbf{a}$ and setting it to zero yields

$$\nabla_{\mathbf{a}}\{f(\mathbf{a})\} = 2\mathbf{a} + \mathbf{X}^T\boldsymbol{\lambda} = 0 \tag{7.240}$$

or

$$\hat{\mathbf{a}} = -(1/2)\mathbf{X}^T\boldsymbol{\lambda}. \tag{7.241}$$

Solving for $\boldsymbol{\lambda}$ yields

$$\boldsymbol{\lambda} = -2(\mathbf{X}\mathbf{X}^T)^{-1}\mathbf{y}, \tag{7.242}$$

and then using (7.242) in (7.241) to yield

$$\hat{\mathbf{a}} = \mathbf{x}^T(\mathbf{X}\mathbf{X}^T)^{-1}\mathbf{y}. \tag{7.243}$$

From (7.240), the solution of $\hat{\mathbf{a}}$ given by (7.241) is a stationary point of $f(\mathbf{a})$. Furthermore, since $\nabla_{\mathbf{a}}^2\{f(\mathbf{a})\} = 2 > 0$, and $f(\mathbf{a})$ is a quadratic form in $\mathbf{a}$, then the local minimum solution of $\hat{\mathbf{a}}$ is the global minimum of $f(\mathbf{a})$. Thus, $\hat{\mathbf{a}}$ of (7.243) (denoted by $\hat{\mathbf{a}}_0$ in (7.122)), yields the minimum energy solution among all possible solutions of the under-determined system of equations satisfying (7.239). □

7.B Proof of Theorem 7.3 of Section 7.4.1

Since $\boldsymbol{\Psi}$ is an unbiased estimator,

$$\mathrm{E}_{\mathbf{X}}\left\{(\boldsymbol{\Psi} - \boldsymbol{\theta})^T\right\} = \int_{\mathbf{x}} p(\mathbf{x})(\boldsymbol{\Psi} - \boldsymbol{\theta})^T d\mathbf{x} = 0. \tag{7.244}$$

Denote the $m \times 1$ gradient $\nabla_{\theta}\{\cdot\}$ and the $m \times m$ Hessian $\nabla_{\theta}^2\{\cdot\}$ operators by

$$\nabla_{\theta}\{\cdot\} = \left[\frac{\partial\{\cdot\}}{\partial\theta_1}, \ldots, \frac{\partial\{\cdot\}}{\partial\theta_m}\right]^T, \tag{7.245}$$

$$\nabla_{\theta}^2\{\cdot\} = \begin{bmatrix} \dfrac{\partial^2\{\cdot\}}{\partial\theta_1^2} & \cdots & \dfrac{\partial^2\{\cdot\}}{\partial\theta_1\partial\theta_m} \\ \vdots & \ddots & \vdots \\ \dfrac{\partial^2\{\cdot\}}{\partial\theta_m\partial\theta_1} & \cdots & \dfrac{\partial^2\{\cdot\}}{\partial\theta_m^2} \end{bmatrix}. \tag{7.246}$$

Upon taking the gradient of (7.244) and assuming the regularity condition is satisfied [interchanging the order of the differentiation and integral in the first line of (7.247)] we have

$$\begin{aligned}\nabla_{\theta}\left\{\mathrm{E}_{\mathbf{X}}\left\{(\boldsymbol{\Psi} - \boldsymbol{\theta})^T\right\}\right\} &= \int_{\mathbf{x}} \nabla_{\theta}\{p(\mathbf{x})(\boldsymbol{\Psi} - \boldsymbol{\theta})^T\}d\mathbf{x} \\ &= \int_{\mathbf{x}} \nabla_{\theta}\left\{p(\mathbf{x})\right\}(\boldsymbol{\Psi} - \boldsymbol{\theta})^T d\mathbf{x} + \int_{\mathbf{x}} p(\mathbf{x})\nabla_{\theta}\left\{(\boldsymbol{\Psi} - \boldsymbol{\theta})^T\right\} d\mathbf{x} = 0. \end{aligned} \tag{7.247}$$

Since $\nabla_\theta \left\{(\boldsymbol{\Psi}-\boldsymbol{\theta})^T\right\} = -\mathbf{I}_m$ and $\nabla_\theta \{p(\mathbf{x})\} = [\nabla_\theta\{\ln p(\mathbf{x})\}]\, p(\mathbf{x})$, (7.247) becomes

$$\int_{\mathbf{x}} [\nabla_\theta \{\ln p(\mathbf{x})\}]\,(\boldsymbol{\Psi}-\boldsymbol{\theta})^T p(\mathbf{x})d\mathbf{x} = \mathbf{I}_m \int_{\mathbf{x}} p(\mathbf{x})d\mathbf{x} = \mathbf{I}_m. \tag{7.248}$$

Now, let $\mathbf{c}$ and $\mathbf{d}$ be two arbitrary $n \times 1$ vectors. In (7.248), use $\mathbf{c}^T$ to multiply on the left-hand side of the first and third expressions and use $\mathbf{d}$ to multiply on the right-hand side of these expressions, resulting in

$$\int_{\mathbf{x}} \mathbf{c}^T[\nabla_\theta \{\ln p(\mathbf{x})\}]\left[(\boldsymbol{\Psi}-\boldsymbol{\theta})^T\mathbf{d}\right] p(\mathbf{x})d\mathbf{x} = \mathbf{c}^T\mathbf{d}. \tag{7.249}$$

Consider the Schwarz Inequality of

$$\left(\int_{\mathbf{x}} f(\mathbf{x})h(\mathbf{x})p(\mathbf{x})d\mathbf{x}\right)^2 \leq \int_{\mathbf{x}} (f(\mathbf{x}))^2 p(\mathbf{x})d\mathbf{x} \int_{\mathbf{x}} (h(\mathbf{x}))^2 p(\mathbf{x})d\mathbf{x}. \tag{7.250}$$

Now, take $f(\mathbf{x}) = \mathbf{c}^T[\nabla_\theta \{\ln \mathrm{p}(\mathbf{x})\}]$ and $h(\mathbf{x}) = (\boldsymbol{\Psi}-\boldsymbol{\theta})^T\mathbf{d}$, resulting in $(f(\mathbf{x}))^2 = \mathbf{c}^T[\nabla_\theta \{\ln \mathrm{p}(\mathbf{x})\}][\nabla_\theta \{\ln \mathrm{p}(\mathbf{x})\}]^T\mathbf{c}$ and $(h(\mathbf{x}))^2 = \mathbf{d}^T(\boldsymbol{\Psi}-\boldsymbol{\theta})(\boldsymbol{\Psi}-\boldsymbol{\theta})^T\mathbf{d}$. Then square both sides of (7.249) and use (7.250) to yield

$$\begin{aligned}&\mathbf{c}^T\left[\int_{\mathbf{x}} [\nabla_\theta \{\ln p(\mathbf{x})\}][\nabla_\theta \{\ln p(\mathbf{x})\}]^T p(\mathbf{x})d\mathbf{x}\right]\mathbf{c}\mathbf{d}^T\left[\int_{\mathbf{x}} (\boldsymbol{\Psi}-\boldsymbol{\theta})(\boldsymbol{\Psi}-\boldsymbol{\theta})^T p(\mathbf{x})d\mathbf{x}\right]\mathbf{d}\\ &= \mathbf{c}^T\mathbf{I}_{\mathrm{m}}\mathbf{c}\mathbf{d}^T\mathrm{E}_{\mathbf{X}}\left\{(\boldsymbol{\Psi}-\boldsymbol{\theta})(\boldsymbol{\Psi}-\boldsymbol{\theta})^T\right\}\mathbf{d} \geq (\mathbf{c}^T\mathbf{d})^2.\end{aligned} \tag{7.251}$$

In particular, substitute $\mathbf{c} = \mathbf{I}_{\mathrm{m}}^{-1}\mathbf{d}$ or $\mathbf{c}^T = \mathbf{d}^T\mathbf{I}_{\mathrm{m}}^{-1}$ in (7.251) to yield

$$\mathbf{d}^T\mathbf{I}_{\mathrm{m}}^{-1}\mathbf{I}_{\mathrm{m}}\mathbf{I}_{\mathrm{m}}^{-1}\mathbf{d}\mathbf{d}^T\mathrm{E}_{\mathbf{X}}\left\{(\boldsymbol{\Psi}-\boldsymbol{\theta})(\boldsymbol{\Psi}-\boldsymbol{\theta})^T\right\}\mathbf{d} \geq (\mathbf{d}^T\mathbf{I}_{\mathrm{m}}^{-1}\mathbf{d})^2$$

or

$$\mathbf{d}^T\mathrm{E}_{\mathbf{X}}\left\{(\boldsymbol{\Psi}-\boldsymbol{\theta})(\boldsymbol{\Psi}-\boldsymbol{\theta})^T\right\}\mathbf{d} \geq \mathbf{d}^T\mathbf{I}_{\mathrm{m}}^{-1}\mathbf{d}. \tag{7.252}$$

Since the expression in (7.252) is valid for an arbitrary $n \times 1$ vector $\mathbf{d}$, then we must have $\mathrm{E}_{\mathbf{X}}\left\{(\boldsymbol{\Psi}-\boldsymbol{\theta})(\boldsymbol{\Psi}-\boldsymbol{\theta})^T\right\} \geq \mathbf{I}_{\mathrm{m}}^{-1}$, which shows the CR bounds given by (7.148) and (7.149a) are valid.

Since $\int_{\mathbf{x}} p(\mathbf{x})d\mathbf{x} = 1$, taking the gradient on both sides of this equation and interchanging the order of differentiation and integration yields

$$\nabla_\theta\left\{\int_{\mathbf{x}} p(\mathbf{x})d\mathbf{x}\right\} = \int_{\mathbf{x}} \nabla_\theta \{p(\mathbf{x})\}\, d\mathbf{x} = \int_{\mathbf{x}} \nabla_\theta \{\ln p(\mathbf{x})\}\, p(\mathbf{x})d\mathbf{x} = 0. \tag{7.253}$$

Taking the gradient on both sides of (7.253) and interchanging the order of differentiation and integration yields

$$\begin{aligned}\mathbf{I}_{\mathrm{m}} &= \mathrm{E}_{\mathbf{X}}\left\{[\nabla_\theta \{\ln p(\mathbf{X})\}][\nabla_\theta \{\ln p(\mathbf{X})\}]^T\right\}\\ &= \int_{\mathbf{x}} [\nabla_\theta \{\ln p(\mathbf{x})\}][\nabla_\theta \{\ln p(\mathbf{x})\}]^T p(\mathbf{x})d\mathbf{x}\\ &= -\int_{\mathbf{x}} \nabla_\theta^2 \{\ln p(\mathbf{x})\}\, p(\mathbf{x})d\mathbf{x} = \mathrm{E}_{\mathbf{X}}\left\{\nabla_\theta^2 \{\ln p(\mathbf{X})\}\right\},\end{aligned} \tag{7.254}$$

which shows that (7.149a) is identical to (7.149b). □

References

[1] M. Schwartz and L. Shaw. *Signal Processing: Discrete Spectral Analysis, Detection, and Estimation*. McGraw-Hill, 1975.

[2] S. D. Silvey. *Statistical Inference*. Halsted Press, 1970.

[3] S. Kotz, T. J. Kozubowski, and K. Podgórski. *The Laplace Distribution and Generalizations*. Birkhäuser, 2001.

[4] H. Wold. A study in the analysis of stationary time series. Ph.D. Dissertation, Uppsala, 1938.

[5] A. N. Kolmogorov. Interpolation and extrapolation of stationary sequences. *Izvestia Acad. Nauk., USSR* 5, 1941.

[6] N. Wiener. *Extrapolation, Interpolation, and Smoothing of Stationary Time Series*. Wiley, 1949.

[7] H. W. Bode and C. E. Shannon. A simplified derivation of least-squares smoothing and prediction theory. In *Proc. IRE* 38, 1950.

[8] L. A. Zadeh and J. R. Ragazzini. An extension of Wiener's theory of prediction. *J. Appl. Physics* 21, 1950.

[9] U. Grenander and M. Rosenblatt. *Statistical Analysis of Stationary Time Series*. Wiley, 1957.

[10] W. B. Davenport Jr. and W. L. Root. *An Introduction to the Theory of Random Signals and Noise*. McGraw-Hill, 1958.

[11] D. Middleton. *An Introduction to Statistical Communication Theory*. McGraw-Hill, 1960.

[12] A. M. Yaglom. *An Introduction to the Theory of Stationary Random Functions*. Prentice-Hall, 1962.

[13] A. Papoulis. *Probability, Random Variables and Stochastic Processes*. McGraw-Hill, 1st ed., 1965; 4th ed., 2002.

[14] R. Deutsch. *Estimation Theory*. Prentice-Hall, 1965.

[15] N. E. Nahi. *Estimation Theory and Applications*. Wiley, 1969.

[16] A. P. Sage and J. L. Melsa. *Estimation Theory and Applications in Communications and Control*. McGraw-Hill, 1971.

[17] M. D. Srinath and P. K. Rajasekaran. *An Introduction to Statistical Signal Processing with Applications*. Wiley, 1979.

[18] N. Mohanty. *Random Signals Estimation and Identification – Analysis and Applications*. Van Nostrand Reinhold, 1986.

[19] J. M. Mendel. *Lessons in Digital Estimation Theory*. Prentice-Hall, 1987.

[20] S. M. Kay. *Fundamentals of Statistical Signal Processing – Estimation Theory*. Prentice-Hall, 1993.

[21] R. E. Kalman. A new approach to linear filtering and prediction problems. *J. Basic Engineering* 82D, 1960.

[22] R. E. Kalman. New results in linear filtering and prediction theory. In *Proc. 1st Symp. of Engineering Applications of Random Function Theory and Probability*. Wiley, 1961, pp. 270–388.

[23] P. Swerling. First order error propagation in a statewise smoothing procedure for satellite observations. *J. Astronautical Sciences* 6, 1959.

[24] S. M. Bozic. *Digital Kalman Filtering*. J. Wiley, 1979.

[25] A. Jazwinski. *Stochastic Processes and Filtering Theory*. Academic Press, 1970.

[26] C. T. Leondes, ed. *Theory and Applications of Kalman Filtering*. AGARD Monograph, 1970.

[27] A. Gelb. *Applied Optimal Estimation*. MIT Press, 1974.

[28] T. P. McGarty. *Stochastic Systems and State Estimation*. Wiley, 1974.

[29] P. S. Maybeck. *Stochastic Models, Estimation, and Control*, volume 1. Academic Press, 1979.

[30] G. J. Bierman. *Factorization Methods for Discrete Sequential Estimation*. Academic Press, 1977.

[31] A. V. Balakrishnan. *Kalman Filtering Theory*. Springer-Verlag, 1984.

[32] C. F. Gauss. Theoria motus corporum coelestium in sectionibus conicis solem ambientium. Published in 1809 (LS work completed in 1795).

[33] P. S. Laplace. Théorie analytique des probabilités (1812). http://en.wikipedia.org/wiki/Pierre-Simon_Laplace.

[34] Galileo Galilei. Dialogo dei massimi sistemi. 1632.

[35] K. Pearson. Contributions to the mathematical theory of evolution, 1894.

[36] R. A. Fisher. On an absolute criterion for fitting frequency curves, 1912.

[37] R. A. Fisher. On the mathematical foundations of theoretical statistics, 1921.

[38] H. Cramér. *Mathematical Methods of Statistics*. Princeton University Press, 1946.

[39] S. S. Wilks. *Mathematical Statistics*. Wiley, 1962.

[40] C. R. Rao. *Linear Statistical Inference and Its Applications*. Wiley, 1965.

[41] M. T. Wasan. *Parametric Estimation*. McGraw-Hill, 1970.

[42] H. W. Sorenson. *Parameter Estimation - Principles and Problems*. M. Dekker, 1980.

[43] P. J. Huber. *Regression Statistic*. Wiley, 1981.

[44] P. J. Rousseeuw and A. M. Leroy. *Robust Regression and Outlier Detection*. Wiley, 2003.

[45] Y. Dodge, ed. *Statistical Data Analysis Based on the L1-Norm and Related Method*. North-Holland, 1987.

[46] P. Bloomfield and W. L. Steiger. *Least Absolute Deviation*. Birkhauser, 1983.

[47] E. J. Candès and M. B. Wakin. An introduction to compressive sampling. *IEEE Signal Proc.* 25, 2008.

[48] R. G. Baraniuk, E. Candès, R. Nowak, and M. Vetterli, eds. *Special Issue on Compressive Sampling. IEEE Signal Proc.* 25, 2008.

[49] S. Boyd and L. Vandenberghe. *Convex Optimization*. Cambridge University Press, 2004.

[50] P. M. Halmos and L. J. Savage. Applications of the radon-nikodym theorem to the theory of sufficient statistics. *Annals Mathematical Statistics*, 1949.

[51] C. R. Rao. *Information and the Accuracy Attainable in the Estimation of Statistical Parameters*. Bulletin Calcutta Math. Soc., 1945.

[52] H. L. Van Trees. *Detection, Estimation, and Modulation Theory*, volume 1. Wiley, 1969.

[53] P. R. Kumar and P. Varaiya. *Stochastic Systems*. Prentice-Hall, 1986.

[54] N. Levanon. *Radar Principles*. Wiley, 1988.

[55] L. L. Scharf. *Statistical Signal Processing*. Addison-Wesley, 1991.

[56] R. D. Gitlin, J. F. Hayes, and S. B. Weinstein. *Data Communication Principles*. Plenum, 1992.

[57] D. H. Johnson and D. E. Dudgeon. *Array Signal Processing*. Prentice-Hall, 1993.

[58] R. N. McDonough and A. D. Whalen. *Detection of Signals in Noise*. Academic Press, 2d ed., 1995.

[59] R. J. Sullivan. *Radar Foundations for Imaging and Advanced Concepts*. Scitech, 2004.

[60] T. A. Schonhoff and A. A. Giordano. *Detection and Estimation Theory*. Pearson, 2006.

[61] T. P. Bayes and R. Price. An essay towards solving a problem in the doctrine of chances. *Phil. Trans. Royal Soc., London* 53, 1763.

Problems

7.1 Let $\{s(t), -\infty < t < \infty\}$ be a zero-mean real-valued stationary process with a covariance function $R(t)$.

(a) Find $\hat{a}$ that minimizes $\varepsilon^2 = \mathrm{E}\{[s(t+\tau) - as(t)]^2\}$. Find $\varepsilon^2_{\min}$.
(b) Let $R(t) = \delta(t)$, $-\infty < t < \infty$. Repeat part (a).
(c) Let $R(t) = e^{-\alpha|t|}$, $\alpha > 0$. Consider the optimum linear estimator using any linear combinations of $\{s(u), -\infty < u \leq t\}$. Show this optimum linear estimator is no better than just the optimum linear estimator $\hat{a}s(t)$ considered in part (a).

7.2

(a) Let X be a uniform r.v. defined on $[0, 1]$ and the r.v. $Y = X^2$.
   1. Find the minimum mean-square estimate $\hat{Y}$ given the r.v. $X = x$. Find the associated mean-square error $\varepsilon^2_{\min}$.
   2. Find the minimum mean-square estimate $\hat{X}$ given the r.v. $Y = y$. Find the associated mean-square error $\varepsilon^2_{\min}$.

(b) Let X be a uniform r.v. defined on $[-1, 1]$ and the r.v. $Y = X^2$.
   1. Find the minimum mean-square estimate $\hat{Y}$ given the r.v. $X = x$. Find the associated mean-square error $\varepsilon^2_{\min}$.
   2. Find the minimum mean-square estimate $\hat{X}$ given the r.v. $Y = y$. Find the associated mean-square error $\varepsilon^2_{\min}$.

7.3 Consider the classical discrete-time Wiener filtering problem. Let the observed random sequence be modeled by $\{X(n) = S(n) + V(n), -\infty < n < \infty\}$, where the signal sequence $\{S(n), -\infty < n < \infty\}$ and the noise sequence $\{V(n), -\infty < n < \infty\}$ are two real-valued zero-mean uncorrelated wide-sense stationary sequences with autocorrelation sequences $R_S(n)$ and $R_V(n)$ and power spectral densities $S_S(f)$, $-1/2 \leq f < 1/2$, and $S_V(f)$, $-1/2 \leq f < 1/2$, respectively.

(a) By using the Orthogonal Principle, find the expression for the optimum sequence $\{h(n), -\infty < n < \infty\}$, for the minimum linear mean-square estimator $\hat{S}(n) = \sum_{-\infty}^{\infty} h(k)X(n-k)$, $-\infty < n < \infty$ of $S(n)$ using all the $\{X(n), -\infty < n < \infty\}$. Do not attempt to solve for the $h(k)$ directly.
(b) By taking the Fourier transform of both sides of the equation involving $h(k)$ in part (a), find the optimum transfer function $H(f) = \sum_{k=-\infty}^{\infty} h(k)\exp(-i2\pi kf)$, $-1/2 \leq f < 1/2$.
(c) Now, let $R_S(n) = a^{|n|}\sigma_S^2$, $-\infty < n < \infty$, $-1 < a < 1$, and $V(n)$ be a white sequence with $\sigma_V^2 = 1$. Find the autocorrelation sequences $R_{SX}(n)$ and $R_X(n)$, and psd's $S_{SX}(f)$ and $S_X(f)$, and the optimum transfer function $H(f)$ explicitly.
(d) Find the minimum mean-square-error $\varepsilon^2_{\min} = \mathrm{E}\{(S(n) - \hat{S}(n))^2\}$ expressed first as a function of $R_S(n)$ and $h(n)$, second as a function of $S_S(f)$ and $H(f)$, and third as a function of $S_S(f)$ and $S_V(f)$.

7.4 Consider the classical continuous-time Wiener prediction/filtering/smoothing problem. This continuous-time problem is analogous to the discrete-time problem considered in Problem 7.3. Let the observed random process be modeled by $\{X(t) = S(t) + V(t),\ -\infty < t < \infty\}$, where the signal process $\{S(t),\ -\infty < t < \infty\}$ and the noise process $\{V(t),\ -\infty < t < \infty\}$ are two real-valued zero mean uncorrelated wide-sense stationary processes with autocorrelation functions $R_S(t)$ and $R_V(t)$ and power spectral densities $S_S(f),\ -\infty < f < \infty$, and $S_V(f),\ -\infty < f < \infty$, respectively.

(a) By using the Orthogonal Principle, find the expression for the optimum linear filter response function $\{h(t),\ -\infty < t < \infty\}$, for the minimum linear mean-square estimator

$$\hat{S}(t+a) = \int_{-\infty}^{\infty} h(\tau)X(t-\tau)\,d\tau,\ -\infty < t < \infty,$$

of $S(t)$ using all the $\{X(t),\ -\infty < t < \infty\}$. If $a > 0$, this is called the classical Wiener prediction problem. If $a = 0$, this is called the classical Wiener filtering problem. If $a < 0$, this is called the classical Wiener smoothing problem. Do not attempt to solve for the $h(t)$ directly.

(b) By taking the Fourier transform of both sides of the equation involving $h(t)$ in part (a), find the optimum transfer function $H(f),\ -\infty < f < \infty$.

(c) Suppose

$$R_S(\tau) = \exp(-|\tau|),\ -\infty < \tau < \infty$$

and

$$R_V(\tau) = 2\delta(\tau),\ -\infty < \tau < \infty.$$

Find the psd's $S_S(f),\ -\infty < f < \infty$, and $S_X(f),\ -\infty < f < \infty$, and the optimum transfer function $H(f)$, explicitly.

(d) Find the minimum mean-square-error

$$\varepsilon^2_{\min} = \mathrm{E}\left\{(S(t+a) - \int_{-\infty}^{\infty} h(\tau)X(t-\tau)\,d\tau)^2\right\}$$

expressed first as a function of $R_S(t)$ and $h(t)$, second as a function of $S_S(f)$ and $H(f)$, and third as a function of $S_S(f)$ and $S_V(f)$.

7.5 Consider the classical continuous-time casual Wiener prediction/filtering/smoothing problem. Let the observed random process be modeled by $\{X(\tau) = S(\tau) + V(\tau),\ -\infty < \tau \le t\}$, where the signal process $\{S(t),\ -\infty < t < \infty\}$ and the noise process $\{V(t),\ -\infty < t < \infty\}$ are two real-valued zero-mean uncorrelated wide-sense stationary processes with autocorrelation functions $R_S(t)$ and $R_V(t)$ and power spectral densities $S_S(f),\ -\infty < f < \infty$, and $S_V(f),\ -\infty < f < \infty$, respectively. By using the Orthogonal Principle, find the expression for the optimum linear causal impulse response function $\{h(t),\ 0 \le t < \infty\}$, for the minimum linear mean-square estimator

$$\hat{S}(t+a) = \int_0^\infty h(\tau)X(t-\tau)\,d\tau,\ -\infty < t < \infty,$$

of $S(t)$ using only the $\{X(\tau),\ -\infty < \tau \le t\}$. This equation is called the Wiener–Hopf equation. Without additional conditions on the two random processes, it is difficult to find the explicit solution of the optimum linear causal $h(t)$ impulse response function. If the two random processes have rational power spectral densities for $S_S(f)$ and $S_V(f)$, then the solution of the optimum linear causal $h(t)$ impulse response function (or its transfer function $H(f)$) can be obtained using a quite involved spectral factorization method (which we will not consider here).

7.6 We have considered various linear estimation problems under the mean-square-error criterion. Under this criterion (with all the r.v.'s having zero means), all the statistical information needed to solve the normal equation is provided by the correlation (i.e., second moment) values of the r.v.'s. Now, consider the use of the mean-quartic criterion to address the "simplest linear estimation" of the r.v. Y by aX. (Note: quartic means the fourth power.) As usual, we assume both Y and aX have zero means. Now, we have

$$\begin{aligned}
\varepsilon^4 &= \mathrm{E}\left\{(Y-aX)^4\right\} = \mathrm{E}\left\{(Y-aX)^2(Y-aX)^2\right\} = \mathrm{E}\left\{(Y^2-2aXY+a^2X^2)^2\right\} \\
&= \mathrm{E}\left\{Y^4 - 4aXY^3 + 2a^2X^2Y^2 + 4a^2X^2Y^2 - 4a^3X^3Y + a^4X^4\right\} \\
&= \mathrm{E}\left\{Y^4 - 4aXY^3 + 6a^2X^2Y^2 - 4a^3X^3Y + a^4X^4\right\} \\
&= \mathrm{M}(0,4) - 4a\mathrm{M}(1,3) + 6a^2\mathrm{M}(2,2) - 4a^3\mathrm{M}(3,1) + a^4\mathrm{M}(4,0),
\end{aligned} \tag{7.255}$$

where we denote the (m, k)th mixed moment of $\{X, Y\}$ by $\mathrm{M}(m,k) = \mathrm{E}\{X^mY^k\}$. We note, all the mixed moments satisfy $m + k = 4$ due to the "quarticness" of the error criterion. The first question is how do we evaluate these five $\mathrm{M}(m, k)$ mixed moments? In the mean-square criterion problem, using this mixed moment notation, we only had to deal with $\mathrm{M}(2,0) = \sigma_X^2$, $\mathrm{M}(1,1) = \mathrm{E}\{XY\}$, and $\mathrm{M}(0,2) = \sigma_Y^2$. In order to evaluate these mixed moments in (7.255), we need to have the joint pdf of X and Y. Thus, we assume now the r.v.'s X and Y are jointly Gaussian with zero means and a correlation matrix given by

$$\mathbf{R} = \begin{bmatrix} 1 & 0.5 \\ 0.5 & 1 \end{bmatrix}. \tag{7.256}$$

Then the joint pdf of X and Y is given by

$$p_{X,Y}(x,y) = \frac{1}{(2\pi)|\mathbf{R}|^{1/2}} \exp\left(-\frac{[x,y]\mathbf{R}^{-1}[x,y]^T}{2}\right). \tag{7.257}$$

Then its characteristic function $\phi_{X,Y}(v_1, v_2)$ is given by

$$\phi_{X,Y}(v_1,v_2) = \int_{-\infty}^{\infty}\int_{-\infty}^{\infty} p_{X,Y}(x,y)\exp(ixv_1 + iyv_2)dxdy = \exp\left(\frac{[v_1,v_2]\mathbf{R}[v_1,v_2]^T}{2}\right). \tag{7.258}$$

However, it is also known that the mixed moments can be obtained from the characteristic function from the expression

$$\mathrm{M}(m,k) = \mathrm{E}\left\{X^m Y^k\right\} = (-i)^{m+k} \frac{\partial^{m+k}\phi_{X,Y}(v_1,v_2)}{\partial v_1^m \partial v_2^k}\bigg|_{v_1=v_2=0}. \tag{7.259}$$

Now, using (7.259), the five mixed moments in (7.255) are available, then find the optimum $\hat{a}_4$ that minimizes ε^4. What is the optimum $\hat{a}_2$ under the mean-square-error criterion? Now, we appreciate why the mean-square error criterion from the computational point of view is preferable compared to the mean-quartic-error criterion.

Hint: Consider the direct evaluation of (7.258) for

$$\phi_{X,Y}(v_1,v_2),\quad \frac{\partial\phi_{X,Y}(v_1,v_2)}{\partial v_1},\quad \frac{\partial\phi_{X,Y}(v_1,v_2)}{\partial v_2},\quad \frac{\partial^2\phi_{X,Y}(v_1,v_2)}{\partial v_1^2},\quad \frac{\partial^3\phi_{X,Y}(v_1,v_2)}{\partial v_1^3},$$

$$\frac{\partial^4\phi_{X,Y}(v_1,v_2)}{\partial v_1^4},\quad \frac{\partial^4\phi_{X,Y}(v_1,v_2)}{\partial v_1^4}\bigg|_{v_1=v_2=0},\quad \frac{\partial^4\phi_{X,Y}(v_1,v_2)}{\partial v_2\partial v_1^3},\quad \frac{\partial^4\phi_{X,Y}(v_1,v_2)}{\partial v_2\partial v_1^3}\bigg|_{v_1=v_2=0},$$

$$\frac{\partial^3\phi_{X,Y}(v_1,v_2)}{\partial v_2\partial v_1^2},\quad \frac{\partial^4\phi_{X,Y}(v_1,v_2)}{\partial v_2^2\partial v_1^2},\quad \frac{\partial^4\phi_{X,Y}(v_1,v_2)}{\partial v_2^2\partial v_1^2}\bigg|_{v_1=v_2=0}.$$

7.7 Consider a zero mean wide-sense stationary random sequence $S(k)$, $-\infty < k < \infty$, with an autocorrelation function $R_S(k) = 4 \times (1/2)^{-|k|}$, $-\infty < k < \infty$. Let this sequence $S(k)$ be observed in the presence of an additive zero mean white sequence $V(k)$, $-\infty < k < \infty$, with a σ_V^2. Namely, denote the observed random sequence by $X(k) = S(k) + V(k)$, $-\infty < k < \infty$. Use the Kalman filtering method of Section 7.2.3.2 to find the steady state solution of the Kalman gain p as a function of the SNR.

7.8 Using M statistically independent samples x_i, $i = 1, \ldots, M$, of a Gaussian process with mean μ and unknown variance σ^2:

(a) show that the maximum-likelihood estimate of σ^2 is $\hat{\sigma}^2 = (1/M)\sum_{i=1}^{M}(x_i - \mu)^2$. Hint: Start with the pdf of $\{x_i,\ i = 1, \ldots, M\}$ involving $V = (1/M)\sum_{i=1}^{M}(x_i - \mu)^2$.
(b) show that V is a sufficient statistic. Hint: Start with V has a Gamma distribution with degree M.
(c) show that σ^2 is also an efficient estimator.

7.9 A single sample is observed from $X = A + W$, where A is a deterministic parameter to be estimated and W is a r.v. with the pdf given by $p_W(u)$. Show that the Cramér–Rao bound for the unbiased estimate $\hat{A}$ of A is given by $\mathrm{var}(\hat{A}) \geq \left\{\int_{-\infty}^{\infty} \frac{\left(\frac{dp_W(u)}{du}\right)^2}{p_W(u)} du\right\}^{-1}$.

Evaluate this estimate for the Laplacian pdf $p_W(u) = \frac{1}{\sqrt{2}\sigma} e^{-\sqrt{2}|u|/\sigma}$, and compare the result to the Gaussian case.

7.10 The realizations of $X_n = ar^n + W_n$, $n = 0, \ldots, N-1$, are observed, where W_n are white Gaussian with variance σ^2, and $0 < r < 1$ is known. Find the CR bound for

a. Show that an unbiased efficient estimator exists, and find its variance. What happens to the variance as $N \to \infty$?

7.11 The realizations of $X_n = a + W_n$, $n = 1, \ldots, N$, are observed, where $\mathbf{W} = [W_1, \ldots, W_N]^T \sim \mathcal{N}(\mathbf{0}, \mathbf{C})$, and $\mathbf{C} = (C_{ij})_{i,j=1,\ldots,N}$ is an arbitrary covariance matrix (positive definite, symmetric). Find the CR bound for a. Does an unbiased and efficient estimator exist? If so, what is its variance?

7.12 Consider the classical Chebyshev polynomial of the first kind, $\{p_k(x), -1 < x < 1, k = 0, 1, \ldots\}$, where $p_0(x) = 0$, and $p_1(x) = x$, and the other terms are defined by the three-term relationship given by $p_{k+1}(x) = 2xp_k(x) - p_{k-1}(x)$, $k = 1, 2, \ldots$. Furthermore, these polynomials form a complete orthogonal system with the weight function $\rho(x) = 1/(\pi\sqrt{1-x^2})$, $-1 < x < 1$, such that

$$\int_{-1}^{1} p_k(x)p_j(x)\rho(x)dx = \begin{cases} 1, & j = k = 0 \\ 1/2, & j = k > 0, \\ 0, & j \neq k. \end{cases}$$

Consider a function $f(x) \in \mathbb{L}_1((-1, 1), \rho(x)) \cap \mathbb{L}_2((-1, 1), \rho(x))$. That is, $f(x)$ satisfies $\int_{-1}^{1} |f(x)|\rho(x)dx < \infty$ and $\int_{-1}^{1} |f(x)|^2\rho(x)dx < \infty$. Then $f(x)$ has a complete orthogonal expansion given by

$$f(x) = \sum_{k=0}^{\infty} b_k p_k(x), \tag{7.260}$$

where the expansion coefficients are given by

$$b_0 = \int_{-1}^{1} f(x)\rho(x)dx, \tag{7.261}$$

$$b_k = 2\int_{-1}^{1} f(x)p_k(x)\rho(x)dx,\ k = 1, 2, \ldots. \tag{7.262}$$

Now, consider the function $f(x) = \exp(-x^2/2)/(\sqrt{2\pi}\rho(x))$, $-1 < x < 1$, and its complete orthogonal expansion of (7.260). From (7.261), the b_0 coefficient is given by

$$b_0 = \int_{-1}^{1} f(x)\rho(x)dx = \int_{-1}^{1} \left(\exp(-x^2/2)/\sqrt{2\pi}\right) dx = 1 - 2\text{Q}(1) = 0.6827, \tag{7.263}$$

and from (7.262), the b_k coefficients are given by

$$b_k = 2\int_{-1}^{1} \left(\exp(-x^2/2)/\sqrt{2\pi}\right) p_k(x)dx,\ k = 1, 2, \ldots. \tag{7.264}$$

(a) Tabulate b_k $k = 0, 1, \ldots, 50$ using (7.263) and (7.264). Note, from (7.264), the Gaussian pdf in the round bracket is an even function and the $p_k(x)$ function for odd integral values of k is an odd function, over the symmetric interval form $(-1, 1)$, then all b_k with odd integral values of k must have the values of zero. Hint: You can use either symbolic integration methods (e.g., Mathematica, Matlab, Maple, etc.) or numerical integration methods to find these coefficients to four significant decimal places. Do your tabulated b_k with odd integral values of k have zero values?

(b) Since half of the coefficients of b_k are zero, perhaps we can use the compressive sensing method of finding these coefficients by treating this as the solution of a sparse linear system of equations. Rewrite the expansion of (7.260) with only $L+1$ number of terms in the sum evaluated at N values of $f(x_i)$, $-1 < x_i < 1$, $i = 1, \ldots, N$, as

$$\mathbf{f} = \begin{bmatrix} f(x_1) \\ f(x_2) \\ \vdots \\ \vdots \\ f(x_N) \end{bmatrix} = \begin{bmatrix} 1 & p_1(x_1) & p_2(x_2) & \ldots & p_L(x_1) \\ 1 & p_1(x_2) & p_2(x_2) & \ldots & p_L(x_2) \\ \vdots & \vdots & \vdots & \vdots & \vdots \\ 1 & p_1(x_N) & p_2(x_N) & \ldots & p_L(x_N) \end{bmatrix} \begin{bmatrix} \hat{b}_0 \\ \hat{b}_1 \\ \vdots \\ \vdots \\ \hat{b}_L \end{bmatrix} = \boldsymbol{\Psi}\hat{\mathbf{b}}, \tag{7.265}$$

where $\mathbf{f}$ is a known $N \times 1$ vector, $\boldsymbol{\Psi}$ is a known $N \times (L+1)$ matrix, and $\hat{\mathbf{b}}$ is an unknown $(L+1) \times 1$ vector whose solution we seek. In the CS method, introduce the $M \times N$ dimensional random matrix $\tilde{\boldsymbol{\Phi}}$, where each component $\tilde{\Phi}_{i,j} \sim \mathcal{N}(0, 1/N)$. Orthonormalize all the rows of $\tilde{\boldsymbol{\Phi}}$ to be denoted by $\boldsymbol{\Phi}$ (i.e., $\boldsymbol{\Phi} = \mathrm{orth}\left(\tilde{\boldsymbol{\Phi}}^T\right)^T$. In the CS method, the desired sparse solution $\hat{\mathbf{b}}$ is found as the minimum of $\hat{\mathbf{b}}$ under the l_1 criterion subject to the constraint of $\boldsymbol{\Phi}\boldsymbol{\Psi}\hat{\mathbf{b}} = \boldsymbol{\Phi}\mathbf{f}$. This minimization under the l_1 criterion subject to a constraint can be performed using various optimization programs including the Matlab cvx program. For your computations, take $N = L + 1 = 2000$ and $M = N/2 = 1000$. Tabulate $\hat{b}_k$, $k = 0, 1, \ldots, 50$. Do your tabulated b_k with odd integral values of k have zero values? How do the values of $\hat{b}_k$, $k = 0, 1, \ldots, 50$ compare to the values of b_k, $k = 0, 1, \ldots, 50$ in part (a) of this problem?

(c) Now, use the classical least-square method to find the $\hat{\mathbf{b}}$ by minimizing the residual of the norm $(\mathbf{f} - \boldsymbol{\Psi}\hat{\mathbf{b}}, 2)$ under the l_2 norm. Tabulate this set of $\hat{b}_k$, $k = 0, 1, \ldots, 50$. How does this set of $\hat{b}_k$ compare to those obtained in part (a) and part (b) of this problem?

7.13 We observe N independent, identically distributed samples with an exponential density,

$$p(x; \lambda) = \begin{cases} \lambda e^{-\lambda x}, & x \geq 0, \\ 0, & x < 0. \end{cases}$$

Find the maximum likelihood estimator for λ (make sure it maximizes the likelihood function).

7.14 Let the observed samples be realizations taken from

$$X(n) = \begin{cases} W(n), & 0 \le n_0 - 1, \\ s(n - n0) + W(n), & n_0 \le n \le n_0 + M - 1, \\ W(n), & n_0 + M \le n \le N - 1, \end{cases}$$

where $W(n)$ are i.i.d. zero-mean Gaussian r.v.'s with variance σ^2. Verify that the ML estimate of n_0 is obtained by maximizing $\sum_{n=n_0}^{n_0+M-1} x(n)s(n - n_0)$.

8 Analytical and simulation methods for system performance analysis

In this chapter we consider various analytical and simulation tools for system performance analysis of communication and radar receiver problems. In Section 8.1, we treat the analysis of receiver performance with Gaussian noise, first using the closure property of Gaussian vectors under linear operations. We then address this issue without using this closure property. Section 8.2 deals with the analysis of receiver performance with Gaussian noise and other random interferences caused by intersymbol interferences (ISI) due to bandlimitation of the transmission channel. Section 8.2.1 introduces the evaluation of the average probability of error based on the moment bounding method. Section 8.3 considers the analysis of receiver performance with non-Gaussian noises including the spherically invariant random processes (SIRP). By exploiting some basic properties of SIRP, Section 8.3.1 obtains a closed form expression for the receiver. We determine the average probability of error for the binary detection problem with additive multivariate t-distributed noise (which is a member of SIRP). Section 8.3.2 again uses some properties of SIRP to model wireless fading channels with various fading envelope statistics. By using Fox H-function representations of these pdfs, novel average probability of error expressions under fading conditions can be obtained. Section 8.3.3 treats the probabilities of a false alarm and detection of a radar problem with a robustness constraint. Section 8.4 first shows a generic practical communication/radar system, which may have various complex operations, making analytical evaluation of system performance in many cases difficult. Section 8.4.1 introduces the use of Monte Carlo (MC) simulation. Section 8.4.2 shows that the use of importance sampling can reduce the number of samples in MC simulations yet still meet system performance requirements. Section 8.5 includes a brief conclusion. Section 8.6 provides some comments on various background information and possible engineering applications of materials considered in this chapter. In Appendix 8.A, we will first introduce some basic concepts on pseudo-random (PR) number generation. Then in Appendix 8.A.1, we use Matlab rand.m for the PR generation of an i.i.d. sequence having a continuous uniformly distributed pdf on $[0, 1]$. In Appendix 8.A.2, we use Matlab randn.m for the PR generation of an i.i.d. sequence having a zero mean and unit variance Gaussian distributed pdf on $(-\infty, \infty)$. In Appendix 8.A.3, an i.i.d. sequence having arbitrary cdf $F(\cdot)$ can be obtained by using the inverse transformation $F^{-1}(\cdot)$ operating on an i.i.d. sequence generated by rand.m. Appendix 8.B includes Theorem 8.3 needed in Section 8.3.2 for the study of wireless fading channel.

8.1 Analysis of receiver performance with Gaussian noise

Consider the binary detection problem, where the received scalar r.v. X is modeled by

$$X = \begin{cases} N, & H_0, \\ s + N, & H_1, \end{cases} \tag{8.1}$$

where N is a scalar Gaussian r.v. with a zero mean and variance σ^2 under both hypotheses and s is a known positive number under H_1. From Chapter 2 we know several optimum decision rules are all based on the LR test, which for this problem results in a threshold test. Denote the decision region for H_1 by $\mathbb{R}_1 = \{x \in \mathbb{R} : x \geq \gamma_0\}$ and the decision region for H_0 by $\mathbb{R}_0 = \{x \in \mathbb{R} : x \leq \gamma_0\}$, where γ_0 is a parameter that depends on s, σ^2, and the specific decision rule. Then the probability of a false alarm P_{FA} is given by

$$\begin{aligned} P_{\text{FA}} = P(x \in \mathbb{R}_1 | H_0) &= \int_{\gamma_0}^{\infty} \frac{e^{-x^2/2\sigma^2}}{\sqrt{2\pi}\sigma} dx = \int_{\gamma_0/\sigma}^{\infty} \frac{e^{-x^2/2}}{\sqrt{2\pi}} dx = \mathrm{Q}(\gamma_0/\sigma) \\ &= 1 - \Phi(\gamma_0/\sigma), \end{aligned} \tag{8.2}$$

and the probability of detection P_{D} is given by

$$\begin{aligned} P_{\text{D}} = \mathrm{P}(x \in \mathbb{R}_1 | H_1) &= \int_{\gamma_0}^{\infty} \frac{e^{-(x-s)^2/2\sigma^2}}{\sqrt{2\pi}\sigma} dx = \int_{(\gamma_0 - s)/\sigma}^{\infty} \frac{e^{-t^2/2}}{\sqrt{2\pi}} dt \\ &= \mathrm{Q}((\gamma_0 - s)/\sigma) = 1 - \Phi((\gamma_0 - s)/\sigma), \end{aligned} \tag{8.3}$$

where $\Phi(\cdot)$ is the cumulative distribution function of a zero mean and unit-variance Gaussian r.v. and $\mathrm{Q}(\cdot)$ is the complementary Gaussian cumulative distribution function.

In (8.1), if N is a Laplacian r.v. of zero mean, then for a positive-valued threshold γ_0, its probability of a false alarm P_{FA} is given by

$$P_{\text{FA}} = \int_{\gamma_0}^{\infty} \frac{\alpha e^{-\alpha x}}{2} dx = (1/2) e^{-\gamma_0 \alpha}, \tag{8.4}$$

and the probability of detection P_{D} is given by

$$P_{\text{D}} = \int_{\gamma_0}^{\infty} \frac{\alpha e^{-\alpha(x-s)}}{2} dx = 1 - \int_{-\infty}^{\gamma_0} \frac{\alpha e^{\alpha x}}{2} dx = 1 - (1/2) e^{-\gamma_0 \alpha}. \tag{8.5}$$

Example 8.1 In both the Gaussian and the Laplacian cases, define SNR(dB)$=$ $20 \log_{10}(s/\sigma)$ or $s/\sigma = 10^{\text{SNR(dB)}/20}$. First, we set the variance of the Laplacian r.v. $\sigma_{\text{L}}^2 = 2/\alpha^2$ to the variance σ^2 of the Gaussian r.v. to 1, then $\alpha = \sqrt{2}$. Furthermore, we constrain $P_{\text{FA}}^{\text{L}} = P_{\text{FA}}$. Then $\gamma_0 = 1.6804$. Figure 8.1 shows a plot of the Laplacian P_{D}^{L} and the Gaussian P_{D} versus SNR(dB) (for $s \geq \gamma_0$), where the dashed curve is for the Laplacian case and the solid curve is for the Gaussian case. For SNR less than about 7.5 dB, these two probabilities of detection are about the same. For higher SNR, since a Gaussian pdf of the r.v. X (with its mean about s under H_1) decreases much more rapidly than that of the Laplacian r.v., since the probabilities of both pdf's from s to ∞ are equal to $1/2$, but the probability of the Gaussian pdf from γ_0 to s is greater than the

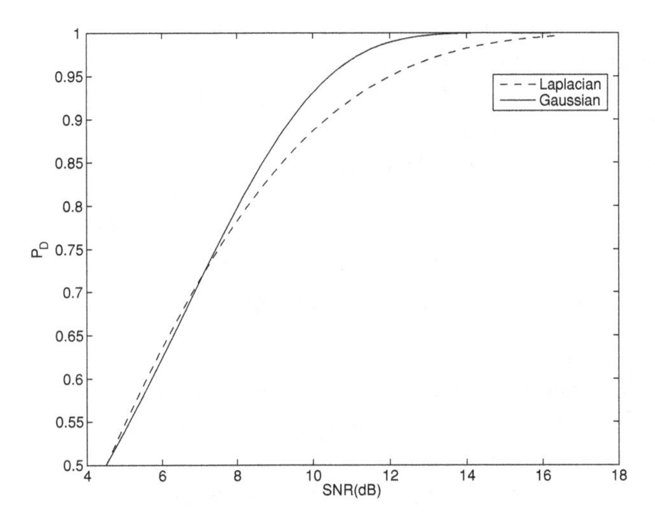

Figure 8.1 Plots of the Laplacian P_D^L (dashed curve) and the Gaussian P_D (solid curve) versus SNR(dB).

probability of the Laplacian pdf. Thus, the Gaussian p_D is greater than the Laplacian P_D^L for large values of SNR. □

Next, consider the binary detection of known real-valued binary signal vectors in WGN as given in (4.1) of Chapter 4. In particular, we take the anti-podal signal vectors $\mathbf{s}_0$ under H_0 and $\mathbf{s}_1$ under H_1 of length n to satisfy $\mathbf{s}_1 = \mathbf{s} = -\mathbf{s}_0$ having equal energy $E = \|\mathbf{s}\|^2$. Then the data vector is modeled by

$$\mathbf{X} = \begin{cases} -\mathbf{s} + \mathbf{N}, & H_0 \\ \mathbf{s} + \mathbf{N}, & H_1 \end{cases} \tag{8.6}$$

where $\mathrm{E}\{\mathbf{N}\} = \mathbf{0}$ and the $\mathrm{E}\{\mathbf{N}\mathbf{N}^T\} = \sigma^2 \mathbf{I}_n$. Then using a LR test for H_0 and H_1 with the LR constant $\Lambda_0 = 1$, from (4.1) the decision regions $\mathbb{R}_1$ and $\mathbb{R}_0$ are then given by

$$\mathbb{R}_1 = \left\{ \mathbf{x} : \{2\mathbf{x}^T\mathbf{s} \geq 0\} \right\}, \ \mathbb{R}_0 = \left\{ \mathbf{x} : \{2\mathbf{x}^T\mathbf{s} < 0\} \right\}. \tag{8.7}$$

From (4.12), we know the sufficient statistic given by

$$\Gamma = \mathbf{X}^T(\mathbf{s}_1 - \mathbf{s}_0) = 2\mathbf{X}^T\mathbf{s} \tag{8.8}$$

is a scalar Gaussian r.v. whose mean value for our problem under H_0 is given by $\mu_{\Gamma_0} = -2E$, under H_1 is given by $\mu_{\Gamma_1} = 2E$, and having a variance of $\sigma_\Gamma^2 = 4\sigma^2 E$ under both hypotheses. Thus, the probability of a false alarm is the probability of a Gaussian r.v. Γ of mean $\mu_{\Gamma_0} = -2E$ and variance $\sigma_\Gamma^2 = 4\sigma^2 E$, integrated over $(0, \infty)$, given by

$$\begin{aligned} P_{\mathrm{FA}} &= \int_0^\infty \frac{1}{\sqrt{2\pi\sigma_{\Gamma_0}^2}} \exp(-(\gamma - \mu_{\Gamma_0})^2/2\sigma_{\Gamma_0}^2) d\gamma \\ &= \int_{\|\mathbf{s}\|/\sigma}^\infty \frac{1}{\sqrt{2\pi}} \exp(-t^2/2) dt = \mathrm{Q}(\|\mathbf{s}\|/\sigma) = \mathrm{Q}(\sqrt{E/\sigma^2}). \end{aligned} \tag{8.9}$$

Of course, the result in (8.9) is identical to the P_{FA} expression of (4.19) of Chapter 4 for the two anti-podal signal vectors with $\Lambda_0 = 1$.

However, suppose we are not aware of the concept of a sufficient statistic Γ of (8.8), including the fact that Γ is a scalar Gaussian r.v. In other words, suppose we can only use explicitly the property of the decision region $\mathbb{R}_1$ given by (8.7). Under H_0, we can rewrite $\mathbb{R}_1$ as

$$\begin{aligned}\mathbb{R}_1 &= \left\{\mathbf{x} : \{2\mathbf{x}^T\mathbf{s} \geq 0\}\right\} = \left\{\mathbf{x} : \{\mathbf{x}^T\mathbf{s} \geq 0\}\right\} \\ &= \left\{\mathbf{n} : \{(-\mathbf{s}+\mathbf{n})^T\mathbf{s} \geq 0\}\right\} = \left\{\mathbf{n} : \{-(\mathbf{s}^T\mathbf{s}) + \mathbf{n}^T\mathbf{s} \geq 0\}\right\} = \left\{\mathbf{n} : \{\mathbf{n}^T\mathbf{s} \geq E\}\right\}.\end{aligned} \tag{8.10}$$

Then the P_{FA} is the n-dimensional integral of an n-th order Gaussian pdf of zero mean and covariance $\sigma^2\mathbf{I}_n$ integrated over the n-dimensional region $\mathbb{R}_1$ of (8.10) given by

$$P_{FA} = \iiint_{\mathbf{n}^T\mathbf{s}\geq E} \frac{1}{(2\pi)^{n/2}} \exp(-\mathbf{n}^T\mathbf{n}/2\sigma^2)d\mathbf{n}. \tag{8.11}$$

In general, since $\mathbf{s}$ is an arbitrary vector of length n of energy E, the integral of (8.11) does not seem to have an analytically tractable solution due to the integration region defining $\mathbb{R}_1$.

Example 8.2 Consider the solution of finding P_{FA} of (8.11) for the binary anti-podal detection problem of (8.6) based upon simulation. Let $\mathbf{s}$ be an arbitrary vector of length n of energy E. Consider M realizations of the Gaussian vector $\mathbf{n}_{(m)} = \text{randn}(n, 1)$, $m = 1, \ldots, M$. Denote the total number of times, Count, when $(\mathbf{s}^T\mathbf{n}_{(m)}) \geq E$. Then the simulated $\tilde{P}_{FA}$ is given by

$$\tilde{P}_{FA} \approx \text{Count}/M. \tag{8.12}$$

In (8.6)–(8.11), set $\sigma^2 = 1$, $n = 10$, and $E = 4$. Then the SNR(dB)$= 10\log_{10}(E/\sigma^2) =$ 6.02 dB. From (8.9), the analysis shows $P_{FA} = \text{Q}(2) = 0.02275$.

Case a. Take the specific $\mathbf{s} = \sqrt{4/10} * \text{ones}(10, 1)$ vector of length $n = 10$ and energy $E = 4$. Consider an initial seed of randn('state', 19) for each M. The second column of Table 8.1 shows the simulated $\tilde{P}_{FA}$ by using (8.12) for Case a for various values of M.

Case b. We use the same parameters as in Case a, except now the initial seed uses randn('state', 2000). The third column of Table 8.1 shows the simulated $\tilde{P}_{FA}$ by using (8.12) for Case b for various values of M.

Case c. We use the same parameters as in Case a, except $\mathbf{s}$ is chosen in a PR manner by using rand('state', 19), $\mathbf{z} = \text{rand}(10, 1)$, $\mathbf{s} = (2/\text{norm}(\mathbf{z})) * \mathbf{z}$, which still has an energy of $E = 4$. The fourth column of Table 8.1 shows the simulated $\tilde{P}_{FA}$ by using (8.12) for Case c for various values of M.

For this simple anti-podal deterministic signal vectors in WGN detection problem, we note that having the analytical theory showing that the sufficient statistic Γ was a scalar Gaussian r.v. allowed us to evaluate explicitly the P_{FA} and P_D values in terms of the

Table 8.1 Three cases of simulated $\tilde{P}_{FA}$ vs. the number of simulation terms M at SNR $= 6.02$ dB.

M	$\tilde{P}_{FA}$(Case a)	$\tilde{P}_{FA}$(Case b)	$\tilde{P}_{FA}$(Case c)
100	0.03000	0.01000	0.01000
1,000	0.02800	0.02500	0.01800
10,000	0.02260	0.02320	0.02290
100,000	0.02253	0.02266	0.02323
1,000,000	0.02279	0.02270	0.02305
5,000,000	0.02276	0.02270	0.02279
10,000,000	0.02274	0.02271	0.02270

Table 8.2 Two cases of simulated $\tilde{P}_{FA}$ vs. the number of simulation terms M at SNR $= 12.0$ dB and SNR $= 14.0$ dB.

M	$\tilde{P}_{FA}$ (12.0 dB)	$\tilde{P}_{FA}$ (14.0 dB)
10,000	0	0
100,000	1.000E − 5	0
1,000,000	3.800E − 5	1.000E − 6
5,000,000	3.540E − 5	4.000E − 7
10,000,000	3.460E − 5	3.000E − 7

Q(·) function expression. Without the analytical theory, we need to revert to brute-force simulation over an n-dimensional space. For this problem, with a large number M of simulation terms, the simulation results of all three cases are essentially close to the theoretical result. These three cases show that the simulation results are essentially not dependent on the initial seed (particularly for large values of M). These results also show, in the presence of WGN, the actual values of the signal vector $\mathbf{s}$ (as long as the energy of $\mathbf{s}$ is the same) are not important in determining the system performance. □

Example 8.3 We take all the parameters of Case a in Example 8.2 except now we first consider $E = 16$ yielding a SNR(dB) $= 12.0$ dB and then consider $E = 25$ yielding a SNR(dB) $= 14.0$ dB. For the 12.0 dB case, (8.9) shows $P_{FA} = 3.1671\text{E} - 5$. For the 14.0 dB case, (8.9) shows $P_{FA} = 3.8665\text{E} - 7$. Now, evaluate the simulated $\tilde{P}_{FA}$ results using (8.12) for these two cases. As seen in Table 8.2, for a system with a very small P_{FA}, one needs a very large number of simulation runs to obtain an accurate result. For the 12.0 dB case, using $M = 1\text{E}6$, 5E6, and 1E7 number of simulation runs, the evaluated corresponding $\tilde{P}_{FA}$ results in Table 8.2 may give us a reasonable belief that the true P_{FA} perhaps may have an approximate value of about 3E − 5. On the other hand, for the 14.0 dB case, using $M = 1\text{E}6$, 5E6, and 1E7 number of simulation runs, the evaluated corresponding $\tilde{P}_{FA}$ results do not seem to give us a hint that the true P_{FA} may have an approximate value of 3.8E − 7. Indeed, for a true $P_{FA} = 3.8665E-7$, we need to obtain about 39 error events over a simulation run of $M = 100\,000\,000 = 1E8$, in order to obtain an approximate $\tilde{P}_{FA} = 39/1E8 = 3.9E-7$. If we used only $M = 10\,000\,000 = 1E7$ simulation runs, then our result of $\tilde{P}_{FA} = 3E-7$ was due to only 3 error events, which certainly does not give us much confidence in the accuracy of that simulated result. In conclusion, as the desired simulated result takes on a small value, the number of simulation runs must increase. To obtain a reasonably accurate simulated $\tilde{P}_{FA}$, intuitively one wants to use a minimum M_{min} number of required simulation runs to be from 10 to 100 times larger than $1/P_{FA}$. For the 14.0 dB case, $1/P_{FA} = 2.7E-6$. Then based on the above approximate expression, we need to have $2.7E7 < M_{min} < 2.7E8$ number of simulation runs. Indeed, using only $M = 1E7$ runs for the 14.0 dB case, the estimated value of $\tilde{P}_{FA} = 3.000E-7$ in Table 8.2 can not be considered to be a reliable estimate of the true P_{FA}.Of course, the brute-force approach of increasing the number of simulation runs has its practical limitations due to excessive simulation times and possible limitation due to computer memory requirements. □

8.2 Analysis of receiver performance with Gaussian noise and other random interferences

Consider a baseband binary communication system over a linear time-invariant band-limited channel modeled by

$$X(t) = \sum_{j=-\infty}^{\infty} A_j h(t - jT) + N(t), \quad -\infty < t < \infty, \tag{8.13}$$

where $\{A_j, -\infty < j < \infty\}$ is an i.i.d. binary information data sequence taking values of ± 1 with equal probability, $h(\cdot)$ is the linear time-invariant impulse response, and $N(t)$ is a zero mean WGN with a two-sided spectral density of $N_0/2$, and the binary data are transmitted every T seconds. The receiver performs sampling at $t = kT + t_0$, yielding

$$X(kT + t_0) = \sum_{j=-\infty}^{\infty} A_j h(kT - jT + t_0) + N(kT + t_0), \quad -\infty < k < \infty. \tag{8.14}$$

For simplicity of notation, denote

$$X_k = X(kT + t_0), \; h_{j-k} = h((j-k)T + t_0), \; N_k = N(kT + t_0). \tag{8.15}$$

Suppose the receiver makes a decision for the binary information data A_0 by using only the sampled value of X_k at $k = 0$, given by

$$X_0 = A_0 h_0 + \sum_{\substack{j=-\infty \\ j \neq 0}}^{\infty} A_j h_j + N_0, \tag{8.16}$$

where t_0 is taken such that the magnitude of $h_0 = h(t_0)$ is the maximum of all $|h(t)|$, $-\infty < t < \infty$. Without loss of generality, we assume h_0 is positive valued and denote

$$Z = \sum_{\substack{j=-\infty \\ j\neq 0}}^{\infty} A_j h_j, \tag{8.17}$$

as a r.v. due to the random behavior of the binary information data A_j for values of $j \neq 0$, and N_0 is a Gaussian r.v. of zero mean variance σ^2. Since h_j is the sampled value of a linear system impulse response function, $|h_j|$ generally decreases (not necessarily monotonically) from its peak value of h_0 as j approaches $\pm\infty$. Thus, in practice with only a finite number of terms in the sum of (8.17), the intersymbol interference (ISI) r.v. Z of (8.17) can be modeled more realistically as

$$Z = \sum_{j=-J_1,\ldots,-1}^{1,\ldots,J_2} A_j h_j, \tag{8.18}$$

with only J_1 terms to the left and J_2 to the right of the $j = 0^{\text{th}}$ term [1]. Denote the total number of terms in (8.18) as $J = J_1 + J_2$. We also note, since $\{A_j,\ j = -J_1, \ldots, -1,\ 1, \ldots, J_2\}$ is a binary sequence of J terms, then the r.v. Z can take on 2^J possible realizations. The total distortion D is defined by

$$D = \sum_{j=-J_1,\ldots,-1}^{1,\ldots,J_2} |h_j|. \tag{8.19}$$

Assuming the decision rule is given by

$$X_0 = A_0 h_0 + \sum_{j=-J_1,\ldots,-1}^{1,\ldots,J_2} A_j h_j + N_0 \begin{cases} \geq 0, \text{ Decide } A_0 = 1, \\ \leq 0, \text{ Decide } A_0 = -1, \end{cases} \tag{8.20}$$

then the probability of a false alarm with ISI is given by

$$\begin{aligned} P_{\text{FA}}(\text{ISI}) &= \text{P}(\text{Decide } A_0 = 1 | A_0 = -1) = P(-h_0 + Z + N > 0) \\ &= \text{P}(Z + N > h_0) = \text{P}(N > h_0 - Z) = \text{P}(N > h_0 + Z) \\ &= \text{E}_Z\{\text{P}(N > h_0 + z | Z = z)\} = \text{E}_Z\{\text{Q}(h_0 + Z)/\sigma\} \\ &= \frac{1}{2^J} \sum_{j=1}^{2^J} \text{Q}((h_0 + z_j)/\sigma), \end{aligned} \tag{8.21}$$

where z_j is one of 2^J realizations of the ISI r.v. Z of (8.18).

Example 8.4 Consider a sampled linear time-invariant impulse response sequence given in Table 8.3. First, take $J_1 = 2$ and $J_2 = 4$ (i.e., assume $h_{-3} = h_5 = 0$). Then $J = 6$ and thus there are $2^6 = 64$ sums in (8.21) that need to be averaged to obtain P_{FA}. Then from (8.19), the total distortion $D = 0.75$. The first column of Table 8.4 shows SNR(dB)

Table 8.3 Sampled impulse response sequence.

h_{-3}	h_{-2}	h_{-1}	h_0	h_1	h_2	h_3	h_4	h_5
−0.001	0.05	−0.2	1	−0.25	0.13	0.1	−0.02	0.002

Table 8.4 Probability of a false alarm with no ISI P_{FA}(No ISI) and with ISI P_{FA}(ISI) with $J = 6$ and $J = 8$ vs. SNR.

SNR(dB)	P_{FA}(No ISI)	P_{FA}(ISI) $J = 6$	P_{FA}(ISI) $J = 8$
6.02 dB	2.275E − 2	5.305E − 2	5.305E − 2
12.04 dB	3.167E − 5	8.911E − 3	8.913E − 3
15.56 dB	9.866E − 10	2.233E − 3	2.233E − 3

$= 20 \log_{10}(h_0/\sigma)$, the second column shows the ideal $P_{FA}(\text{No ISI}) = Q(h_0/\sigma)$, the third column shows P_{FA}(ISI) with $J = 6$, and the fourth column shows P_{FA}(ISI) with $J = 8$.

Clearly, while P_{FA}(No ISI) decreases rapidly with increasing SNR, P_{FA}(ISI) with $J = 6$ decreases quite slowly due to the degradation of the ISI with a quite large value of $D = 0.75$. Now, taking $J = 8$ (including the effects of $h_{-3} = -0.001$ and $h_5 = 0.002$), then $D = 0.753$. As expected since the total distortion under $J = 8$ is very close to that under $J = 6$, we note the values of P_{FA}(ISI) with $J = 8$ are essentially identical to those values of P_{FA}(ISI) with $J = 6$. This also shows the original assumption of taking $J = 6$ was reasonable. □

However, in severe bandlimited scenarios, the number of significant impulse response sequence terms J can be very long. In Example 8.4, with either $J = 6$ or $J = 8$, the number of terms of the $Q(\cdot)$ function to be averaged in (8.21) was $2^6 = 64$ or $2^8 = 256$, which are quite computationally acceptable. However, when J becomes very large, it is not feasible to compute 2^J number of terms in (8.21). Thus, some other method is needed to tackle this problem.

8.2.1 Evaluation of P_e based on moment bound method

Consider the Chebyshev impulse response function $h(t)$ given by (8.22), sampled at $t = -20T : 20T$, with $h(0) = 1$ [2].

$$h(t) = 0.4023 \cos(2.839|t|/T - 0.7553) \exp(-0.4587|t|/T) + 0.7162 \cos(1.176|t|/T - 0.1602) \exp(-1.107|t|/T). \quad (8.22)$$

The significant impulse response sequence terms are from $J_1 = 20$ to $J_2 = 20$ resulting in $J = 40$ and a total distortion of $D = 0.274$. While the total distortion of this impulse response function is fairly low, its overall length is very high. Clearly, the brute-force averaging of $2^{40} \approx 10^{12}$ number of terms in (8.21) is not computationally feasible. While there are several possible methods to evaluate the average probability of error P_e of a binary detection problem, we will consider the use of the moment space method [2].

From probability theory, statistical moments are well known and useful. We know for the r.v. X, the mean $E\{X\}$ is the first moment, $E\{X^2\}$ is the second moment, and $\sigma^2 = E\{(X - E\{X\})^2\}$ is the variance. Then $E\{g(X)\}$ is defined as the generalized moment of X w.r.t. the deterministic function $g(\cdot)$. We note the average probability of error P_e in the binary detection problem given in (8.13)–(8.21), when A_j has an equal probability of $\frac{1}{2}$ of taking ± 1, is given by

$$P_e = \mathrm{P}(\text{Decide } A_0 = 1 | A_0 = -1) \times \frac{1}{2} + \mathrm{P}(\text{Decide } A_0 = -1 | A_0 = 1) \times \frac{1}{2}$$

$$= E_Z\{Q((h_0 + Z)/\sigma)\} \tag{8.23}$$

$$= E_Z\{[Q((h_0 + |Z|)/\sigma) + Q((h_0 - |Z|)/\sigma)]/2\}. \tag{8.24}$$

We note, P_e as given by either (8.23) or (8.24) is a generalized moment of either $g(x) = Q((h_0 + x)/\sigma)$ or $g(x) = [Q((h_0 + |x|)/\sigma) + Q((h_0 - |x|)/\sigma)]/2$. As shown in (8.21), the expectation of $E\{\cdot\}$ can be evaluated by brute force as the average of 2^J terms, when 2^J is sufficiently small. When 2^J is too large, instead of performing the average by brute force with 2^J terms, we may be able to obtain tight upper and lower bounds of the generalized moment of P_e using some other appropriate upper and lower bounds of generalized moments based on the Moment Space Isomorphism Theorem.

Theorem 8.1 ***Moment Space Isomorphism Theorem*** *[2, 3]*
Let Z be a r.v. with a pdf $p_Z(z)$ defined over a finite closed interval $\mathbb{I} = [a, b]$. Let $k_z(z)$, $k_2(z)$, ..., $k_N(z)$ be a set of N continuous functions defined on $\mathbb{I}$. Then the generalized moment of the r.v. Z generated by the function $k_i(z)$ is denoted by

$$m_i = \int_{\mathbb{I}} k_i(z) p_Z(z) dz = E_Z\{k_i(Z)\},\ i = 1, \ldots, N. \tag{8.25}$$

We denote the N-th moment space by

$$\mathcal{M} = \{\mathbf{m} = [m_1, m_2, \ldots, m_N] \in \mathbb{R}^N : m_i = \int_{\mathbb{I}} k_i(z) p_Z(z) dz,\ i = 1, \ldots, N\}, \tag{8.26}$$

where $p_Z(z)$ ranges over all the possible pdfs defined on $\mathbb{I}$. Then $\mathcal{M}$ is a closed, bounded, and convex set. Now, let C be the curve $\mathbf{r} = (r_1, r_2, \cdots, r_N)$ traced out in $\mathbb{R}^N$ by $r_i = k_i(z)$, $i = 1, \cdots, N$, for $z \in \mathbb{I}$. Let $\mathcal{H}$ be the convex hull of C. Then these two convex bodies are identical

$$\mathcal{M} = \mathcal{H}. \tag{8.27}$$

□

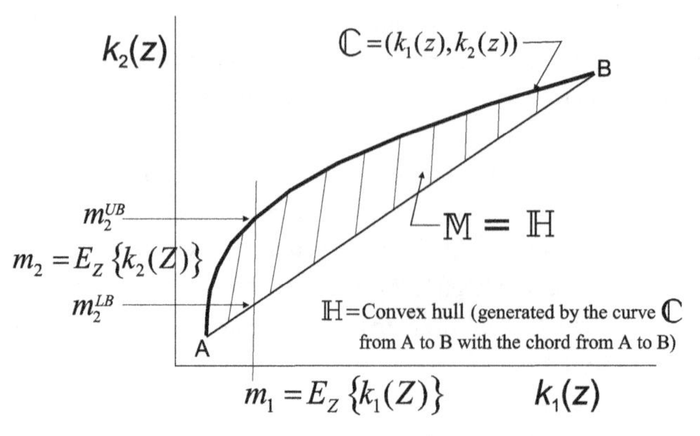

Figure 8.2 Convex body $\mathcal{M} = \mathcal{H}$ in two dimensions generated from the convex hull of the curve $\mathcal{C} = (k_1(z), k_2(z))$ from A to B with the chord from A to B and the evaluation of the upper and lower bounds of m_2 from the evaluated m_1 based on the Isomorphism Theorem.

To illustrate the Isomorphism Theorem of (8.27), consider the specific case of $N = 2$, with the curve $\mathcal{C}$ generated parametrically by some $(k_1(z), k_2(z))$ as shown in Fig. 8.2. Generate its convex hull $\mathcal{H}$ using the curve $\mathcal{C}$ from A to B and the chord from A to B by the shaded region in Fig. 8.2. Suppose we want to evaluate $m_2 = \mathrm{E}_Z\{k_2(Z)\}$, where $k_2(z)$ is specified by a problem of interest, but we are not able to do so readily. However, we are able to find a function $k_1(z)$ so that we can evaluate $m_1 = \mathrm{E}_Z\{k_1(Z)\}$ readily. Then the Isomorphism Theorem using (8.27) shows that for the given value of m_1, as indicated by the vertical dashed line in Fig. 8.2, intersecting the convex body of $\mathcal{M}$ (which is identical to the convex body of $\mathcal{H}$), thus yields the upper bound m_2^{UB} and the lower bound m_2^{LB} to the desired $m_2 = \mathrm{E}_Z\{k_2(Z)\}$. In other words, from the known computable value of $m_1 = \mathrm{E}_Z\{k_1(Z)\}$, we are able to obtain m_2^{LB} and m_2^{UB} satisfying

$$m_2^{\mathrm{LB}} \leq m_2 = \mathrm{E}_Z\{k_2(Z)\} \leq m_2^{\mathrm{UB}}, \tag{8.28}$$

without needing to perform the actual statistical generalized moment evaluation of $m_2 = \mathrm{E}_Z\{k_2(Z)\}$.

While $k_2(z)$ is specified by the problem of interest, we are completely free to choose $k_1(z)$. Of course, the goodness of the moment bounding method as seen in Fig. 8.2 is a function of the thickness of the convex body where the vertical line (given by the value of m_1) intersects the convex body. Ideally, we would like the convex body to be as thin as possible. Of course, if $k_1(z) = k_2(z)$, then the convex body will be a finite length chord of zero thickness, so $m_2^{\mathrm{LB}} = m_2^{\mathrm{UB}} = m_2 = m_1$. But since we were not able to evaluate $m_2 = \mathrm{E}_Z\{k_2(Z)\}$ equally, we were not able to evaluate $m_1 = \mathrm{E}_Z\{k_1(Z)\}$. In the following Example 8.5 and Example 8.6, we illustrate the use of the Isomorphism Theorem to obtain the upper and lower bounds of P_{e} for the Chebyshev impulse response function of (8.22) with $D = 0.274$ and $J = 40$. We note, Matlab has the program convhull.m that yields the convex hull of a curve defined by P points in two dimensions of the form $\{k_1(z_i),\ k_2(z_i),\ z_i \in \mathbb{I},\ i = 1, \ldots, P\}$.

Table 8.5 Variance and exponential lower and upper bounds vs. SNR(dB).

SNR(dB)	Variance LB	Exponential LB	Exponential UB	Variance UP
4 dB	5.77E − 2	5.76E − 2	5.78E − 2	5.77E − 2
8 dB	6.75E − 3	6.69E − 3	6.82E − 3	6.84E − 2
12 dB	6.14E − 5	6.29E − 5	6.84E − 5	1.05E − 4
16 dB	1.29E − 9	3.58E − 9	4.44E − 9	1.04E − 7
20 dB	5.01E − 21	1.25E − 17	1.56E − 17	1.30E − 14

Example 8.5 Variance bound [2]. Take $k_1(z) = z^2/D$ and $k_2(z) = [\mathrm{Q}((h_0 + |Z|)/\sigma) + \mathrm{Q}((h_0 - |Z|)/\sigma)]/2$, for $-D \leq z \leq D$. By direct analysis, the upper and lower envelope of convex body $\mathcal{M}$ can be found and thus upper and lower bounds of $m_2 = \mathrm{E}_Z\{k_2(Z)\}$ can be expressed explicitly in terms of

$$m_1 = \mathrm{E}\{Z^2/D\} = \sigma_Z^2/D = (1/D) \sum_{j=-20,\,\ldots,\,-1}^{1,\ldots,20} h_j^2. \tag{8.29}$$

Details on the evaluations of these bounds are omitted here [2], but their values for five SNR values are given in the second and fifth columns in Table 8.5. We note the variance lower and upper bounds are quite close for low SNR values, but these bounds become quite close for large SNR values. □

Example 8.6 Exponential Bound [2].

In order to obtain tightness between the lower and upper bounds of the convex body, one needs to find a $k_1(z)$ that approximates $k_2(z)$ well over $[-D, D]$, but still having a $m_1 = \mathrm{E}_Z\{k_1(Z)\}$ that can be readily evaluated. We know $\mathrm{Q}(x)$ is well approximated by $(2\pi x^2)^{-1/2}\exp(-x^2/2)$, for $x \geq 3$. Now, we choose $k_1(z) = \exp(c(h+z))$, for some real-valued parameter c to be chosen carefully, and $k_2(z) = \mathrm{Q}((h+z)/\sigma)$, for $-D \leq z \leq D$. Since $\mathrm{Q}(x)$ decreases exponentially as $-ax^2$, and $k_1(x)$ decreases exponentially as $-cx$, we would expect $k_1(x)$ can approximate $k_2(x)$ for large values of x (at high SNR values) much better than a $k_1(x)$ that has powers of x (such as in the case of the variance bound). More importantly, now we can evaluate $m_1 = \mathrm{E}_Z\{k_1(Z)\}$ as given by

$$\begin{aligned} m_1 &= \mathrm{E}\left\{\exp\left(c(h_0 + \sum_{j=-20,\ldots,-1}^{1,\ldots,20} A_j h_j)\right)\right\} = \exp(ch_0) \prod_{j=-20,\ldots,-1}^{1,\ldots,20} \mathrm{E}\{\exp(cA_j h_j)\} \\ &= \exp(ch_0) \prod_{j=-20,\ldots,-1}^{1,\ldots,20} \cosh(ch_j), \end{aligned} \tag{8.30}$$

readily. The actual lower and upper envelopes of $\mathcal{M}$ are fairly involved functions of the parameters c and σ (which is a function of the SNR). In particular, for a given σ, we can find an optimum $\hat{c}$ to minimize the thickness between the lower and upper envelope.

Again, details on the solutions of the $\hat{c}$ are omitted here but are given in [2]. The lower and upper bounds on P_e using the optimized $\hat{c}$ for four values of SNRs are given in the third and fourth columns of Table 8.5. As can be seen, the exponential lower and upper bounds of P_e remain very tight even for large SNR values. These exponential lower and upper bounds remain very tight essentially over all four values of SNRs, unlike the behaviors of the lower and upper bounds for the variance bounds.

We note, the moment space bounding method based on the Isomorphism Theorem has been utilized to evaluate bounds on average error probability for binary communication systems with ISI using N moments with $N > 2$ [4], with uncertainties in the moments [5, 6], interferences from other users in a Gold-coded CDMA system [7]; for co-channel interferences [8], and for bounds on probability of missed detection in cognitive radio [9].

8.3 Analysis of receiver performance with non-Gaussian noises

8.3.1 Noises with heavy tails

The use of a Gaussian noise model in communication and radar systems is physically justified based on standard assumptions of the consequence of Central Limit Theorem on the additions of many small independent random disturbances of approximately similar contributions. Of course, a Gaussian noise model is also attractive due to its analytical tractability in the analysis of the receiver performance of such a system. It turns out that there are various conditions in which the Gaussian noise model is not valid and the pdfs of the associated noise process have heavy tails. A univariate Gaussian pdf decreases in the limit for a large value of x as $C_1 \exp(-ax^2)$, for some positive C_1 and a. A univariate pdf for a large value of x behaving as $C_2 \exp(-ax^p)$, for some positive C_2 and $p < 2$, is often said to be heavy-tailed (HT). In particular, HT pdfs include all those tails for large values of x decrease algebraically as $C_3 x^{-q}$, for some positive C_3 and q. Figure 8.3a shows the absolute value of 10 000 realizations of a PN Gaussian sequence of norm equal to 10 having a mean of zero and standard deviation $\sigma = 0.1$. From Gaussian theory, the probability of the absolute value larger than $3\sigma = 0.3$ is approximately 2.7×10^{-3}. For 10 000 samples, we would then expect about 27 such events. From these 10 000 realizations, Fig. 8.3a, we actually observed 24 such events. From theory, the probability of the absolute value larger than $4\sigma = 0.4$ is 6.6×10^{-5}. For 10 000 samples, we would then expect about 0.66 such event. From these 10 000 realizations, we actually observed one such event. The conclusion is that for a Gaussian process, most of the realizations are within few σ of the mean of the process. In other words, there are very few large valued "rare events" for a Gaussian process. Figure 8.3b shows 10 000 realizations of a PN Weibull distributed sequence having a norm of ten and Fig. 8.3c shows 10 000 realizations of a PN Student t-distributed sequence also having a norm of ten. For these two HT pdfs, we observed many more "rare events" having very large values.

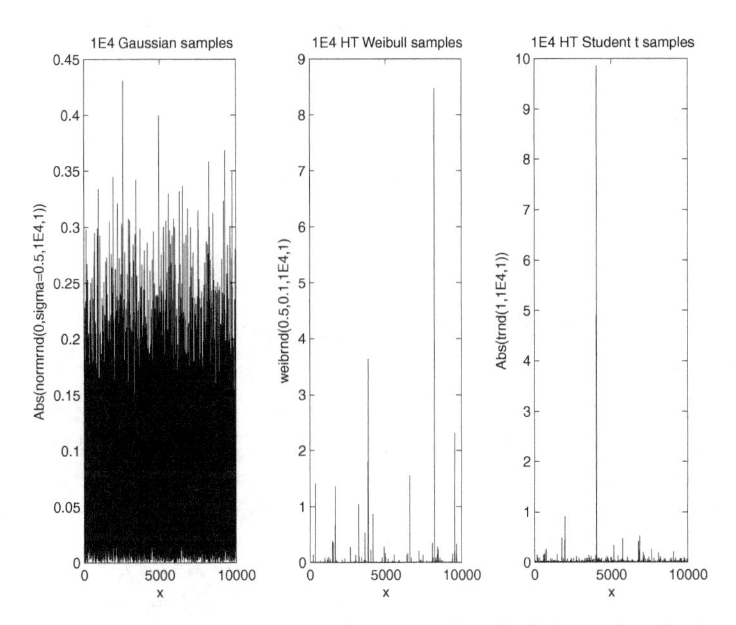

Figure 8.3 Plots of 10 000 PR Gaussian (a), PR Weibull (b), and PR Student t (c) realizations.

An $n \times 1$ Gaussian random vector $\mathbf{X}$ is fully characterized by its quadratic form $Q_n(\mathbf{x}, \boldsymbol{\mu}, \mathbf{R}) = (\mathbf{x} - \boldsymbol{\mu})^T \mathbf{R}^{-1}(\mathbf{x} - \boldsymbol{\mu})$, where $\boldsymbol{\mu}$ is the $n \times 1$ mean vector and $\mathbf{R}$ is an $n \times n$ positive-definite covariance matrix. An $n \times 1$ random vector $\mathbf{X}$ having a pdf of the form $p_{\mathbf{X}}(\mathbf{x}) = D_n h(Q_n(\mathbf{x}, \boldsymbol{\mu}, \mathbf{R}))$ for a scalar positive-valued $h(\cdot)$ and a positive-valued normalization constant D_n is said to be a spherically invariant random vector (SIRV). A random process $\{X(t), -\infty < t < \infty\}$ is called spherically invariant (SIRP) if any set of n sampled r.v.'s taken from $X(t)$ forms a SIRV $\mathbf{X}$.

Theorem 8.2 ***Spherically Invariant Random Process (SIRP) Representation*** *[10]. A necessary and sufficient condition for a random process to be a SIRP is its n-dimensional pdf satisfies*

$$p_{\mathbf{X}}(\mathbf{x}) = (2\pi)^{-n/2} |\mathbf{R}|^{-1/2} h((\mathbf{x} - \boldsymbol{\mu})^T \mathbf{R}^{-1}(\mathbf{x} - \boldsymbol{\mu})),$$
$$h(r) = \int_0^\infty v^{-n} \exp(-r/(2v^2)) p_V(v) dv, \tag{8.31}$$

where $p_V(\cdot)$ is an arbitrary one-dimensional pdf defined on $(0, \infty)$. □

A SIRV is characterized by $(n, \boldsymbol{\mu}, \mathbf{R}, p_V)$.

Example 8.7 Let $p_V(v) = \delta(v - 1)$, then (8.31) shows $h(r) = \exp(-r/2)$ and

$$p_{\mathbf{X}}(\mathbf{x}) = (2\pi)^{-n/2} |\mathbf{R}|^{-1/2} \exp\{-(\mathbf{x} - \boldsymbol{\mu})^T \mathbf{R}^{-1}(\mathbf{x} - \boldsymbol{\mu})/2\}, \tag{8.32}$$

which yields the well-known n-th order Gaussian pdf. □

Example 8.8 Let

$$p_V(v) = 2\exp\{-1/(\beta v^2)\}v^{-(2\alpha+1)}/(\Gamma(\alpha)\beta^\alpha),\ 0 \le v < \infty,\ 0 < \alpha,\ 0 < \beta, \quad (8.33)$$

then (8.31) becomes

$$p_{\mathbf{X}}(\mathbf{x}) = \frac{\Gamma(\alpha + n/2)}{(2\pi)^{n/2}\Gamma(\alpha)|\mathbf{R}|^{1/2}\left(\frac{1}{\beta} + \mathbf{x}^T\mathbf{R}^{-1}\mathbf{x}\right)^{\alpha+n/2}}, \quad (8.34)$$

which is the n-th order generalized multivariate t-distributed pdf with 2α degrees of freedom [10]. □

By using other one-dimensional $p_V(\cdot)$ pdf's in (8.31), various explicit multivariate pdf's can also be obtained.

Now, consider the detection of a known $n \times 1$ vector $\mathbf{s}$ in the presence of an $n \times 1$ multivariate SIRV $\mathbf{N}$ modeled by

$$\mathbf{X} = \begin{cases} \mathbf{N}, & H_0, \\ \mathbf{s} + \mathbf{N}, & H_1, \end{cases} \quad (8.35)$$

where $\mathbf{s}$ is a $n \times 1$ deterministic vector of energy $\|\mathbf{s}\|^2 = E$ and $\mathbf{N}$ is an arbitrary $n \times 1$ zero mean SIRV characterized by $(n, \mathbf{0}, \mathbf{R}, p_V)$ having a multivariate pdf $p_{\mathbf{N}}(\cdot)$ of the form given by (8.31). Then under the LR test with a unity LR constant Λ_0,

$$\Lambda(\mathbf{x}) = p_{\mathbf{X}}(\mathbf{x}|H_1)/p_{\mathbf{X}}(\mathbf{x}|H_0) = p_{\mathbf{N}}(\mathbf{x} - \mathbf{s})/p_{\mathbf{N}}(\mathbf{x}) \begin{cases} \ge 1 \Rightarrow \text{Decide } H_1, \\ \le 1 \Rightarrow \text{Decide } H_0, \end{cases} \quad (8.36)$$

or equivalently

$$\int_0^\infty \left[\exp\{-(\mathbf{x} - \mathbf{s})^T\mathbf{R}^{-1}(\mathbf{x} - \mathbf{s}\}/(2v^2)) - \exp\{-\mathbf{x}^T\mathbf{R}^{-1}\mathbf{x}/(2v^2)\}\right] v^{-n} p_V(v) dv$$
$$\begin{cases} > 0 \Rightarrow \text{Decide } H_1, \\ < 0 \Rightarrow \text{Decide } H_0, \end{cases} \quad (8.37)$$

or equivalently

$$\exp\{-(\mathbf{x} - \mathbf{s})^T\mathbf{R}^{-1}(\mathbf{x} - \mathbf{s})/(2v^2)\} \begin{cases} > \exp\{-\mathbf{x}^T\mathbf{R}^{-1}\mathbf{x}/(2v^2)\} \Rightarrow \text{Decide } H_1, \\ < \exp\{-\mathbf{x}^T\mathbf{R}^{-1}\mathbf{x}/(2v^2)\} \Rightarrow \text{Decide } H_0. \end{cases} \quad (8.38)$$

Taking the $\ln(\cdot)$ on both sides of (8.38) yields

$$\Gamma = \mathbf{x}^T\mathbf{R}^{-1}\mathbf{s} \begin{cases} > \mathbf{s}^T\mathbf{R}^{-1}\mathbf{s}/2 = \gamma_0 \Rightarrow \text{Decide } H_1, \\ < \mathbf{s}^T\mathbf{R}^{-1}\mathbf{s}/2 = \gamma_0 \Rightarrow \text{Decide } H_0. \end{cases} \quad (8.39)$$

The sufficient statistic of $\Gamma = \mathbf{X}^T\mathbf{R}^{-1}\mathbf{s}$ of (8.39) shows the LR receiver with a unity LR constant Λ_0 for the binary detection of a known vector in SIRV noise problem of (8.35) is just a correlation receiver similar to that of a binary colored Gaussian noise detection problem. From SIRV theory [10], if $\mathbf{X}$ is a SIRV characterized by $(n, \boldsymbol{\mu}, \boldsymbol{\rho}, p_V)$

and $\mathbf{L}$ is a $n \times m$ deterministic matrix such that $\mathbf{L}^T\mathbf{L}$ is non-singular, then the $1 \times m$ vector $\mathbf{Z} = \mathbf{X}^T\mathbf{L}$ is also a SIRV characterized by $(m, \mathbf{L}^T\boldsymbol{\mu}, \mathbf{L}^T\boldsymbol{\rho}\mathbf{L}, p_V)$. In other words, the closure property of linear operations on Gaussian random vectors yielding another Gaussian random vector is also valid for SIRV. Thus, for $\Gamma = \mathbf{X}^T\mathbf{R}^{-1}\mathbf{s}$, take $\boldsymbol{\rho} = \mathbf{R}$ and $\mathbf{L} = \mathbf{R}^{-1}\mathbf{s}$. Then under H_0, Γ is a SIRV scalar r.v. with the mean $\mu_\Gamma = 0$ and the variance $\sigma_\Gamma^2 = \mathbf{s}^T\mathbf{R}^{-1}\mathbf{R}\mathbf{R}^{-1}\mathbf{s} = \mathbf{s}^T\mathbf{R}^{-1}\mathbf{s} = 2\gamma_0$. We note $2\gamma_0$ is a form of SNR. The pdf of Γ under H_0 is given by

$$\begin{aligned} \mathrm{P}_\Gamma(x|H_0) &= \int_0^\infty \frac{\exp\{-x^2/(2\cdot 2\gamma_0\cdot v^2)\}}{\sqrt{2\pi\cdot 2\gamma_0}} v^{-1} p_V(v)dv \\ &= \int_0^\infty \frac{\exp\{-x^2/(2(2\gamma_0 v^2))\}}{\sqrt{2\pi(2\gamma_0 v^2)}} p_V(v)dv. \end{aligned} \tag{8.40}$$

Then the probability of a false alarm has an analytically closed form given by

$$\begin{aligned} P_{\text{FA}} &= \int_{\gamma_0}^\infty \mathrm{P}_\Gamma(x|H_0)dx = \int_{\gamma_0}^\infty \int_0^\infty \frac{\exp\{-x^2/(2(2\gamma_0 v^2))\}}{\sqrt{2\pi(2\gamma_0 v^2)}} p_V(v)dvdx \\ &= \int_0^\infty \int_{\gamma_0}^\infty \frac{\exp\{-x^2/(2(2\gamma_0 v^2))\}}{\sqrt{2\pi(2\gamma_0 v^2)}} dx p_V(v)dv \\ &= \int_0^\infty \int_{\gamma_0/\sqrt{2\gamma_0 u^2}}^\infty \frac{\exp\{-t^2/2\}}{\sqrt{2\pi}} dt p_V(v)dv \\ &= \int_0^\infty \mathrm{Q}\left(\sqrt{\frac{\gamma_0}{2v^2}}\right) p_V(v)dv. \end{aligned} \tag{8.41}$$

Example 8.9 Consider the binary detection problem of (8.35) where $\mathbf{s}$ is an $n \times 1$ deterministic vector and $\mathbf{N}$ is an $n \times 1$ Gaussian random vector with a zero mean and a positive-definite correlation matrix $\mathbf{R}$. Then as shown in Example 8.7, $p_V(v) = \delta(v-1)$. Thus, the probability of a false alarm expression of (8.41) is now given by

$$\begin{aligned} P_{\text{FA}} &= \int_0^\infty \mathrm{Q}\left(\sqrt{\frac{\gamma_0}{2v^2}}\right)\delta(v-1)dv = \mathrm{Q}\left(\sqrt{\frac{\gamma_0}{2}}\right) \\ &= \mathrm{Q}\left(\sqrt{\frac{\mathbf{s}^T\mathbf{R}^{-1}\mathbf{s}}{4}}\right) = \mathrm{Q}\left(\frac{1}{2}\sqrt{\mathbf{s}^T\mathbf{R}^{-1}\mathbf{s}}\right), \end{aligned} \tag{8.42}$$

which is equal to the P_{FA} expression of the binary detection of a known vector in CGN of (4.111) with $\Lambda_0 = 1$ and $\mathbf{s}_0 = \mathbf{0}$ [10]. □

Example 8.10 Consider the binary detection problem of (8.35) where $\mathbf{s}$ is an $n \times 1$ deterministic vector and $\mathbf{N}$ is an $n \times 1$ n-th order generalized multivariate t-distributed pdf given by (8.34) with $\alpha = 1$. Then, as in Example 8.8, take $p_V(v) = (2/(\beta v^3))\exp(-1/(\beta v^2))$, $0 \le v < \infty$, $0 < \beta$. Then, using this $p_V(v)$ in (8.41) with a

Table 8.6 Comparison P_{FA}'s for Gaussian and t-distributed cases and simulated t-distributed $\tilde{P}_{FA}$ for $N = 1E3$, $1E4$, $1E5$, and $1E6$ samples.

SNR(dB)	Gaussian P_{FA}	t-Distribution $\tilde{P}_{FA}$ $(N = 1E3)$	$\tilde{P}_{FA}$ $(N = 1E4)$	$\tilde{P}_{FA}$ $(N = 1E5)$	$\tilde{P}_{FA}$ $(N = 1E6)$	P_{FA}
−4.00	1.197E − 1	1.600E − 1	1.825E − 1	1.801E − 1	1.800E − 1	1.800E − 1
2.04	9.321E − 3	7.200E − 2	7.210E − 2	7.166E − 2	7.166E − 2	7.147E − 2
5.56	2.087E − 4	3.400E − 2	3.850E − 2	3.852E − 2	3.557E − 2	3.588E − 2
8.06	1.268E − 6	1.900E − 2	2.040E − 2	2.133E − 2	2.109E − 2	2.116E − 2
10.0	2.033E − 9	1.110E − 2	1.470E − 2	1.419E − 2	1.386E − 2	1.386E − 2

change of variables and [20, p. 303], we obtain

$$\begin{aligned} P_{FA} &= \int_0^\infty Q\left(\sqrt{\frac{\gamma_0}{2v^2}}\right)(2/(\beta v^3))\exp\{-1/(\beta v^2)\}dv \\ &= \frac{1}{\beta}\int_0^\infty Q\left(\sqrt{\frac{\gamma_0 t}{2}}\right)\exp\{-t/\beta\}dt \\ &= \frac{1}{2}\left[1-\sqrt{\frac{\beta \mathbf{s}^T\mathbf{R}^{-1}\mathbf{s}}{\beta \mathbf{s}^T\mathbf{R}^{-1}\mathbf{s}+8}}\right]. \end{aligned} \tag{8.43}$$

In (8.43), if $\gamma_0 \to 0$, then $P_{FA} \to 1/2$, while if $\gamma_0 \to \infty$, then $P_{FA} \to 0$. □

Example 8.11 Consider the binary detection problem of (8.35) and compare the P_{FA} of Example 8.9 for the Gaussian noise vector $\mathbf{N}$ with the P_{FA} of Example 8.10 for the t-distributed noise vector $\mathbf{N}$ with the same $\mathbf{s}$ vector and the same $\mathbf{R}$ matrix. Here we take randn('state', 49), $n = 10$, $\mathbf{s} = \text{randn}(n, 1)$, $\mathbf{s} = A * \mathbf{s}/\text{norm}(\mathbf{s})$, $A = 2, 4, 6, 8, 10$, $\mathbf{R} = \text{toplitz}(\mathbf{r})$, $\mathbf{r} = 0.5^{[0:1:n-1]}$, and $\alpha = \beta = 1$. Then SNR(dB)$= 10 * \log_{10}(\|\mathbf{s}\|^2/\text{Trace}(\mathbf{R}))$. In Table 8.6, the first column shows the SNR(dB), the second column shows the analytical evaluation of the Gaussian P_{FA} obtained from (8.42), the third to sixth columns show the simulated t-distributed $\tilde{P}_{FA}$ for $N = 1E3$, $1E4$, $1E5$, and $1E6$ terms, respectively, and the seventh column shows the analytical evaluation of the t-distributed P_{FA} obtained from (8.43). From Table 8.6, we note the P_{FA} of the Gaussian noise case has a much smaller value compared to that of the P_{FA} of the t-distributed noise case for the same SNR (particularly for large values of SNR). This result is not too surprising, since the t-distributed pdf is a "heavy-tailed" pdf with a much slower decrease toward the tail of the pdf and thus more probability mass shows up at the tail region. As can be seen from Table 8.6, for the t-distributed case, increasing SNR is not very effective in improving the performance of the detector (e.g., in achieving a lower P_{FA}). Thus, performances of communication and radar systems with heavy-tailed noises may behave much worse than predicted using the standard Gaussian noise model. This ill effect of heavy-tailed noises may have catastrophic consequences in practice. □

8.3.2 Fading channel modeling and performance analysis

Consider the study of fading channels with variation of the amplitudes of the received carrier waveforms due to random propagation effects in wireless communication and radar systems [11, 12]. In a narrowband receiver, the background noise (due to Central Limit Theorem) can be represented as a narrowband Gaussian process expressed as $Y(t) = Y_{\mathrm{I}}(t)\cos(2\pi ft) - Y_{\mathrm{Q}}(t)\sin(2\pi ft)$, where $Y_{\mathrm{I}}(t)$ and $Y_{\mathrm{Q}}(t)$ are two zero mean independent low-pass Gaussian processes. Then its envelope is denoted by $R_Y(t) = (Y_{\mathrm{I}}(t)^2 + Y_{\mathrm{Q}}(t)^2)^{1/2}$, which has a Rayleigh pdf. Recently, [13–15] proposed to use an SIRP to model propagation fading on the received narrowband process given by $\{X(t) = V \bullet Y(t), -\infty < t < \infty\}$. Then, $X(t)$ is also a zero mean narrowband process and its envelope can be expressed as $R_X(t) = (X_{\mathrm{I}}(t)^2 + X_{\mathrm{Q}}(t)^2)^{1/2} = ((VY_{\mathrm{I}}(t))^2 + (VY_{\mathrm{Q}}(t))^2)^{1/2} = V \bullet R_Y(t)$, where $\bullet$ denotes a multiplication. To simplify our notation, we suppress the variable t and denote the original envelope of the Gaussian process Y by R (since it is Rayleigh distributed), and the fading SIRP envelope by X. Then, we have

$$X = V \bullet R, \tag{8.44}$$

where V is the same univariate r.v. having a pdf $p_V(v)$ used to define the n-dimensional SIRV in (8.31) of Theorem 8.2. If the SIRP model is to be useful to characterize a fading channel statistic, then we must be able to show that for any of the known $p_X(\cdot)$ envelope pdf's (e.g., Weibull, Nakagami-m, etc.) and $p_R(\cdot)$ being a Rayleigh pdf, then there must be a $p_V(\cdot)$ pdf satisfying (8.44). But from elementary probability theory, if V and R in (8.44) are two independent univariate r.v.'s, then their pdfs must satisfy

$$p_X(x) = \int_0^\infty p_V(v) p_R(x/v)(1/v) dv, \ 0 < x < \infty. \tag{8.45}$$

The question is, given an arbitrarily known fading $p_X(\cdot)$ envelope pdf and the Rayleigh pdf $p_R(\cdot)$, how does one explicitly find the $p_V(\cdot)$ pdf satisfying the integral equation of (8.45)? Finding this $p_V(\cdot)$ is not only needed to show that $p_X(\cdot)$ is the envelope of a SIRP process, but more importantly, as shown shortly in Section 8.3.2.1, various performance measures of fading communication systems such as average error probability can be evaluated when $p_V(\cdot)$ is explicitly available [16, 17].

One possible way to attempt to find the $p_V(\cdot)$ pdf given $p_X(\cdot)$ and pdf $p_R(\cdot)$ is to formulate the problem using the Mellin transform. The Mellin transform of any $p(x)$ univariate pdf defined on $[0, \infty)$ is given by

$$\mathrm{M}\{f(x)\} = \int_0^\infty p(x) x^{s-1} dx = \mathrm{E}\left\{X^{s-1}\right\} = F(s). \tag{8.46}$$

By raising both sides of (8.44) to the $(s-1)$ power and taking expectations, (8.46) shows

$$\mathrm{M}\{p_X(x)\} = \mathrm{M}\{p_V(v)\} \bullet \mathrm{M}\{p_R(r)\}. \tag{8.47}$$

Then

$$\mathrm{M}\{p_V(v)\} = \mathrm{M}\{p_X(x)\}/\mathrm{M}\{p_R(r)\}, \tag{8.48}$$

or

$$p_V(v) = \mathrm{M}^{-1}\{\mathrm{M}\{p_X(x)\}/\mathrm{M}\{p_R(r)\}\} = \mathrm{M}^{-1}\{F_X(s)/F_R(s)\}. \tag{8.49}$$

Example 8.12 In order to use (8.49) to find $p_V(\cdot)$, consider the Weibull pdf $p_X(x) = abx^{b-1}\exp(-ax^b)$, $0 \le x < \infty$, $0 < a$, $0 < b$, and its Mellin transform $F_X(s) = a^{-(s-1)/b}\Gamma((s+b-1)/b)$. The Rayleigh pdf $p_R(r) = 2ar\exp(-ar^2)$, $0 \le r < \infty$, $0 < a$, has the Mellin transform of $F_R(s) = a^{-(s-1)/2}\Gamma((s+1)/2)$. While we can find $\{F_X(s)/F_R(s)\} = [a^{-(s-1)/b}\Gamma((s+b-1)/b)]/[a^{-(s-1)/2}\Gamma((s+1)/2)]$, it is not clear how to find the Inverse Mellin of this expression to obtain $p_V(\cdot)$. □

In Example 8.12, if we replaced $p_X(x)$ by the Nakagami-m pdf, we would end up with an expression $\{F_X(s)/F_R(s)\}$ in which it is also not clear how to find the Inverse Mellin of this expression to obtain $p_V(\cdot)$. Clearly, we need another approach to solve this problem. The Fox H-function representation theory can be used to solve this problem.

8.3.2.1 Fox H-function representation method for finding $p_V(\cdot)$

The Fox H-function [18] represents a systematic approach to the characterization of various previously known and new mathematical special functions. H-functions include all generalized hypergeometric functions, confluent hypergeometric functions, etc. The Fox H-function $H(z)$ is defined as the Inverse Mellin transform $\mathrm{M}^{-1}\{\cdot\}$ of $h(s)$ denoted by

$$\begin{aligned} H(z) &= H_{p,q}^{m,n}\left[z \,\middle|\, \begin{matrix} (a_1, A_1), \ldots, (a_p, A_p) \\ (b_1, B_1), \ldots, (b_p, B_p) \end{matrix}\right] \\ &= \mathrm{M}^{-1}\{h(s)\} = \frac{1}{2\pi}\oint h(s)z^{-s}ds, \end{aligned} \tag{8.50}$$

where $\mathrm{M}^{-1}\{\cdot\}$ is the Inverse Mellin transform of $h(s)$ with the notation of

$$h(s) = \frac{\prod_{j=1}^{m}\Gamma(b_j + B_j s)\prod_{j=1}^{n}\Gamma(1 - a_j - A_j s)}{\prod_{j=m+1}^{q}\Gamma(1 - b_j - B_j s)\prod_{j=n+1}^{p}\Gamma(a_j + A_j s)}. \tag{8.51}$$

In (8.51), $\Gamma(\cdot)$ is the gamma function. In other words, $h(s)$ and its Inverse Mellin transform $H(z)$ are completely specified by $m, n, p, q, \{a_j, j = 1, \ldots, p\}, \{A_j, j = 1, \ldots, p\}, \{b_j, j = 1, \ldots, q\}$, and $\{B_j, j = 1, \ldots, q\}$ parameters. In the Non-Line-Of-Sight (NLOS) fading scenario [16, 17], most of the seven known fading envelope pdf's can be represented as a generalized gamma pdf $p_X^{(1)}(x)$ given by

$$p_X^{(1)}(x) = \frac{\beta_1' a_1'^{\alpha_1'/\beta_1'} x^{-\alpha_1'-1}\exp(-a_1' x^{\beta_1'})}{\Gamma(\alpha_1'/\beta_1')} = \frac{a_1'^{1/\beta_1'}}{\Gamma(\alpha_1'/\beta_1')} H_{0,1}^{1,0}\left[a_1'^{1/\beta_1'}x \,\middle|\, \begin{matrix} \text{------} \\ (\frac{\alpha_1'-1}{\beta_1'}, \frac{1}{\beta_1'}) \end{matrix}\right],$$

$$0 \le x < \infty,\ 0 < a_1' < \infty,\ 0 < \alpha_1' < \infty,\ 0 < \beta_1' < \infty. \tag{8.52}$$

In particular, these seven fading envelopes given in Table 8.7 have specific Fox H-function representations.

Table 8.7 Seven NLOS positive-valued univariate pdf's expressed in terms of generalized gamma function with associated H-function parameters.

Case – pdf	a_1'	α_1'	β_1'	k_1	c_1	b_1'	B_1'
1. Gamma			1	$1/\Gamma(\alpha_1')$	a_1'	$(\alpha_1'-1)$	1
2. Rayleigh		2	2	$\sqrt{a_1'}$	$\sqrt{a_1'}$	$1/2$	$1/2$
3. Weibull		$\alpha_1'=\beta_1'$		$a_1'^{1/\alpha_1'}$	$a_1'^{1/\alpha_1'}$	$(\alpha_1'-1)/\alpha_1'$	$1/\alpha_1'$
4. Nakagami-m	m/Ω	$2m$	2	$\sqrt{m/\Omega}/\Gamma(m)$	$\sqrt{m/\Omega}$	$(2m-1)/2$	$1/2$
5. Chi-squared	$1/2$	$v/2$	1	$1/(2\Gamma(v/2))$	$1/2$	$(v/2-1)$	1
6. Half-Gaussian		1	2	$\sqrt{a_1'}/\Gamma(1/2)$	$\sqrt{a_1'}$	0	$1/2$
7. 1-sided expon.		1	1	a_1'	a_1'	0	1

The reason for finding the fading envelope pdf's to have H-function representations is the usefulness of Theorem 8.3 in yielding explicit expression of $p_V(\cdot)$ for each of the seven fading envelope pdf's. Due to the long and detailed solution of $p_V(\cdot)$, we have placed Theorem 8.3 in Appendix 8.B.

For the NLOS scenario, the Rayleigh pdf, $p_R^{(2)}(x)$, is also an example of the generalized gamma pdf that can be expressed as

$$p_R^{(2)}(x) = 2a_2''x\exp(-a_2''x^2) = \sqrt{a_2''}H_{0,1}^{1,0}\left[\sqrt{a_2''}\left|\begin{matrix}---\\(1/2,\ 1/2)\end{matrix}\right.\right],\ 0\le x<\infty,\ 0<a_2''. \tag{8.53}$$

Now, using $p_X^{(1)}(x)$ of (8.52) as the $p_X^{(1)}(x)$ expression in (8.102) and $p_R^{(2)}(x)$ of (8.53) as the $p_R^{(2)}(x)$ expression in (8.103) in Theorem 8.3 in Appendix 8.B, the $p_V(\cdot)$ expression of (8.104) corresponding to any one of the seven fading envelope pdf's in Table 8.7 has the explicit form of

$$p_V(x) = \frac{a_1'^{1/\beta_1'}}{\sqrt{a_2''}\Gamma(\alpha_1'/\beta_1')}H_{1,1}^{1,0}\left[\frac{a_1'^{1/\beta_1'}}{\sqrt{a_2''}}x\left|\begin{matrix}\{(1/2,\ 1/2)\}\\\{((\alpha_1'-1)/\beta_1',\ 1/\beta_1')\}\end{matrix}\right.\right]. \tag{8.54}$$

Example 8.13 Consider Case 1 in Table 8.7 with β_1', then $p_X^{(1)}(x)$ of (8.52) becomes a gamma function and $p_V(x)$ of (8.54) is given by

$$p_V(x) = \frac{a_1'}{\sqrt{a_2''}\Gamma(\alpha_1')}H_{1,1}^{1,0}\left[\frac{a_1'}{\sqrt{a_2''}}x\left|\begin{matrix}\{(1/2,\ 1/2)\}\\\{((\alpha_1'-1),\ 1)\}\end{matrix}\right.\right]. \tag{8.55}$$

Consider the explicit evaluation of $p_V(x)$ of (8.55) for the three cases of $\alpha_1' = 1,\ 3/2$, and 2 [19].

Case 1a. Let $\alpha_1' = 1$. Then

$$p_V(x) = cH_{1,1}^{1,0}\left[cx\left|\begin{matrix}\{(1/2,\ 1/2)\}\\\{(0,\ 1)\}\end{matrix}\right.\right] = \frac{c}{\sqrt{\pi}}\exp(-c^2x^2/4),\ 0<x<\infty,\ 0<c, \tag{8.56}$$

which is just a half-Gaussian pdf on $(0, \infty)$. The result of (8.56) was obtained from (8.55), which was obtained using Theorem 8.3 in Appendix 8.B. Given the complexity of the solution of $p_V(x)$ in (8.104), it will be reassuring to confirm that the $p_V(x)$ of (8.56) and $p_R^{(2)}(x)$ of (8.53) do yield the $p_X^{(1)}(x)$ of (8.52), using the integral equation of (8.45). By taking $a_2'' = 1/2$, then $p_R^{(2)}(x)$ of (8.53) yields $p_R^{(2)}(x) = x\exp(-x^2/2)$ and by taking $c = \sqrt{2}$ in $p_V(x)$ of (8.56) yields $p_V(x) = \sqrt{2/\pi}\exp(-x^2/2)$. Then the integral of (8.45) becomes

$$\begin{aligned} p_X^{(1)}(x) &= \int_0^\infty \left(\frac{x}{v}\right) \exp(-(x^2/(2v^2))) \left(\sqrt{\frac{2}{\pi}}\right) \exp(-v^2/2) \left(\frac{1}{v}\right) dv \\ &= \left(\sqrt{\frac{2}{\pi}}\right) x \int_0^\infty \left(\frac{1}{v^2}\right) \exp(-(x^2/(2v^2))) \exp(-v^2/2) dv, \ 0 < x < \infty. \end{aligned} \tag{8.57}$$

From [20, p. 341] we have

$$\int_0^\infty \left(\frac{1}{x^2}\right) \exp(-(a/x^2)) \exp(-\mu x^2) dx = (1/2)\sqrt{\pi/a}\exp(-2\sqrt{a\mu}). \tag{8.58}$$

By taking $a = x^2/2$ and $\mu = 1/2$ in (8.58), then the second line of (8.57) yields

$$p_X^{(1)}(x) = \left(\sqrt{\frac{2}{\pi}}\right) x(1/2)\sqrt{2\pi/x^2}\exp(-2\sqrt{x^2/4}) = \exp(-x), \ 0 < x < \infty. \tag{8.59}$$

With $\beta_1' = \alpha_1' = a_1'' = 1$, then $p_X^{(1)}(x)$ of (8.52) yields $p_X^{(1)}(x) = \exp(-x)$, $0 < x < \infty$, which is identical to the $p_X^{(1)}(x)$ of (8.59).

Case 2b. Let $\alpha_1' = 3/2$. Then

$$\begin{aligned} p_V(x) &= kH_{1,1}^{1,0}\left[cx \left| \begin{matrix} \{(1/2,\ 1/2)\} \\ \{(1/2,\ 1)\} \end{matrix} \right. \right] \\ &= \frac{2c}{\sqrt{\pi}}\left\{ \frac{\sqrt{cx}\,F_1(3/4, 1/2, -(cx)^2/4)}{\Gamma(1/4)} - \frac{(cx)^{3/2} F_1(5/4, 3/2, -(cx)^2/4)}{\Gamma(-1/4)} \right\},\ 0 < x < \infty. \end{aligned} \tag{8.60}$$

The $p_V(x)$ of (8.60) does not seem to be a "known pdf with a name." However, by using symbolic integration in Mathematica, we can show the $p_V(x)$ of (8.60) is positive-valued with unit area for any $0 < c$, and thus is a valid pdf over $0 < x < \infty$. A plot of $p_V(x)$ of (8.60) using Mathematica is given in Fig. 8.4.

Case 2c. Let $\alpha_1' = 2$. Then

$$p_V(x) = cH_{1,1}^{1,0}\left[cx \left| \begin{matrix} \{(1/2,\ 1/2)\} \\ \{(1,\ 1)\} \end{matrix} \right. \right] = \frac{c^3 x^2}{2\sqrt{\pi}} e^{-(cx/2)^2},\ 0 < x < \infty,\ 0 < c. \tag{8.61}$$

$p_V(x)$ in (8.61) is clearly positive-valued over $0 < x < \infty$. In particular, $p_V(x)$ of (8.61) is a valid pdf since its integral is the second moment of a half-Gaussian pdf having the value of 1. □

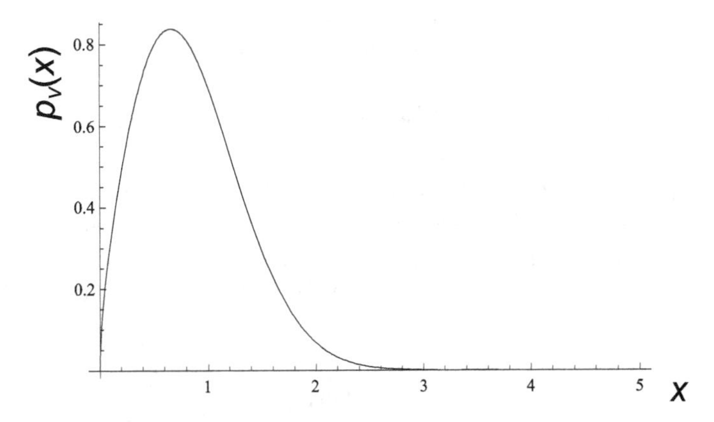

Figure 8.4 Plot of $p_V(x)$ of Case 1b vs. x.

The method provided by Theorem 8.3 in Appendix 8.B for finding $p_V(x)$ in Case 1 can be readily used to find the other six cases of the fading envelope pdf's in the NLOS scenarios [20]. In the LOS scenarios, we need to replace the Rayleigh pdf $p_R^{(2)}(x)$ with the Rician pdf.

8.3.2.2 Evaluation of average error probabilities for SIRP fading channels

Consider the communication problem of a BPSK communication system under the AWGN model with a random fading on the received envelope [21]. The average bit error probability of a BPSK system is well known and is given by $P_e = Q\left(\sqrt{2E/N_0}\right)$ in (4.82). In the presence of a random fading r.v. R on the envelope with a pdf of $p_R(r)$, $0 < r < \infty$, the new P_e under fading is given by

$$P_e = \int_0^{\infty} Q\left(\sqrt{2r^2 E/N_0}\right) p_R(r) dr, \tag{8.62}$$

under the assumption that the realization of R is known at the receiver and the demodulation is ML. When R is a Rayleigh pdf denoted by $p_R(r) = (r/\sigma^2)\exp(-r^2/(2\sigma^2))$, $0 \leq r < \infty$, (8.62) reduces to the well-known average error probability under the Rayleigh fading [22, p. 701] of

$$P_e^{(\text{Rayleigh})}(\bar{\gamma}) = \frac{1}{2}\left[1 - \sqrt{\frac{\bar{\gamma}}{1+\bar{\gamma}}}\right], \tag{8.63}$$

where $\bar{\gamma} = 2\sigma^2 E/N_0$ is the average fading SNR and asymptotically decreases as $1/(4\bar{\gamma})$. Now, assume the random fading r.v. X is a SIRV, then the average error probability under the SIRV fading is given by

$$P_e = \int_0^{\infty} \frac{1}{2}\left[1 - \sqrt{\frac{v^2\bar{\gamma}}{1+v^2\bar{\gamma}}}\right] p_V(v) dv = \int_0^{\infty} P_e^{(\text{Rayleigh})}\left(v^2\bar{\gamma}\right) p_V(v) dv, \tag{8.64}$$

where $p_V(v)$ is the univariate pdf used in (8.44) and (8.31) to define the SIRV.

Example 8.14 Consider the heavy-tailed $p_V(v)$ pdf defined by

$$p_V(v) = \sqrt{\frac{2}{\pi}} v^{-2} \exp\left(-\frac{1}{2v^2}\right), \; 0 < v < \infty. \tag{8.65}$$

We note the $p_V(v)$ of (8.65) for large values of v decreases as $1/v^2$. Thus, this $p_V(v)$ is heavy-tailed and its second moment is infinite (or more correctly does not exist). In order to find its corresponding $p_X(x)$ pdf, take the $p_V(v)$ of (8.65) and the Rayleigh pdf $p_R(r) = 2ar\exp(-ar^2)$, $0 < r < \infty$ of (8.53) (with $a = \frac{1}{2\sigma^2}$), and insert them into (8.45).

$$\begin{aligned}\int_0^\infty p_V(v)p_R(x/v)(1/v)dv &= \int_0^\infty \sqrt{\frac{2}{\pi}} v^{-2} \exp\left(-\frac{1}{2v^2}\right) 2a\left(\frac{x}{v}\right)\exp\left(-a\left(\frac{x}{v}\right)^2\right)dv \\ &= 2ax\sqrt{\frac{2}{\pi}}\int_0^\infty \frac{1}{v^4}\exp(-b/v^2)dv, \; b = \left(\frac{1}{2} + ax^2\right).\end{aligned} \tag{8.66}$$

Using symbolic integration of Mathematica, we obtained

$$\int_0^\infty \frac{1}{v^4}\exp(-b/v^2)dv = \left(\frac{\sqrt{\pi}}{4b^{3/2}}\right). \tag{8.67}$$

By using (8.67) in the second line of (8.66), then

$$p_X(x) = \frac{2ax}{(1+2ax^2)^{3/2}}, \; 0 \le x < \infty. \tag{8.68}$$

We call the envelope fading pdf $p_X(x)$ of (8.68) the "Half-Cauchy-Like" pdf. This pdf is Cauchy-like, since at large values of x, it decreases as $1/x^2$, just like a Cauchy pdf. Thus, the Half-Cauchy-Like pdf (like the Cauchy pdf) is heavy-tailed and its second moment is infinite (or more correctly is not defined).

We can use the $p_V(v)$ pdf of (8.65) in (8.64) to evaluate the average error probability of the fading scenario with the Half-Cauchy-Like pdf [16, 17]. The integral of (8.64) with the $p_V(v)$ pdf of (8.65) does not appear to have a closed form expression. However, by numerical integrations, the average error probabilities of the Half-Cauchy-Like pdf can be evaluated. In Table 8.8, the first column gives the four fading SNR values, the second column is for the P_e given by (8.63) for the Rayleigh fading channel, and the third column is for the P_e given by (8.66) for the Half-Cauchy fading channel. It is well known that the P_e of a BPSK system under Rayleigh fading as given by (8.63) decreases very slowly with increasing SNR. However, it is most surprising that under the Half-Cauchy-Like fading scenarios, even though this $p_V(v)$ pdf has infinite fading power (i.e., infinite second moment) and its corresponding $p_V(v)$ also have infinite second moment, its average error probability is well behaved. In fact, by comparing column three results to column two results in Table 8.8, the P_e of the system with the Half-Cauchy-Like fading cases are even slightly lower than those of the Rayleigh fading cases. Thus, the heavy-tailed fading statistic (even with infinite power) does not seem to have caused catastrophic system degradations. In fact, by using (8.65) in (8.64), we can

Table 8.8 Comparisons of average error probability $P_e(\text{Rayleigh})$ under Rayleigh fading with average error probability $P_e(\text{Half}-\text{Cauchy}-\text{Like})$ under Half-Cauchy-Like fading vs. SNR.

Fading SNR	$P_e(\text{Rayleigh})$	$P_e(\text{Half}-\text{Cauchy}-\text{Like})$
5	0.04356	0.03669
10	0.02327	0.02089
15	0.01588	0.01467
20	0.01204	0.01132

show asymptotically $P_e^{(\text{HCL})}(\bar{\gamma}) \sim 1/(4\bar{\gamma})$, which has the same asymptotic behavior as $P_e^{\text{Rayleigh}}(\bar{\gamma}) \sim 1/(4\bar{\gamma})$ of (8.63). □

8.3.3 Probabilities of false alarm and detection with robustness constraint

Consider a radar system in which n pulses are sent similar to that shown in Fig. 6.9 in Section 6.3.2 [11]. In the absence of a target, the simple Swerling I model yields a return envelope of each pulse to be Rayleigh distributed due to a large number of scatters of about equal strengths. Then the square of the envelope of each pulse X_i has a Central Chi-Square distribution with 2 degrees of freedom given by

$$p_{X_i}(x_i) = (2\sigma^2)^{-1} \exp\left(-x_i/(2\sigma^2)\right), \; i = 1, \ldots, n, \tag{8.69}$$

where σ^2 is the variance of the narrowband zero mean Gaussian I and Q channels of the radar system. This Chi-Square distribution has 2 degrees of freedom, due to one degree of freedom from each of the Gaussian I and Q channels. In the presence of a target, the Swerling Model III is applicable, where the return envelope of each pulse contains a large dominant constant term plus Rayleigh distributed random scatters of about equal strengths. Then the square of the envelope of each pulse X_i has a Non-Central Chi-Square distribution with 2 degrees of freedom given by

$$p_{X_i}(x_i) = (2\sigma^2)^{-1} \exp(-(\lambda + x_i)/(2\sigma^2)) \mathrm{I}_0(\sqrt{\lambda x_i}/\sigma^2), \; i = 1, \ldots, n, \tag{8.70}$$

where $\lambda = 2A^2$, A is the amplitude of the sinusoid of each pulse, and $\mathrm{I}_0(\cdot)$ is the modified Bessel function of order 0.

Consider the radar detector's sufficient statistic Γ as the sum of returns of n pulses given by

$$B = \{\Gamma = X_1 + X_2 + \ldots + X_n \geq T_1\}, \tag{8.71}$$

where T_1 is a threshold taken to control the probability of false alarm. Then Γ is a Central Chi-Squared r.v. of $2n$ degrees of freedom given by

$$p_\Gamma(\gamma) = \left((2\sigma^2)^n (n-1)!\right)^{-1} \gamma^{n-1} \exp\left(\gamma/(2\sigma^2)\right), \; 0 \leq \gamma < \infty. \tag{8.72}$$

Then

$$P_{\mathrm{FA}}^{(B)} = \mathrm{P}(B) = \mathrm{P}(\Gamma \geq T_1) = 1 - \mathrm{chi2cdf}(T_1/\sigma^2, 2n), \tag{8.73}$$

where chi2cdf(x, ν) is the Matlab function of a Central Chi-Square cdf of unit σ and of ν degrees of freedom. However, we note in (8.71), Γ can be greater than or equal to the threshold T_1 even in the absence of a target, if one or more of the X_i happen to have sufficiently large values if those scatters happen to yield large returns (e.g., due to highly reflective surfaces), leading to an error of declaring the presence of a target when it is not present [23]. Thus, for better robustness detection control (in avoiding having only few large contributions of X_i contributing to Γ), in addition to the condition in (8.71), we also impose the constraint of

$$\begin{aligned} C = &\{m \text{ or more of } X_i \geq T_2, \text{ but the other } X_i < T_2, \\ &i = 1, \ldots, n,\ T_2 \leq T_1,\ 1 \leq m \leq n\}. \end{aligned} \tag{8.74}$$

The constraint in (8.74) ensures that at least m of the X_i's are sufficiently large when a target is present. Or equivalently, in the absence of a target, one and up to $(m-1)$ large X_i's will not result in declaring a target is present. Thus, the constraint in (8.74) is a simple form of robustness control on reducing false alarms when a target is not present. Suppose we want the probability of false alarm only based on the condition of C. Then

$$P_{\mathrm{FA}}^{(C)} = \mathrm{P}(C) = \sum_{i=m}^{n} \binom{n}{i} (1 - \mathrm{chi2cdf}(T_2/\sigma^2, 2))^i (\mathrm{chi2cdf}(T_2/\sigma^2, 2))^{n-i}, \tag{8.75}$$

where we assume all the X_i r.v.'s are independent and each has a pdf given by (8.44). Here $\binom{n}{i}$ denotes the binomial coefficient indexed by n and i.

Now, consider the probability of false alarm subject to the constraints of (8.71) and (8.74), where each X_i has the pdf of (8.69), is then given by

$$\begin{aligned} P_{\mathrm{FA}} &= \mathrm{P}(B \cap C) \\ &= \mathrm{P}(\{X_1 + X_2 + \cdots + X_n \geq T\} \cap \\ &\quad \{m \text{ or more of } X_i \geq T_2, \text{ but the other } X_i < T_2, \\ &\quad i = 1, \ldots, n,\ T_2 \leq T_1,\ 1 \leq m \leq n\}). \end{aligned} \tag{8.76}$$

Unfortunately, there does not appear to be any obvious analytical method for the evaluation of the expression of (8.76). Series of numerical integrations (for small values of n and m) or simulations can be used to evaluate (8.75), but these details will not be discussed here [23].

Example 8.15 Figure 8.5 shows the allowable relationship between T_1 and T_2 for the case of $n = 4$ and $m = 2$ for $P_{\mathrm{FA}} = 10^{-6}$. We note, at the upper left corner of the curve in Fig. 8.5, with $T_2 = 0$, then the constraint of (8.74) is not imposed, and thus $P_{\mathrm{FA}} = 10^{-6}$ is attained with $T_1 = 42.701$ consistent with the value given by $\mathrm{P}(B)$ of (8.73). In (8.76), as T_2 increases, then the collection of sample points in event C becomes smaller, thus

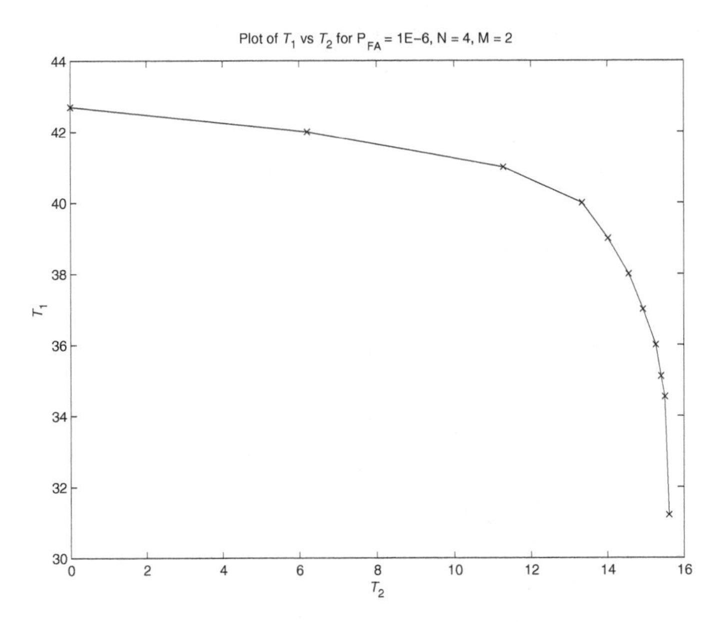

Figure 8.5 Plot of T_1 vs. T_2 for $P_{\mathrm{FA}} = 10^{-6}$ with $n = 4$ and $m = 2$.

forcing the collection of sample points in event B to become larger, which is equivalent to having a smaller value of T_1, as can be seen in Fig. 8.5 [23]. □

In the presence of a target, we want to evaluate the probability of detection subject to the constraints of (8.71) and (8.74), except now each X_i has the pdf of (8.70), which is given by

$$\begin{aligned} P_{\mathrm{D}} &= \mathrm{P}(B \cap C) \\ &= \mathrm{P}(\{X_1 + X_2 + \cdots + X_n \geq T\} \cap \\ &\quad \{m \text{ or more of } X_i \geq T_2,\ i = 1, \ldots, n,\ T_2 \leq T_1,\ 1 \leq m \leq n\}). \end{aligned} \tag{8.77}$$

The evaluation of (8.77) [just as in the evaluation of (8.76)] does not appear to have an analytical solution, and simulations need to be used.

Example 8.16 Figure 8.6 shows the P_{D} obtained by simulations for various allowable values of T_1 and T_2 subject to $P_{\mathrm{FA}} = 10^{-6}$ for various SNR(dB). Here SNR(dB) = $10 \log_{10}(A^2/\sigma^2)$. From Fig. 8.6, we note, for a given SNR = 8 dB, there are five possible P_{D} values. Starting at $T_1 = 42.701$ and $T_2 = 0$ (i.e., with no "robustness constraint"), then $P_{\mathrm{D}} = 0.85$. Imposing larger constraint values on T_2 resulted in $T_1 = 31.214$ and $T_2 = 15.607$, which yielded a much lower $P_{\mathrm{D}} = 0.51$. Thus, robustness in reducing possible false alarm errors also leads to a significant reduction of P_{D} values. In practice, "excessive robustness" may not be a desirable choice. □

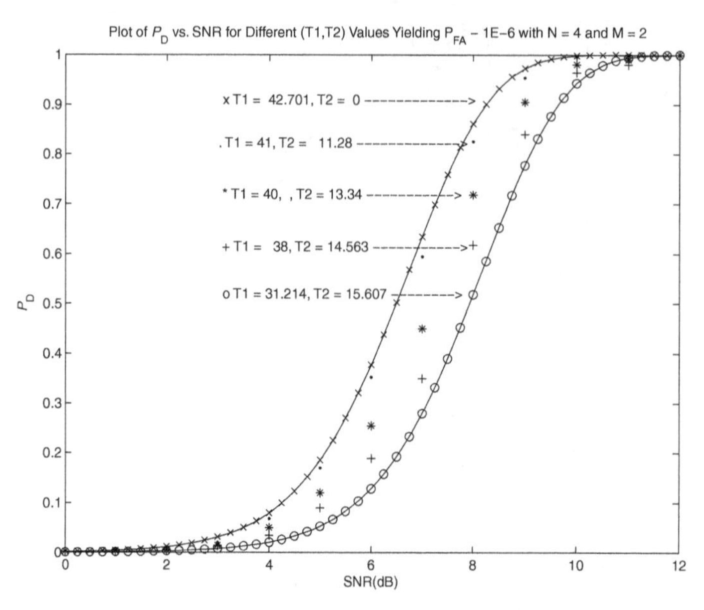

Figure 8.6 Plot of P_D vs. SNR(dB) for different T_1 and T_2 values with $P_{FA} = 10^{-6}$ and $n = 4$ and $m = 2$.

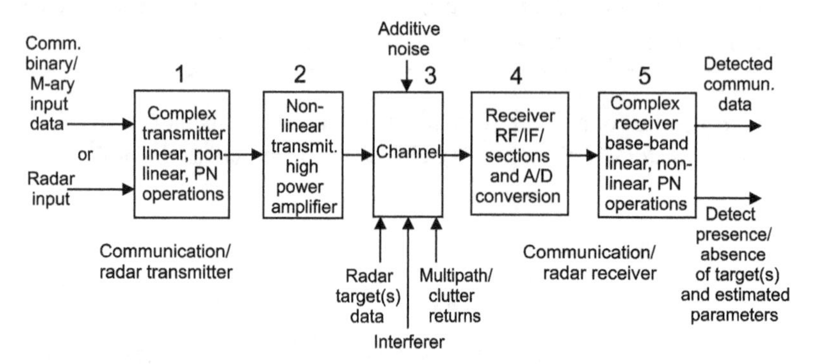

Figure 8.7 A generic communication/radar system with various complex processings.

8.4 Monte Carlo simulation and importance sampling in communication/radar performance analysis

From Examples 8.2 and 8.3 in Section 8.1, we already used the simple simulation method of (8.12) to evaluate the simulated probability of a false alarm $\tilde{P}_{FA}$ for a binary detection problem. Indeed, from previous chapters, we only considered some basic detection and estimation sub-systems of various communication and radar systems. Figure 8.7 shows a high-level block diagram of six blocks of a generic communication and radar system. In particular, we make some comments on various known PR operations as well as unknown random phenomena encountered in these blocks. In a communication system, the binary/M-ary information data should be truly random. Inside sub-system 1, various linear and non-linear source encoding and modulation operations plus known PN sequences are imposed for synchronization and interleaving purposes, and in a CDMA system, various known spreading PN codes are also used. In sub-system 2, for transmitter power efficiency, the high power RF amplifier is normally set near the saturation point of non-linear operating. The channel represented by box 3 contains an additive random noise (often modeled as AWGN), but there may be random interferers leaking from possible adjacent bands, other users' signals in a CDMA system, and potential random intentional or non-intentional waveform. Other than in free-space propagation, various random multipath waveforms may disturb the desired transmitted waveform. In sub-systems 4 and 5, again all sorts of linear and non-linear operations exist in the RF and IF sections, with various additional random effects due to PN sequence de-interleaving, and in a CDMA system with PN despreading effects, and A/D quantization noises, and bit, word, and frame synchronization noises, and decoding and demodulation random errors. In a radar system, most of the random phenomena of a communication system may also be present, except the interference from random clutter may be very intense and in some scenarios, strong random jamming waveforms must be reduced in order to declare the presence or absence of target(s) and associated estimated parameters such as target range(s), size(s), and doppler(s). It is clear that both the communication and radar systems need to be modeled as complex non-linear stochastic dynamical systems. While conceptually various analytical tools (e.g., Volterra stochastic methods) may be used to analyze some of these systems, system performances of most practical complete communication and radar systems can not be evaluated analytically, but must be evaluated using simulation methods [2, 24–26].

8.4.1 Introduction to Monte Carlo simulation

Simulation using pseudo-random (PR) sequences is often given the more colorful name of Monte Carlo (MC) simulation, coined by S. Ulam, in "honor" of seemingly random gambling outcomes at the famed Monte Carlo Casino in the principality of Monaco [27]. In Examples 8.2 and 8.3 of Section 8.1, we already used the simple simulation method of (8.12) to evaluate the simulated probability of a false alarm $\tilde{P}_{\mathrm{FA}}$ for a simple binary detection problem. In Section 8.2, we also considered the use of simulation for the evaluation of the average probability of error for a binary detection problem with ISI. In Section 8.3, we compared the analytical and simulation results for evaluating average probability of errors of Gaussian and t-distributed binary detection problems as well as the probability of a false alarm and probability of detection for a radar problem with robustness constraint. From all these and other problems, it is clear that we want to find efficient MC methods (i.e., reduced number of simulation terms) to yield the desired system performances.

In many engineering and other practical situations, we often want to evaluate the probability of the tail behavior of a pdf modeling some problems of interest. Of course, if the pdf of the r.v. is known explicitly, then this probability can be obtained either analytically (if the cdf of this r.v. has an analytical expression) or by various numerical methods. However, if the internal operations of a system of interest (such as that given in

Fig. 8.7) are too complicated, then MC simulation may be a reasonably simple approach to model this problem and obtain some probability results fairly rapidly. One basic problem in simulation is always to determine how many simulation runs M are needed to achieve the desired accuracy of the problem.

Suppose we want to evaluate the probability of the tail end of a univariate r.v. X with a pdf $p_X(x)$ given by

$$P_e(c) = \int_c^\infty p_X(x)ds = 1 - F_X(c), \tag{8.78}$$

where $F_X(x)$ is the cdf of the r.v. Of course, if $F_X(x)$ has an analytical expression, then P_e is given as the complementary cdf evaluated at c. Otherwise, if $p_X(x)$ is available, then various numerical integration methods can also be used to evaluate the integral in (8.78).

Example 8.17 Suppose we want to find P_e for the following three commonly encountered pdf's in communication and radar detection systems:

$$\text{Gamma pdf:} \quad p_1(x) = \frac{x^{a-1}\exp(-x/b)}{b^a\Gamma(a)}, \ 0 \le x < \infty, \ 0 < a, \ b. \tag{8.79}$$

$$\text{Gaussian pdf:} \quad p_2(x) = \frac{\exp\left(-(x-\mu)^2/(2\sigma^2)\right)}{\sqrt{2\pi\sigma^2}},$$

$$-\infty < x < \infty, \ -\infty < \mu < \infty, \ 0 < \sigma. \tag{8.80}$$

$$\text{Weibull pdf:} \quad p_3(x) = abx^{b-1}\exp(-ax^b), \ 0 \le x < \infty, \ 0 < a, \ b. \tag{8.81}$$

For all three cases, consider $c = 3$ and $c = 5$ to be used in (8.78). For the Gamma pdf, take $a = 1.2$ and $b = 0.4$, for the Gaussian pdf, take $\mu = 0$ and $\sigma^2 = 1$, and for the Weibull pdf, take $a = 1.7$ and $b = 2.7$. Plots of the log of the complementary cdf's for these three pdfs over the region $3 \le x \le 5$ are given in Fig. 8.8. Values of these complementary cdf's at $c = 3$ and $c = 5$ are available. Specifically, $P_e^{(\text{Gamma})}(3) = 9.23 \times 10^{-4}$, $P_e^{(\text{Gamma})}(5) = 6.83 \times 10^{-6}$, $P_e^{(\text{Gaussian})}(3) = 1.35 \times 10^{-3}$, $P_e^{(\text{Gaussian})}(5) = 2.87 \times 10^{-7}$, $P_e^{(\text{Weibull})}(3) = 9.71 \times 10^{-3}$, and $P_e^{(\text{Weibull})}(5) = 1.01 \times 10^{-8}$.

Now, simulation results for these three complementary cdfs for different simulation runs M are given in Table 8.9.

For our specified parameters of these three pdf's, at large value of x, from (8.79), we note the Gamma r.v. decreases approximately as an exponential of $(-2.5x)$, while from (8.80), we note the Gaussian r.v. decreases approximately as an exponential of $(-0.5x^2)$, and finally from (8.81), we note the Weibull r.v. decreases approximately as an exponential of $(-1.7x^{2.7})$. Thus, it is not surprising that at $c = 5$, the Weibull complementary cdf is smaller than that of the Gaussian complementary cdf, which is also smaller than that of the Gamma complementary cdf. From Table 8.9 at $c = 5$, using $M = 10^7$, the Gamma simulation result of $\tilde{P}_e = 6.90 \times 10^{-6}$ seems to be useful to estimate the true $P_e = 6.82 \times 10^{-6}$; while the Gaussian simulation result of $\tilde{P}_e = 3.00 \times 10^{-7}$ is barely able to estimate the true $P_e = 2.87 \times 10^{-7}$; while the Weibull simulation result

Table 8.9 Listing of P_e and $\tilde{P}_e$ for Gamma, Gaussian, and Weibull complementary cdf's for $c = 3$ and $c = 5$ vs. M.

M	Gamma comp. cdf		Gaussian comp. cdf		Weibull comp. cdf	
	$c = 3$	$c = 5$	$c = 3$	$c = 5$	$c = 3$	$c = 5$
P_e	9.23×10^{-4}	6.83×10^{-4}	1.35×10^{-3}	2.87×10^{-7}	9.71×10^{-3}	1.01×10^{-8}
	$\tilde{P}_e$		$\tilde{P}_e$		$\tilde{P}_e$	
10^3	1.00×10^{-3}	0	2.00×10^{-3}	0	1.00×10^{-2}	0
10^4	7.00×10^{-4}	0	1.20×10^{-3}	0	1.02×10^{-2}	0
10^5	8.90×10^{-4}	1.00×10^{-5}	1.49×10^{-3}	0	9.49×10^{-3}	0
10^6	8.84×10^{-4}	4.00×10^{-6}	1.29×10^{-3}	0	9.58×10^{-3}	0
10^7	9.25×10^{-4}	6.90×10^{-6}	1.34×10^{-3}	3.00×10^{-7}	9.67×10^{-3}	0

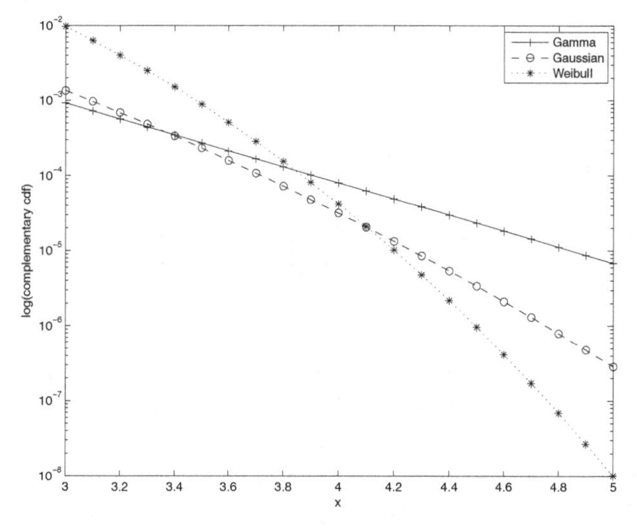

Figure 8.8 Plots of log of the complementary cdf's of the Gamma, Gaussian, and Weibull r.v.'s vs. x.

of $\tilde{P}_e = 0$ is not able to estimate the true $P_e = 1.01 \times 10^{-8}$. These Weibull simulation results seem to indicate that for $c = 5$, the rare events at the tail of the Weibull pdf do not appear at all even with simulation runs of $M = 10^7$. This leads to the issue of trying to improve the efficiency of the MC simulation upon some modification so that more of the rare events at the tail of the pdf actually will show up in the simulation runs. This leads to the use of MC importance sampling simulation method. □

8.4.2 MC importance sampling simulation method

As noted above, the reason why the simulations for the above three (Gamma, Gaussian, and Weibull) cases in Example 8.17, even for quite a large number of simulation runs,

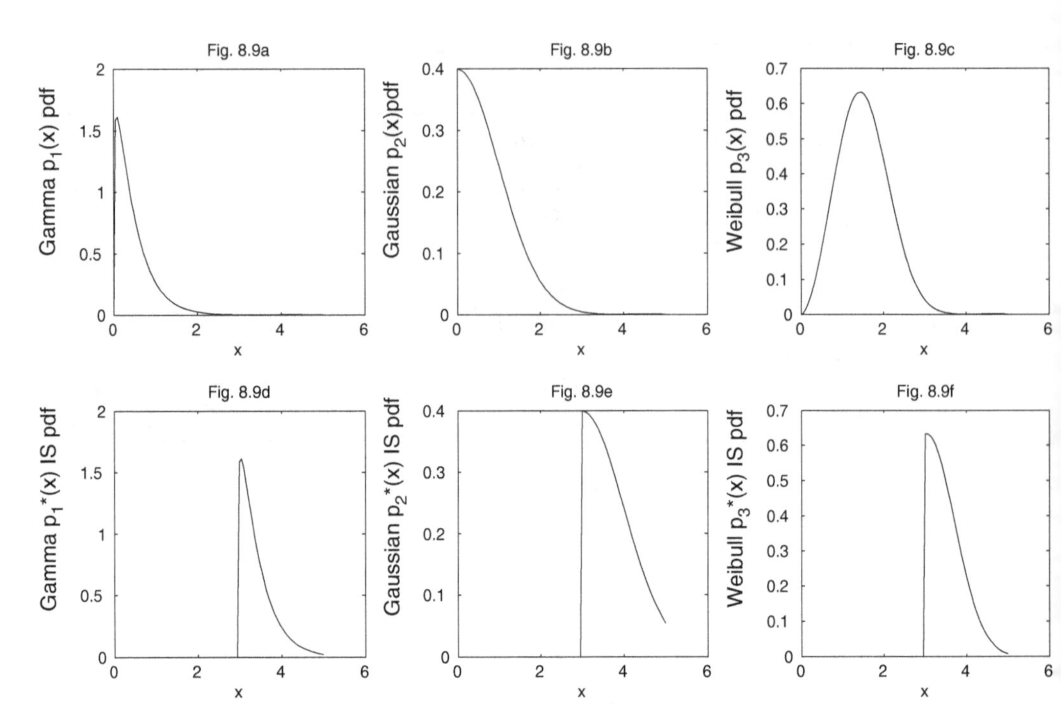

Figure 8.9 Plots of original Gamma pdf (Fig. 8.9a), Gaussian pdf (Fig. 8.9b), Weibull pdf (Fig. 8.9c), and modified Gamma IS pdf (Fig. 8.9d), Gaussian IS pdf (Fig. 8.9e), and Weibull IS pdf (Fig. 8.9f) vs. x.

were not too effective was that the actual number of samples in the simulation appearing in the interval $[c, \infty)$, particularly for $c = 5$, was very low. This can be seen if we plot the pdfs of these cases as shown in the top row of Fig. 8.9 (denoted as Fig. 8.9a, 8.9b, and 8.9c). Intuitively, if we can modify the three pdfs in (8.79)–(8.81), such as simple translations to the right for these three pdfs (as shown in the bottom row of Fig. 8.9 (denoted as Fig. 8.9d, 8.9e, and 8.9f), then more samples will appear in the interval of interest. This is an example of importance sampling (IS) methodology [26, 28–33] in simulation, which we will pursue as follows. Consider the evaluation by simulation of the probability P_R of the r.v. X with a pdf $p_X(x)$ on a subset of the real line denoted by R. An example of such a probability $\mathrm{P}(R)$ is that of $P_e(c)$ given by (8.78). Let the indicator function $I_R(x)$ be defined by

$$I_R(x) = \begin{cases} 1, & x \in R, \\ 0, & x \notin R. \end{cases} \tag{8.82}$$

Then P_R can be written as

$$P_R = \int_R P_X(x)dx = \int_{-\infty}^{\infty} I_R(x)p_X(x)dx = \mathrm{E}_X\{I_R(X)\}. \tag{8.83}$$

In particular, suppose the values of $p_X(x)$ over R are low, which indicates that if we do simulation based on samples taken from $p_X(x)$, the number of simulation samples will also be low. Consider a new IS pdf p_x^*, defined over R, such that (8.83) can be rewritten as

$$P_R = P_R^{\mathrm{IS}} = \int_R \frac{p_X(x)}{p_x^*(x)} p_X^*(x) dx = \int_{-\infty}^{\infty} I_R(x) w(x) p_X^*(x) dx = \mathrm{E}_{p_X^*}\{I_R(X) w(X)\}, \tag{8.84}$$

where the weighting function $w(x)$ is defined by

$$w(x) = \frac{p_X(x)}{p_X^*(x)}, \ x \in R. \tag{8.85}$$

From (8.84) and (8.85), it is clear that p_X^* must be non-zero over R in order for $w(x)$ to be well defined, but otherwise p_X^* can have an arbitrary value over R. However, in order for this IS pdf p_X^* to be useful, it must have much larger values than p_X over R, so that the values of $w(x)$ over R will have values much less than one. This also means that many more samples will appear if the simulation runs are based on that of p_X^* rather than the original p_X over R. From the last expression of (8.83), we know P_R is just the average of $I_R(X)$ using the p_X pdf. If we just replace the p_X pdf in (8.83) with the p_X^* pdf, then we would have counted too many simulation samples from the p_X^* pdf. Thus, from the last expression of (8.84), P_R^{IS} shows that if we use samples from the p_X^* pdf, then we need to reduce each averaged value by an attenuation factor of $w(x)$ for all sampled values in R.

Now, using the indicator function $I_R(x)$ notation of (8.82), the simulation formula of (8.12) can be rewritten as

$$\tilde{P}_{\mathrm{FA}} = \frac{1}{M} \sum_{m-1}^{M} [I_R(x_m)] \,|_{x_m \in \{\text{PR realizations from } p_X\}}. \tag{8.86}$$

Then the simulation formula for $\tilde{P}_R^{\mathrm{IS}}$ using sample realization taken from the P_X^* pdf is given by

$$\tilde{P}_R^{\mathrm{IS}} = \frac{1}{M} \sum_{m-1}^{M} [I_R(x_m) w(x_m)] \,|_{x_m \in \{\text{PR realizations from } p_X^*\}}. \tag{8.87}$$

Example 8.18 Let us consider some possible IS pdf's for the three Gamma, Gaussian, and Weibull pdfs of Example 8.17 by translations [31]. From Fig. 8.9a, we note the peak of the Gamma pdf p_1 of (8.79) is located near $x = 0.07$. For the $c = 3$ case, the region R is given by $[3, \infty)$. One simple new IS pdf p_1^* is then obtained by just taking the original Gamma pdf p_1 of (8.79) and translating it by 2.93 (i.e., 3 – 0.07), as shown in Fig. 8.9d. Similarly, since the peak of the Gaussian pdf of (8.80) with a zero mean and unit variance is located at $x = 0$, the new IS p_2^* is then obtained by just taking the original Gaussian pdf p_2 of (8.80) and translating it by 3, as shown in Fig. 8.9e. Finally, since the peak of the Weibull pdf of (8.81) is located near $x = 1.8$, the new IS p_3^* is then obtained by just taking the original Weibull pdf p_3 of (8.81) and translating it by 1.2, as shown in Fig. 8.9f. For the $c = 5$ case, then p_1^* takes p_1 and translates it by 4.93 to the right, p_2^* takes p_2 and translates it by 5 to the right, and p_3^* takes p_3 and translates it by 3.2 to the right. Using simulations based on these IS pdfs, results corresponding to those of Table 8.9 are given in Table 8.10.

Table 8.10 Listing of P_e and $\tilde{P}_e^{IS}$ for Gamma, Gaussian, and Weibull complementary cdf's for $c = 3$ and $c = 5$ vs. M.

M	Gamma comp. cdf		Gaussian comp. cdf		Weibull comp. cdf	
	$c = 3$	$c = 5$	$c = 3$	$c = 5$	$c = 3$	$c = 5$
P_e	9.2325E − 4	6.8274E − 4	1.3499E − 3	2.8665E − 7	9.7097E − 3	1.0127E − 8
	$\tilde{P}_e$		$\tilde{P}_e$		$\tilde{P}_e$	
1E3	9.086E − 4	6.719E − 6	1.402E − 3	3.048E − 7	9.009E − 3	8.745E − 9
1E4	9.086E − 4	6.719E − 6	1.335E − 3	2.833E − 7	9.786E − 3	1.010E − 8
1E5	9.181E − 4	6.790E − 6	1.341E − 3	2.837E − 7	9.703E − 3	1.010E − 8
1E6	9.220E − 4	6.819E − 6	1.346E − 3	2.856E − 7	9.710E − 3	1.013E − 8
1E7	/	/	1.349E − 3	2.864E − 7	9.703E − 3	1.012E − 8

By comparing the results of Table 8.10 to those of Table 8.9, we note that with only $M = 1E4$ IS simulation runs, the IS simulated results are within 1% of the true values for all the three pdf's for $c = 3$ and $c = 5$. These results demonstrate the efficiencies of the IS simulation method over the conventional MC simulation method. □

In Example 8.18, the new IS pdfs for all three cases were obtained by translating certain ad hoc amounts to the right from the original three pdfs. Now, we want to study some basic properties of the IS simulation method and find the optimum $\hat{p}_X^*$ IS pdf for a given p_X pdf.

First, we want to show $\tilde{P}_R^{IS}$ is an unbiased estimator. From

$$E_{P_X^*}\{\tilde{P}_R^{IS}\} = \frac{1}{M}\sum_{m=1}^{M} E_{P_X^*}\{I_R(X)w(X)\} = \frac{1}{M}\sum_{m=1}^{M} P_R^{IS} = P_R^{IS}, \tag{8.88}$$

it is clear that $\tilde{P}_R^{IS}$ is an unbiased estimator. Next, consider the variance σ_{IS}^2 of the IS estimator P_R^{IS} given by

$$\sigma_{IS}^2 = \int_{-\infty}^{\infty}\left(I_R(x)\frac{p_X(x)}{p_x^*(x)} - P_R\right)^2 P_X^*(x)dx = \int_{-\infty}^{\infty} I_R(x)w^2(x)p_X^*(x)dx - P_R^2. \tag{8.89}$$

Then the conventional MC simulation variance σ_{MC}^2 is given by

$$\sigma_{MC}^2 = \int_{-\infty}^{\infty}(I_R(x) - P_R)^2\, p_X(x)dx = \int_{-\infty}^{\infty} I_R(x)p_X(x)dx - P_R^2. \tag{8.90}$$

Then σ_{IS}^2 becomes zero, if the optimum $\hat{p}_X^*$ IS pdf is taken to be

$$\hat{p}_X^*(x) = \begin{cases} p_X(x)/P_R, & x \in I_R(x), \\ 0, & x \notin I_R(x). \end{cases} \tag{8.91}$$

To show $\hat{p}_X^*$ is the optimum IS pdf, take $\hat{p}_X^*$ of (8.91) and substitute it for $\hat{p}_X^*$ of (8.90), to yield $\sigma_{IS}^2 = 0$. We should note the expression for the optimum IS pdf in (8.91) $\hat{p}_x^*$ is not usable, since it requires the knowledge of the true probability P_R.

Table 8.11 Minimum IS simulation variance $\hat{\sigma}^2_{\text{IS}}$ and MC simulation variance σ^2_{MC} for $c = 3$ and $c = 5$.

$\hat{\sigma}^2_{\text{IS}}$		σ^2_{MC}	
$c = 3$	$c = 5$	$c = 3$	$c = 5$
6.17E − 6	4.71E − 13	1.35E − 3	2.87E − 7

Example 8.19 Consider the Gaussian pdf $p_2(x)$ of (8.90) with a mean of zero and a unit variance used to evaluate $\widetilde{P}_R^{\text{IS}}$ in Example 8.18. At that time, we obtained the ad hoc c by arbitrarily translating the original zero mean Gaussian pdf $p_2(x)$ to the right by an amount c. Now, with the availability of the variance σ^2_{IS} expression of (8.78), we can ask what is the optimum translation parameter t, in the class of all IS pdfs of the form $\{p_2(x - t), -\infty < t < \infty\}$ for a specified value of c defining $R = [c, \infty)$? Using $p_2(x) = (2\pi)^{-1/2} \exp(-x^2/2), -\infty < x < \infty\}$, and $p_x^*(x) = p_2(x - t), -\infty < x < \infty$ in (8.89) yields

$$\sigma^2_{\text{IS}} = \exp(t^2) \int_{-\infty}^{\infty} I_R(x) \exp(-(x + t)^2/2) dx - P_R^2. \tag{8.92}$$

Then

$$\begin{aligned}\frac{\partial \sigma^2_{\text{IS}}}{\partial t} &= e^{\frac{t^2}{\sigma^2}} \left\{ (2t/\sigma^2) \int_{(c+t)/\sigma}^{\infty} \frac{1}{\sqrt{2\pi}} e^{-\frac{u^2}{2}} du + (1/\sigma) \int_{(c+t)/\sigma}^{\infty} \frac{1}{\sqrt{2\pi}} (-u) e^{-\frac{u^2}{2}} du \right\} \\ &= \exp(t^2/\sigma^2) \left\{ (2t/\sigma^2) \text{Q}((c + t)/\sigma) - (1/\sigma\sqrt{2\pi}) \right\} \\ &\approx e^{\frac{t^2}{\sigma^2}} \left\{ 2t/(\sigma(c + t)\sqrt{2\pi}) e^{-\frac{(c+t)^2}{2\sigma^2}} - (1/\sigma\sqrt{2\pi}) e^{-\frac{(c+t)^2}{2\sigma^2}} \right\} = 0. \end{aligned} \tag{8.93}$$

Equation (8.93) yields

$$\frac{2\hat{t}}{c + \hat{t}} = 1 \Leftrightarrow \hat{t} = c. \tag{8.94}$$

Furthermore, since

$$\left. \frac{\partial \sigma^2_{\text{IS}}}{\partial t^2} \right|_{t=c} > 0, \tag{8.95}$$

this shows the local minimum of the IS simulation variance is attained with the optimum translation parameter of $\hat{t} = c$ [31].

Table 8.11 shows the minimum IS simulation variance $\hat{\sigma}^2_{\text{IS}}$ of (8.92) using the optimum translation parameter of $\hat{t} = c$ and the conventional MC simulation variance σ^2_{MC} of (8.90) for the two cases of $c = 3$ and $c = 5$. □

We note, optimum translation of the original pdf to the desired $\hat{P}_x^*(\cdot)$ can be extended to vector-valued simulation pdfs for evaluation of average probability of error of communication systems with ISI [31] as well as to coded communication systems [34].

8.5 Conclusions

In Section 8.1, we considered the analysis of linear receiver performance with Gaussian noises. The performances of such receivers are simple to evaluate by exploiting the fact that the sufficient statistic was Gaussian due to the closure property of Gaussian processes and vectors under linear operations. If this closure property of Gaussian processes was not available (or not known), then brute-force simulations had to be used to evaluate the receiver performances. In Section 8.2, we first considered a binary communication system with random ISI effect. For short length ISI disturbances, brute-force statistical enumerations could be used. However, for long-length ISI problems, we proposed the use of the moment bound method to yield upper and lower bounds of the average probability of error of the system. Section 8.3 dealt with analysis of receiver performances under non-Gaussian noise disturbances. In Section 8.3.1, we introduced the SIRP with its explicit n-dimensional pdf's characterized by an arbitrary univariate pdf over $(0, \infty)$. Explicit closed form expressions for the average probability of error of systems with SIRP noises were then obtained. In Section 8.3.2, we studied the receiver envelope pdfs of wireless communication fading channels modeled as SIRP. Mellin transform and H-function representation methods were used to find the univariate pdfs characterizing the SIRP. This univariate pdf allowed us to evaluate the average probability of error of the fading communication system in a novel manner. Section 8.3.3 treated the probabilities of a false alarm and detection of a radar system with a robustness constraint. Section 8.4 considered the need to use MC simulation for complex communication and radar system performance evaluations when analytical methods are not possible. In particular, the concept of importance sampling in MC was introduced to reduce the number of simulation runs without losing the desired system performance. In practice, proper use of IS in simulation appears to be highly problem dependent. Appendix 8.A reviews some basic PR sequence generations needed for simulations. Appendix 8.B provides Theorem 8.3 needed in characterizing fading channel pdfs.

8.6 Comments

In Section 8.1, we explored the concept of sufficient statistic first introduced by R. A. Fisher for hypothesis testing and estimation [35]. Closure property of Gaussian processes under linear operations was probably known to many statisticians as documented by Cramér [36]. Early publications (in the 1950s and 1960s) in radar by Woodward [37] and in communications by Middleton [38] all exploited the Gaussian sufficient statistical properties readily to evaluate system performances. This case showed that the full theoretical understanding of a model can be immensely useful in avoiding the need for extensive brute-force simulations for performance analysis. In the 1960s and 1970s, there were considerable interests in communications over bandlimited channels with ISI. Lucky et al. [1] was one of the early books dealing with these problems. Since then, it is known that a detection rule based on the single value of the observed x_0 in (8.20) is sub-optimum for deciding the information data a_0. Maximum likelihood sequence estimation (i.e., Viterbi algorithm) over the observed data sequence centered about x_0 provides an optimum equalization procedure to combat the ill effects of the ISI. Of course, it is beyond the scope of this chapter to consider such solutions. The moment space method used in Section 8.2 (credited here to Dresher [3]) appeared to have been discovered independently by various mathematicians and operations research workers, although not necessarily fully documented. As commented at the end of that section, the moment space method has been used besides treating the ISI problem.

In Section 8.3, we considered the detection issues of non-Gaussian random processes and noises. Gaussian processes and noises are analytically tractable and justified physically in ideal freespace propagation conditions. Since large fractions of the realizations of Gaussian processes are clustered around a few σ's from their mean values, assuming the Gaussian processes are stationary, then long-term data collections can estimate the desired autocorrelation matrix $\mathbf{R}$ well. Thus, the linear Gaussian colored noise receiver derived from theory generally performed well in practice. As shown in Fig. 8.3(b) and Fig. 8.3(c), many heavy-tailed random processes have rare large-valued events (thus they are called "rare events"). Similarly, Hall [39] studied very low frequency RF propagation channels with atmospheric noises from lightning disturbances and other rare highly spiky-valued events. Over a sufficiently long observation period, he showed these noises may be modeled by t-distributed statistics. As is shown in Example 8.7 and Example 8.8 of Section 8.3.1, SIRP included Gaussian processes and t-distributed processes, and thus provided a general approach to study receiver detection and performance analysis. As shown in Example 8.10 in Section 8.3.1, system performances of t-distributed (and probably other heavy-tailed (HT)) noise processes are significantly worse than those of comparable Gaussian noise processes of equivalent SNR. A major reason for this low performance problem is that the large-valued rare events biased the resulting pdf, in that much of the time this pdf is not modeling well most of the outcomes of the HT processes. Thus, there may be many decision errors much of the time. If adaptive data collection procedures (such as deleting some of the large rare events appropriately) were used, then the resulting pdfs formed from these observed data models may model most of the realizations fairly well much of the time, and thus the system performances of such detectors may improve significantly. For the given t-distributed noise model of (8.34), SIRP theory correctly yielded the LR (i.e., optimum) receiver of (8.39). Yet the receiver of (8.39) is far from optimum, since the statistic obtained from a long-term averaging, resulting in a t-distribution, is far from optimum in modeling much of the time the actual distribution of the process. Yet, the receiver of (8.39) based on SIRP theory does provide a useful lower bound on the system performance when these rare events in the HT processes are averaged without some appropriate pre-processing (i.e., without some sorts of rare events removal).

Section 8.3.2 provided a unified approach for received envelope statistics based on the SIRP model. For each observed NLOS fading envelope pdf, a corresponding univariate pdf was derived from the SIRP theory using Mellin transform and Fox H-function representations. While the intermediate steps of using H-representations seemed to be complicated, the final resulting univariate pdfs are actually quite explicit. This

univariate r.v. (due to multipath and scattering propagation effects) modulating the Rayleigh envelope led to the observed fading envelope. The performances of various communication systems under fading were obtained using these univariate pdfs. In particular, even when the fading envelope r.v. has infinite power (i.e., infinite second moment as shown in Example 8.14 of Section 8.3.2), the system performance was shown to be well behaved compared to the well-known Rayleigh fading case. Of course, for both the Rayleigh as well as the half-Cauchy-like fading envelope cases, the system performances are very poor. It is most interesting to note asymptotically that $P_{\mathrm{FA}} \sim 1/(2\beta\gamma_0)$ for the t-distributed problem has a very similar form to that of the $P_{\mathrm{e}}^{\mathrm{Rayleigh}} \sim 1/(4\bar{\gamma}_0)$ for the Rayleigh fading problem and that of the $P_{\mathrm{e}}^{(\mathrm{RHL})} \sim 1/(4\bar{\gamma}_0)$ for the half-Cauchy-like fading problem, where $2\gamma_0$ and $\bar{\gamma}_0$ are both some forms of SNR. In practice, spatial diversity is used to remedy the low fading system performance problems, by using several receivers at different physical locations. In many scenarios, it is not too likely to have severe fadings simultaneously at all these locations. Modern MIMO schemes not only achieve spatial diversity at the transmitters and receivers, but also achieve enhanced data rate transmissions with appropriate codings. Of course, again it is beyond the scope of this chapter to consider such solutions.

Section 8.3.3 studies a standard radar detection problem with an additional constraint to provide a simplified model. This simplified model assumes all reflecting returns are independent. Modern radar systems may be able to provide robustness control in discriminating against large reflecting returns that are not independent by using more sophisticated methods. The geometry of the n-dimensional volume needed for system evaluations (either P_{FA} or P_{D}) does not appear to lead to analytical solutions. Thus, brute-force MC simulations (or IS MC) methods need to be used. The robustness control method proposed here may also be useful in cognitive radio problems.

In Section 8.4, we provided some motivations on the need to use MC simulations for complex system performance evaluations. In particular, we showed the use of an IS scheme based on optimized translation of the pdfs of interest yielded an efficient IS method for reducing the number of simulation runs.

8.A Generation of pseudo-random numbers

Before we consider the concept of the generation of a sequence of pseudo-random numbers, we can ask, how can we tell whether a sequence of numbers is random? Suppose we are given the following short sequence {4, 14, 23, 34}; can we "predict" the next integers? One possible solution is, "No," since the next number is "42," which is the name of a station in a north–south IRT 7th Avenue subway train route in New York City. Clearly, this sequence of numbers is not random.

On the other hand, the concept of generating random outcomes is quite simple, as demonstrated by a sequence of tossing of a coin. We can list the outcome of a "head" (to be denoted by the symbol "H" or the number "1") and a "tail" (to be denoted by the symbol "T" or the number "0"). Then a possible realization of a sequence of random outcomes may be denoted by $\{H, H, T, H, T, \ldots\}$ or $\{1, 1, 0, 1, 0, \ldots\}$. If given a sequence of 1's and 0's, from the toss of a coin, why do we believe the outcomes are "random"? In other words, what characterizes a "random sequence"?

Given a coin toss of length n, we may define the probability of a "1" by P("1") and the probability of a "0" by P("0") by using: P("1") = Number of "1"s $/n$ and P("0") = Number of "0"s$/n$, using the "relative frequency concept" to obtain the "probability" of an event.

Various algorithms have been proposed to perform testing for "randomness." Specifically, consider the following "run-length" testing for randomness of a binary {0, 1} sequence.

1. Test for randomness of order 1.
Count the number of "1"s and the number of "0"s. If their relative frequencies are equal (or close) to $\frac{1}{2}$ the sequence passes the "test for order 1."

2. Test for randomness of order 2.
Check the relative frequencies of blocks of length 2 of the form "00," "01," "10," and "11." If they are all equal (or close) to $1/4$, then it passes the "test of order 2".

⋮

k. Test for randomness of order k.
Check for relative frequencies of blocks of length k. There are $2k$ such blocks and their relative frequencies should be $1/2k$. If they are all equal (or close) to $1/2k$, then it passes the "test of order k".

Clearly, we need more systematic ways to generate "random sequences" than flipping coins. In practice, particularly when we are interested in a very long sequence of random outcomes, a more systematic method for generating a random sequence is needed. Various algorithmic methods (i.e., specific formulas) for the pseudo-random (PR) generation of numbers have been proposed. These generators are called PR because such sequences behave like a random sequence, such as meeting the test for randomness of order k, for some large positive integer k, but at the same time are generated from a deterministic algorithm. Thus, a PR sequence is a deterministic sequence but possesses some random sequence properties. One of the earliest PR generators proposed by John von Neumann was the "middle square" method.

Example 8.20 Start with a three-digit number, such as "157." After squaring, take "464" out of $157^2 = 24649$. Then take "573" in $24649^2 = 607573201$, and so on, etc. It turns out the "middle square" method is not very good for generating PR numbers. □

8.A.1 Uniformly distributed pseudo-random number generation

Consider the Power Residue Method (also called the Multiplicative Congruence Method) that uses the algorithm

$$z_k = (\alpha \times z_{k-1}) \bmod M, \; x_k = z_k/M, \; k = 1, 2, \ldots, \tag{8.96}$$

Table 8.12 Sample mean, sample variance, and sample standard deviation of simulated PR uniform rand.m generated results for sequences of different length n.

n	Sample Mean	Sample Variance	Sample Stand. Dev.
10	6.7882e − 001	9.8672e − 002	3.1412e − 001
100	5.1244e − 001	8.8645e − 002	2.9773e − 001
1000	5.0069e − 001	8.3333e − 002	2.8868e − 001
10000	5.0171e − 001	8.3505e − 002	2.8897e − 001
100000	5.0089e − 001	8.3277e − 002	2.8858e − 001
1000000	4.9956e − 001	8.3278e − 002	2.8858e − 001

where z_0 is called the "seed" integer in $\mathbb{M} = \{0, \ldots, M-1\}$ for some prime integer M, α is an integer in $\mathbb{M}$, to generate PR uniformly distributed numbers $\{x_k, k = 1, 2, \ldots\}$ on $[0, 1]$. Here $x \bmod M$ means finding the reminder in $\mathbb{M}$ after dividing x by the largest possible positive multiples of M.

The Matlab generator rand.m produces an i.i.d. sequence of PR uniformly distributed r.v. on $[0, 1)$. An immediate question is how well do the n outcomes of $x = \text{rand}(1, n)$ compare to a uniformly distributed r.v. on $[0, 1)$. Given a r.v. X, its sample mean is defined by

$$\bar{x} = \frac{1}{n}\sum_{i=1}^{n} x_i, \tag{8.97}$$

which uses the n realizations $\{x_i, i = 1, \ldots, n\}$ to approximate the true (ensemble averaged) mean defined by

$$\mu = \int_0^1 x dx = 1/2. \tag{8.98}$$

The *sample variance* defined by

$$S^2 = \frac{1}{n-1}\sum_{i=1}^{n}(x_i - \bar{x})^2 \tag{8.99}$$

provides an approximation to the variance

$$\sigma^2 = m_2 - \mu^2 = 1/3 - 1/4 = 1/12 = 0.08333. \tag{8.100}$$

The sample standard deviation is defined by $S = \sqrt{S^2}$ and provides an approximation to the standard deviation of $\sigma = 0.2887$. In Table 8.12, we show the sample mean, sample variance, and sample standard deviation from a PR uniform rand.m generated results for sequences of length $n = 10$, $n = 100$, $n = 1000$, $n = 10\,000$, $n = 100\,000$, and $n = 1\,000\,000$ terms.

Thus, as $n \to \infty$, all three simulated parameters converge to the corresponding ensemble-averaged parameters. Analysis on the rate of convergence of these parameters to the $n \to \infty$ case are known, but will not be considered here. We also note, all these simulation results used an initial seed of "98" by setting rand('state',98) before running rand(1, n) for each n. If we use a different initial seed, the simulation results

Table 8.13 Sample mean, sample variance, and sample standard deviation of simulated zero mean and unit variance PR Gaussian sequences of different length n.

n	Sample Mean	Sample Variance	Sample Stand. Dev.
10	$-1.1042e-001$	$3.6457e+000$	$1.9094e+000$
100	$1.6703e-001$	$1.3548e+000$	$1.1639e+000$
1000	$2.7533e-002$	$1.0029e+000$	$1.0015e+000$
10000	$-1.3583e-002$	$1.0061e+000$	$1.0031e+000$
100000	$-1.1374e-003$	$1.0045e+000$	$1.0022e+000$
1000000	$-6.8987e-004$	$1.0003e+000$	$1.0002e+000$

will be different. Of course, as n becomes larger the simulation results become less dependent on the initial seed.

8.A.2 Gaussian distributed pseudo-random number generation

Various algorithms for generating PS Gaussian r.v. are known. Homework Problem 8.1 shows one such well-known algorithm. The Matlab generator randn.m produces an i.i.d. sequence of PR Gaussian distributed r.v. of zero mean and unit variance on $(-\infty, \infty)$. An immediate question is how well do the n outcomes of randn$(1, n)$ of this generator compare to the Gaussian r.v. on $(-\infty, \infty)$. As before, we list the sample mean, sample variance, and sample standard deviation obtained from randn$(1, n)$ as a function of n, for sequences of length $n = 10$, $n = 100$, $n = 1000$, $n = 10\,000$, $n = 100\,000$, and $n = 1\,000\,000$ terms as shown in Table 8.13.

As in the previous PR simulation cases for uniformly distributed r.v., as $n \to \infty$, all three simulated parameters of the Gaussian cases converge to the corresponding ensemble-averaged Gaussian parameters. Analysis on the rate of convergence of these parameters to the $n \to \infty$ case are known, but will not be considered here. We also note, all these simulation results used an initial seed of "98" by setting randn('state',98) before running randn$(1, n)$ for each n. If we use a different initial seed, the simulation results will be different. Of course as before, as n becomes larger the simulation results become less dependent on the initial seed.

8.A.3 Pseudo-random generation of sequences with arbitrary distributions

Now, consider the PR generation of an i.i.d. sequence with an arbitrary cdf using the transformation method applied to an i.i.d. sequence of PR uniform r.v.s on $[0, 1)$. Denote the i.i.d. PR uniform sequence by $\{U_i, i = 1, 2, \ldots\}$ defined on $[0, 1)$. Then the sequence $\{X_i, i = 1, 2, \ldots\}$ defined by

$$X_i = F^{-1}(U_i),\ i = 1, 2, \ldots, \tag{8.101}$$

is the desired sequence of i.i.d. r.v. with a cdf of $F(x)$. This algorithm in (8.101) becomes intuitively obvious by considering Fig. 8.10 given below. From Fig. 8.10, each

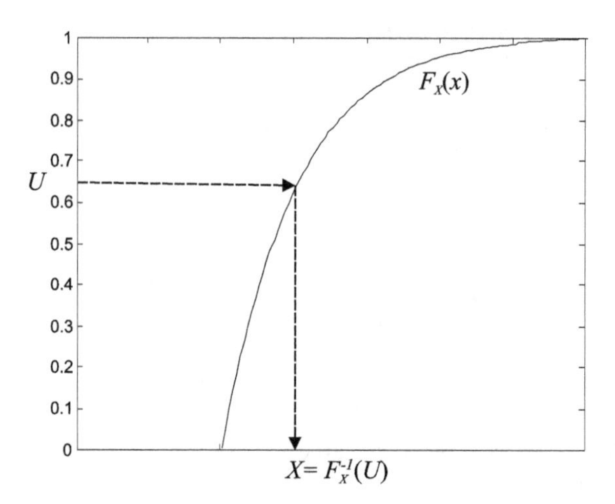

Figure 8.10 Inverse transformation $F^{-1}(U_i)$ of an i.i.d. sequence of PR uniform sequence U_i to generate an i.i.d. sequence X_i of PR sequence with a $F(\cdot)$ cdf.

$0 \leq U_i < 1$, upon the inverse $F^{-1}(.)$ mapping, generates a PR r.v. X_i having the desired $F(\cdot)$ cdf.

Example 8.21 Suppose we want to generate a PR exponential r.v. X with a cdf of $F(x) = 1 - \exp(-x)$, $0 \leq x < \infty$ (or a pdf of $f(x) = \exp(-x)$, $0 \leq x < \infty$), using Matlab.

1. Set the seed =2007. That is, set rand('seed',2007).
2. Generate 10 000 U_i's. Use $\mathbf{U} = \text{rand}(1, 10000)$.
3. Use the inverse transformation $\mathbf{X} = F^{-1}(\mathbf{U})$ to generate 10 000 X_i's. Use $\mathbf{X} = -\log(1 - \mathbf{U})$.

For $F(x) = 1 - \exp(-x)$, $0 \leq x < \infty$, its $\lambda = 1$. Thus, its ensemble-averaged mean and ensemble-averaged variance both are equal to 1. By computations, the sample mean of **X** yields 0.9786 and the sample standard yields 0.9549. Plots of the simulated cdf of **X** using 200 bins and the analytical expression of an exponential cdf are shown in Fig. 8.11.

From the sample mean and sample variance, the transformed **X** appears to be a PR sequence of an exponential r.v. with a λ of approximately unit value. □

8.B Explicit solution of $p_V(\cdot)$

Theorem 8.3 *Let $X = R \bullet V$, where R and V are independent r.v.'s. Let all three of the pdf's have H-function representations. If*

$$p_X^{(1)}(x) = k_1 H_{p_1,q_1}^{m_1,n_1}\left[c_1 x \left| \begin{matrix} \{(a_1', A_1'), \ldots, (a_{p_1}', A_{p_1}')\} \\ \{(b_1', B_1'), \ldots, (b_{q_1}', B_{q_1}')\} \end{matrix} \right.\right] \tag{8.102}$$

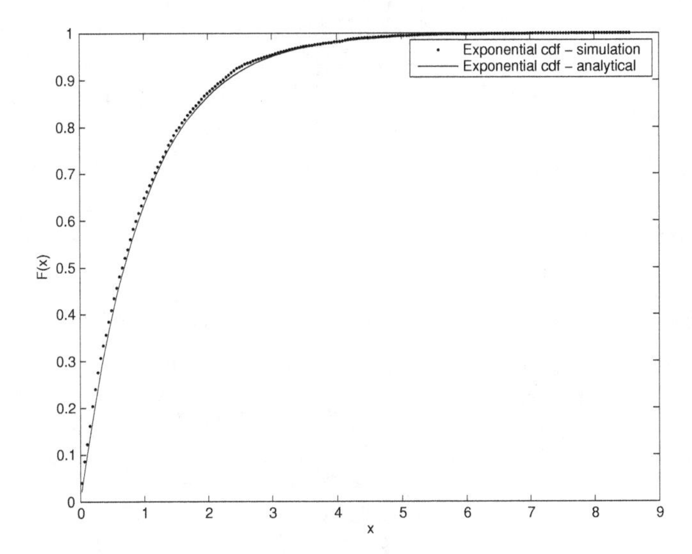

Figure 8.11 Simulated cdf of an exponential r.v. obtained from an inverse transformation of an exponential cdf of 5000 PR uniform samples using 200 bins compared to the analytical expression of an exponential cdf.

and

$$p_X^{(2)}(x) = k_2 H_{p_2,q_2}^{(R)m_2,n_2}\left[c_2 x \,\middle|\, \begin{matrix} \{(a''_1, A''_1), \ldots, (a''_{p_2}, A''_{p_2})\} \\ \{(b''_1, B''_1), \ldots, (b''_{q_2}, B''_{q_2})\} \end{matrix}\right] \tag{8.103}$$

are known, then $p_V(x)$ can be expressed explicitly in terms of the parameters of $p_X^{(1)}(x)$ and $p_R^{(2)}(x)$ in (8.104). Proof: The explicit closed form expression of $p_V(x)$ is given below.

$$
\begin{aligned}
&p_V(x) \\
&= \mathrm{M}^{-1}\left\{\frac{\mathrm{M}\left\{k_1 H_{p_1,q_1}^{m_1,n_1}\left[c_1 x \,\middle|\, \begin{matrix}\{(a'_1, A'_1), \ldots, (a'_{p_1}, A'_{p_1})\} \\ \{(b'_1, B'_1), \ldots, (b'_{q_1}, B'_{q_1})\}\end{matrix}\right]\right\}}{\mathrm{M}\left\{k_2 H_{p_2,q_2}^{m_2,n_2}\left[c_2 x \,\middle|\, \begin{matrix}\{(a''_1, A''_1), \ldots, (a''_{p_2}, A''_{p_2})\} \\ \{(b''_1, B''_1), \ldots, (b''_{q_2}, B''_{q_2})\}\end{matrix}\right]\right\}}\right\} \\
&= \mathrm{M}^{-1}\left\{\frac{k_1 \dfrac{\prod_{j=1}^{m_1}\Gamma(b'_j+B'_j s)\prod_{j=1}^{n_1}\Gamma(1-a'_j-A'_j s)}{\prod_{j=m_1+1}^{q_1}\Gamma(1-b'_j-B'_j s)\prod_{j=n_1+1}^{p_1}\Gamma(a'_j+A'_j s)} c_1^{-s}}{k_2 \dfrac{\prod_{j=1}^{m_2}\Gamma(b''_j+B''_j s)\prod_{j=1}^{n_2}\Gamma(1-a''_j-A''_j s)}{\prod_{j=m_2+1}^{q=2}\Gamma(1-b''_j-B''_j s)\prod_{j=n_2+1}^{p_2}\Gamma(a''_j+A''_j s)} c_2^{-s}}\right\} \\
&= \mathrm{M}^{-1}\left\{\frac{k_1 \prod_{j=1}^{m_1}\Gamma(b'_j + B'_j s)\prod_{j=1}^{n_1}\Gamma(1-a'_j - A'_j s)\prod_{j=m_2+1}^{q_2}\Gamma(1-b''_j - B''_j s)\prod_{j=n_2+1}^{p_2}\Gamma(a''_j + A''_j s) c_1^{-s}}{k_2 \prod_{j=m_2+1}^{q_1}\Gamma(1-b'_j - B'_j s)\prod_{j=n_1+1}^{p_1}\Gamma(a'_j + A'_j s)\prod_{j=1}^{m_2}\Gamma(b''_j + B''_j s)\prod_{j=1}^{n_2}\Gamma(1-a''_j - A''_j s) c_2^{-s}}\right\}
\end{aligned}
$$

$$
= \mathrm{M}^{-1}\left\{ \frac{k_1 \left\{ \prod_{j=1}^{m_1} \Gamma(b'_j + B'_j s) \prod_{j=n_2+1}^{p_2} \Gamma(a''_j + A''_j s) \right\} \left\{ \prod_{j=1}^{n_1} \Gamma(1 - a'_j - A'_j s) \prod_{j=m_2+1}^{q_2} \Gamma(1 - b''_j - B''_j s) \right\} c_1^{-s}}{k_2 \left\{ \prod_{j=m+1}^{q_1} \Gamma(1 - b'_j - B'_j s) \prod_{j=1}^{n_2} \Gamma(1 - a''_j - A''_j s) \right\} \left\{ \prod_{j=n_1+1}^{p_1} \Gamma(a'_j + A'_j s) \prod_{j=1}^{m_2} \Gamma(b''_j + B''_j s) \right\} c_2^{-s}} \right.
$$

$$
= k H_{p,q}^{m,n}\left[cx \,\middle|\, \begin{matrix} \{(a_1,\ A_1),\ \ldots,\ (a_p,\ A_p)\} \\ \{(b_1,\ B_1),\ \ldots,\ (b_q,\ B_q)\} \end{matrix} \right],
$$

$$
k = k_1/k_2,\ c = c_1/c_2,\ m = m_1 + p_2 - n_2,\ n = n_1 + q_2 - m_2,\ p = p_1 + q_2,\ q = q_1 + p_2,
$$

$$
a = \{a_1, \ldots, a_p\} = \{\{a_1{}', \ldots, a_{n_1}{}'\}, \{b_{m_2+1}{}'', \ldots, b_{q_2}{}''\}\}, \{\{a_{n_1+1}{}', \ldots, a_{p1}{}'\}, \{b_1{}'', \ldots, b_{m_2+1}{}''\}\}
$$

$$
A = \{A_1, \ldots, A_p\} = \{\{A_1{}', \ldots, A_{n_1}{}'\}, \{B_{m_2+1}{}'', \ldots, B_{q_2}{}''\}\}, \{\{A_{n_1+1}{}', \ldots, A_{p1}{}'\}, \{B_1{}'', \ldots, B_{m_2+1}{}''\}\}
$$

$$
b = \{b_1, \ldots, b_q\} = \{\{b_1{}', \ldots, b_{m_1}{}'\}, \{a_{n_2+1}{}'', \ldots, a_{p_2}{}''\}\}, \{\{b_{m_1+1}{}', \ldots, b_{q1}{}'\}, \{a_1{}'', \ldots, a_{n_2+1}{}''\}\}
$$

$$
B = \{B_1, \ldots, B_q\} = \{\{B_1{}', \ldots, B_{m_1}{}'\}, \{A_{n_2+1}{}'', \ldots, A_{p_2}{}''\}\}, \{\{B_{m_1+1}{}', \ldots, B_{q1}{}'\}, \{A_1{}'', \ldots, A_{n2+1}{}''\}\}
\tag{8.104}
$$

References

[1] R. W. Lucky, J. Salz, and E. J. Weldon Jr. *Principles of Data Communication*. McGraw-Hill, 1968.

[2] K. Yao and R. M. Tobin. Moment space upper and lower error bounds for digital systems with intersymbol interference. *IEEE Trans. Inf. Theory* 22(1), 1976.

[3] M. Dresher. Moment spaces and inequalities. *Duke Math. J.* 20, 1953.

[4] K. Yao and E. Biglieri. Multidimensional moment error bounds for digital communication systems. *IEEE Trans. Inf. Theory* 26(4), 1980.

[5] E. Biglieri. Probability of error for digital systems with inaccurately known interference. *IEEE Trans. Inf. Theory* 30(2), 1984.

[6] B. H. Cheng, L. Vandenberghe, and K. Yao. Semidefinite programming bounds on the probability of error of binary communication systems with inexactly known intersymbol interference. *IEEE Trans. Inf. Theory* 51(8), 2005.

[7] K. Yao. Error probability of asynchronous spread spectrum multiple access communication systems. *IEEE Trans. Commun.* 25(8), Aug. 1977.

[8] R. M. Tobin and K. Yao. Upper and lower error bounds for coherent phase-shift-keyed systems with cochannel interference. *IEEE Trans. Commun.* 25(2), Feb. 1977.

[9] E. Biglieri. Effect of uncertainties in modeling interference in spectrum sensors. *Electr. Ltrs.* 47(13), June 2011.

[10] K. Yao. A representation theorem and its applications to spherically-invariant random processes. *IEEE Trans. Inf. Theory* 19(5), 1973.

[11] J. V. DiFranco and W. L. Rubin. *Radar Detection*. Artech House, 1980.

[12] W. C. L. Lee. *Mobile Communications Engineering*, 2d ed. McGraw-Hill, 1998.

[13] A. Abdi, H. A. Barger, and M. Kaveh. Signal modeling in wireless fading channel using spherical interference in nakagami fading. In *Proc. Int. Conf. Acoustics, Speech, Signal Processing (ICASSP)*, pp. 2997–3000, 2000.

[14] V. A. Aalo and J. Zhang. Average error probability of optimum combining with a co-channel interference in nakagami fading. In *Conf. Record of IEEE Wireless Communication and Networking*, pp. 376–381, 2000.

[15] K. Yao. Spherically invariant random processes: Theory and applications. In V. K. Bhargava et al., ed. *Communications, Information and Network Security*, pp. 315–332. Kluwer Press, 2002.

[16] K. Yao, M. K. Simon, and E. Biglieri. Statistical modeling of fading channel statistics based on SIRP and H-functions. In *Proc. IEEE International Conference on Communications, Circuits and Systems (ICCCAS)*, pp. 265–268, 2004.

[17] K. Yao, M. K. Simon, and E. Biglieri. Unified theory on wireless communication fading statistics based on SIRP. In *Proc. IEEE 5th Workshop on Signal Processing Advances in Wireless Communication*, pp. 135–139, 2004.

[18] C. Fox. The g and h-functions as symmetrical Fourier kernels. *Trans. American Mathematical Society*, pp. 395–429, 1961.

[19] K. Yao. Unified SIRP fading statistical modeling and turbulence effects. Progress in Electromagnetics Research Symposium (PIERS 20, Cambridge) Vu graphs presentation, June 2011. 35 p.

[20] I. S. Gradshteyn and I. M. Ryzhik. *Table of Integrals, Series, and Products*. Academic Press, 1965.

[21] M. K. Simon and M. S. Alouini. *Digital Communication over Fading Channel*. Wiley, 2000.

[22] S. Benedetto and E. Biglieri. *Principles of Digital Transmission with Wireless Applications*. Kluwer, 1999.

[23] K. Yao. Evaluation of radar detection p_{FA} and p_{D} with constraints. Personal notes, 2002.

[24] R. L. Mitchell. *Radar Signal Simulation*. Artech House, 1976.

[25] P. Balaban, E. Biglieri, M. C. Jerichim, and K. S. Shanmugan, ed. *Special Issue on Computer-Aided Modeling, Analysis, and Design of Communication Systems II. IEEE J. Sel. Areas Commun.*, 1988.

[26] M. C. Jeruchim, P. Balaban, and K. S. Shanmugan. *Simulation of Communication Systems*. Plenum Press, 1992.

[27] S. Ulam. On the Monte Carlo methods. In *Proc. 2nd Symposium on Large Scale Digital Calculating Machinery*, pp. 207–212, 1951.

[28] J. M. Hammersley and D. C. Handscomb. *Monte Carlo Methods*. Methuen, 1964.

[29] K. S. Shanmugan and P. Balaban. A modified Monte Carlo simulation technique for the evaluation of error rate in digital communication systems. *IEEE Trans. Commun.* 28(11), Nov. 1983.

[30] R. Y. Rubinstein. *Simulation and the Monte Carlo Method*. Wiley, 1981.

[31] D. Lu and K. Yao. Improved importance sampling technique for efficient simulation of digital communication system. *IEEE J. Sel. Areas Commun.* 6(1), Jan. 1988.

[32] F. M. Gardner and J. D. Baker. *Simulation Techniques–Models of Communication Signals and Processes*. Wiley, 1997.

[33] R. Srinivasan. *Importance Sampling–Applications in Communications and Detection*. Springer, 2002.

[34] J. C. Chen, D. Lu, D. J. S. Sadowsky, and K. Yao. On importance sampling in digital communications, Part I: Fundamentals. *IEEE J. Sel. Areas Commun.* 11(3), Apr. 1993.

[35] R. A. Fisher. On the mathematical foundations of theoretical statistics. *Phil. Trans. of the Royal Soc.* 222A, 1922.

[36] H. Cramér. *Random Variables and Probability Distributions*. Cambridge University Press, 1937.

[37] P. M. Woodward. *Probability and Information Theory, with Applications to Radar*. Pergamon Press, 1953.

[38] D. Middleton. *An Introduction to Statistical Communication Theory*. McGraw-Hill, 1960.
[39] H. M. Hall. A new model for impulsive phenomena–application to atmospheric noise communication channel. Technical Report SEL-66-052, Stanford University, 1966.

Problems

8.1 One simple way to verify the elements of a PR sequence $\{U_n,\ n = 1, \ldots, N\}$ do not have strong adjacent neighbor influence is to plot (U_n, U_{n+1}), for $n = 1, \ldots, N$. Consider the use of the uniform generator rand in Matlab. Specifically, take rand(state, 49), $U =$ rand$(1, N)$, with $N = 501$. Then plot $(U(1:500), U(2{:}501))$. If the sequence values of U do not have strong adjacent neighbor influence, what would you expect the spread of these 500 points in the plot?

8.2 Repeat Problem 8.1 by using the linear congruential generator (B.J.T. Morgan, *Elements of Simulation*, Chapman and Hall, 1984, p. 61), defined by

$$X_{n+1} = 781 * X_n + 387(\text{mod } 1000), \tag{8.105}$$

$$U_n = X_n/1000,\ n = 0, \ldots, 501.$$

Plot (U_n, U_{n+1}), for $n = 1, \ldots, 500$, as in Problem 8.1. From the spread of these 500 points in the plot, what can you say about the influence of the adjacent neighbors for the sequence generated by (8.105)?

8.3 Consider the generation of a uniformly distributed r.v. on (0, 1). Let us use the multiplicative congruence method (IMSL, GGUBFS). Specifically, consider

$$s_{n+1} = 7^5 * s_n \text{mod}(2^{31} - 1),\ n = 0, \ldots, \tag{8.106}$$

$$x_n = s_n/(2^{31} - 1),\ n = 1, \ldots, . \tag{8.107}$$

s_0 is the seed between 1 and $2^{31} - 1 = 2\,147\,483\,646$ and x_n is the PR uniform r.v. over (0, 1). Write a Matlab code for this PR uniform generator. The first thing to do in checking whether the algorithm you coded is correct is to take $s_0 = 20052005$, then you should obtain $x_1 = 0.934395521848647$. If you do not obtain this x_1 value, then debug your program until you get this number. Take $s_0 = 49$, and obtain x_n, $n = 1, \ldots, 502$. Plot (x_n, x_{n+1}), for $n = 1, \ldots, 500$ and plot (x_n, x_{n+2}), for $n = 1, \ldots, 500$, to check on the correlations among x_n and x_{n+1} as well as among x_n and x_{n+2}.

8.4 Consider the generation of two sequences of independent Gaussian r.v. of zero mean and unit variance based on the Box–Muller (1958) and Marsaglia–Bray (1064) method.

(a) Start with independent uniformly distributed r.v. U_1 and U_2 on (0, 1). Use the method of $S_i = 7^5 S_{i-1}\text{mod}(2^{31} - 1)$, $U_i = S_i/(2^{31} - 1)$ to generate the U_i and U_{i+1}. Obtain independent uniformly distributed r.v. V_1 and V_2 on $(-1, 1)$ by taking $V_1 = 2U_1 - 1$ and $V_2 = 2U_2 - 1$. (Keep V_1 and V_2 in floating point format.)
(b) Compute $S = V_1^2 + V_2^2$.
(c) If $S \geq 1$, go back to (a) and use two new pairs of U_i and U_{i+1}'s, and repeat.

(d) If $S < 1$, take $X_1 = V_1\sqrt{(-2\ln S)/S}$, $X_2 = V_2\sqrt{(-2\ln S)/S}$ where X_1 and X_2 are claimed to be two independent Gaussian r.v. of zero mean and unit variances.
(e) Take $s_0 = 49$. Using steps (a) to (d) to generate a sequence of 1000 x_1's denoted by $\mathbf{x}_1$ and 1000 x_2s denoted by $\mathbf{x}_2$.
(f) Compare the empirical cdf of the $\mathbf{x}_1$ sequence with the cdf of the Gaussian r.v. of zero mean and unit variance. How do these two cdf's compare to each other?
(g) A simple test for the correlation of the realizations of $\mathbf{x}_1$ and $\mathbf{x}_2$ is to plot $x_1(n)$ versus $x_2(n)$, $n = 1, \ldots, 1000$, to look for regular patterns in this plot. Are there any obvious regular patterns in this plot?

8.5 Consider the two PR Gaussian sequences of $\mathbf{x}_1$ and $\mathbf{x}_2$ of Problem 8.4. A simple method to verify the "Gaussianness" of $\mathbf{x}_1$ is to plot the empirical cdf of $\mathbf{x}_1$ versus the standard Gaussian cdf as done in Problem 8.4(f). A more quantitative method is to use the Kolmorogov–Smirnov (KS) test to compare both of these sequences with the standard Gaussian cdf. Consider the use of the Matlab kstest2($\mathbf{x}_1, \mathbf{x}_2$) function for the KS test. Find and learn an elementary understanding of the KS test, and provide a brief explanation of the KS test result for these two sequences.

8.6 Consider the analysis and simulation of a scalar binary communication system modeled by

$$X_i = \begin{cases} 1 + N_i, & H_1 \\ -1 + N_i, & H_0 \end{cases}, \quad i = 1, \ldots, n, \tag{8.108}$$

where the data $A_i = 1$, under H_i, and $A_i = -1$, under H_0, each with equal probability of $1/2$, and the noise N_i, $i = 1, \ldots, n$, are a sequence of i.i.d. Gaussian r.v.'s of zero mean and unit variance. For each received x_i, the receiver declares H_i if $x_i \geq 0$, and declares H_0 if $x_i < 0$.

(a) Using analysis, find the average error probability P_e of this system.
(b) Start with $s_0 = 2003$. Use steps (a) to (d) of Problem 8.4 to generate 1000 x_1s and 1000 x_2s. Let $n_i = x_1(i)$, and if $x_2(i) > 0$, set $a_i = 1$, and if $x_2(i) \leq 0$, set $a_i = -1$, for $i = 1, \ldots, n$. Count the number of decision errors, C_e, for $a_i = 1$ and $a_i = -1$. Then the simulated average error probability is given by $\tilde{P}_e = C_e/1000$. Your $\tilde{P}_e$ need not be identical to P_e, but should not be too different.

8.7 Duplicate the results of Example 8.2 for all three cases.

8.8 Consider a binary communication system with only a one-term ISI interference disturbance modeled by $X = A_0h_0 + A_1h_1 + N$, where A_0 and A_1 would be two independent r.v.'s taking ± 1 values with equal probability of $1/2$, assume h_0 and h_1 with $h_0 > h_1 > 0$, and the noise N is a Gaussian r.v. with a zero mean and unit variance. Suppose the information data A_0 of interest takes the value of -1. Consider a decision threshold set at 0. Show the error probability P_e^{ISI} with ISI is given by $P_e^{\text{ISI}} = \frac{1}{2}\text{Q}(h_0 - h_1) + \frac{1}{2}\text{Q}(h_0 + h_1)$, while in the absence of ISI (i.e., $h_1 = 0$), the error probability P_e is given by $P_e = \text{Q}(h_0)$. Furthermore, regardless of the actual values of

h_0 and h_1 (but assuming $h0 > h1 > 0$), show $P_e^{ISI} > P_e$ is always true. (Hint: What is the behavior of the Q(x) function for $0 \le x < \infty$?)

8.9 Duplicate the results of Example 8.4 for both $J = 6$ and $J = 8$.

8.10 Use the moment space bounding method of Section 8.2.1 to find the upper and lower error probabilities of the ISI problem of Example 8.4 for $J = 6$ at SNR = 6.02 dB by finding m_2^{UB} and m_2^{LB} as the values of the intersection of the vertical chord taking the value of m_1 with the convex hull. Repeat this evaluation for SNR = 12.04 dB. Why are the m_2^{UB} and m_2^{LB} values much closer for the SNR = 6.02 dB case compared to the m_2^{UB} and m_2^{LB} values for the SNR = 12.04 dB case?

8.11 Consider Case 1a of Example 8.13, where

$$p_R^{(2)} = x \cdot \exp(-x^2/2), \ 0 < x < \infty, \tag{8.109}$$

and

$$p_V(x) = \sqrt{2/\pi}\exp(-x^2/2), \ 0 < x < \infty. \tag{8.110}$$

Explicit analytical evaluation using (8.57)–(8.59) showed

$$p_X^{(1)}(x) = \exp(-x), \ 0 < x < \infty. \tag{8.111}$$

Use simulation to verify if a sequence of r.v.'s $\{R_i, \ i = 1, \ldots, N\}$ having a Rayleigh distribution given by $p_R(2)(x)$ of (8.109) and a sequence of r.v.'s $\{V_i, \ i = 1, \ldots, N\}$ independent of R_i having a half-Gaussian distribution $p_V(x)$ of (8.110), then the sequence of r.v.'s $\{X_i = R_i \times Vi, \ i = 1, \ldots, N\}$ has the distribution given by (8.111). Specifically, take randn('state', 19) and generate $N = 5000$ Rayleigh r.v.'s whose realizations are denoted by $\mathbf{r}$ = raylrnd(1, 5000). Similarly, take randn(state, 47) and generate $N = 5000$ half-Gaussian r.v.'s whose realizations are denoted by $\mathbf{v}$ = abs(normrnd(0, 1, 1, 5000)). Form $\mathbf{x} = \mathbf{r}. * \mathbf{v}$. Estimate the mean of $\mathbf{x}$. Plot the histogram of $\mathbf{x}$ and show this histogram approximates $p_X^{(1)}(x)$ of (8.111). Plot the sample cdf of x and compare to the cdf corresponding to the pdf $p_X^{(1)}(x)$ of (8.111).

8.12 (*) In a radar system, consider the transmission of the waveform

$$s_0(t) = A_0 \cos(2\pi f_0 t), \ 0 \le t \le T, \tag{8.112}$$

where both A_0 and f_0 are some deterministic parameters. Suppose the returned waveform is modeled by

$$X(t) = A\cos(2\pi f_0 t + \theta) + N(t), \ 0 \le t \le T, \tag{8.113}$$

where A is now modeled as a Rayleigh r.v. due to random scatterings from the fluctuating radar target with the pdf given by

$$p_A(a) = (2a/\bar{E})\exp(-a^2/\bar{E}), \ 0 \le a < \infty, \tag{8.114}$$

where $\bar{E}$ is the averaged received energy, Θ is a uniform r.v. on $[0, 2\pi)$, and $N(t)$ is a zero mean w.s. stationary white Gaussian process with a power spectral density of $N_0/2$.

(a) Consider the non-coherent receiver derivation in Section 6.1, where A was assumed to be a deterministic constant and Θ was a uniform r.v. on $[0, 2\pi)$. Denote the conditional LR function using (6.9) and (6.16) to obtain

$$\Lambda(r|a) = \exp(-E/N_0)\mathrm{I}_0(2ar/N_0),\ 0\ eqr < \infty,\ 0 \leq a < \infty, \tag{8.115}$$

conditioned on the r.v. A taking the value of a. Then the LR function is given by

$$\Lambda(r) = \int_0^\infty \Lambda(r|a)p_A(a)da. \tag{8.116}$$

Show r is still a sufficient statistic even when A is a fluctuating r.v.

(b) Evaluate the probability of detection

$$P_\mathrm{D} = \int_0^\infty p_\mathrm{D}(a)p_\mathrm{A}(a)da, \tag{8.117}$$

where $p_A(a)$ is given by the Rayleigh pdf of (8.114) and the conditional probability of detection from (6.33) yields

$$p_\mathrm{D}(a) = \mathrm{Q}\left(\sqrt{2E/N_0}, \gamma_0\right),\ E = a^2T/2,\ \gamma_0 = \sqrt{-(N_0T/2)\ln(P_\mathrm{FA})}. \tag{8.118}$$

Hint: After substituting (8.118) in (8.117), P_D becomes a double integral. Inside this double integral of (8.117), one of the integrals has the form of the l.h.s. of (8.119). By using this closed form expression on the r.h.s. of (8.119) to perform the other integral of (8.117), it yields the desired explicit solution of P_D.

$$\int_0^\infty t\mathrm{I}_0(at)\exp(-bt^2)dt = \frac{\exp(a^2/(4b))}{2b}. \tag{8.119}$$

Index

a posteriori probability 72, 138
additive noise (AN) 7, 18
additive white Gaussian noise, AWGN 8
alternative hypothesis 85
amplitude shift keyed, ASK 188
antenna gain 13, 33
anti-podal 12, 13, 102, 108, 162, 170, 273
Apery's constant 128
augmented 176
autocorrelation function 26, 27, 47, 48, 50, 55, 56, 63, 104, 118, 119, 120, 122, 124, 125, 128, 144, 146, 147
  matrix 109, 110, 216, 223, 226, 305
  sequence 23, 34, 56, 216
average cost 67, 72, 76, 150, 256
average error probability, average probability of error 10, 13, 28, 71, 101, 103, 112, 155, 168, 185, 271, 279, 297, 303, 304

bandlimited 26, 104, 276, 304
bandpass-filter-envelope-detector 204
base station 2, 16, 29
basis 149, 185
Bayes 78, 91, 97, 98, 253, 259
  criterion viii, 66, 103, 141, 147, 149, 150, 185
  estimator ix, 256, 259
Bernoulli r.v. 39, 94
Bessel function 193, 203, 293
binary detection ix, 8, 11, 13, 88, 112, 138, 190, 202, 272, 279, 285, 296, 297
  deterministic signals 28, 66, 149
  hypothesis test, binary hypothesis testing 79, 81, 85, 88, 97, 105, 138, 146, 186
  phase-shift-keyed (BPSK) modulation, binary-phase-shifted-keyed, BPSK, binary phase-shifted-keying 13, 108, 149, 162
  frequency-shift-keyed, Binary frequency-shifted-keying, BFSK 109, 158, 161
block mean square estimation 226
  operation 226
Bochner–Khinchin theorem 49
Brownian motion random process 120

carrier synchronization 189
causal 146, 265, 266
cdf 36
CDMA system 282, 297
cell under test (CUT) 209
cell-averaging-constant false alarm rate, CA-CFAR ix, 208, 209
Central Limit Theorem (CLT) 39, 97, 282, 287
Chebyshev inequality 240, 241, 242
Chi-square distribution, Chi-square r.v. 93, 289, 293
classification ix, xi, 149, 171, 172, 173
Cochannel interferences 282
coherent detection 200, 201
  integration ix, 190, 201, 202, 205, 206, 210
  receiver 188, 189, 200, 201
colored Gaussian noise, CGN 110
  matched filter 131
communication systems vii, 2, 17, 35, 139, 149, 157, 163, 186, 187, 190, 282, 287, 303, 306, 312, 313
complementary Gaussian distribution function 9, 92, 130, 205, 272
complete orthonormal, CON 116, 154
completeness 117
composite hypothesis 85, 86, 87
concave function 73, 74
conditional expectation 214, 219
  pdf 41, 42, 43
  probability 5, 35, 37, 38, 56, 65, 88, 150, 156, 228, 250
consistency 215, 238, 258
consistent 241
constant false alarm rate, CFAR 190, 207
continuous r.v. 37, 88
  time r.p. 48, 55
  time waveform 105, 106, 250
  time white noise process 104
convex optimization xi, 147, 182, 186, 187, 189, 237, 263
correction term 229
correlation 6, 23, 24, 28, 47, 100
  receiver vii, 3, 7, 28, 136, 138, 139, 156, 284
cost function 66, 72, 255, 256, 257

covariance matrix 8, 40, 41, 42, 51, 56, 58, 62, 97, 110, 113, 130, 132, 134, 138, 139, 142, 143, 144, 219, 224, 268, 283
Cramer–Rao bound (CRB) 215, 243, 259
critical value 87
CVX 181, 186, 187, 189, 237, 269

decision region 65, 67, 69, 70, 71, 72, 73, 75, 78, 81, 93, 185, 187, 188, 192, 194, 213, 272, 273, 274
detection i, ii, xi, 1, 2, 3, 5, 10, 12, 28, 65
determinant 40, 51, 97
deterministic 3, 28, 38, 51, 53, 88, 97, 104, 108, 110, 112, 113, 116, 117, 123, 127, 130, 137, 138
diagonalization 115
discrete r.v. 37, 62
discrete-time WSS random processes 35, 54
discriminant function 172, 174, 175, 176, 180, 184
Doppler 209, 214, 297
double-threshold detection 205
  receiver 190, 210

efficiency 86, 215, 238, 258, 259
efficient 91, 240, 252
eigenfunction 144, 145
eigenvalue 113, 118, 120, 121, 122
  problem 113, 114, 115, 118, 119, 120, 122
eigenvector 100, 113, 114, 115, 118, 121, 123
energy 6, 8, 100, 106, 108, 116, 124, 127, 128, 129, 138, 146, 149, 153, 154, 156, 160, 162, 163, 165, 168, 236
energy per bit (E) 15, 165
ensemble averaging 39, 53, 308
envelope 192, 194, 204, 206, 209, 210, 271, 287, 288, 289
  detector 194, 195, 210
equivalent temperature 14
ergodic 53, 54, 64
ergodicity 53, 54, 216
estimate 2, 17, 18, 19, 22, 24, 27, 100, 130, 139, 201, 210, 214, 215, 216, 217, 219, 238, 248
estimation robustness 21
estimator 29, 30, 144, 215, 225, 226, 228, 229, 230, 231, 234, 238
exponential r.v. 39, 310

factorization theorem 79, 94, 242, 244
false alarm 65, 80, 85, 90, 94, 160, 271
first-order autoregressive sequence 228
Fisher 2, 31, 65, 91, 243
  information matrix 243
Fourier series expansion 54, 117
  transform 48, 49, 50, 55, 146
Fox H-function 271, 288, 305
frequency-shifted-keying, FSK 109

Galileo 2, 20, 258
gamma distribution, gamma distribution 209
Gauss xi, 2, 20, 21, 57, 259
Gaussian pdf 8, 40, 41, 42
  random process, Gaussian r.p. 51, 104, 106, 119
  variable, Gaussian r.v. 39
  vector 40, 41, 51, 62, 116, 172, 224, 283
Gauss–Markov theorem 234
general orthogonal principle 222
geometry of signal vectors 7, 185
geo-synchronous satellite 16, 32
gold-coded CDMA system 282
Gram–Schmidt orthonormalization method, GSOM 150, 151
Gram–Schmidt, GS 150

heavy-tailed (HT) 282, 286, 305
hypothesis testing 80, 105

importance sampling (IS) 299, 300
impulse response function 50, 136, 144
independence 38, 41
independent 37, 40, 41, 46
independent and identically distributed (i.i.d.) 46, 83
information rate 13
inner product 99, 100, 106
integrate-and-dump 137
interfering power 27, 28
intersymbol interferences 271, 277
inverse Fourier transform 50, 51, 55
isomorphism Theorem 279, 280, 282

joint pdf 38
  probability cumulative distribution function 37

Kalman filter 227, 229, 230, 262, 263
  gain 229, 230, 267
Kantorovich inequality 133, 139, 140
Karhunen–Loeve (KL) expansion 117
Karish–Kuhn–Tucker, KKT 147
Kolmogorov 2, 31, 45, 57, 60, 258, 262
Kotel'nikov 29, 31, 92, 141

Lagrangian multiplier, Lagrange multiplier 20, 21, 241, 259
Laplace 2, 31, 57, 60, 259
Laplacian r.v. 39, 234, 272
least-absolute-error (LAE) 233
  criterion 19, 232
  estimation 230
least-square (LS) 230, 233
  criterion 231
  estimation, least-square-error estimation 29
  error, LSE 20, 21

likelihood function 3, 4, 84, 242, 247, 248, 249, 250, 269
ratio (LR) function 4, 67, 72
test, LR test 82, 84
linear 106
discriminant 174, 175, 176
dynamical system 227
machine 185
mean square estimation 29, 226, 258
programming, LP 182
receiver 190, 204, 205, 304
time-invariant operation 51, 56
time-invariant, LTI 50, 136, 276
linearly independent 149, 150, 152, 153, 154, 235, 236
separable 149, 176, 179, 180, 183, 189
LR constant 67, 71, 75, 76, 77, 97, 98, 99, 101, 103, 138, 147, 273

Maclaurin power series 189
Marcum-Q function 199, 207
Markov covariance matrix 134
sequence 228
M-ary amplitude shift keyed, M-ASK 188
detection 149
deterministic signals 149
modulation 149, 185
matched filter 131, 135, 136, 143, 146, 156, 173, 188, 194, 202, 205, 210, 212
receiver 135, 136, 157
Matlab xi, 20, 30
maximum a posteriori (MAP) decision 155
estimator 155, 188
maximum likelihood (ML) 3, 247
decision 3
estimator 215
mean 39
mean vector 40, 42, 51
square (MS) 23, 214, 215
square error (MSE) 23, 24, 214
square error criterion 23, 89, 257, 258
square estimation 215, 218, 221, 230, 258
median 88
Mellin transform 287, 288
Mercer theorem 119, 128, 137, 144
Middleton 29, 31, 91, 141, 211
Minimax criterion 72, 74
minimum distance decision rule 149, 166, 188
receiver 157
minimum mean-square-error (MMSE) 24, 26, 27, 29
probability of error criterion 149
variance 234, 238, 241
modified Bessel function 193, 203
moment 23, 39, 238
space isomorphism theorem 279
space method 279, 305
Monte Carlo importance sampling 299
simulation 271, 297
most efficient 244
most efficient estimator 244
most powerful test 80, 91
most sparseness 237
multi-dimensional (variate) cdf 38, 218

Nakagami-m pdf 213, 288
narrowband Gaussian process 41, 114, 287
N-dimensional probability density function, N-dimensional pdf 45, 46, 51
Neyman–Pearson (NP) criterion 75, 97
Neyman–Pearson, NP 65, 139
noise figure 14, 33
noise spectral density 14, 108, 146
non-coherent detection 195, 201
integration 207
non-linear mean square estimation 229
optimization 21, 29, 186, 259
non-parametric test 90
non-singular covariance matrix 40, 58, 133
norm 30, 99, 100, 106
normal equation 216, 223, 233
normality 114
normalized correlation 102, 107, 236, 251
null hypothesis 85
Nyquist sampling interval 104
rate 104

observation equation 228, 230, 232
omni-directional antenna 13
optimum decision regions 67, 138, 142, 144, 159
predictor 229
orthogonal 102, 109, 151, 159, 162, 169, 170, 196, 215, 220
matrix 115, 133
principle 214, 215, 222, 225, 226, 228, 258, 264
projection (projection) 221
signals 109, 169, 170
transformation 113, 115
orthonormal 116, 120, 122, 138, 140, 151, 152, 153, 154
series expansion 116, 149, 152
set 120, 150, 238
orthonormality 114, 116, 253
over-determined system of equations 258, 259

parameter estimation 215, 258
pattern recognition 171, 173
perceptron 177, 178, 186
phase error 189
synchronization 201
phase-shifted-keying, PSK 162

PN sequence 148, 297
Poisson formula 189
  r.v. 39, 62, 163
population 238
positive-definite 110
  semidefinite 110
post-detection integration 202, 204, 205, 206
power of the test 86, 89, 258
  spectral density (psd), power spectral density function 14, 48, 104, 125
prior probability 67, 72, 73, 156, 271
probability 36
  cumulative distribution function (cdf) 36
  density function (pdf) 36
  of correct decision 164, 167
  of detection 81, 86, 101, 107, 109, 131, 139, 295
  of error 5, 10, 71, 101, 150, 279
  of false alarm 80, 81, 85, 107
propagation loss 13, 16, 32
pseudo-number generation 271, 306, 307, 309

Q function 9, 16, 92, 100
quadra-amplitude-modulation, QAM 149, 166, 167
quadratic receiver 190, 205
quadrature phase-shifted-keying, QPSK, Quadrature-PSK, 4-PSK 149, 214

random event 36
  experiment 35
  noise 40, 297
  noise sequence 109
  processes (r.p.) 43, 44, 45
  sample 238
  sequence (r.s.) 54
  variable (r.v.) 36
  vector 38
rare events 282, 299, 305
Rayleigh pdf 199, 200, 206, 212, 287
  r.v. 39, 196
RC low-pass filter 50
realizations 3, 18, 38, 45, 46, 53, 54
receiving operational curve (ROC) 76
recursive estimator 228
residual 231, 235
Rice 198, 210
Rician pdf 198, 199, 200, 206, 291
Riemann integration 105
  zeta function 128
robustness constraint 293, 295

sample 23, 35, 44, 45, 84, 206, 207
  mean 87, 216, 232, 238, 308
  point 35, 45
  space 44, 238
  variance 239, 308
Schwarz inequality 131, 133, 139, 143, 261
second order random process, SORP 117, 118
sequential likelihood ratio test (SLRT) 80
  test 81
series expansion 54, 116, 123, 137, 152
sidelobe canceller (SLC) 28
signal-to-noise ratio (SNR) 3, 7, 15, 108, 143, 147, 199
significance level 85, 87, 88
simple hypothesis testing 65, 85
singular detection 127, 129, 139
sinusoidal waveform 14, 210
sparseness 236
spectral decomposition 134, 140
spherically invariant random processes 283, 312
square integrable 116
  law detector 209
state 227, 229, 230, 258
stationarity 46, 53, 223
stationary of order *n* 46, 47, 53
statistic 45, 79, 88, 238
strictly stationary 47, 53
student t 282, 283
sufficiency 215, 238
sufficient 242
  statistic 72, 79, 103, 105, 111, 112, 116, 127, 135, 137, 159, 204, 242, 244, 273, 304
support vector machine, SVM 149, 182
symbol 164, 165
synchronization 189, 201, 297

tapped delayed line 209
Taylor series linear approximation 246
theoretical data compression 80
thermal noise 14, 139
time averaging 53
  bandwidth product 108, 109
transfer function 50, 55

UAV 16, 29
unbiased estimator 234, 238, 240, 242, 243, 244, 302
uncorrelated 40, 41, 118, 119
under-determined system of equations 235, 236, 237
uniform r.v. 39
uniformly most powerful (UMP) test 85, 86, 87

variance 39
vector measurements 77

Weibull distribution 288, 289, 298, 299, 300, 301, 302

white Gaussian noise (WGN) 8, 98, 104, 110
  process 14, 190
white matched filter 131
  noise, WN 98
  process (w.p.) 50
whitening filter 112, 113, 115
whitening matrix 113, 115, 116
wide-sense stationary, WSS 47, 54
  random process, WSSRP 118
Wiener 2
  random process 120
wireless fading channel 271
Woodward 2, 31, 304

Yao 312, 313

CPSIA information can be obtained
at www.ICGtesting.com
Printed in the USA
LVHW100028190819
628100LV00008B/106/P